Membrane Science and Engineering

Membrane Science and Engineering

Editor: Hannah Palmer

CALLISTO REFERENCE

www.callistoreference.com

Callisto Reference,
118-35 Queens Blvd., Suite 400,
Forest Hills, NY 11375, USA

Visit us on the World Wide Web at:
www.callistoreference.com

ISBN: 978-1-64116-113-8 (Hardback)

Cataloging-in-Publication Data

Membrane science and engineering / edited by Hannah Palmer.
    p. cm.
Includes bibliographical references and index.
ISBN 978-1-64116-113-8
1. Membranes (Biology). 2. Membranes (Technology). I. Palmer, Hannah.
QH601 .M46 2019
571.64--dc23

Table of Contents

Preface

Membrane science and engineering is the scientific study of the transport of substances between two fractions through a permeable membrane. It is used as a mechanical separation method for separating gaseous or liquid streams. The separation method requires no application of heat like distillation and crystallization. It is of significance in food, biotechnology and pharmaceutical industries. Important applications include the filtration of water using reverse osmosis, treatments in food industry, petrochemical vapor recovery and electrolysis for chlorine production. The medical applications of membrane technology are present in hemodialysis among many others. Research in membrane technology explores new techniques of wastewater treatment, energy recovery, etc. There has been rapid progress in this field and its applications are finding their way across multiple industries. This book covers in detail some existing theories and innovative concepts revolving around membrane science and engineering. Students, researchers, experts and all associated with this field will benefit alike from this book.

Various studies have approached the subject by analyzing it with a single perspective, but the present book provides diverse methodologies and techniques to address this field. This book contains theories and applications needed for understanding the subject from different perspectives. The aim is to keep the readers informed about the progresses in the field; therefore, the contributions were carefully examined to compile novel researches by specialists from across the globe.

Indeed, the job of the editor is the most crucial and challenging in compiling all chapters into a single book. In the end, I would extend my sincere thanks to the chapter authors for their profound work. I am also thankful for the support provided by my family and colleagues during the compilation of this book.

Editor

Casting Membrane of Acrylamide/Polymethacrylic acid and Reinforced by PAC for Application in Fuel Cell Unit

Abdel-Hady EE[1], El-Toony MM[2]* and Abdel-Hamed MO[1]

[1]Physics Department, Faculty of Science, Menia University El-Menia City, Egypt
[2]National center for radiation research and technology, Atomic energy authority, Egypt, 3 Ahmad El-Zomr street, P.O. Box 29- Nasr City, Cairo, Egypt. 11370

Abstract

Gamma irradiation poses very important role for casting the polymer membranes. Acrylamide was hydrolyzed by phosphoric acid, while it plays a potential role for coordination of poly methacrylic acid and hydrolyzed acrylamide (acrylic acid) around them. Adding PAC in different ratios for membrane reinforcement besides adsorbing excess of hydrogen proton in aqueous media which enhances proton conductivity. Thermal characterization of the membranes was carried out using thermal gravimetric analysis, while their mechanical properties were investigated by measuring hardness and Chemical description was discussed by studying FTIR and ion exchange capacity (IEC). Morphological characteristics were performed by scan electron microscope. Crystallinity was reviewed using X-ray diffraction which attempts the homogeny of the casted membrane. Electrical resistivity of the membrane was measured resulted in 7 ohm/cm while it reduced to 2.7 ohm by temperature rising to 80°C which confirmed the availability of the membrane in fuel cell usage.

Keywords: Casting; Phosphoric acid; Acrylamide; Methacrylic acid; PAC; Fuel cell; Electrical resistivity.

Introduction

Wetting of solids by polymer solutions is important in many aspects for polymer processing, including polymer blending, coating of films and production of reinforced polymer composites [1–12]. The influence of the surface on the formation of cast polymer films has mostly been investigated for Langmuir–Blodgett films [1–3], while the influence of wetting phenomena on polymer complex suspensions has been studied intensively in the paint, ink and coating literature [4–6]. The influence of wettability on phase separation of polymer blends [7–10] and co-polymers [11,12] has also been intensively studied. In membrane technology, wetting plays an important role in the preparation of flat-sheet membranes via phase inversion, more in specific during the casting of the membrane dope solution on a solid surface. While industrial production of flat-sheet membranes mostly takes place on woven/non-woven polymeric support layers, a variety of supporting materials is used in lab-scale experiments. Wetting/dewetting is in this respect often reflected in film shrinkage, either during casting and/or during immersion in the coagulation bath. During the immersion process, stresses are built up in the cast polymer films as a consequence of combined wetting/dewetting phenomena and polymer solidification. This induces the shrinkage of the formed membrane and eventually its release from the support. During the earlier stage of the membrane casting process, shrinkage can also be observed whenever the polymer film has a tendency to dewet the support surface. This is in contrast to the behavior of the same thin film cast on a wettable support, where the thin film remains immobile on the substrate. Numerous studies have been performed on acid-doped membranes concerning its proton conductivity [13–17], thermal stability [18], water drag coefficient [19,20], methanol crossover [21], and fuel cell tests [15,19]. The proton conductivity of polymer membranes increased when the acid doping level increased, while the mechanical strength of the membranes decreased significantly at the same time [19,21]. Therefore, a doping level of around 5 (5 mol of H_3PO_4 per mole of polymer repeat unit) is reported to be the maximum to maintain reasonable mechanical strength of polymer membranes for PEMFC applications. Lifetime tests of polymer based fuel cells showed that mechanical degradation of the membranes under operational conditions is one of the major reasons for the performance degradation of phosphoric acid-doped polymer fuel cells [22,23]. Optimization of casting solvent is very important for membrane fabrication as it affects the consequent separation performance. Macro-phase separation can be observed for blended membranes cast from polymer/solvent solution, and asymmetric membranes are prepared by phase separation methods [24]. Different solvent pairs result in different membrane morphology and performance. For symmetric (dense) membrane also, fabrication conditions can markedly affect permeation properties [25]. The membrane behavior of a given polymer strongly depends on both the intrinsic material properties and the membrane fabrication conditions. Moreover, practical applications, it is essential to understand the consequences of membrane fabrication conditions, e.g. the effect of solvent selection on the physical structure of the membranes must be carried out. Under the program of the New Energy and Industrial Technology Development Organization, Sanyo Electric Corp., Toshiba Fuel Cell Power System Corp., Matsushita Electric Industrial Corp., Ebara Ballard Corp. and Aishin Seiki Corp. have been examining the possibility of commercializing a 1kW-class PEMFC residential system based on polymer membranes that operates in the temperature region 70–80°C. Such low-temperature operation is an advantage of this type of fuel cell, but it also causes some problems, including low electrical efficiency [26,27] and CO poisoning of the platinum catalyst [28,29]. Furthermore, polymer electrolytes only exhibit high proton conductivities at high humidities. High-temperature PEMFCs (HT-

*Corresponding author: El-Toony MM, National center for radiation research and technology, Atomic energy authority, Egypt, E-mail: Toonyoptrade@yahoo.com

PEMFCs) have the potential to solve the problems inherent with LT-PEMFCs. In order to realize such a HT-PEMFC, a number of basic polymers have been investigated for the preparation of acid–base electrolytes, including polybenzimidazole (PBI) [30], polyethylene oxide (PEO) [31], polyvinyl alcohol (PVA) [32,33], polyacrylamide (PAAM) [34,35], and polyethylenimine (PEI) [36].

In this work casting of hydrolyzed acrylamide with polymethacrylic acid powdered, inserted through PAC. Different gamma irradiation doses were used for membrane compatiblization and composite preparation. Characterization of the membranes using FTIR, thermal gravimetric analysis (TGA), X-ray diffraction (XRD) and Scanning Electron Microscope (SEM). Some parameters were measured such as Ion Exchange Capacity (IEC) and water uptake prior to investigate electrical resistivity which reduced with raising temperature up to 80°C.

Experimental

Materials

- Commercial polymethacylic acid powder and powdered activated carbon (PAC) were purchased from OPTCo, Egypt.

- Reagent grade, Acrylamide of purity 98.5% was purchased from Cytec co., Italy, while phosphoric acid of purity 99.8% were supplied by Merck, Germany, other chemicals such as solvents, alkali acids etc., were reagent grade.

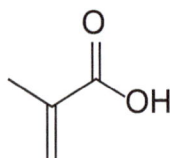

Methacrylic acid- (Picture1) NH_2-C(=O)-CHCH$_2$ –Acrylamide (Picture 2)

Activated carbon as seen at SEM-(Picture 3)   PAC as seen by naked eye-(Picture 4)

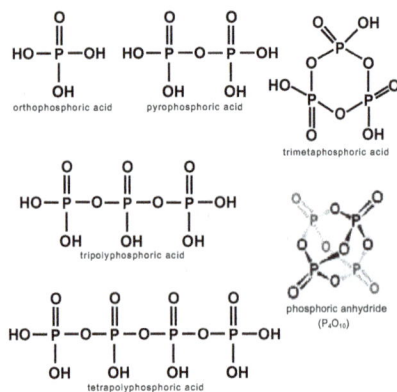

Series of polyphosphoric acids-(Picture 5)

Preparation of grafted membrane

10% of Acrylamide was solvated in double distilled water, with addition 1% of phosphoric acid, 1% polymethacrylic acid and

powdered activated carbon for different ratios 0.5%, 1% and 2% to the acrylamide used. The content of the ceiled vial was heated at 80°C for 3 hours. Pouring the viscous solution into clean and dry glass petridish in different thickness was performed. The petridish content was exposed to γ-rays from a ^{60}Co source with different doses. Evaporation of the irradiated soft gelatinous mixture in ambient temperature for 48 hours till attaining casted membrane was done. It detached carefully from the petridish while it kept in dried ceiled vial.

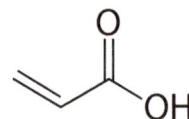

Acrylic acid-(Picture 6)

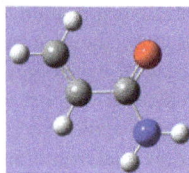

Ammonium polyphosphate is an inorganic salt of polyphosphoric acid and ammonia containing both chains and possibly branching. Its chemical formula is $[NH_4PO_3]_n$ showing that each monomer consists of an orthophosphate radical of a phosphorus atom with three oxygens and one negative charge neutralized by an ammonium anion leaving two bonds free to polymerize. In the branched cases some monomers are missing the ammonium anion and instead link to three other monomers. The properties of ammonium polyphosphate depend on the number of monomers in each molecule and to a degree on how often it branches. Shorter chains are more soluble and less thermally stable, but short polymer chains (e.g. pyro-, tripoly-, and tetrapoly-) show increasing solubility with increasing chain length.

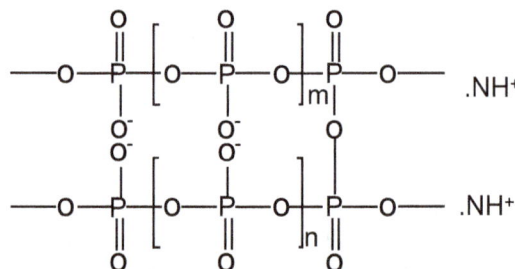

Different contribution of Ammonium phosphate-(Picture 7)

NH4+ -O-P(=O)(OH)2 = NH6PO4 = NH4H2PO4 = ammonium dihydrogen phosphate =

.............OH

....+....-...|

H4N...O-P=O

.............|

............OH

NH4+ -O-P(=O)(-OH)-O- +(NH4)+ = N2H9PO4 = (NH4)2HPO4 = ammonium hydrogen phosphate =

............OH

...+....-...|

H4N...O-P=O

.............|

............O...NH4

...............-....+

Triammonium phosphate (NH4)3PO4 (including 3x water of crystallization)

.............-....+

............O...NH4

...+....-...|

H4N...O-P=O x3 H2O

.............|

............O...NH4

..............-....+

Theoretical approach

Acrylamide was hydrolyzed by water in presence of phosphoric acid, to acrylic acid while poly ammonium phosphate was synthesized. Orientation of acrylic acid and poly methacrylic acid around the ammonium phosphate polymer chain facilitate transfer of hydrogen proton through the net matrix. PAC has a potential role for sorbing any migrated hydrogen protons which enhance their conductivity. Beside the previous mentioned role played by PAC that, it has highly electrical properties which facilitates the exchanging of hydrogen protons leading to amplification of hydrogen proton conductivity. These casted membranes have high water swelling properties due to their introduced hydrophilic properties beside their affinity to making permanent hydrogen bond with water.

Water uptake studies

Respective water uptake/swelling behavior of different ratios of grafted membranes and foam were studied in water as a function of different doses (KGy), and different composition. Swollen polymer was wiped off with tissue paper and then weighed immediately to know the swelling percent or water uptake percent, which was calculated as:

$$\text{Water uptake\%} = \frac{\text{Weight of swollen grafted membrane} - \text{Weight of dry grafted membrane}}{\text{Weight of dry grafted membrane}} \times 100 \quad (1)$$

Ion Exchange Capacity (IEC)

IEC of the polymer membranes was measured using a typical titration method. The dried membrane in the protonic form was immersed into 25.00 mL of 3 M NaCl solution for 24 h. A large excess of Na^+ ions in the solution ensured nearly complete ion exchange. Then, 10.00 mL of the 3 M NaCl solution containing the released H^+ ions was titrated against 0.05 M NaOH solution using phenolphthalein as indicator. The IEC was calculated using the following equation:

$$\text{IEC}_{exp} = \frac{0.05 \times V\,\text{NaOH} \times n}{W\text{dry}} \left(\text{m.equiv.}\,/\,g\right) \quad (2)$$

Where V NaOH (mL) is the volume of the 0.05 M NaOH solution used for titration. n is the factor corresponding to the ratio of the amount of NaCl solution taken to immerse the polymer (25.00 mL) to the amount used for titration (10.00 mL), which is 2.5, W_{dry} (g) is the dry weight of the polymer electrolyte membrane in the protonic form.

Ac Impedance Measurement

Proton conductivity measurements of casted membranes were carried out. Proton conductivity measurements were derived from AC impedance spectroscopy measurements over a frequency range of 50 to 10^6 Hz with an oscillating voltage of 50–500 mV, using a system 3532 Hioiki bridge LCR hitester. Each membrane sample was cut into sections 2.5 cm × 2.5 cm prior to being mounted in the cell. The cell was placed in a temperature controlled container open to air by a pinhole where the sample was equilibrated at 100% RH at ambient atmospheric pressure and clamped between two electrodes. The proton conductivities of the samples were measured in the longitudinal direction and were calculated from the impedance data, using the following relationship:

$$\sigma = \frac{l}{RS} \quad (3)$$

where σ is the proton conductivity (in S/cm), l is the distance between the electrodes used to measure the potential (l=1cm), S is the membrane cross-sectional surface area (membrane width × membrane thickness) for protons to transport through the membrane (in cm²) and R is derived from the low intersection of the high frequency semicircle on a complex impedance plane with the Re (Z) axis.

Scientific Equipments

FTIR characterization: The functional groups of the grafted membrane were studied using Mattson 1000, Pye-Unicam, England.

Scanning Electron Microscope: Investigation and magnification of the polymer surface was carried out by SEM, JEOL-JSM-5400; Japan.

Thermal Gravimetric Analysis: Shimadzu TGA -50, Japan was used to characterize the thermal stability of the casted membrane.

X-ray diffraction: X-ray diffraction pattern was recorded in the range of diffraction angle θ on philipe pw 1730, x-ray generated equipped with scintillation counter. The diffraction pattern were run with nickel filter (Cuka), λ =1.54A°. The X-ray diffractogram was obtained using the following experimental condition filament current 28 mA, voltage 40 Kv, scanning speed 20 mm/min.

Results and Discussion

Characterization of castesd membrane

FTIR: Poly phosphoric acid ammonium salt shows strong bands at 1040–910 cm⁻¹ theoretically while it appeared at 1074.5 cm⁻¹ that belong to asymmetric stretching vibrations of the P–OH group and at 1242 cm⁻¹ that corresponds to P-O stretching. At higher poly phosphoric acid ammonium salt ratio, the P–O–H vibration at 937 cm⁻¹ becomes stronger, indicating the existence of excess acidic protons. Additionally, phosphoric acid units give rise to broad bands with medium intensity at 1714 cm⁻¹ and 2903–2842 cm⁻¹ as seen in (Figure 1a, 1b, 1c and 1d). The intensity of these peaks increased in parallel with the concentration of phosphoric acid grafted through the net matrix while as PAC increased its concentration the peak intensity tend to be less. The band broadening around 3174 cm⁻¹ is due to the hydrogen bonding network which is necessary for proton conduction.

Thermal gravimetric analysis: The increase powdered activated carbon into the casted membrane improved their thermal stability. The thermogram could be divided into 5 divisions. First division described the membrane stability upon raising the temperature (working temperature of the membrane). This division has a temperature range up to 100°C, at which no loss of weight has been seen. Second division explained the slow weight decrease upon raising the temperature. The loss of weight through which was 6% for 1% PAC while it increased to

8% by raise the PAC content to 3% which may be due to promotion of volatile gasses by PAC insertion into the casted membrane. The range of temperature through this division was 100-220°C for 3% PAC and from 100-200°C for 1% PAC casted membrane. The third division described the convex zone regarding to temperature axe. The weight loss through PAC/ casted membrane was 17% for 3% PAC and 15% for 1% to a range of 220- 330°C while it reduced for 3% PAC of the membrane to a range of 220-230°C. The fourth division described the abrupt change of weight by temperature raising up, via which 1% PAC membrane, the temperature range was 320-370°C while it expanded to a range of 320-380°C for 3% PAC-membrane. The weight loss through this division showed similar loss of weight for 1 and 3% PAC /membrane which was 13%. The fourth division showed a weight reduction to 52% from the original weight for 1% PAC-membrane while it was 50% for 3% PAC-membrane. This loss of weight has performed up to 550°C.

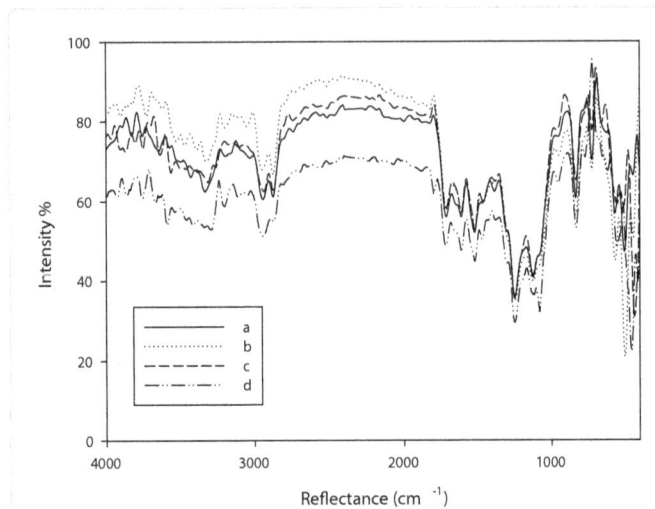

Figure 1: FTIR of polymethacrylic acid and poly phosphoric acid ammonium salt blend at different Gamma irradiation dose (KGy).

a) 10 KGy irradiation dose and 0.1% PAC
b) 20 KGy irradiation dose and 0.1% PAC
c) 20 KGy irradiation dose and 0.2% PAC
d) 20 KGy irradiation dose and 0.3% PAC

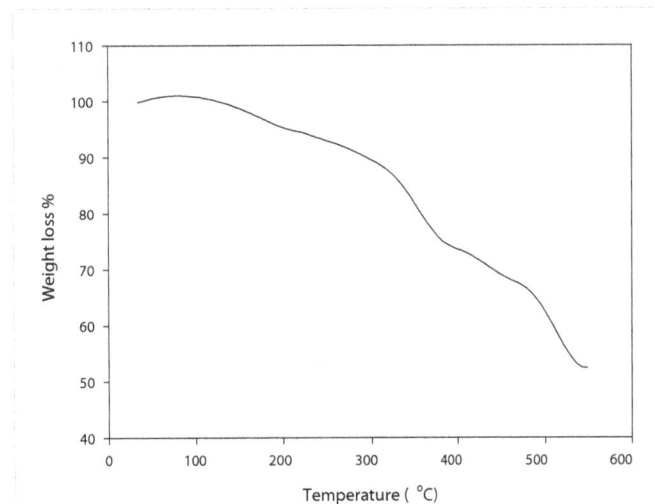

Figure 2: Thermogram of casted membrane of polymethacrylic acid/ polyacrylamide inserted in 1% PAC at 20 KGy irradiation dose.

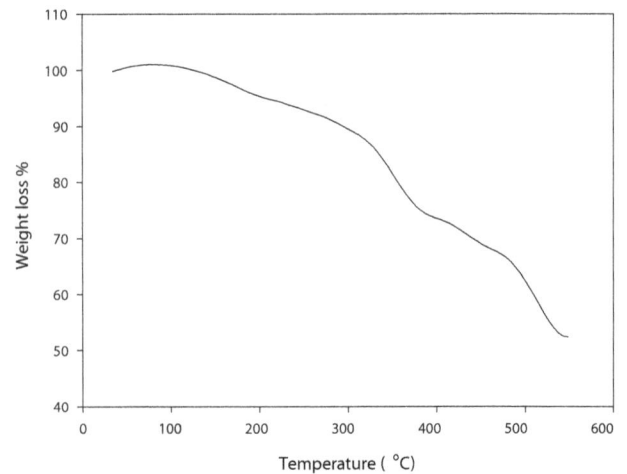

Figure 3: Thermogram of casted membrane of polymethacrylic acid / polyacrylamide inserted in 3% PAC at 20 KGy irradiation dose.

Figure 4: Scan electron microscope of polymethacrylic acid and poly phosphoric acid ammonium salt blend at different Gamma irradiation dose (KGy).

a) 10 KGy irradiation dose and 0.1% PAC
b) 20 KGy irradiation dose and 0.1% PAC
c) 20 KGy irradiation dose and 0.2% PAC
d) 20 KGy irradiation dose and 0.3% PAC

Scanning electron microscope: Figure 4a showed irregular scattering of black aggregates while their distribution is over the net matrix which may explain by more or less reaction of PAC with polymer. It can be seen dark zones distributed over white one which could pointed to less cross-linking of poly methacrylic acid and poly phosphoric acid ammonium salt.

Irradiation plays a very important role for homogenizing polymers comprises the sample. All the particles are appeared to be tightly closed, regular distribution of carbon aggregates in (Figure 4b) which is explained by their diffusion onto deepest layer of matrix. Whitish zones have seen and disappearance of black and white may due to cross-linking between poly phosphoric acid ammonium salt and poly methacrylic acid.

By increasing the PAC content as seen in (Figure 4c) it can seen larger content of PAC aggregate diffused into the net polymer while exceeding scattered regularly over the membrane surface. No contact among the PAC aggregates which seen as isolated islands over the sample. Large porous has noticed which may allow larger passage of hydrogen protons and increase the chance of water uptake.

In Figure 4d the cross-linking between the two blended polymers have been obviously seen while increase the PAC aggregates forming larger islands and in some cases resemble continuous series of big islands.

X-ray diffraction: The cast membrane showed highest crystallininty at 10 KGy gamma irradiation dose and 2% PAC content. While the membrane crystallinity decrease with increasing the irradiation dose to 20 KGy as it seen in (Figure 5b). By increasing the carbon content to 2% the maximum amorphousity has been aroused as it illustrated in (Figure 5c). The hump existed through which confirmed disappearance of the cast crystallinity. The amorphicity start to decrease by increasing the PAC content to 3% as it shown in (Figure 5d).

Water uptake

Water uptake is an important property for PEMs which needs to be seriously considered, because the loss of water is anticipated to bring two negative impacts on cell performance: decreased proton conductivity and degraded membrane-catalyst interface [37–39]. To verify the improved hygroscopic property of membranes by incorporating PAC particles, the water uptakes were measured. According to the Eikerling's theory, there are two kinds of water in PEMs [40,41]: one is bound water which solvates the function groups; the other is bulk water which fills the micro-pores. .An increase in the water uptake of membranes with the increase in the irradiation dose (KGy) was observed in (Figure 6). The date of non cross-linked membranes has been also studied. The increase in swelling corresponds well with the higher content of cross-linked phosphoric acid ammonium salt groups, which in turn increase the hydrophilicity in the membrane. The most interesting aspects is that; carbon content increase the water uptake till certain extent over which less water uptake which may due to reactivity of carbon content towards hydrogen proton which disrupted their ease of motion.

Ion exchange capacity

The ion exchange capacities were determined by acid–base titration and are plotted in (Figure 7). It was found that the ion exchange capacity increases with an increase in the degree of carbon content to certain percent over which it start to decreased. Carbon content played important role in reinforcement of casted membrane, more phosphoric acid groups are incorporated within the membranes, which led the membranes become more hydrophilic and enable to absorb more water in order to enhance the facilitation of the proton (H⁺) mobility through the membranes. Figure 7 listed the IEC values of the studied membranes showed the highest IEC value of 0.93 mmolg⁻¹ compared with the recast Nafion® of 1.5 mmolg⁻¹ .The increased IEC value illustrated that a little more acid sites resulting. From acidic poly phosphoric acid ammonium salt resulted and hydrolyzed acrylamide (acrylic acid) particles existed in the membrane composite. The enhanced acid property was expected to promote the membrane proton conductivity at various conditions [42–44].

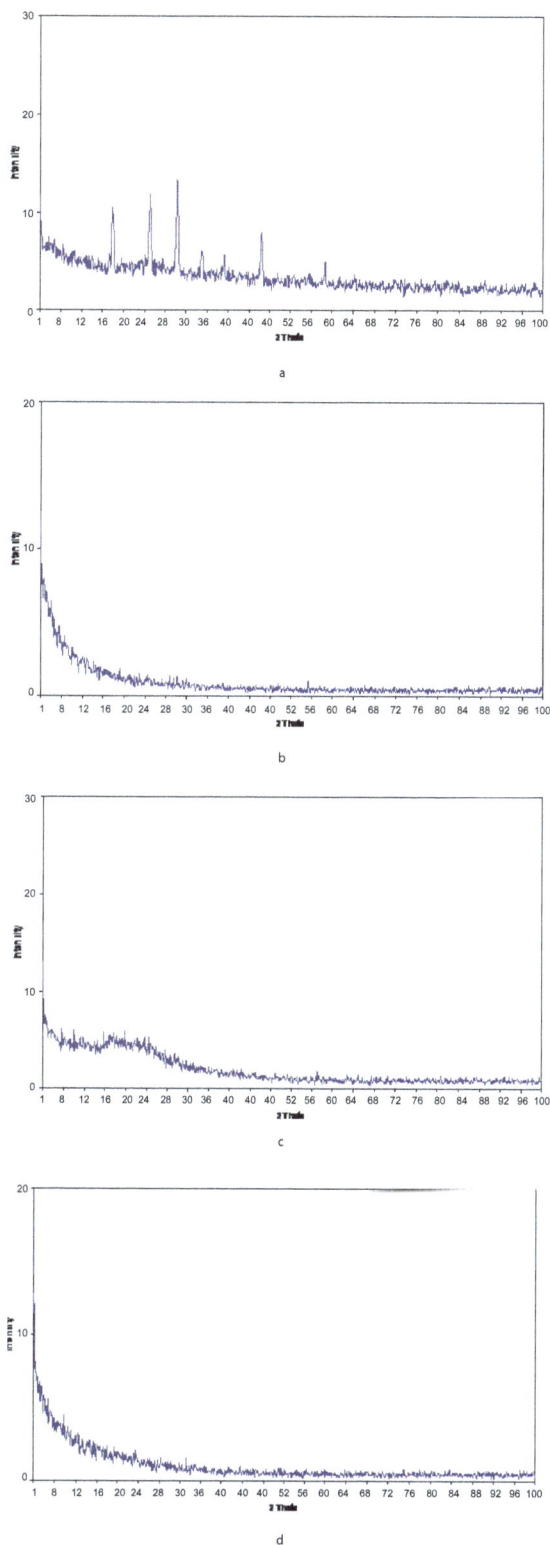

Figure 5: X-ray diffraction of polymethacrylic acid and poly phosphoric acid ammonium salt blend at different Gamma irradiation dose (KGy).

a) 10 KGy irradiation dose and 0.1% PAC
b) 20 KGy irradiation dose and 0.1% PAC
c) 20 KGy irradiation dose and 0.2% PAC
d) 20 KGy irradiation dose and 0.3% PAC

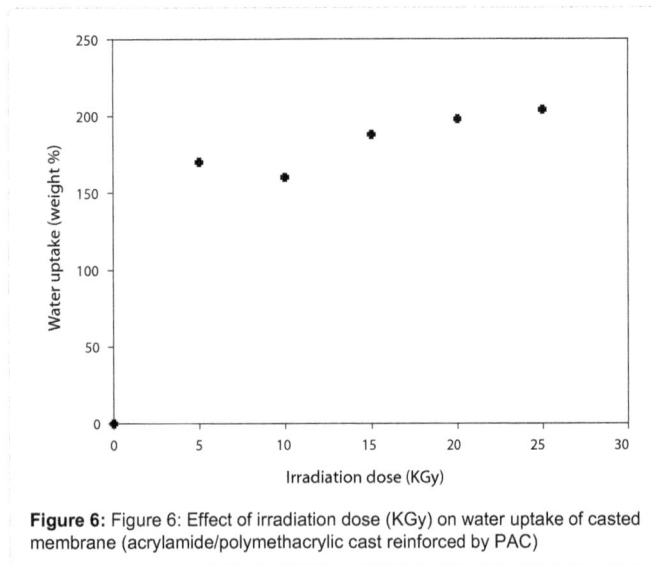

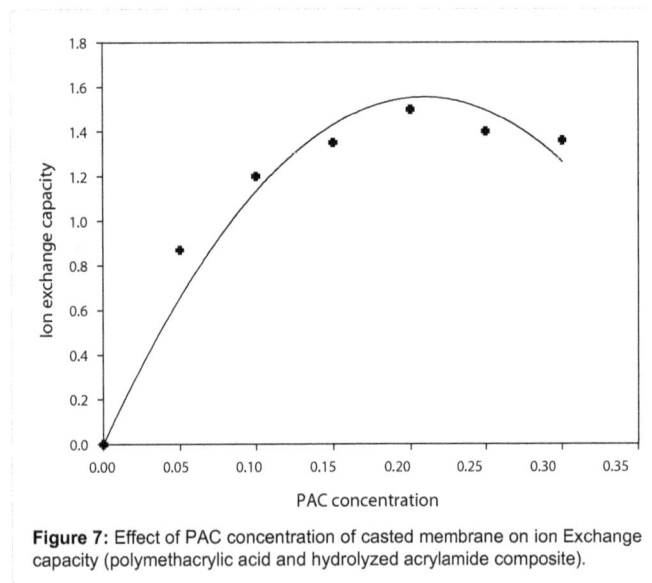

Figure 6: Figure 6: Effect of irradiation dose (KGy) on water uptake of casted membrane (acrylamide/polymethacrylic cast reinforced by PAC)

Figure 7: Effect of PAC concentration of casted membrane on ion Exchange capacity (polymethacrylic acid and hydrolyzed acrylamide composite).

Ac impedance

The proton conductivity of casted membrane (polymethacrylic acid and poly phosphoric acid ammonium salt blend) with various carbon content measured at room temperature is illustrated in (Figure 9). The specific resistivity is decreased with decreasing carbon content. The resistivity decreases from 17 to 7 ohm cm^{-1} as the carbon percent decreased from 0.3 to 0.05%. This is due to the ionic conductivity is a function of IEC, solvent uptake and hydration number of the membranes was found to be strongly dependence upon the less carbon content which play important role in reinforcement of casted membrane, more phosphoric acid groups are incorporated within the membranes, which conduct the membranes become more hydrophilic and enable to absorb more water in order to enhance proton (H$^+$) mobility through the membranes.

Raising the temperature leads to increasing up the kinetic energy of hydrogen proton to be diffused easily through the membrane's active sites. Temperature expands the pore size of the synthesized membrane and so enhancing the proton passage through which. Temperature has

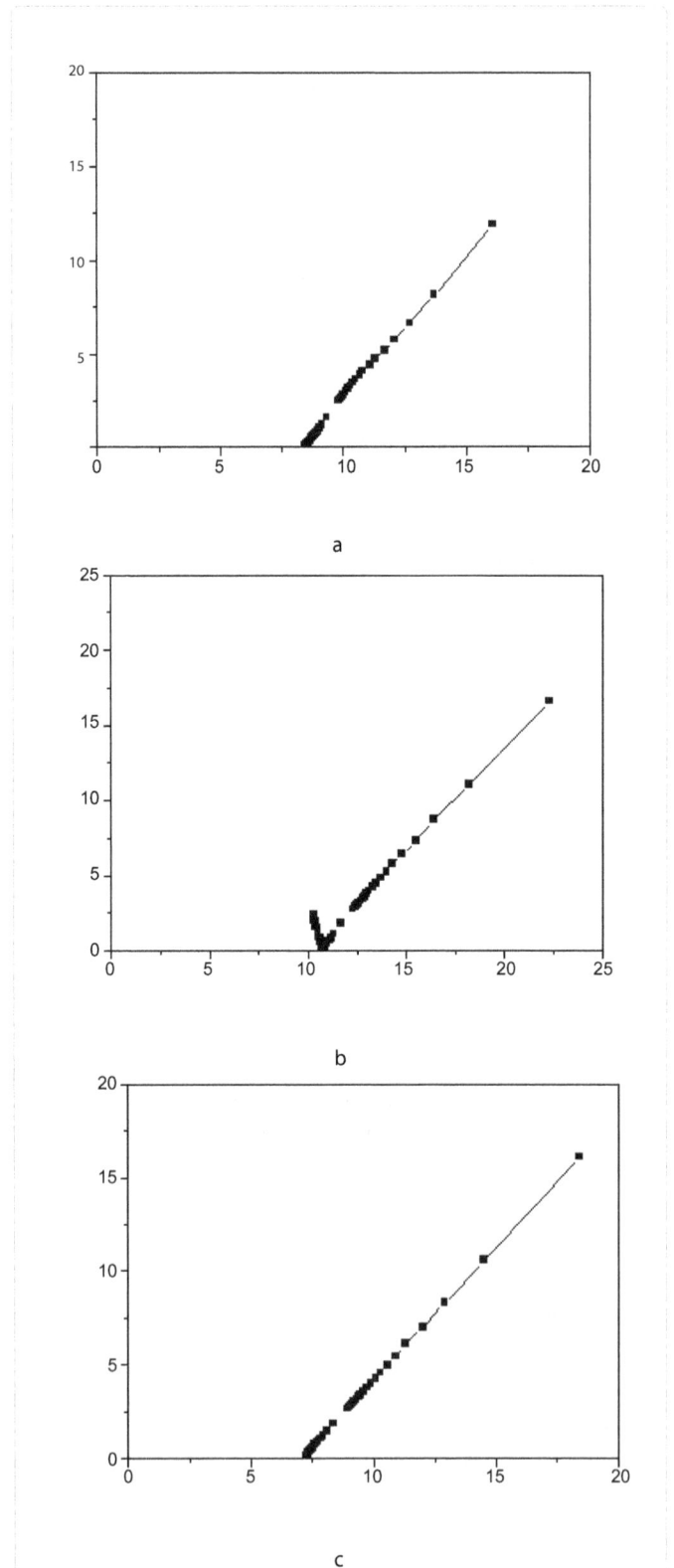

a

b

c

Figure 8: Nyquist plots of the electrodes of the casted membrane (polymethacrylic acid and hydrolyzed acrylamide composite) at different Frequencies

a) 10 KGy irradiation dose and 0.1% PAC
b) 20 KGy irradiation dose and 0.1% PAC
c) 20 KGy irradiation dose and 0.2% PAC

a potential role in activate the poly phosphoric ammonium salt and hydrolyzed acrylamide hanged through the membrane net matrix which increase their chance for accept and donate the hydrogen proton. All these factors increasing the electrical conductivity of the membrane as it could be seen in the (Figure 10) decreasing the EC with the temperature is linear relationship in the range of 25 to 80°C for the casted membrane [45].

In order to better understand some of the key factors involved in design of a composite membrane, some quantitative models of the conductivity are obtained. One of these models is based on the dusty-fluid model [46,47]. It is claimed that the proton diffusion obstruction presented in the polymer matrix is viewed as an additional frictional interaction with large immobile dust or gel particles. Within this framework, the inorganic additive is simply considered as an additional dust species immobilized within the polymer matrix. If the incorporated components are nonconductive (unlike carbon particles), the proton diffusive resistance will be increased and vice versa. So the objective of increased PEM conductivity can be achieved by the

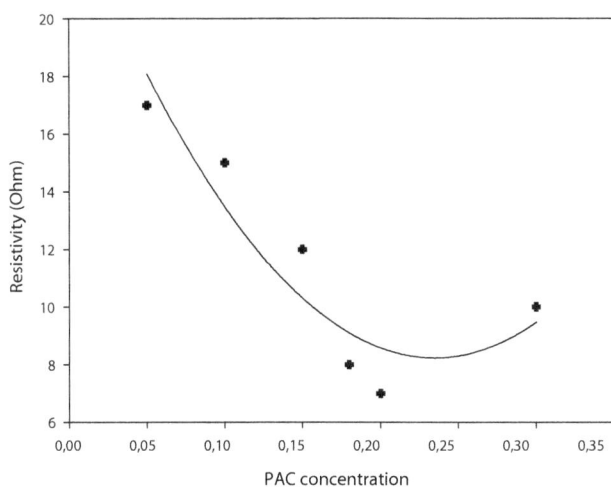

Figure 11: Effect of carbon content on the hardness of the casted membrane (polymethacrylic acid and hydrolyzed acrylamide composite).

presence of acidic additives (like carbon particles) [46]. The presence of acid sites on the surface of carbon particles increases the total numbers of acid sites within the PEM, which correspondingly increases the available numbers of charge carriers and offers more effective proton transfer channels to enhance the proton conductivity.

Hardness testing and durability

Evaluation of the mechanical properties of casted membrane with variation of carbon content which inserted through them for the purpose of reinforcement. Hardness measurements showed increase their value by increasing the carbon content to 2% while it decrease with small extent over which till 3%. These results attempt the availability of the membrane and their durability through all the carbon content range.

Conclusion

This work aims to synthesis low cost and efficient polymer membranes used in fuel cell. Acrylamide was hydrolyzed with phosphoric acid and mixed with polymethacrylic acid to prepare the cast composite. Powdered activated carbon (PAC) was inserted into the cast mixture for membrane reinforcement and has a role for sorption excess of hydrogen proton. The cast composite was irradiated with different doses (10 and 20 KGy) for compatiblization and blend formation. Characterization of the membranes using FTIR, Thermal gravimetric analysis (TGA), X ray diffraction (XRD) recommend their usage in fuel cell application. Measurements of some impacts such as water uptake and ion exchange capacity and hardness were carried out. Electrical resistivity investigations confirmed the membranes usage into fuel cell. Raising the temperature to 80°C resulted in reduction of resistivity to 2.7 ohm which well competes with commercial membranes (Nafion®).

References

1. Peach S, Robert D, Polak, Franck C (1996) Characterization of partial monolayers on glass using friction force microscopy. Langmuir 12: 6053-6058.

2. Li Z, Tolan M, Höhr T, Kharas D, Qu S (1998) Polymer thin films on patterned Si surfaces. Macromolecules 31: 1915-1920.

3. Muller BP, Stamm M (1998) Correlated roughness, long-range correlations, and dewetting of thin polymer films. Macromolecules 31: 3686-3692.

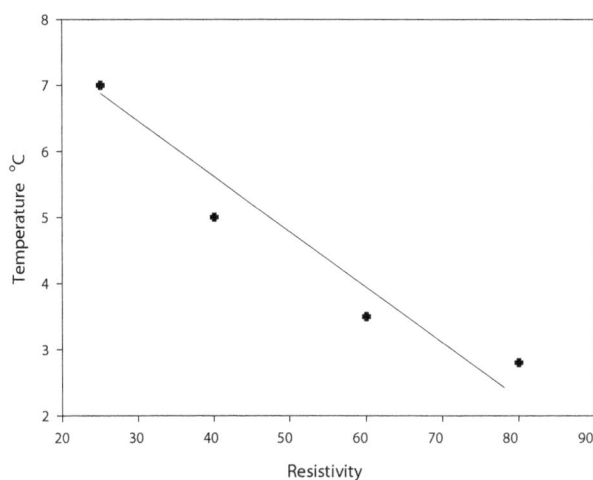

Figure 9: Effect of PAC % on Resistivity of the casted membrane (polymethacrylic acid and hydrolyzed acrylamide composite).

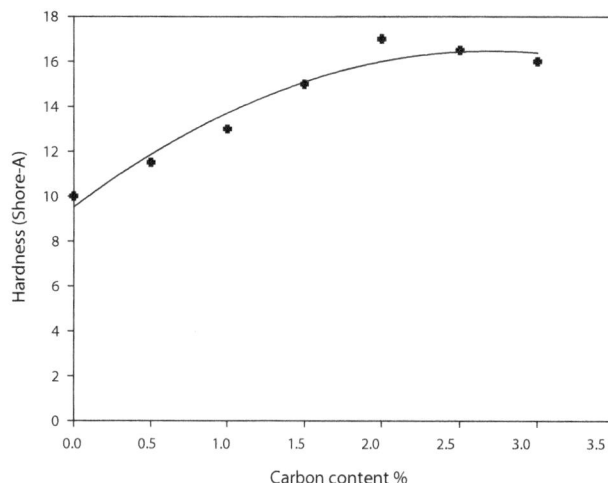

Figure 10: Effect of temperature on the resistivity of the casted membrane (polymethacrylic acid and hydrolyzed acrylamide composite).

4. Patton TC (1979) Paint flow and pigment dispersion. In: A Rheological Approach to Coating and Ink Technology. Wiley-Interscience Publication/John Wiley & sons, USA.

5. Bierwagen GP (1975) Surface dynamics of defect formation in paint films. Prog Org Coat 3: 101-113.

6. Hansen CM (1972) Surface dewetting and coatings performance. J Paint Technol 44: 57-60.

7. Hayashi M, Ribbe A, Hashimoto T, Weber M, Heckmann W (1998) The influence of wettability on the morphology of blends of polysulfones and polyamides. Polymer 39: 229-308.

8. Machell JS, Greener J, Contestable BA (1990) Optical properties of solvent cast polymer films. Macromolecules 23: 186-194.

9. Muller P, O'Neill SA, Affrossman S, Stamm M (1998) Phase separation and dewetting of weakly incompatible polymer blend films. Macromolecules 31: 5003-5009.

10. Schikk´en K, Claesson PM, Malmsten M (1997) Properties of poly(ethylene oxide)–poly(butylene oxide) diblock copolymers at the interface between hydrophobic surfaces and water, J Phys Chem B 101: 4238-4252.

11. Eskilsson K, Tiberg F (1998) Interfacial behavior of triblock copolymers at hydrophilic surfaces. Macromolecules 31: 5075-5083.

12. Wainright JS, Wang JT, Weng D, Savinell RF, Litt M (1995) Acid-doped polybenzimidazole: a new polymer electrolyte. J Electrochem Soc 142: L121-L123.

13. Savinell RF, Litt MH (1996) US Patent 5,525,436.

14. Wang JT, Savinell RF, Wainright J, Litt M, Yu H (1996) A H_2/O_2 fuel cell using acid doped polybenzimidazole as polymer electrolyte. Electrochim Acta 41: 193-197.

15. Bouchet R, Siebert E (1999) Proton conduction in acid doped polybenzimidazole. Solid State Ionics 118: 287-299.

16. Kawahara M, Morita J, Rikukawa M, Sanui K, Ogata N (2000) Synthesis and proton conductivity of thermally stable polymer electrolyte: poly(benzimidazole) complexes with strong acid molecules. Electrochim Acta 45: 1395-1398.

17. Samms SR, Wasmus S, Savinell RF (1996) Thermal stability of Nafion in simulated fuel cell environment. J Electrochem Soc 143: 1225-1232.

18. Li Q, He R, Jenson JO, Bjerrum NJ (2004) PBI based polymer membrane for high temperature fuel cell-preparation, characterization and fuel cell demonstration. Fuel Cells 4: 147-159.

19. Qingfeng L, Hjuler HA, Bjerrum NJ (2001) Phosphoric acid doped polybenzimidazole membranes: physic chemical characterization and fuel cell applications. J Appl Electrochem 31: 773-779.

20. Liu G, Zhang H, Hu J, Zhai Y, Xu D, et al. (2006) Studies of performance degradation of a high temperature PEMFC based on H_3PO_4-doped PBI. J Power Sources 162: 547-552.

21. Zhai Y, Zhang H, Liu G, Hu J, Yi B (2007) Degradation study on MEA inH3PO4/PBI high-temperature PEMFC life test. J Electrochem Soc 154: B72-B76.

22. Wijmans JG, Smolders CA (1986) In Bungay PM, Lonsdale HK, de Pinho MN (Eds.) Synthetic Membranes: Science, Engineering and Application. NATO ASI Series, Series C: Mathematical and Physical Sciences.

23. Brandrup J, Immergut EH (Eds) (1989) Polymer Handbook 3rd ed. Wiley, New York, USA.

24. Omata T (2007) J Fuel Cell Technol 7: 6.

25. Omata T (2008) Proceedings of the 4th International Hydrogen & Fuel Cell Expo, FC-2.

26. Shi W, Hou M, Shao Z, Hu J, Hou Z, et al. (2007) A novel proton exchange membrane fuel cell anode for enhancing CO tolerance. J Power Sources 174: 164-169

27. Wan CH, Zhuang QH (2007) Novel layer wise anode structure with improved CO-tolerance capability for PEM fuel cell. Electro Chim Acta 52: 4111-4123.

28. Xing B, Savadogo O (2000) Hydrogen/Oxygen Polymer Electrolyte Membrane Fuel Cells (PEMFCS) Based on Alkaline-Doped Polybenzimidazole (PBI). Electrochem Commun 2: 697-702.

29. Donoso P, GoreckiW, Berthier C (1988) NMR, conductivity and neutron scattering investigation of ionic dynamics in the anhydrous polymer protonic conductor $PEO(H_3PO_4)_x$. Solid State Ionics 28-30: 969-974.

30. Weeks SP, Zupancic JJ, Swedo JR (1988) Proton conducting interpenetrating polymer networks. Solid State Ionics 31: 117-125.

31. Vargas MA, Vargas RA, Mellander BE (1999) More studies on the $PVAl+H_3PO_2+H_2O$ proton conductor gels. Electrochim Acta 45: 1399-1403.

32. Lassègues JC, Desbat B, Trinquet O, Cruege F, Poinsignon C (1989) From model solid-state protonic conductors to new polymer electrolytes. Solid State Ionics 35: 17-25.

33. Rogriguez D, Jegat C, Trinquet O, Grondin J, Lassègues JC (1993) Proton conduction in poly (acrylamide)-acid blends. Solid State Ionics 61: 195-202.

34. Daniel MF, Desbat B, Cruege F, Trinquet O, Lassègues JC (1988) Solid state protonic conductors: Poly (ethylene imine) sulfates and phosphates. Solid State Ionics 28-30: 637-641.

35. Adjemian KT, Dominey R, Krishnan L, Ota H, Majsztrik P, et al. (2006) Function and Characterization of Metal Oxide-Nafion Composite Membranes for Elevated Temperature H_2/O_2 PEM Fuel Cells. Chem Mater 18: 2238-2248.

36. Thampan TM, Jalani NH, Choi P, Datta R (2005) Systematic approach to design of higher temperature composite proton-exchange membranes. J Electrochem Soc 152: A316-A325.

37. Kanamura K, Morikawa H, Umegaki T (2003) Observation of interface between Pt electrode and nafion membrane. J Electrochem Soc 150: A193-A198.

38. Eikerling M, Kornyshev AA, Stimming U (1997) Electro physical properties of polymer electrolyte membranes: A random network model. J Phys Chem B 101: 10807-10820.

39. Bia C, Zhanga H, Zhanga Yu, Zhuc X, Maa Y, et al. (2008) Fabrication and investigation of SiO2 supported sulfated zirconia/Nafion® self-humidifying membrane for proton exchange membrane fuel cell applications. Journal of Power Sources 184: 197-203.

40. Zhai Y, Zhang H, Hu J, Yi B (2006) Preparation and characterization of sulfated zirconia (SO_4^{2-}/ZrO_2)/Nafion composite membranes for PEMFC operation at high temperature/low humidity. J Membr Sci 280: 148-155.

41. Zhang Y, Zhang H, Zhai Y, Zhu X, Bi C (2007) Investigation of self-humidifying membranes based on sulfonated poly (ether ether ketone) hybrid with sulfated zirconia supported Pt catalyst for fuel cell applications. J Power Sources 168: 323-329.

42. Zhang Y, Zhang H, Zhu X, Bi C (2007) J Phys Chem B 111: 6391.

43. Abdel-Hady EE, Abdel-Hamed MO, El-Toony MM, El-Sharkawy MR (2011) Comparative study between sulfonation and phosphoration for commercial PTFE grafted with Styrene for fuel cell application. J Membra Sci Technol 1: 3.

44. Thampan TM, Jalani NH, Choi P, Datta R (2005) Systematic approach to design of higher temperature composite proton-exchange membranes. J Electrochem Soc 152: A316-A325.

45. Mason EA, Malinauskas AP (1983) Gas transport in porous media: The dusty gas mode. Elsevier, Amsterdam, Netherlands.

2

Investigation of Inlet Boundary Conditions on Capillary Membrane with Porous Wall during Dead-End Backwash

Hussam Mansour* and Wojciech Kowalczyk

Chair of Mechanics and Robotics, University of Duisburg-Essen, Lotharstr, Duisburg, Germany

Abstract

The capillary membrane technology has become one of the effective methods for producing drinking water. The membrane lifetime and permeability are significantly affected by operating and backwash conditions. To enhance the backwash process, the flow in the porous wall and the pressure drop inside the capillary membrane were investigated numerically. For this purpose, 3D model describing steady-state laminar flow inside the capillary membrane operated in dead-end mode was simulated. The influence of various boundary conditions on both the flow pattern inside the capillary membrane and the characteristic of the membrane were studied. Hereby, the pressure drop in the module and the axial as well as radial velocity profile were estimated with the consideration of the membrane fouling. The calculation of permeate flux contributes to increase the backwash performance and minimize energy consumption. The method of coupling Navier-Stokes equation for the free flow and Darcy-Forchheimer approach for the prediction of the flow in the porous membrane is proposed in the current study. The CFD model was validated by comparing the numerical results with the experimental data. A very good agreement was achieved.

Keywords: Ultrafiltration; Porous capillary; Dead-end; Backwash; CFD

Nomenclature

u: velocity (m/s)

p: pressure (Pa)

υ: kinematic viscosity (m²/s)

μ: dynamic viscosity (kg/m.s)

ρ: density (kg/m³)

D: viscous resistance (1/m²)

F: inertial resistance (1/m)

Introduction

Capillary membrane filtration process applied in water treatment makes use of the porous structure of the membrane. For the backwash process operated in dead-end mode, the driving force is generated by exposing the capillary to high outer pressure that forces the water to flow through the porous wall in outside-in direction. The magnitude of the driving force depends on many factors, among others, on the pressure drop in the capillary membrane. Predicting the behavior of the flow adjacent to the porous wall is essential to calculate the pressure drop and loss in the capillary and in the porous wall, respectively. Consequently, the energy consumption can be reduced, the permeate flux can be estimated and the performance of the backwash process can be improved.

The complexity of interaction between the free and porous flow regimes highlights the demand to develop a CFD model which can optimize the backwash operating parameters. Thus, the CFD model provides a library of boundary conditions that predicts the corresponding pressure drop in the capillary and the axial and radial velocity profile with the consideration of the membrane fouling. Furthermore, the model investigates the influence of the membrane characteristic and the operating parameters on the performance of the backwash process.

The flow inside a tube with porous wall has been for long time under investigation. The early models to solve Navier-Stokes equation for a laminar flow in porous tube were proposed in Berman [1] and Yuan [2] and extended by Bernales [3]. Berman [1] presented a solution to the two-dimensional model in laminar flow in channels with uniform wall suction. This study was extended to cylindrical coordinates to investigate suction as well as injection in flow along a porous channel [2]. Also there is a description of the flow in the porous wall for a circular tube with uniform suction [4-6] and for different permeability [7]. Further investigation was performed for a tube with variable wall suction [8]. Karode [9] derived analytical solution for the pressure drop in a rectangular slit and cylindrical tube with porous walls for a constant permeability. This expression was in an agreement with the solution proposed by Berman [1]. Another expression for the pressure drop and velocity was derived by Kim [10]. It is obtained by applying the perturbation theory to a dead-end cylindrical porous tube. More recently [11], one dimensional model is introduced to determine the fluid in a porous channel with wall suction or injection based on the analytical solution proposed by Berman and Yuan [1, 2].

Linking Darcy and Stokes equation [12] contributes to understanding of interfacial phenomena in coupled free and porous flow regimes applicable to Cross flow filtration. Various methods for coupling the free flow and the flow in permeable wall models were introduced and the strength and weakness of each method were discussed [13]. An analytical solution for the coupling between the transmembrane pressure and velocity in tubular membrane is proposed and validated with direct numerical simulation [14]. With Stokes

*Corresponding authors: Hussam Mansour, Chair of Mechanics and Robotics, University of Duisburg-Essen, Lotharstr. Duisburg, Germany
E-mail: hussam.mansour@uni-due.de

equation free flow dynamics and with Darcy equation non-isothermal, non-inertial and incompressible flow in a porous medium is modeled. Darcy's law was extended to non-Stokes flow and the resulting relation between pressure gradient and mass flux in porous media is presented [15].

In order to provide better insight into the flow and pressure drop in membrane channel with porous wall, a CFD model was developed which offers more detailed information about the physical phenomena accompanied with filtration process [16, 17]. Another work predicts the pressure drop and permeates flux in the hollow fiber using a simplified model equation [18].

Therefore, the optimization of the backwash process is associated with developing a numerical model which describes accurately the pressure drop under various boundary conditions and the flow through the porous wall of a capillary membrane in ultrafiltration module. Furthermore, adjusting the correct boundary conditions for further simulations of backwash process needs the knowledge of the velocity distribution in vicinity of the membrane surface.

Governing Equations

The fluid dynamic behavior of an incompressible, laminar and viscous flow in steady state is described by continuity Equation 1 and momentum Equation 2. The porous wall is simplified as homogeneous and isotropic medium. In order to incorporate the flow through the porous wall in the considered model, a source term S is added to the momentum equation [19]. This term is distributed in two parts, a viscous loss term D and an inertial loss term F. These two terms represent the properties of the fluid and the porous wall. The pressure drop has a proportional relation to the velocity in the viscous term and squared velocity in the inertial term whose effects become dominant with the increase of the velocity. The viscous parameter is calculated as the inverse of the permeability whereas the inertial term was not considered (F=0).

$$\nabla u = 0 \tag{1}$$

$$u\nabla u = \frac{1}{\rho}\nabla p + \upsilon\nabla^2 u + S \tag{2}$$

$$S = -(\mu D + \frac{1}{2}|u|F)u \tag{3}$$

Numerical Model

A single phase, steady-state, laminar, incompressible flow in a capillary membrane with a porous wall was carried out (Figure 1). The capillary length is 980 mm and has an inner and outer diameter of 1.4 mm and 2.3 mm, respectively. The equations above will be solved in a computational domain discretized in 11 Mio structured control volume. This high mesh resolution covers the whole computational domain and ensures mesh independent solution. Further mesh refinement has no influence on the solution and will only increase computational efforts and time. The element size is uniform distributed in the capillary inside the capillary membrane and can be seen in Figure 2.

The meshing is generated in ANSYS-ICEM with *mesh extension and read in OpenFOAM by the command fluentMeshToFoam. RenumberMesh command is used to rearrange the numbering of the cells in all time directories which will reduce the computational efforts.

Once the mesh is generated, the governing equation in the previous

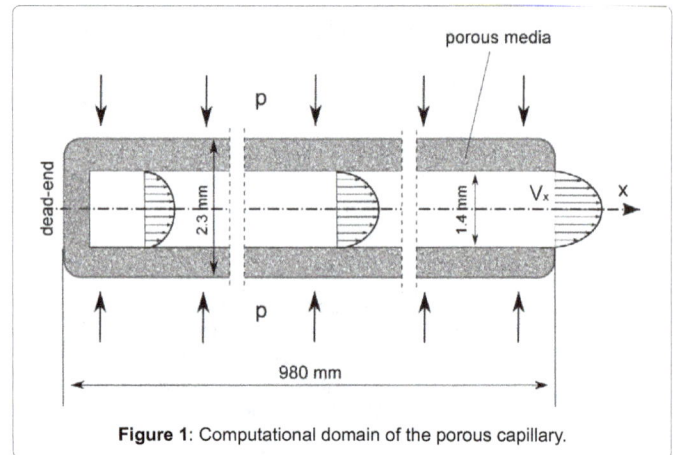

Figure 1: Computational domain of the porous capillary.

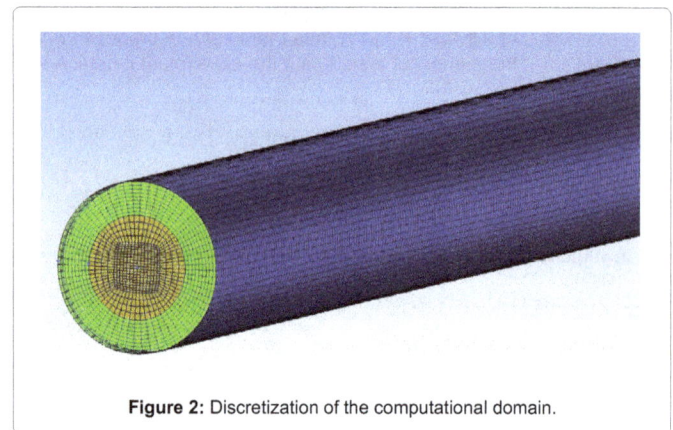

Figure 2: Discretization of the computational domain.

section is discretized by Gauss theorem to convert the volume integral to surface integral. For discretization of the convection term of the yielding governing equations is done by the upwind scheme which ensures boundedness and stability of the solution with an acceptable accuracy. Gauss linear scheme is used for the discretization of the diffusion term. The source term which is considered for the flow in a porous wall is treated explicitly in term of the temporal discretization. The resulting algebraic equations are solved based on the Semi-Implicit Method for Pressure Linked Equations called SIMPLE algorithm [20] to find a converged solution. The under-relaxation factors required to improve the stability of the solution are chosen for the velocity and pressure as 0.3 and 0.7, respectively. The overall iteration is assumed to be converged when the residual for each equation is below than $10e^{-5}$.

Boundary Conditions

The porous wall is homogeneous, isotropic and characterized by its viscous resistance D in equation 3 or permeability ε. Moreover, the used fluid water is assumed to be incompressible, isothermal and with a constant density and viscosity of 1000 kg/m and $10e^{-6}$ m²/s, respectively. The flow is laminar since a Reynolds number of less than 1000 is calculated for all the simulated conditions.

For each parameter such as pressure and permeability, a steady state simulation is carried out resulting the steady state flux, pressure drop and axial as well as radial velocity distribution. The boundary conditions are adjusted in such a way; they match the operating conditions of the experiment.

For all the simulations, the driving force for the water is the pressure gradient. Therefore, the pressure has a constant value at the inlet namely operating pressure and an atmospheric pressure at the outlet. No slip velocity at the dead-end is set which is treated as a wall. The components of the velocity gradient are considered to be zero. Different permeability values for the porous wall are taken into account. This variation of the capillary permeability considers the formation of the fouling layer since this layer has a porous structure.

Results and Discussion

The flow through the porous wall and inside the capillary membrane operated in dead-end mode was simulated. The operating parameters for the numerical simulation match those used in the experiment in terms of the applied pressure and the capillary permeability. The range of the boundary conditions applied in this numerical simulation is explained in the Table 1.

This range is adapted for the experiments used commonly for the UF in dead-end capillary membrane. The validation of the results is realized by comparing the flux at the outlet between the experimentally measured and simulated values. Figure 3 shows a good agreement between the simulation and the experiment results with very small deviations due to the accuracy of the measurements. Both simulation

Parameter	Simulated range
Permeability [liter/(m² h bar)]	80-400
Pressure [bar]	2-4

Table 1: Range of the studied boundary conditions.

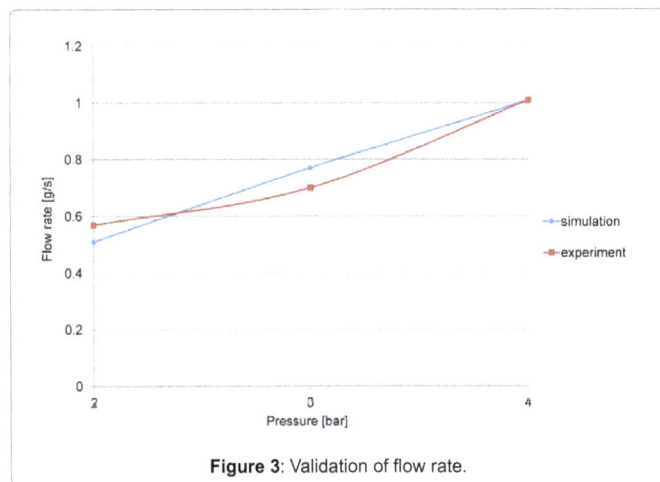

Figure 3: Validation of flow rate.

and experiment show that the flow rate is directly proportional to the applied pressure. An increase in the outer pressure on the capillary is accompanied with increase in the water flow rate at the outlet.

Figure 4 shows the axial and radial velocity profiles at laminar flow condition along different lines placed at different distances from the capillary axis. Considering the flow rate balance, the axial velocity increases from zero at the dead-end towards the outlet where the maximum value is reached. For the flow at distance of r=0.7 mm along the interface between the porous wall and the capillary surface the axial velocity is very small and refers to negligible tangential variation of the flow in the porous wall. However, the radial velocity at the interface and in the porous wall has a constant value along the capillary. It can be supposed that the velocity in the porous wall is composed only of radial component whereas inside the capillary the dominant component is the radial velocity.

Evaluating the axial velocity at three control lines inserted in different positions across the capillary and porous wall is illustrated in Figure 5. The velocity profile changes with the distance over the capillary length. The fluid accelerates towards the outlet resulting different maximum velocity and subsequently different Reynolds number. The developing velocity profile within the capillary membrane can be observed at the control lines. It also can be seen that the velocity through the porous wall has no axial component. This observation confirms the normal penetrating of the water to the capillary in the presence of the permeate wall. Here r* is in the abscissa which denotes the normalized diameter of the capillary where zero is on the axis. The dimensionless form simplifies the presentation of the results independent of the scale.

The pressure distribution along the three control lines across the capillary is shown in Figure 6. Over the cross section, the pressure in the capillary indicates a constant value whereas in the porous wall this value reaches its maximum.

The effect of various boundary conditions such as the operating pressure and membrane permeability on the flow inside the capillary is studied in the next section.

Effect of the Operating Pressure

Since the pressure gradient is the driving force for the backwash process, determining the pressure drop inside the capillary gains more interest.

The numerical results show that the pressure changes along the capillary length for various operating pressure. As it is expected, the pressure inside the capillary increases when the applied pressure

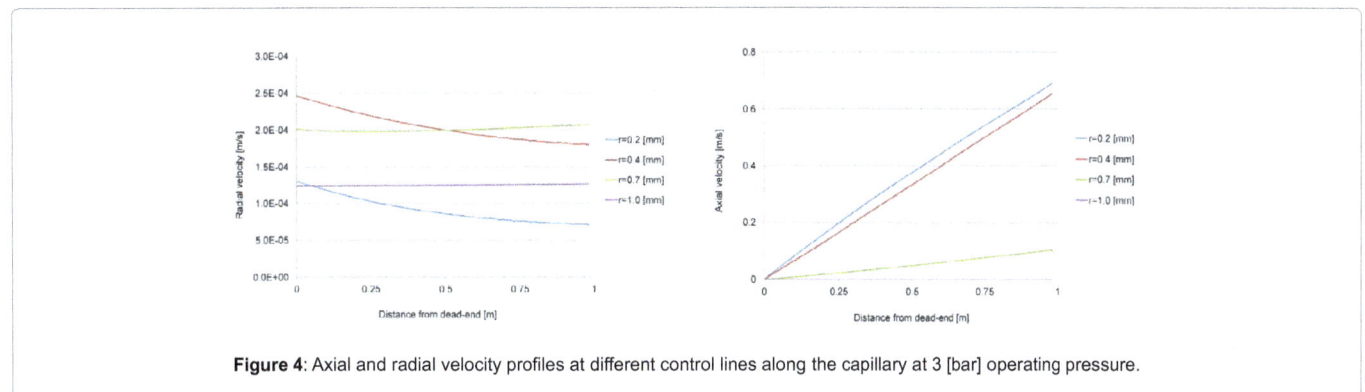

Figure 4: Axial and radial velocity profiles at different control lines along the capillary at 3 [bar] operating pressure.

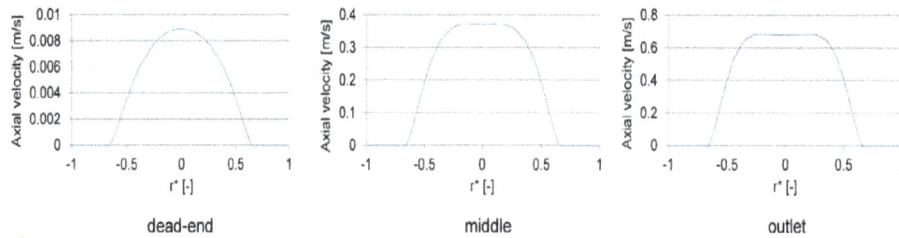

Figure 5: Axial velocity across the capillary at three control lines, at dead-end, in the middle of the capillary and at the outlet for 3 [bar] operating pressure.

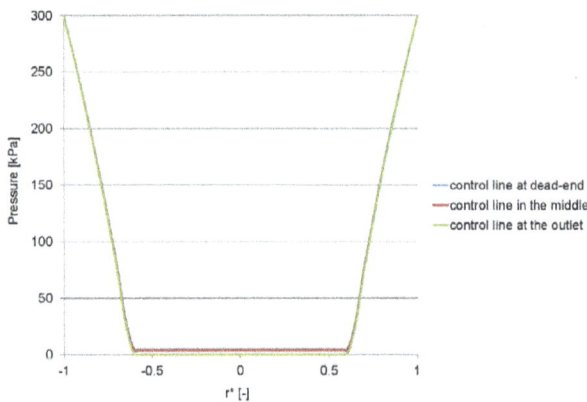

Figure 6: Pressure across the capillary at three control lines, at dead-end, in the middle of the capillary and at the outlet for 3 [bar] operating pressure.

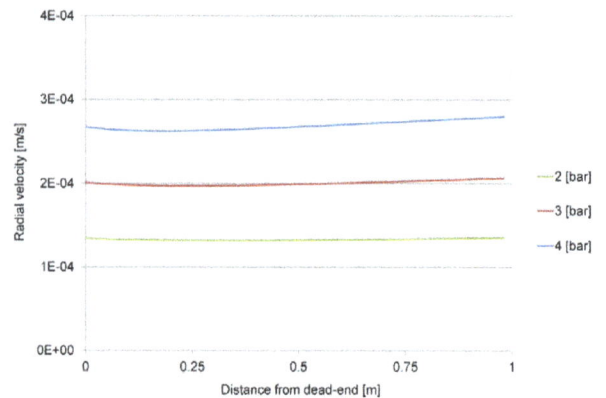

Figure 8: Radial velocity at the interface between porous wall and capillary surface r=0.7 [mm] for different operating pressure and permeability of 160 [liter/ (m² h bar)].

The radial velocity profile is independent of the operating pressure and has a constant value over the capillary length. In Figure 8 the radial velocity is plotted against the capillary length at the interface between the porous wall and capillary surface r=0.7 mm for different operating pressure. The profile variation from the dead-end to the outlet is very small and the velocity considered as constant. This assumption of uniform normal-wall velocity is found to be a reasonable approach and is applicable for further multiphase simulation when the permeable wall is neglected.

For the axial velocity, increasing the pressure gradient, as referred in Figure 9 leads to significant changes in the driving force which controls flow rate and the axial velocity. Reynolds number recorded from the axial velocity in the capillary at the outlet is 700, 980 and 1260 for 2, 3 and 4 bar operating pressure, respectively.

Effect of the Membrane Permeability

The fouling layer decreases membrane permeability and forms additional resistance for the water flow through the capillary. The total membrane resistance results from the sum of all involved resistance in series model namely, the membrane resistance, irreversible fouling resistance and cake layer resistance. TTherefore, the thickness of the fouling layer has a significant influence on the pressure drop inside the capillary and the operating pressure. Hereby, the variation of the membrane permeability due to the formation of fouling layer is taken into account.

Table 2 summarizes the flow rate resulting from various vales of the membrane permeability at constant operating pressure. The flow

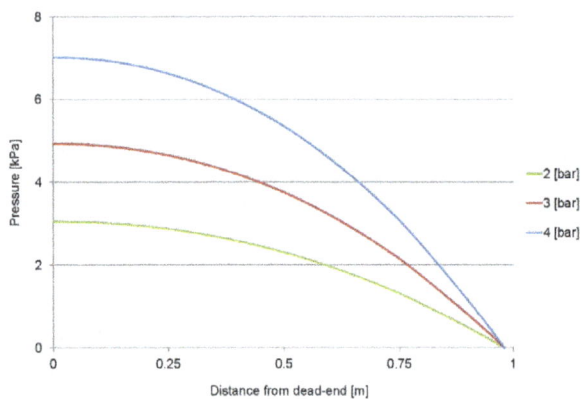

Figure 7: Pressure at the capillary axis for different operating pressure.

increases. The flow rate reduces towards the dead-end, thus increasing the pressure inside the capillary. The pressure starts with value zero at the outlet and increases to its maximum at the dead-end. Increasing the operating pressure improve the removal of suspended particles because the pressure gradient inside the capillary increases that will has consequences on the axial velocity (Figure 7).

At a small operating pressure the Reynolds number indicates laminar flow. However, the flow is fluctuating between laminar and turbulence when the pressure increases. This enhances the energy loss in the capillary and leads to more operating costs.

The results of the simulations show that, during the backwash water enters the capillary perpendicular to the capillary wall independent of the encountered conditions, such as the fouling layer and operating pressure. Moreover, the membrane permeability has a

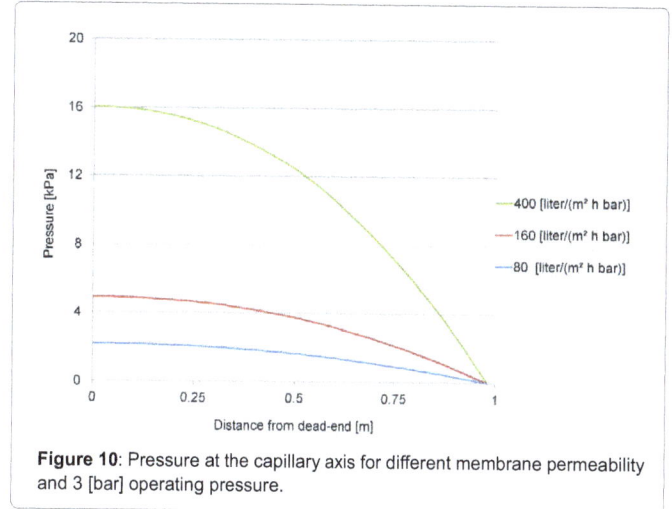

Figure 9: Axial velocity at the capillary axis for different operating pressure and permeability of 160 [liter/ (m² h bar)].

Permeability [liter/(m² h bar)]	Flux [g/s]
80	3.87e-7
160	7.7e-7
400	1.87e-6
800	3.43e-6

Table 2: Flux at different permeability and 3 [bar] operating pressure.

rate through the porous wall is increasing with the destruction of the fouling layer which changes the resistance towards allowing more water to penetrate the capillary membrane.

In Figure 10, the relation between the fouling layer and the pressure inside the capillary is presented. The pressure and related driving force are higher when the membrane permeability declines. It is assumed that the formation of the fouling layer is a modification of the membrane permeability. Thus, the removal of the deposited particles increases the pressure since it enhances the membrane permeability.

Based on the driving force inside the capillary, the estimated axial velocity at the capillary axis shows a proportional relation to the membrane permeability (Figure 11) the more the membrane permeability, the higher the axial velocity caused by the higher pressure gradient.

For a high permeability, the flow can pass through the porous wall easily and the radial velocity is increasing. However, the profile remains constant over the capillary length apart from the permeability value (Figure 12). These small changes in the radial velocity from the dead-end to the outlet have no considerable influence on the flow entering the capillary. Thus, again the assumption of a normal-wall velocity is an acceptable approach and can be used for further simulation without the consideration of the flow in the porous wall.

Conclusions

The effect of different boundary conditions on the backwash process is investigated. Simulations were carried out for capillary membrane model operated in dead-end mode.

It was found that converting the operating pressure into a constant flux over the capillary length is acceptable approach for the definition of the boundary condition. This assumption has a significant impact on further simulation for the considered model.

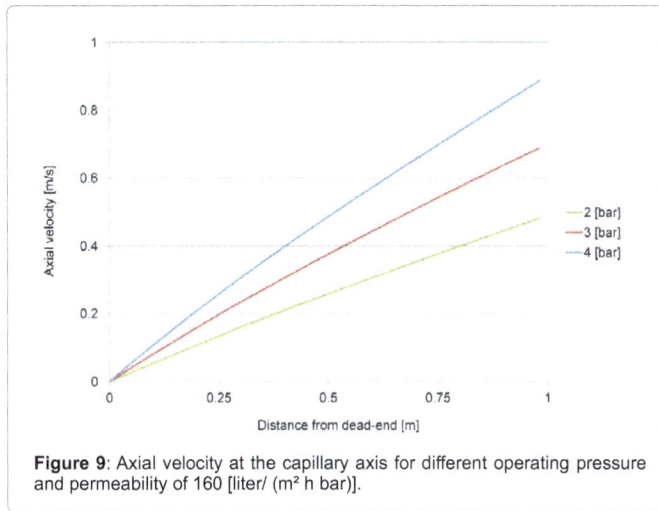

Figure 10: Pressure at the capillary axis for different membrane permeability and 3 [bar] operating pressure.

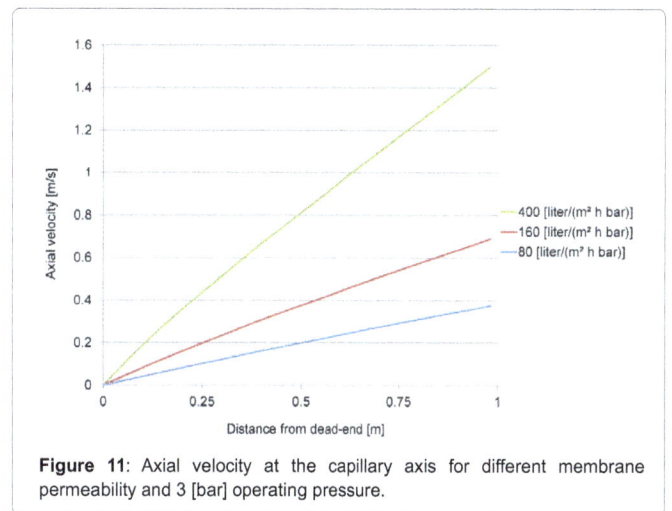

Figure 11: Axial velocity at the capillary axis for different membrane permeability and 3 [bar] operating pressure.

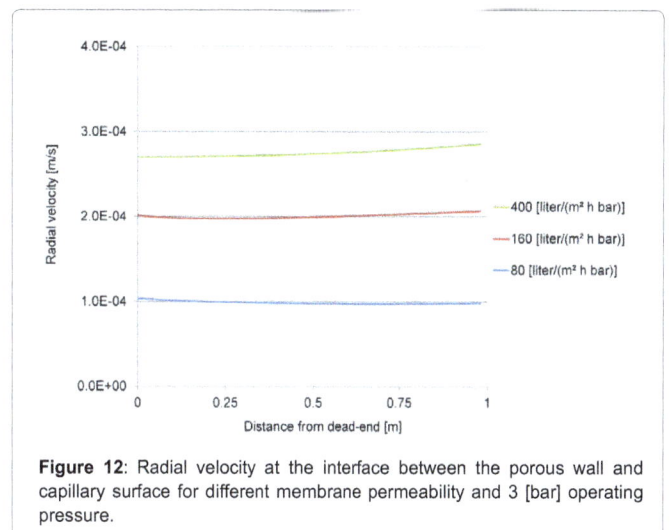

Figure 12: Radial velocity at the interface between the porous wall and capillary surface for different membrane permeability and 3 [bar] operating pressure.

proportional relation to the flux penetrating the porous wall, i.e. the flow rate increases when the permeability increases. This permeability represents the packing density of the retained particles on the capillary surface and can accordingly vary. In addition, increasing the operating pressure leads to higher flux at the same membrane permeability. The previously highlighted relation was subsequently confirmed by experimental validation which shows a good agreement with the numerical simulation for predicting the flux at the outlet for different boundary conditions.

Acknowledgements

The Deutsche Forschungsgemeinschaft (DFG) is gratefully acknowledged for the financial support as well as Professor Rolf Gimbel and Mrs. Anik Keller from Chair of Process Engineering and Water Technology, University of Duisburg-Essen for the experimental validation.

References

1. Berman AS (1953) Laminar flow in channels with porous walls. Journal of Applied Physics 24: 1232-1235.

2. Yuan SW (1955) Laminar Pipe Flow with Injection and Suction Through a Porous Wall. Trans. ASME 78: 719-724.

3. Bernales B, Haldenwang P (2014) Laminar flow analysis in a pipe with locally pressure-dependent leakage through the wall. European Journal of Mechanics, B/Fluids 43: 100-109.

4. Quaile JP, Levy EK (1975) Laminar Flow in a Porous Tube With Suction. Journal of Heat Transfer 97: 66-71.

5. Terrill RM, Thomas PW (1969) On laminar flow through a uniformly porous pipe. Applied Scientific Research 21: 37-67.

6. Terrill RM (1983) Laminar Flow in a Porous Tube. Journal of Fluids Engineering, Transactions of the ASME 105: 303-307.

7. Terrill RM, Shrestha GM (1966) Laminar flow through a channel with uniformly porous walls of different permeability. Applied Scientific Research 15: 440-468.

8. Galowin LS, Fletcher LS, Desantis MJ (1974) Investigation of Laminar Flow in a Porous Pipe with Variable Wall Suction. AIAA Journal 12: 1585-1589.

9. Karode S (2001) Laminar flow in channels with porous walls. Journal of Membrane Science 191: 237-241.

10. Kim AS, Lee YT (2011) Laminar flow with injection through a long dead-end cylindrical porous tube: Application to a hollow fiber membrane. AIChE Journal 57: 1997-2006.

11. Oxarango L, Schmitz P, Quintard M (2004) Laminar flow in channels with wall suction or injection: a new model to study multi-channel filtration systems. Chemical Engineering Science 59: 1039-1051.

12. Hanspal NS, Waghode AN, Nassehi V, Wakeman RJ (2009) Development of a predictive mathematical model for coupled stokes/Darcy flows in cross-flow membrane filtration. Chemical Engineering Journal 149: 132-142.

13. Nassehi V (1998) Modelling of combined Navier–Stokes and Darcy flows in crossflow membrane filtration. Chemical Engineering Science 53: 1253-1265.

14. Tilton N, Martinand D, Serre E, Lueptow RM (2012) Incorporating Darcy's law for pure solvent flow through porous tubes: Asymptotic solution and numerical simulations. AIChE Journal 58: 2030-2044.

15. Teng H, Zhao TS (2000) An extension of Darcy's law to non-Stokes flow in porous media. Chemical Engineering Science 55: 2727-2735.

16. Wiley DE, Fletcher DF (2003) Techniques for computational fluid dynamics modelling of flow in membrane channels. Journal of Membrane Science 211: 127-137.

17. Pak A, Mohammadi T, Hosseinalipour SM, Allahdini V (2008) CFD modeling of porous membranes. Desalination 222: 482-488.

18. Ghidossi R, Daurelle JV, Veyret D, Moulin P (2006) Simplified CFD approach of a hollow fiber ultrafiltration system. Chemical Engineering Journal 123: 117-125.

19. Forchheimer P (1901) Wasserbewegung durch Boden. Zeitschrift des Vereines Deutscher Ingenieuer 45: 1782-1788.

20. Ferziger JH, Peric M (2002) Computational Methods for Fluid Dynamics. Springer London, Limited.

3

New Ion Selective Sensitive Electrode of Pd (II) as Multisensor Based on IRA-410 via Low Cost Oxidation Reduction Process

AT Kassem*, N El Said, and HF Aly

Hot Labs. Center, Atomic EnergyAuthority, P.C. 13759, Cairo, Egypt

Abstract

A novel ion selective IRA-410 membrane disc sensor for Pd (II) ions has been prepared and studied. This electrode has a wide linear dynamic range from 10^{-1} to 2.5×10^{-6} mol^{-1} with a Nernstian slope of 16.5 ± 0.2 mV decade^{-1} and low detection limit of 1.6×10^{-6} mol l^{-1}. It has a fast response time (<1 s) and good selectivity with respect to different metal ions. IRA-410 based electrode was suitable for aqueous solutions of pH range from (1.0- 9.0). It can be used for about 10 months with complex with Pd (II) was calculated by using segmented sandwich membrane method. The formation constant of ionophore of IRA-410 and its Pd (II)-Complex is examined using Fourier-transform infrared analysis and elemental analysis techniques. The proposed electrode has been used successfully as an indicator electrode in potentiometric determination in aqueous nitrate and /or chloride media.

Keywords: Membrane sensor; Nernstian slope; IRA-410 based electrode; Ionophore complex; Aqueous and nitrate media.

Introduction

Separation of palladium by strongly basic anion exchangers IRA-410 and IRA-900 from intermediate radioactive liquid waste in chloroacetic acid/nitrate medium containing thirteen elements have been achieved. Different conditions have been studied, the effect of NaNO$_3$ as salt content, chloroacetic acid and hydrogen ion concentration have been investigated. Selective recovery of palladium from the [ILLW] solution was achieved using the column technique. The selectivity increased by using chloroacetic acid/nitrate than in nitrate medium. The elution of palladium was carried out via reduction with formic acid [1-3] where the rate of the reduction process was increased by decreasing the formic acid concentration till 50% followed by dissolving the separated palladium by nitric acid. Selective efficient method is used for separation of palladium [4-12] by strongly basic anion exchangers IRA-410 and IRA-900 from intermediate radioactive nitrate medium different conditions for exchange behaviour of palladium from (ILLYO solutions containing number of elements were investigated by batch technique. Selective recovery of palladium from (ILL) solution was achieved using column technique the elution [13] of palladium was carried out via reduction with formic acid. Ion-selective electrode based potentiometry has become a well-established electro-analytical technique. In this technique the most exciting and fastest growing area of research is the use of ion sensitive membrane electrodes for analysis of wastewater containing heavy metals. Using this approach the applicability of the potentiometric method has been greatly extended [14-17] enabling the simple and accurate determinations of many heavy metal ions and has led to a search for suitable materials that can be used for preparation of sensitive and selective ion-sensors, chemical sensors or more commonly ion-selective electrodes ((ISEs) [18,19]. Ion selective electrode based on palladium (II) dichloride acetylthiophene fenchone azine (I) has been developed.

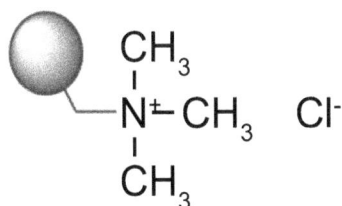

Figure 1: Structure: Amberlite IRA- 410 in chloride medium.

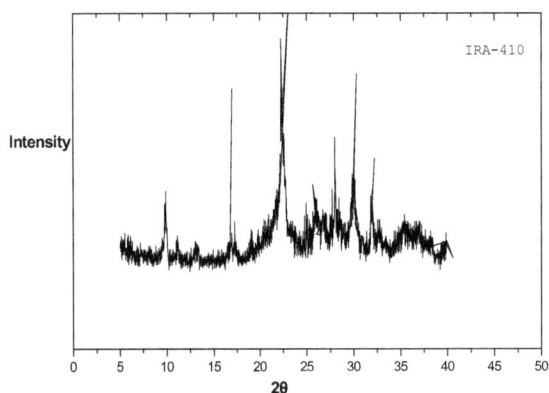

Figure 2: XRD Pattern for PVC on pd after.

Experiment

Reagents and chemicals

The strongly basic anion exchanger was used as previously described [4]. The plasticizers were obtained from Aldrich (Milwaukee, WI). While poly vinyl chloride powder (PVC) were obtained from Fluka (Buchs, Switzerland). The chloride salts of all cations studied (Figure 1).

XRD and SEM characterization of Amberlite-410

As it is shown in Figures 2, 3a and 3b, Pd (II) was chosen as the

*Corresponding author: AT Kassem, Hot Labs. Center, Atomic Energy Authority, P.C. 13759, Cairo, Egypt, E-mail: nessemsalam@gmail.com

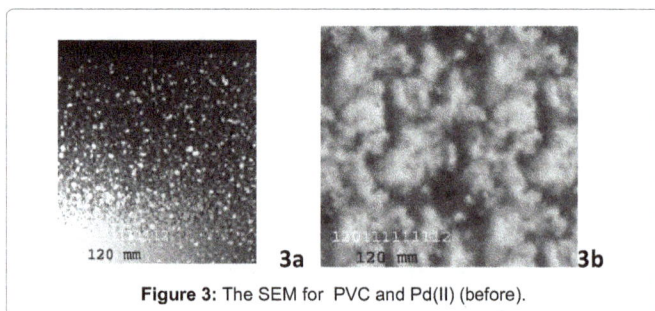

Figure 3: The SEM for PVC and Pd(II) (before).

Resin	C%	N%	O%	Cl%
IRA-410	85	12	2.33	0.67

Table 1: Elemental analysis results after addition the IRA-410.

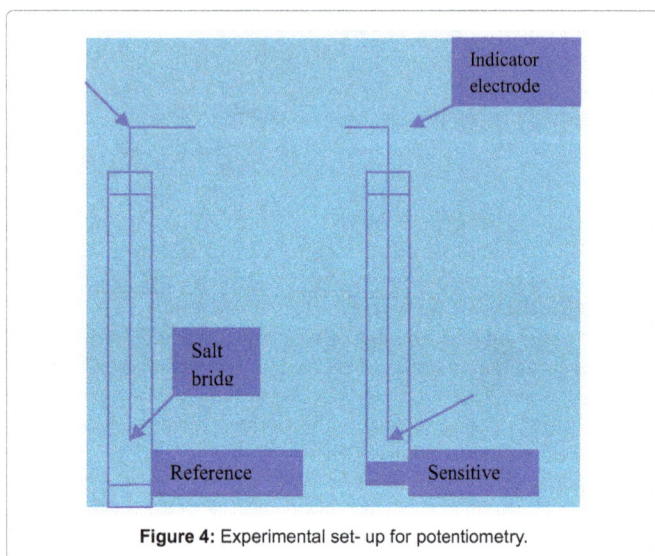

Figure 4: Experimental set- up for potentiometry.

Membrane	PVC	Composition % Plasticizer	NBPP	Additive	Slope/mV decade^{-1})
1	60	2,NaTFPB	----------	2,KTClPB	12.5 ± 0.2 mV
2	60	2,NaTFPB	----------	2,KTClPB	28.4 ± 0.2 mV
3	60	2,NaTFPB	0.8	4,KTClPB	14.6 ± 0.2 mV
4	60	4,NaTFPB	1.2	4,KTClPB	9.8 ± 0.2 mV
5	60	2,NaTFPB	1.6	3,KTClPB	12.4 ± 0.2 mV

Table 2: Optimization of membrane ingredients.

appropriate sorbent for the sorption of Amberlite-410 among all the sorbents tested in this work. X-ray powder diffraction (XRD) characterization showed that the natural samples of Amberlite-410, Figure 2: XRD pattern of acid-treated. Amberlite-410 with 2MHCl almost pure. Elemental content of the mineral was revealed using energy dispersive X-ray spectroscopy (EAR). The percentages of the elements are given in Table 1. The values given correspond to an average of Data points selected randomly on the surface of Amberlite-410. A scanning electron microscope (SEM) micro image is provided in Figures 2, 3a and 3b. This Figures show typical Amberlite-410 crystals with sizes varying up to several μm.

Powder XRD pattern of Pd nanoparticles is shown in Figure 2. The d-spacing corresponding to XRD lines is 2.236, A°. These d-spacing values correspond to (111), (planes with lattice constant, a = 3.871 A°. This observation confirms the presence of metallic Pd with fcc structure. XRD line corresponding to {111} plane is found to be

unusually intense. SEM images of black particles are shown in Figure 3a and 3b. Aggregates of irregular-shaped particles are observed and the size of Pd particles varies from 8 to 25 nm. Of particles are formed due to the self-assembling nature of palladium tetra chloroacmplox on IRA-410. This self-assembly of the particles also confirms the capping ability of Pd on IRA-410.

Apparatus: Potentiometeric measurements were performed at 25 ± 1°C using a Fisher scientific-computer aided pH/ milivoltmeter (Model 450) with a palladium PVC matrix membrane electrodes in conjunction with a double junction Hg|HgCl$_2$ |KCl (satd.) reference electrode (Cole-Parmer Co., Chicago, Illinois 60648). A Fisher Accumet Model 825 MP pH-meter (accuracy ± 0.00 1 pH) with a glass pH electrode (Fisher electrode no. 13-639-90) was used for all pH measurements). Figure 4: using the KBr technique was used for infrared measurements. (Figure 3a,b) (Figure 4) (Table 2).

Calibration Curve: Figure 5: These sensors exhibit the maximum working concentration range of 1.0×10^{-7} to 1.0×10^{-3} M with a slope of >16 mVdecade-1 of activity (Sensor of Pd (II) ion selective electrode). The PVC-membranes were prepared and aggregation into sensor electrodes using established procedures [20-24]. The prepared membranes contained 1.2 mg ionophore (IRA-410), 0.60 mg lipophilic salt, 60 mg PVC and 10 mg membrane solvent were mixed with 0.5 ml (THF) I mixed together very well and making compact disk with diameter 0.5 cm and thickness 0.2 cm and drying. In the glass dish with an diameter of 30 mm resting on a smooth mould. THF tetrahydrofuran was allowed to evaporate for 48 h standing at room temperature. Transparent PVC membranes were obtained with a thickness of 0.2 mm. A 11 mm diameter piece was cut out from the PVC membrane and attached to a PVC tube by means of PVC–THF solution Figure 6: wavelength(nm)

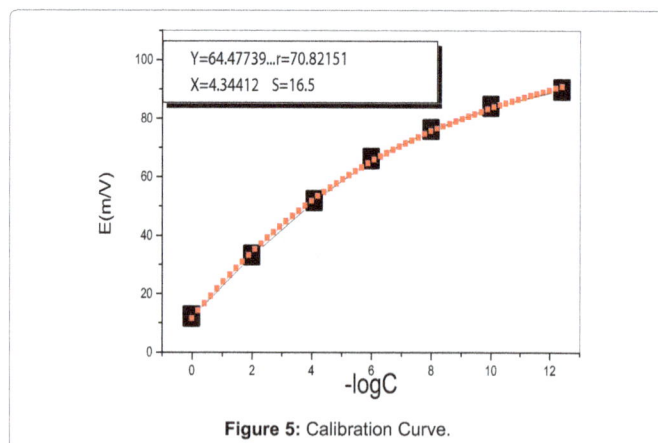

Figure 5: Calibration Curve.

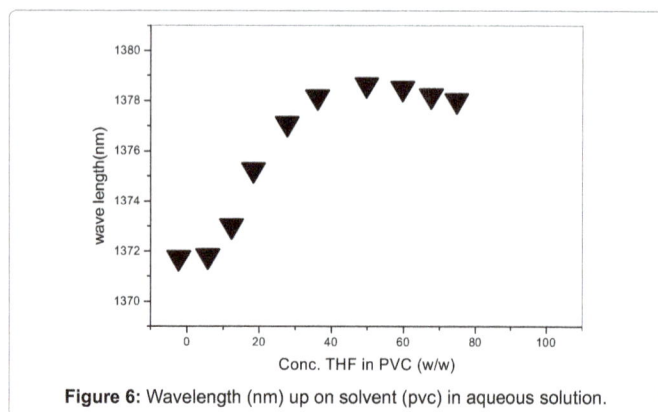

Figure 6: Wavelength (nm) up on solvent (pvc) in aqueous solution.

New Ion Selective Sensitive Electrode of Pd (II) as Multisensor Based on IRA-410 via Low Cost Oxidation Reduction Process

17

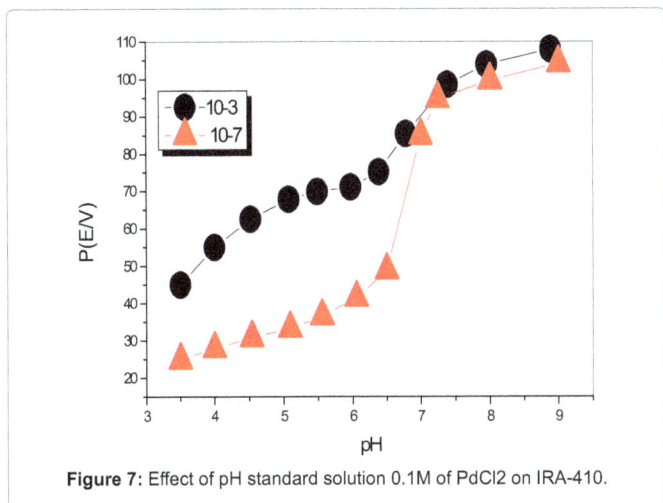

Figure 7: Effect of pH standard solution 0.1M of PdCl2 on IRA-410.

upon solvent(PVC)in aqueous solution as previously described [25]. The PVC tube with the membrane was then incorporated into an $Hg|HgCl_2|KCl$ inner electrode (1.0 mm diameter). The dried tube was filled with internal solution contained 10^{-3} moll^{-1} pdCl$_2$ and 10^{-3} mol l^{-1} KCl. Then, the electrode was conditioned for 1 h in 10^{-3} mol l^{-1} pdCl$_2$ solution [26].

Preparation of PVC–membrane: The PVC-membranes were prepared and aggregation into sensor electrodes using established procedures (20-24). The prepared membranes contained 1.2 mg ionophore (IRA-410), 0.60 mg lipophilic salt, 60 mg PVC and 10 mg membrane solvent were mixed with 0.5 ml (THF) I mixed together very well and making compact disk with diameter 0.5 cm and thickness 0.2 cm and drying. In the glass dish with a diameter of 30mm resting on a smooth mould. THF tetrahydrofuran was allowed to evaporate for 48 h standing at room temperature. Transparent PVC membranes were obtained with a thickness of 0.2 mm. A 11 mm diameter piece was cut out from the PVC membrane and attached to a PVC tube by means of PVC–THF solution as previously described (25). The PVC tube with the membrane was then incorporated into an $Hg|HgCl_2|KCl$ inner electrode (1.0 mm diameter). The dried tube was filled with internal solution contained 10^{-3} moll^{-1} pdCl$_2$ and 10^{-3} mol l^{-1} KCl. Then, the electrode was conditioned for 1 h in 10^{-3} mol l^{-1} pdCl$_2$ solution [26].

Potential measurement: All potential measurements were performed at 25 ± 1°C using a Fisher scientific-computer aided pH/milivoltmeter (Model 450). The electrochemical system for this electrode can. The performance of the electrode was investigated by measuring its potential in palladium chloride solutions prepared in the concentration range (10^{-1} to 10^{-7}) mol l^{-1} by gradual dilution of stock standard solution 0.1 mol l^{-1} of pdCl$_2$, with triply distilled water. The potentiometric selectivity coefficients (log K Pot Pd, B) were measured using the separation solution method (SSM) and the mixed solution method (MSM) [27,28]. Dependence of pH on electrode response was examined (Figure 7): Adjusting the pH of the measured standard solution with 1×10^{-3} mol^{-1} hydrochloric acid or sodium hydroxide solutions (Figures 6 and 7).

Underwent response Sensors palladium according the equation Nernstian with selectivity similar and the knowledge that the sensors palladium prepare traditional because we prepare palladium disc, and put the bottom of reference electrode .The electrode signal feels palladium in another solutions whether the sample solution sea water or others and, of course, used plastic films made of PVC, which showed

long-term response to senstivity this led to the use of sensors for a long period of more than 6 months and under the continued use of this investigation.

Results and discussion

Dynamic response time

Dynamic response time of the Pd (II) sensor: The determination of palladium (II) by potentiometric titration based on the formation of a water-insoluble ion pairs of PdX_4^{2-} and PdX_3^{-} complexes (X = Cl$^-$, Br$^-$, I$^-$, CN$^-$, SCN$^-$) [29] with a cationic titrant such as alkyl ammonium [30-32], alkylphosphonium [33-35] and crystal violet [36] were published. An important requirement for preparation of an ion selective sensor is that membrane electroactive material should have high lipophilicity and strong affinity for a target metal ion and poor affinity to the others.

It is well known that coordination abilities of ligands containing sulfur atom, are very selective to the transition metal ions. However, most of these show some limitations in their working activity range, selectivity, response time, pH range and lifetime. Thus, the development of reliable sensing ion selective sensors for palladium ion is considerable importance for environment and human health. To improve the analytical selectivity, it is essential to search novel carrier compounds that would interact with palladium ion with high selectivity. Because of the ligand that contain sulphur highly selective for Pd (II), classified as a "soft" Lewis acid. The use of IRA-410 as an ionophore is reported in the construction of a Palladium (II)-PVC membrane electrode and their characteristic and properties of selective electrode were studied. The novelty comes from that it can be used in aqueous and solid phases. The washing was used by elution of palladium absorbed palladium via reduction of palladium by formic acid. It is used for reducing of palladium producing palladium metal and sensing it in metallic palladium. The reduced palladium can be used as sensor hydrogen.

Time response, when total solution effect of the time according the equation (1)

$$y(t) = \{Kt + K(t_0 - T)_{\frac{-t}{e^\tau}}\} \tag{1}$$

Where Kt represent yp (t)...... $K(t_0\text{-}T)_e^{-t/r}$ yH(t)

The first term from equation describe the particular solution the end term is homogeneous solution. The dynamic response time of the Pd (II) (ISE)s selective electrode , of the most important factors .study

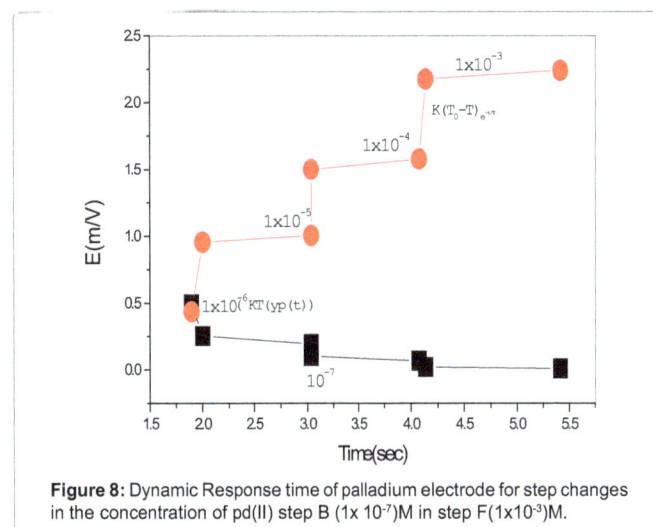

Figure 8: Dynamic Response time of palladium electrode for step changes in the concentration of pd(II) step B (1x 10^{-7})M in step F(1x10^{-3})M.

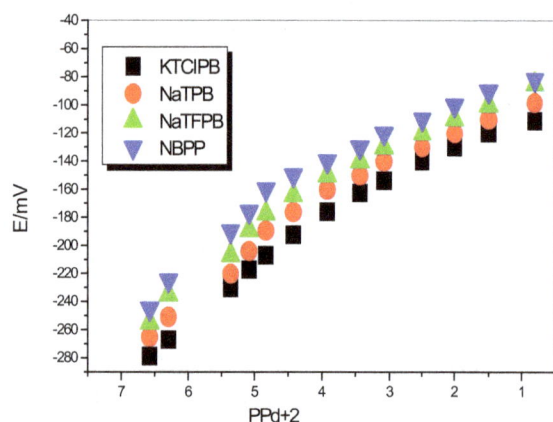

Figure 9: The Potential responses of pd(II) membrane ISEs prepared with different types of potential.

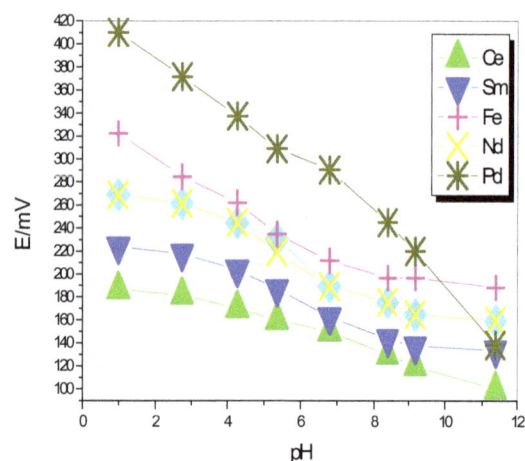

Figure 10: Effect of pH on the potential responses of pd(II) ion-selective electrode.

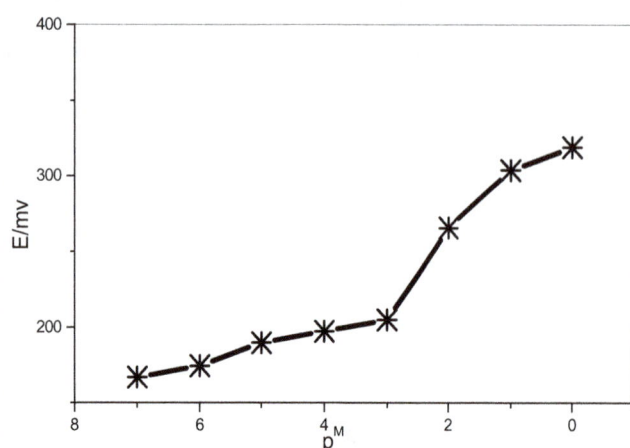

Figure 11: Potential response of various ion-selective membrane based on IRA-410.

the practical for response time of the ion selective electrode recorded by changing of the Pd(II) concentration in solution from (1.0×10^{-7} to

1.0×10^{-3})M. The potentials versus time traces are shown in Figure 8. As can be seen, the whole concentration range of plasticized membrane electrode reaches its equilibrium responses in a very short time (<1s)

Membrane composition

Due to some similarities between the functional groups of IRA-410 with those of the previously reported ligating molecules for lanthanide ions, and especially for palladium ion [2–6] as well as its negligible water solubility, we decided to examine the suitability of IRA-410 as potential ionophore in constructing some lanthanide ion-selective electrodes. Our preliminary solution studies revealed that, NBPP forms a quite stable complex with Pd (II) ion, which readily precipitates out from dioxan solution. While the extent of complexation of transition metal ions as well as other lanthanide such as Fe (III), Nd (III) and Sm (III) with IRA-410 was found to be much lower, as was examined by segmented sandwich membrane method [37]. Subsequently, the ligand IRA-410 was tested as an ionophore for the preparation of a variety of ions, mono, di, and trivalent metal ion-selective electrodes. The potential response of various ion-selective electrodes based on the proposed ionophore is shown in Figure 10. As expected, among different cations tested, Pd (II) with the most sensitive response seems to be suitably determined with the PVC membrane based on ligand NBPP. While the response slope of the other ion-selective electrodes are much lower the values expected from Nernstian equation, although in a limited concentration range. The response of membrane electrode depends on some parameters such as plasticizer–PVC ratio, amount of ionophore and additive used since; the nature of plasticizer influences the dielectric constant of the membrane phase both the mobility of ionophore molecules and the state of ligands [38-41]. It was expected to play a key role in the determining the ion-selective electrode characteristics. Polar plasticizers lead to the lowering of the membrane resistance as compared with polar plasticizers, which contain other functional groups with potential coordination sites which might [42] compete with carrier Thus, several solvents such as THBE, EHBS, DOP, o-NDPE, o-NPPE, and were tested (Figures 9-11). In fact, the Pd (II) ion-selective electrode based on IRA-410 better than the other examined mediators. It has a good Nernstian slope of 19.5 ± 0.2 mV decade-1 over a wide of concentration range from 10^{-1} to 2.5×10^{-6} mol l^{-1}, with detection limit of 1.6×10^{-6} mol l^{-1}. On the other hand, THBE, EHBS, DOP, o-NDPE and o-NPPE solvents give non-Nernstian slopes of 12.5, 28, 14.6, 9.8, and 12.4, respectively.

Response of different anions: In preliminary experiments, various PVC-membrane ion-selective electrodes with the synthesized ion pair were prepared and tested for different anions. The potential response of the electrode for different anions is shown in Figure 9: The results exhibited significantly high selectivity to palladium ion over other anions. Hence, ion pair was selected as a carrier for preparation of palladium selective electrode.

Preparation of sandwich membrane

A potentiometric method to determine ionophore complex formation constants in solvent polymeric membrane phases, it requires membrane potential measurements on two-layer sandwich membranes, where only one side contains the ionophore. If both membrane segments have the same ionic strength, it is convenient to assume that the activity coefficients for the complexed and uncomplexed ions are approximately equal. In that case, they can be omitted and the complex constant is related to the potential .This relationship allows for the convenient determination of formation constants of ionophore complexes within the membrane phase on the basis of transient membrane potential

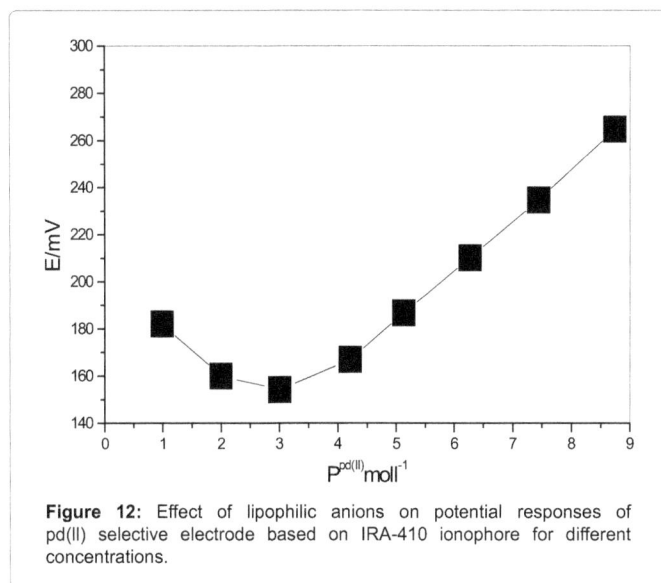

Figure 12: Effect of lipophilic anions on potential responses of pd(II) selective electrode based on IRA-410 ionophore for different concentrations.

Interfering ions	$\log K =_{p(II)}^{MPM}$
Ce^{3+}	−3.00
Sm^{3+}	−2.89
Fe^{3+}	−3.02
Nd^{3+}	−3.10

Table 3: Comparison of the selectivity coefficients of different pd(II) electrodes.

measurements on two-layer sandwich membranes can be if ion pairing neglected. Ion-selective electrode membranes were cast by mixing the recorded membrane components to give a total cocktail mass of 181.8 mg in 2 ml of THF. The solvent THF allowed evaporating overnight. A series of disks were cut with a cork borer from the parent membrane. These disks were conditioned overnight in each of metal chloride 0.01 M salt solutions shown in Figure 10 of Ce (III), Sm (III), Fe (III) and Nd (III). All membrane electrode potential measurements were performed at laboratory ambient temperature in unstirred salt solution (identical to the conditioning and inner filling solution) versus Ag/AgCl reference electrode. Sandwich membrane was made by pressing two individual membranes (ordinary one without ionophore and one with the same components and additional ionophore) together immediately after blotting them individually, dry with tissue paper. The combined segmented membrane was then rapidly mounted into the electrode body and immediately measured. The time required from making the membrane sandwich contact until final membrane potential measurement was less than 1 min.

Sample preparation

Determination of palladium sample by Potentiometeric titrations: THF (15.0 mL) was placed in the titration vessel and the required volume of the investigated acid and two drops of the indicator solution were added. The indicator electrode, either H2/Pd or glass electrode and a SCE as the reference one were immersed in the investigated solution and connected to a pH-meter. The solution was then titrated with standard solution of (potassium hydroxide or sodium methylate) and the potential was read after each addition of titrant. The test solution was stirred magnetically under a continuous stream of dry nitrogen (Figures 9-11).

The optimization of permselectivity of membrane sensors

The optimization of permselectivity of membrane sensors is known to be highly dependent on the incorporation of addition a membrane components. In fact it has been demonstrated that, the presence of lipophilic negatively charged additives improves the potentiometric behavior of certain cation-selective electrodes by reducing the ohmic resistance and improving the response behavior and selectivity [43,44]. Some of the lipophilic ions such as, potassium tetrakis (4-chlorophenyl) borate (KTClPB), sodium tetra phenyl borate (NaTPB), sodium tetrakis (1-imidazolyl borate) (NaIB) and sodium tetrakis (4-fluorophenyl) borate dehydrate (NaTFPB), were tested (Figure 12). It has been found that, the suitable lipophilic additive which improves the sensitivity of Pd (II) electrode was KTClPB with a good Nernstian slope of 19.5 ± 0.2 mVdecade-1. While the other lipophilic ions have slopes of 21, 13.6 and 9, respectively (Figure 11). Shown that the potential response of various ion-selective membranes based on IRA-410. The amount of ionophore has effect on the electrode sensitivity. So that, amounts of NBPP carrier (0.8, 1.2 and 1.6 mg) were examined. The results indicate that, the membrane containing 1.20 mg NBPP ionophore exhibits a good Nernstian slope of 19.5 ± 0.2 mV decade-1 and high selectivity of Pd (II) ion.

Effect of internal solution, response time and pH

The working of membrane electrode in relation to variation of reference solutions was investigated. It was found that, the variation of the concentration of the internal solution (10^{-1} to 10^{-4} mol^{-1} of KCl solution) causes significant effect on corresponding potential response. However a solution of 10^{-3} moll^{-1} KCl mixed with 10^{-3} mol l^{-1} [pdCl$_4$]$^{-4}$ would be used as a suitable internal solution, it had a good slope 19.5 ± 0.2 mV decade-1. The detection limit, taken at the point of intersection of the extrapolated linear segment of the calibration curve, was 1.6×10^{-6} moll^{-1}. The static response time of the membrane electrode thus obtained was <10 s. The sensing behavior of the electrode remained unchanged when the potential recorded from low to high concentrations or vice versa. The life time of the present electrode was at least 3 months. During this time, the detection limit of the electrode remained almost constant and the slope of the response decrease from 16.5 ± 0.2 to 16.1 ± 0.15 mV decade^{-1}. After this time the electrochemical behavior of the electrode gradually deteriorates. The effect of pH on the response of the electrode was studied over the pH range from 1 to 11 at different concentrations (10^{-2} to 10^{-5} moll^{-1}) of Pd (II) solution. The pH of solutions was adjusted with either HCl or NaOH solutions. Potential remains constant at pH range from 4 to 8 (Figure 12). The increase of potential below pH~ 4 may be ascribed to interference by H+ ion and the decrease of potential above pH= 8 may be due to formation of some hydroxyl complex of the pd(II) ions in solution from hydrolysis of palladium chloride. The performance of the electrode was assessed in partially non-aqueous media using ethanol–water mixture; it is observed that the electrode functions well in presence of up to 10% (v/v) non-aqueous (alcoholic) content. Higher alcoholic content disturbs the functioning of system (Figure 12).

Electrode selectivity

The influence of interfering ions on the response behavior of ion-selective membrane electrode is usually described in terms of selectivity coefficient log Potpalladium B. The potentiometric selectivity coefficients log Pot palladium B of palladium electrode were evaluated by (SSM) and (MSM)(27,30,31). The resulting values of the selectivity coefficients are summarized in Table 3. It is evident from the selectivity coefficients data, that the sensor exhibits a high performance for Pd

(II) ion compared with alkali, alkaline-earth, transition and heavy metal ions. Comparison of the main analytical features of some the previously described pd(II) ion-selective [18-22] electrodes) with the proposed Pd (II) electrode revealed that, the present electrode exhibited a better selectivity, especially in the presence of Hg (II) and Fe (III). This participation of these functional groups in the binding with pd(II) ions. Elemental analysis of NBPP and its Pd (II) complex was examined and showed that, the formation of NBBP–pd(II) complex for a membrane segment may form with the complex stoichiometry n =1. Membrane potential values emf for the examined metal salt solutions of Ce (III), Sm (III), Fe (III) and Nd (III) were deter-mined by subtracting the cell potential for a membrane without ionophore from that of the sandwich membrane. The determined formation constants (log β ILn) for the examined different complexes were recorded in Table 3. A careful analysis of the data in, reveals that pd(II) has significant cation-binding characteristics. A comparison between the potentiometric behaviors of the proposed electrode with the previously reported ISEs for Pd (II).

Determination of cations in some pharmaceutical samples.

Samples containing 50 ml of distilled water and stirred for 3 h in thermostat adjusted to 25 ± 1°C and allowed to stand to 2 that the same temperature before potential measurement. The palladium selective electrode and reference electrode were immersed in the precipitate solution. The concentration of soluble pd(II) ions was measured and Solubility product was calculated for cerium oxalate was 2.2×10^{-26} while reference was 3.2×10^{-26} and for cerium phosphate was 2.3×10^{-23} while reference was 1×10^{-23} with small deviation of reference (Table 3).

Determination of palladium in some water and radioactive waste samples

As discussed before palladium cation can be determined in water and radioactive waste samples

Reaction mechanism

The formation of $[PdCl_4]^{-2}$ species are suggested to be bounded to the resin. This concept is supported by the considerable stability of the formed palladium tetrachloro complex ($PdClO_4 = 15$). The complex formation of palladium tetrachloro complexes explains the fast loading of palladium on the resin phase as a brown zone which may be formed as a tetrachloro complex. It, also, explains the difficulty of elution which is attributed to the stability of $[PdCl_4]_2$ complex. Therefore, the separation of the complexed palladium from the solid phase can be carried out via reduction using the weak complexing agent; formic acid [1]. To explain the reduction steps of palladium chloride by formic acid, the reduction using ethylene was consulted [2] in comparison to the following reduction reaction case. In this respect the following steps are suggested [1]. Sepaon Mechanism: The formation of a complex for a metal ion M⁺ with the following steps is also suggested;

The following steps are also suggested

a. $HCOOH + R[PdCl_4]^2 \xrightarrow{K_1} R[PdCl_3(H000H)] + Cl$

b- $R[PdCl_3(H000H)] + H_2O \xrightarrow{K_2} R(Pd(HCOOH)(OH_2)Cl_2) + Cl$

c- $R(Pd(HCOOH)(OH_2)Cl_2) \xrightarrow{\sim K_3} R[Pd(HCOOH)(OH)Cl_2] + H^+$

d- Step (c) indicates that the rate of the reduction process is reversibly proportional to the H⁺ concentration which explains its increase by decreasing the formic acid concentration.

The complexed species formed in step (c) are suggested to go via possible rearrangements through an addition of an OH group [14]; as in (d):

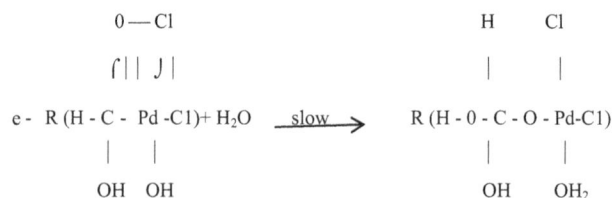

$$e\text{-}\ R(H\text{-}C\text{-}Pd\text{-}Cl) + H_2O \xrightarrow{\ slow\ } R(H\text{-}O\text{-}C\text{-}O\text{-}Pd\text{-}Cl)$$

Step (d) is suggested to be the rate determining step where the addition of the hydroxide ion to the coordinated formic acid occurs. This step is followed by a proposed fast reaction as in (e):

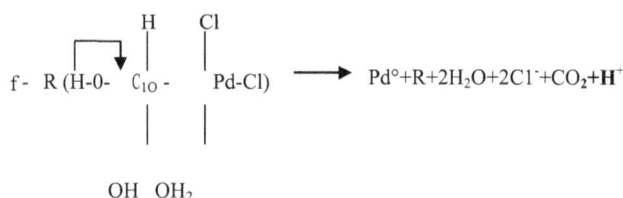

$$f\text{-}\ R(H\text{-}O\text{-}\ C_{10}\text{-}\ Pd\text{-}Cl) \longrightarrow Pd^\circ + R + 2H_2O + 2Cl^- + CO_2 + H^+$$

Conclusion

New ion selective sensitive electrode of Pd (II) as multisensor based on IRA-410 via low Cost oxidation reduction process as membrane disc sensor from. It can be used as sensor in HCl solution as $(PdCl_4)^2$ in, Pdo sorbed on solid phase IRA-410 and eluted as $Pd(NO_3)^2$ in nitrate medium. This electrode has a wide linear dynamic range d low detection limit of 1.6×10^{-6} mol l^{-1}. It has a fast response time (<1 s) and good selectivity with respect to different metal ions in pH range from (1.0-9.0). It can be used for about 10 months with complex with Pd (II) was calculated by using segmented sandwich membrane method. The proposed electrode has been used successfully as an indicator electrode in potentiometric determination in aqueous nitrate and /or chloride media.

References

1. El-Said N, Mekhail A, Khalifa SM, Aly HF (1996) Selective separation of palladium from simulated intermediate radioactive waste nitrate by IRA-410 and IRA-900 anion exchanger. J Radioanalytical and Nuclear Chemistry 208: 243-255.

2. El-Said N, Siliman AM, El-Sherif E, Borai EH (2002) Separation of palladium from simulated intermediate radioactive waste/chloroacetic acid/nitrate medium by IRA-410 and IRA-900 anion exchangers. J Radioanalytical and Nuclear Chemistry 251: 285-292.

3. El-Said N, Mekhael AS, Khalifa SM, Aly HF (1995) Separation of strontium from simulated waste in K2SO4/Nitrate medium containing Sr2+, Eu3+, Ce3+, Pd2+, Rh3+, Ru3+, UO22+, Fe3+, Cr3+, Ni2+, Al3+, Ca2+ and Cs+, by strongly basic anion exchanger. J Radioanalytical and Nuclear Chemistry 208: 257-270.

4. Amini MK, Shahrokhian S, Tangestaninejad S (1999) Thiocyanate-selective electrodes based on nickel and iron phthalocyanines. J Anal Chim Acta 402.

5. Shamsipur M (2001) A dinuclear zinc complex with (E)-4-dimethylamino-N'-(2-hydroxybenzylidene)benzohydrazide. J Anal Chem 73.

6. Ganjali M, Poursaberi T (2001) Using PVC ion-selective electrodes for the potentiometric flow injection analysis. J Anal Chem 370.

7. Khan AA, Inamuddin (2004) Applications of Hg(II) sensitive polyaniline Sn(IV) phosphate composite cation-exchange material in determination of Hg2+ from aqueous solutions and in making ion-selective membrane electrode. Sensors and Actuators B: Chemical 120: 10-18.

8. Khan AA (2003) A Sensitive Cigarette Smoke Sensor - International Frequency. Sensors and Actuators B 120.

9. Ahmad M, Hashim N, Ghani SA (2011) Cobalt(II) selective membrane electrode based on palladium(II) dichloro acetylthiophene fenchone azine. J Talanta 87.

10. Dhiman SB, LaVerne JA (2013) Radiolysis of simple quaternary ammonium salt components of Amberlite resin. J Nuclear Materials 436: 8-13.

11. Wren JC (1983) Literature study of volatile radioiodine release from ion-exchange resins during transportation. J Kysela R Pejsa Radiat Phys Chem 22.

12. Devi PSR, Joshi S, Verma R, Lali AM, Gantayet LM (2010) Radiation-induced decomposition of anion exchange resins. J Radiat Phys Chem 79.

13. Iwai Y, Yamanishi T, Hiroki A, Tamada M (2009) Procurement Preparation of ITER Key Components by JADA. J Fusion Sci Tech 56.

14. Pillay KKS (1986) Effects of Ionizing Radiation on Modern Ion Exchange Materials. J Radio Nucl Chem I 102.

15. Pandey PC, Prakash R (1998) Characterization of Electropolymerized Polyindole Application in the Construction of a Solid State, Ion Selective Electrode. J Electrochemical Society 12.

16. Amed MT, Clay PG, Hall GR (1966) Radiation-induced decomposition of ion-exchange resins. Part II. The mechanism of the deamination of anion-exchange resins. J Chem Soc B 102.

17. Qin Y, Peper S, Bakke E (2002) Plasticizer-Free Polymer Membrane Ion-Selective Electrodes Containing a Methacrylic Copolymer Matrix. J Electroanalysis 14.

18. Hall GR, Streat M (1963) The solubility of silver chloride in methanol and water mixtures. J Chem Soc 114.

19. Baidak M, Badali JA, LaVerne (2011) Radiation Chemical Effects with Heavy Ions. J Chem Sci 115.

20. Moody GJ, Oke RB, Thomas JDR (1970) A calcium-sensitive electrode based on a liquid ion exchanger in a poly(vinyl chloride) matrix. J Analyst 95.

21. Cragge GJ, Moody, Thomas JDR (1974) Experiments with the pvc matrix membrane calcium ion-selective electrode. J Chem Educ 51.

22. Ma TS, Hassan SSM (1982) Organic Analysis Using Ion-selective Electrodes: Methods 1. Academic Press, London, UK.

23. Saleh MB, Abdel Gaber AA (2001) Novel Zinc Ion-Selective Membrane Electrode Based on Sulipride Drug. J Electroanalysis 13.

24. Sombatsompop N (2004) Structural changes of PVC in PVC/LDPE melt-blends: Effects of LDPE content and number of extrusions. J Polymer Engineering and Science 44.

25. El-Nemmaa EM, Badawi NM, Hassan SSM (2009) Cobalt phthalocyanine as a novel molecular recognition reagent for batch and flow injection potentiometric and spectrophotometric determination of anionic Surfactants. J Talanta 78.

26. Nataliya M (2013) Nanocrystalline ZnO(Ga): Paramagnetic centers, surface acidity and gas sensor properties. J Sensors Actuators B: Chemical 186.

27. Koteswara (2014) Construction and performance characteristics of polymeric membrane electrode and coated graphite electrode for the selective determination of Fe3+ ion J: Materials Science and Engineering 36.

28. Mi Y, Bakker E (1999) Ion-Selective Electrode Characteristics. J Anal Chem 71.

29. Vytras K (1985) Stabilizing Potential of Solid-Contact Sensors Selective towards Surface-Active Substances. J Ion-Sel Electrode Rev 7.

30. Scibone G, Mantella L, Danesi PR (1970) Liquid anion membrane electrodes sensitive to metal cation concentration. J Anal Chem 42.

31. Yu M, Sednew EM, Rakhmanko GL, Starobinets (1985) Potentiometric titration of palladium(II) halide complex. J Anal Khim 40.

32. Vytrs K, Kalous J, Simickova B, Cerna J, Silena I (1988) Potentiometric titration of palladium(II) halide complex anions based on ion-pair formation. J Anal Chim 357.

33. Stvlgiene S, Tautkus R, Kazlauska OM, Petrukhin Z (1990) Electroanalysis: From Laboratory to Field Versions Lab 56.

34. Kazluskas R, Jankauskas V, Petrukhin (1986) Theory and applications of ion-selective electrodes: Part 7. J Ionnyi Obmen Ionometriya 5.

35. Badr I, Meyerhoff M, Hassan S (1995) Membrane sensors for batch and flow injection potentiometric. J Anal Chem 67.

36. Vasilev VA, Gurev IA, Zavod (1989) Effect of Palladium on Thermal Degradation of Fiban K-1 Lab. 55.

37. Bakker E, Buhlmann P, Pretsch E (1997) Carrier-Based Ion-Selective Electrodes and Bulk Optodes. 1. General Characteristics. J Chem Rev 97.

38. Fakhari AR, Shamsipur M (1997) Development of a highly selective voltammetric sensor for nanomolar detection of mercury ions using glassy carbon electrode modified with a novel ion imprinted polymeric nanobeads and multi-wall carbon nanotube. J Anal Chem 99.

39. Wongsan W (2013) Bifunctional polymeric membrane ion selective electrodes using phenylboronic acid as a precursor of anionic sites and fluoride as an effector: A potentiometric sensor for sodium ion and an impedimetric sensor for fluoride ion. J Electrochimica Acta 111.

40. Kim YJ, Kim J, Kim YS, Lee JK (2013) Erratum to "TiO2-poly(4-vinylphenol) nanocomposite dielectrics for organic thin film transistors" [Org. Electron. 14 (2013) 3406–3414]. Organic Electrons 15: 640.

41. Khan Kabir-ud-din AA, Beg MA (2014) Potentiometric estimation of potassium hydroxotetracyanotungstate(IV) and determination of standard potential. J Electroanalytical Chemistry 20: 239-244.

42. Eugster R (1991) Study of the influence of polymeric membrane composition on the sensitivity of acoustic wave sensors for metal analysis. J Anal Chem 63.

43. Stoica AI (2014) Heavy metals determination from food and pharmaceutical samples Development of sensors, biosensors and immunosensors. J Electrochimica Acta 113.

44. Bakker E (2014) Enhancing ion-selective polymeric membrane electrodes by instrumental control. J Trends in Analytical Chemistry 53.

4

Development of an Asymmetric Ultrafiltration Membrane from Naturally Occurring Kaolin Clays: Application for the Cuttlefish Effluents Treatment

Sonia Bouzid Rekik[1-3]*, Jamel Bouaziz[1], Andre Deratani[2] and Semia Baklouti[3]

[1]*Laboratory of Industrial Chemistry, National School of Engineering, University of Sfax, BP 1173, 3038 Sfax, Tunisia*
[2]*Institut Européen des Membranes (IEM), Université Montpellier 2, Place E. Bataillon, 34095 Montpellier Cedex 5, France*
[3]*Laboratory of Materials Engineering and Environment, National School of Engineering, University of Sfax, BP 1173, 3038 Sfax, Tunisia*

Abstract

This work concerns to the development and characterization of support and ultrafiltration membranes from naturally occurring- kaolin clays as principal components. The preparation and characterization of porous tubular supports, using kaolin powder with corn starch as poreforming agent, were reported. It has been found that the average pore size was about 1 µm while the pore volume was 44% for supports sintered at 1150°C with a flexural strength of about 15 MPa. The deposition of the active layer was performed by slip casting method. The rheological study of various coatings with different concentration of kaolin powder, polyvinyl alcohol (PVA) and water under different conditions regarding temperature and stirring time was done. After drying at room temperature for 24 h, the membrane was sintered at 650°C. The average pore diameter of the active layer was 11 nm and the thickness was around 9 µm. The determination of the water permeability shows a value of 78 l/h.m².bar. This membrane can be used for crossflow ultrafiltration. The application of the cuttlefish effluent treatment shows an important decrease of turbidity, inferior to 1.5 NTU and chemical organic demand (COD), retention rate of about 87%. So, it seems that this membrane is suitable to use for wastewater treatment.

Keywords: Kaolin clays; Sintering; Ceramic supports; Ultrafiltration membrane; Cuttlefish effluent

Introduction

Membrane separation processes extend more and more every day in industrial uses, with new requirements concerning the materials and preparation procedures. Due to their potential application in a wide range of industrial processes such as water and effluent treatments [1-5], drink clarification [6,7], milk pasteurization [8-11], biochemical processing [12], inorganic membrane technology grows in importance.

The use of ceramic membranes has many advantages such as high thermal and chemical stability, pressure resistance, long lifetime, good resistance to fouling, and ease of cleaning [13-17]. The main process to prepare ceramic membranes includes first the obtaining of a good dispersion of small particles and then, the deposition of such dispersion on a support by a slip casting method.

Unfortunately, ceramic membrane fabrication, even though commercially available, still remains highly expensive from a technical and economic point of view due to the use of expensive powders such as alumina [18-22], zirconia, titania and silica [23,24]. To reduce the cost of ceramic membrane fabrication, recent research works are focused on the use of cheaper raw materials such as apatite powder [25], natural raw clay [26-29], graphite [30], phosphates [31,32], dolomite, kaolin [33-37] and waste materials such as fly ash [38-42].

To increase the reliability of the tenacity of these membranes having an asymmetric structure composed of at least two layers which are the support and the active layer, there is still a need for improvement in rheological properties of dip solution required for the preparation of active layer. The control of the rheological parameters of the dip solution is of significant importance for optimizing the final composition of the suspension. Indeed, the rheological properties depend on the physico-chemical characteristics of the raw material used and on the conditions under which the dip solution was prepared. Besides, the active layer thickness and microstructure of the membrane are mainly controlled by the viscosity of the dip solution which depends mainly on the particle size, the nature of the raw material, the addition of polymer and on temperature [43,44].

The properties of the kaolin suspensions in water are largely modified by the presence of polymers. Many studies reported in the literature, treated as subject the rheological behavior of systems like "water- apatite- additive" and "water-clay-additive" where the additives were often a polymer. Most of these studies were devoted to the colloidal and/or rheological properties of these suspensions and on the effect of polymeric additives in order to obtain the optimal composition of the suspension [45-47].

In order to decrease the membrane cost, the present work describes the preparation of both porous support and active layer based on naturally occurring- kaolin clays (chemical structure: $Al_2O_3.2SiO_2.2H_2O$ [48,49]), to elaborate a new asymmetric ceramic ultrafiltration membrane. This membrane is fabricated in different steps.

The first step is the preparation of porous tubular ceramic support using kaolin clays. Corn starch powder was added as pore-forming agent to produce sufficient porosity with acceptable mechanical property. The properties of porous support were discussed as a function of sintering temperature in order to optimize the preparation conditions. Their structural and functional properties are determined by different techniques. The most important parameters used in the characterization

*Corresponding author: Sonia Bouzid Rekik, National School of Engineering, University of Sfax, Sfax, Tunisia, E-mail: bouzidsonia@gmail.com

of these substrates are: surface and internal morphology, mean pore size, pore size distribution, porosity, mechanical and chemical stability.

The second step is the characterization of the rheological properties of a dip solution which will be used to elaborate an ultrafiltration membrane by deposition using slip casting process of the active layer supported by the kaolin support. The adhesion of the layer to the macroporous ceramic support is achieved by capillary suction. The thickness of the layer depends on the physico-chemical properties of medium, as well as the viscosity of the slip. The study reveals also that the thickness of the active membrane layer can be controlled by the percentage of the kaolin powder added to the suspension and the deposition time.

The last step of this work concerns the application of the ultrafiltration elaborated membrane to the treatment of cuttlefish conditioning wastewater. Figure 1 shows a schematic representation of the sequential steps of the support, active layer, and the procedure of asymmetric ceramic membrane preparation.

Materials and Methods

Materials

In this study, both the supports and membranes were prepared from clay. The clay used in the present study is a kaolin Codex (notes as K), it was recommended by the L.P.M Cerina (Laboratoire des Plantes Medicinales, Tunisia). Corn starch powder was used as a pore former.

Characterization of the kaolin powder

The chemical composition of the kaolin powder was determined by spectroscopic techniques, as Xray fluorescence for metals and by atomic absorption for alkaline earth metals.

Phase identification was performed by XRD analysis (Philips X'Pert X-ray) diffractometer) with Cu Kα radiation ($\lambda=1.5406$ A°), and the crystalline phases were identified by reference to the International Center for Diffraction Data cards.

Fourier transforms infrared (FTIR) spectra by the KBr method using an IR spectrometer [Perkin-Elmer spectrum BX]. The spectra were collected for each measurement over the spectral range 600-4000 cm^{-1} with a resolution of 4 cm^{-1}.

Thermogravimetric analysis (TG) and differential thermal analysis (DTA) were carried out from ambient temperature to 1300°C at a rate of 10°C min^{-1} under air, using a setaram SETSYS Evolution 1750.

The particle size distributions of kaolin (K) were determined by the Dynamic Laser Scattering (DLS) technique using water as dispersing medium (Mastersizer 2000, Malvern Instruments).

Supports elaboration

The different steps for preparing ceramic materials is the choice of the nature and particle size of the ceramic powder, the choice of organics additions, the optimal volume of water, the aging time paste, shaping by extrusion, and the time of drying and sintering. The preparation of the porous membranes involved the formulation of a plastic and homogeneous paste that contained the clay ceramic powder mixed with organic additive and water.

The optimal formulation of the ceramic precursor paste was carried out in an empirical way. The selected composition of powders used for the plastic paste preparation was 90 wt.% Kaolin (K) and 10 wt.% Starch as a plasticizer. To dos this, a quantity of 540 g of kaolin and 60

g of starch powders were uniformly dry mixed using a rotary mixer at a speed of 65 rpm for 10 min. The powder mixture was aged with a progressive addition of water (300 ml). Water content was adapted according to the type of clay material used in the formulation. As a result, more homogeneity in the final substrate structure was acquired. Before the extrusion phase, an aging stage of the aqueous suspension is necessary to obtain a good homogeneity and to favor the formation of porosities. This step is required to prepare a paste with rheological properties allowing the shaping by extrusion. To this end, the excess of liquid was eliminated and the obtained paste was kept in a closed plastic bag for 24 h under high humidity environment to avoid premature drying and ensure a homogenous distribution of water and organic additives. After aging, the paste was extruded into the tubular specimens trough an extruder and then the wet pieces are set on stems at room temperature during 24 h to ensure a homogenous drying and to avoid twisting and blending.

Finally, a thermal treatment was carried out in a programmable furnace at different final temperatures. The Type 30400 Automatic and Programmable furnaces are general laboratory and heat treating furnaces. Two steps have been determined. Subsequently, the membrane was heated up to 400°C for 2 h at a heating rate of 2°C/min in order to eliminate the organic additive used as pore-forming agent. In a second step, the support was sintered at a temperature

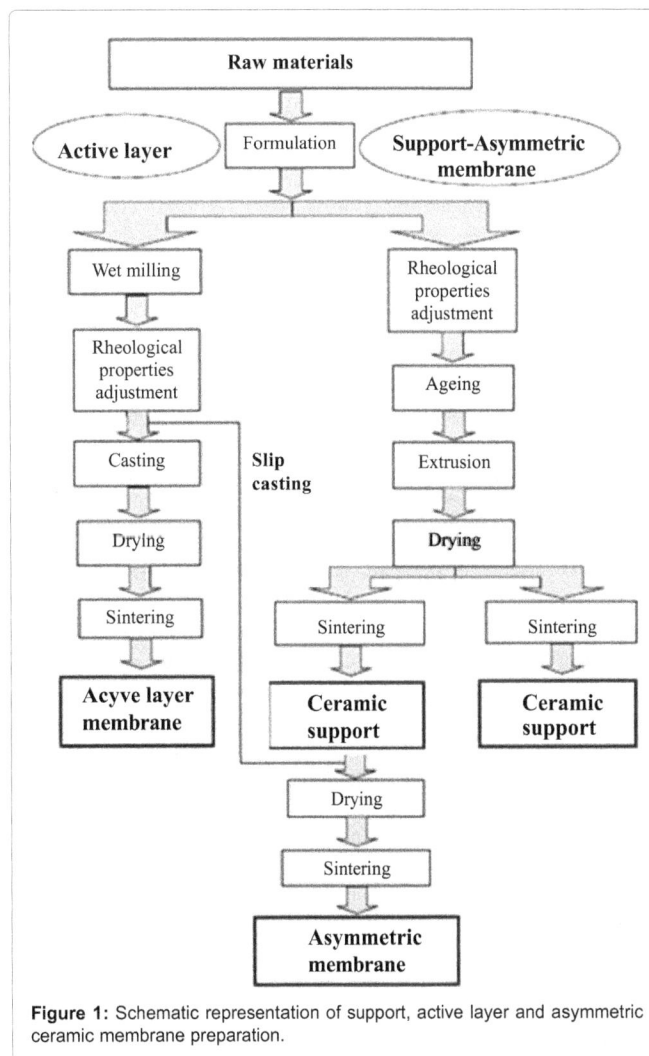

Figure 1: Schematic representation of support, active layer and asymmetric ceramic membrane preparation.

ranging from 1100 to 1250°C with a ramping rate of 5°C min⁻¹ in order to avoid the formation of cracks during sintering of the samples. The final temperature was kept constant for 1 h. After that, the consolidated support was allowed to cool until the environmental temperature was reached.

The temperature-time schedule not only affects the pore diameters and porous volume of the final product but also allows obtaining the final morphology and mechanical strength. Tubular supports were elaborated with external/internal diameter of 16/11 mm and the length of 150 mm.

Supports characterization

The evolution of densification and surface quality of the supports sintered at different temperatures were determined by scanning electron microscopy.

Porosity and pore size distribution were measured by mercury porosimetry. This technique relies on the penetration of mercury into a membrane's pores under pressure [34-36,50,51]. The intrusion volume is recorded as a function of the applied pressure and then the pore size was determined.

The mechanical resistance tests were performed using the three points bending method (LLOYD Instrument) to control the resistance of the membranes fired at different temperatures.

The corrosion tests were carried out using aqueous solutions of nitric acid (pH=2.5) and sodium hydroxide (pH=12.5) at 45 and 80°C, respectively. All the samples were ultrasonically rinsed in distilled water, dried at 110°C and stored in a dryer. The degree of corrosion was characterized by the percentage of the weight loss.

Ultrafiltration membrane elaboration

To decrease the probability of cracking or fissures during the sintering process of the membrane, in the preparation of the active layer the same raw materials of the support were used.

The active ultrafiltration layer from kaolin was prepared by a slip casting process on kaolin support from a mixture of the inorganic powder and a 12 wt.% aqueous solution of polyvinylique alcohol (PVA) (Rhodoviol 25/140 (Prolabo)), which is used as binder and plasticizer. The slip casting process [42,52] was applied to coat the support tube to elaborate ultrafiltration membrane as described in Figure 1.

In the first step, a stable suspension or slip has to be obtained by putting in suspension the mineral powder in water. Then, 12 wt.% aqueous solution of PVA was mixed to the preceding suspension. Desagglomeration of the mineral powder and homogenization of the coating formulation was ensured by mechanical stirring using an ordinary magnetic stirrer at its maximum speed. The coating formulation was poured inside the support for a few minutes at room temperature while the tube was closed at one end. A layer is then formed on the inner side of the porous tube due to capillary suction. Afterwards, the excess was drained out. Drying was realized at room temperature for 24 hours. The sintering temperature was fixed at 650°C for 3 hours. A heating temperature at 250°C for 2 hours is necessary in order to eliminate completely the PVA, present in great quantity in the slip [3]. Relatively slow temperature rate (2°C/min) was needed in order to avoid the formation of cracks on the layer.

Composition and characterization of the slip

Different coating formulations B.n differing by PVA concentration

(1<n<3) and kaolin powder (4<n<8) were used to determine the optimal composition of the final layer.

At the first time, the kaolin powder quantity was kept at 4% when the PVA aqueous solution percentage was varied and then, at the second time, the PVA aqueous solution percentage was fixed to 40% when the kaolin powder percentage was varied (Tables 1 and 2).

The viscosity is an important parameter to determine the slip optimal composition. It was measured just before slip casting by using an Antoon Paar rheometer model MCR301.

Filtration pilot

The laboratory pilot used for the filtration experiments was equipped with a cross-flow filtration system implementing tubular ceramic membrane of 15 cm length. The tubular membrane takes place in a stainless steel carter. The transmembrane pressure (TMP) can reach 8 bars wich corresponds to the ultrafiltration range. It was controlled by an adjustable valve on the retentate side. Temperature was kept at 25°C by a thermal exchange system. The membrane was conditioned by immersion in pure deionized water for a minimum of 24 h before filtration tests. The determination of the water membrane permeability was performed with distilled water.

Effluent characterization

Wastewater samples were taken from the effluents produced by a sea-product freezing factory located in Sfax, Tunisia. The dark colour in this effluent was due to the presence of sepia ink (containing melanin) as suspension particles which have a size range between 56 and 161 nm [31].

A large number of analyses were conducted on each sample and the following parameters were measured: turbidity, Chemical Oxygen Demand (COD), temperature, pH and conductivity. The COD values of raw effluents from the production process ranged between 2000 and 3000 mgL⁻¹ with an average concentration of 2615 mgL⁻¹. The turbidity measured for the raw effluent presents a very high value which is in the range of 335 NTU (Table 3).

The techniques used to analyze collected samples of feed, retentate and permeate are reported below:

- Turbidity: Using a HACH «2100 N Turbidimeter» turbidimeter.

Coating formulation	% of PVA[a]	% of water
B.1	20	76
B.2	30	66
B.3	40	56

a: the percentage of PVA here is relatif to the quantity of the 12-wt% PVA aqueous solution added to the powder-water dispertion.

Table 1: Composition of coating formulation prepared with different percentage of PVA.

Coating formulation	% of kaolin	% of water/PVA aqueous solutions[b]
B.4	2	94
B.5	4	92
B.6	6	90
B.7	8	88
B.8	10	86

b: mixture of 12 wt % PVA aqueous solution and water

Table 2: Composition of coating formulation prepared with varying percentage of kaolin.

Turbidity (NTU)	COD (mg.L⁻¹)	Conductivity (mS.cm⁻¹)	pH
335	2615	204	7

Table 3: Turbidity, COD, conductivity and pH of raw effluents.

- Dissolved organic carbon: Using a «REHROTEST TRS 200 NF T 90-101» COD Analyser.

- Conductivity: Using a «Consort K 911» conductimeter.

Results and Discussion

Powder characterization

The chemical composition of kaolin is given in Table 4, where the main impurities are CaO, K2O, TiO_2 and Fe_2O_3. It reveals that the major components were silica (SiO_2: 47.85%) and aluminium oxide (Al_2O_3: 37.60%).

Phase identification is of great importance before any membrane manufacturing. For example, Figure 2 presents the XRD patterns of the raw kaolin. It can be seen that kaolinite (K) was the major mineral component with a small amount of quartz (Q) and illite (I) impurities. No other components were observed, because the impurities are in so tiny quantity (Table 4) and most of them are probably incorporated into the crystal structure of kaolinite.

The results of FTIR spectroscopy of the starting clay (Figure 3) show the kaolin characteristic bands: Si-O (at around 993, 1024, and 1112 cm⁻¹), Si-O-Al (at around 525, 750, and 795 cm⁻¹), Al-OH (at around 910 cm⁻¹) and OH (at around 3684, 3668, 3651, and 3618 cm⁻¹) [48,49,53].

The particle size distribution of kaolin was determined by the Dynamic Laser Scattering (DLS) technique, and the results shown in Figure 4. This method gave an average particle size in the order of 4 μm.

A total weight loss is observed by TGA to be about 15% of kaolin (Figure 5). In fact, the weight loss consists of two distinct stages: The first one is considered as a slight weight loss between room temperature and 150°C, because of the dehydration of the clay. The second mass loss detected between about 400 and 700°C is mainly due to the phenomenon of dehydroxylation of kaolinite confirmed by DTA which shows an endothermic peak at 560°C leading to the transformation of kaolinite to metakaolinite. A third stage, wich is characterized by an exothermic reaction appeared at about 975°C without any weight loss. The exothermic peak corresponds to the metakaolin-mullite transformation [54].

Supports characterization

For the development of high-quality supports, the following properties are of major importance: pore size distribution, porosity, surface texture, mechanical properties and chemical stability.

Supports morphology: Figure 6 illustrates SEM pictures for the supports sintered at the four different temperatures considered in this work. The optimal sintering temperature was determined by comparing the texture of the different obtained samples.

The ceramic substrates sintered at lower temperature (1100°C) show highly porous structure. Below 1100°C, we detect the presence of intergranular contacts which are large enough to ensure the ceramics cohesion (beginning of sintering). We observe that the formation of grain boundaries is achieved within a narrow temperature range of 1150°C-1200°C. The aspect of the surface is homogeneous and does

not present any macro defect (cracks, etc.). Beyond 1250°C, the densification process (shrinkage) dominates and a relatively open structure is still observed.

The morphology as well as the microstructure of optimal support fired at 1150°C is shown in Figure 6 for the cross section and the surface views. A smooth inner surface is observed which will allow the deposit of a homogeneous membrane layer.

Open porosity and pore size distribution for supports: The evolution of the average pore diameter and the porosity with the temperature of sintering reveals that the porosity decreases from 49 to 35% between 1100°C and 1250°C, while the pore diameter increased from 0.81 to 1.4 μm (Figure 7). The beginning of the material densification occurred when the temperature increases [25,33,36,37,41,55].

At 1150°C, the characteristics of the support are an average pore diameter of 1 μm and a porosity of 44%. Consider Figure 8 which presents a Single (mono) Modal of pore size distribution or

	SiO₂	Al₂O₃	Fe₂O₃	MgO	K₂O	CaO	TiO₂	LOI*
Kaolin (%)	47.85	37.60	0.83	0.17	0.97	0.57	0.74	11.27
* LOI: Loss on Ignition at 1000°								

Table 4: Chemical composition of the used kaolin (wt %).

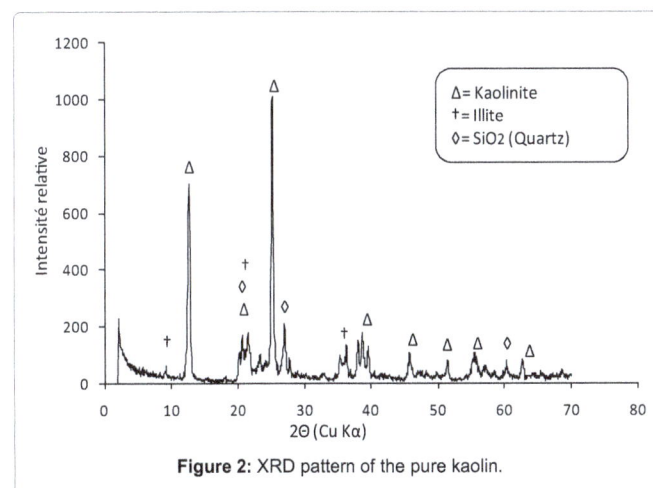

Figure 2: XRD pattern of the pure kaolin.

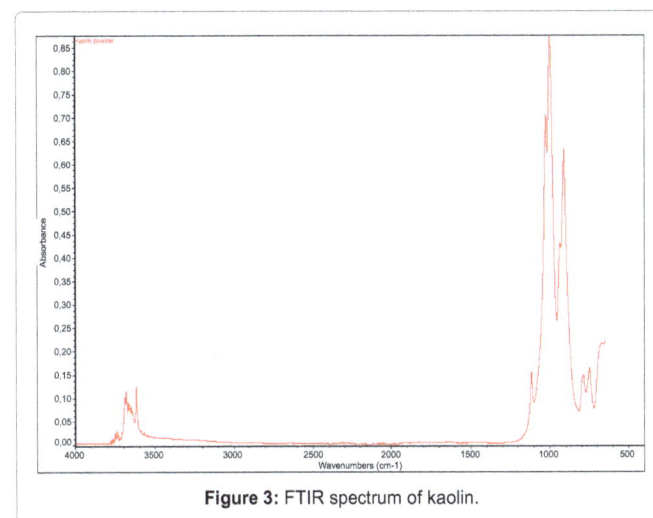

Figure 3: FTIR spectrum of kaolin.

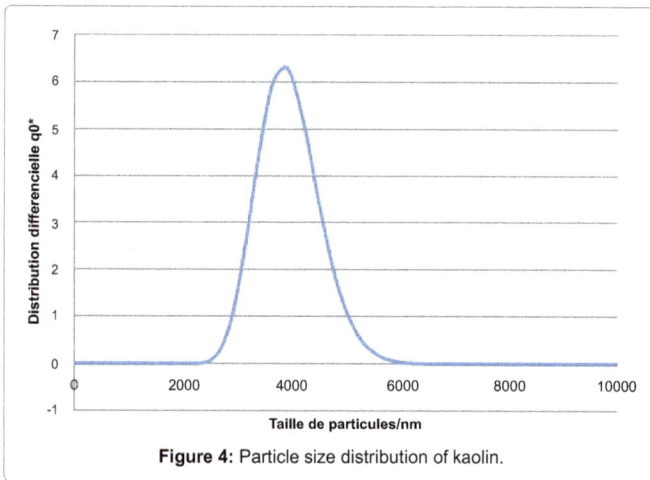

Figure 4: Particle size distribution of kaolin.

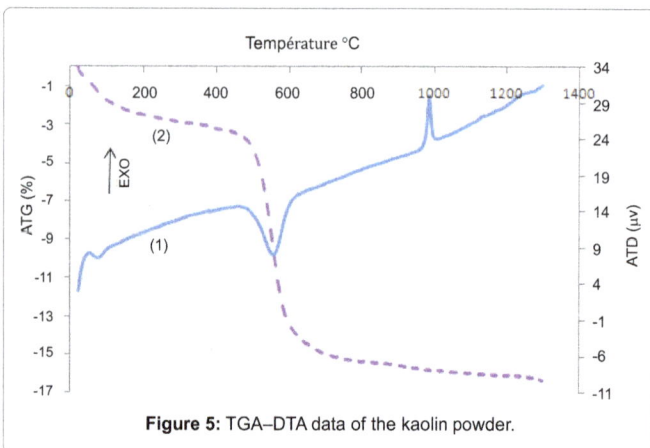

Figure 5: TGA–DTA data of the kaolin powder.

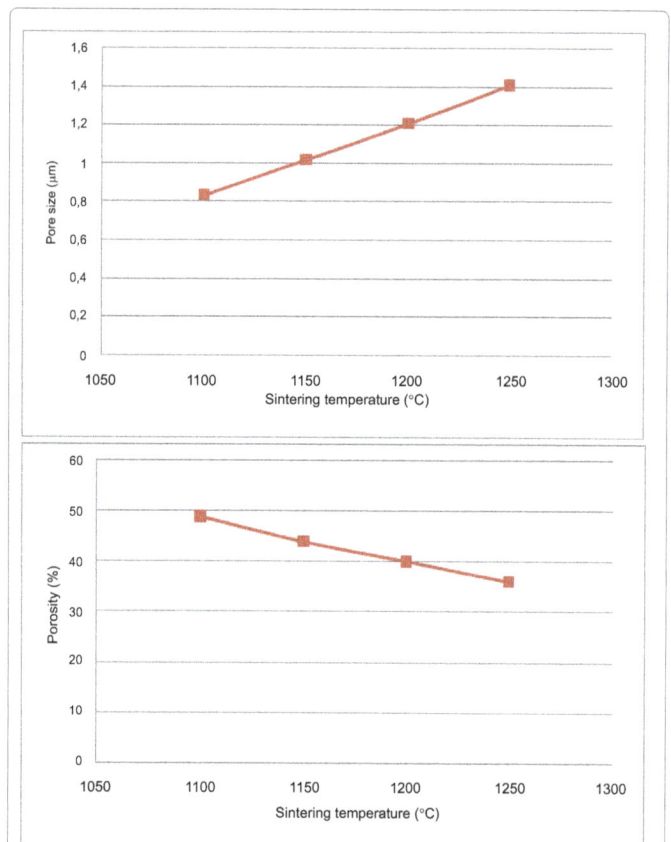

Figure 6: SEM pictures of the elaborated supports fired at various sintering temperature.

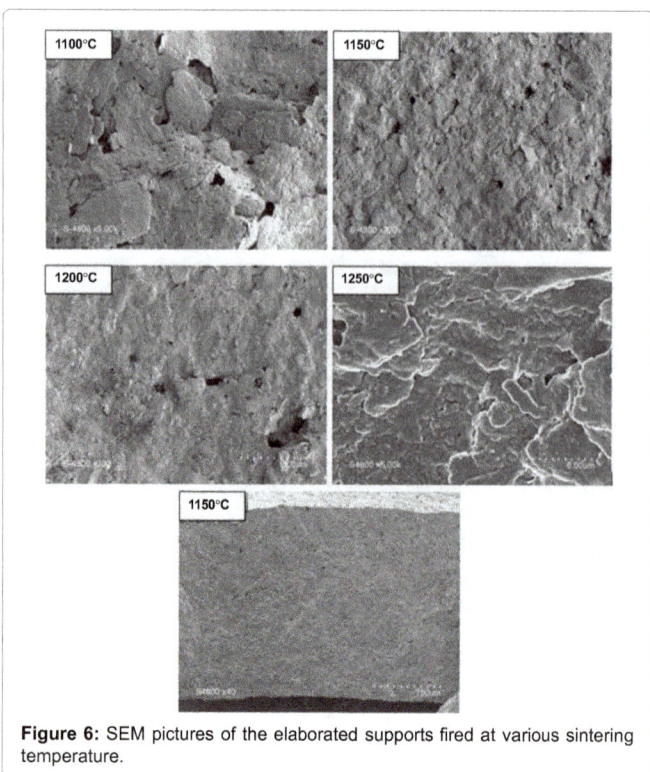

Figure 7: Effect of sintering temperature on average pore size and porosity of supports.

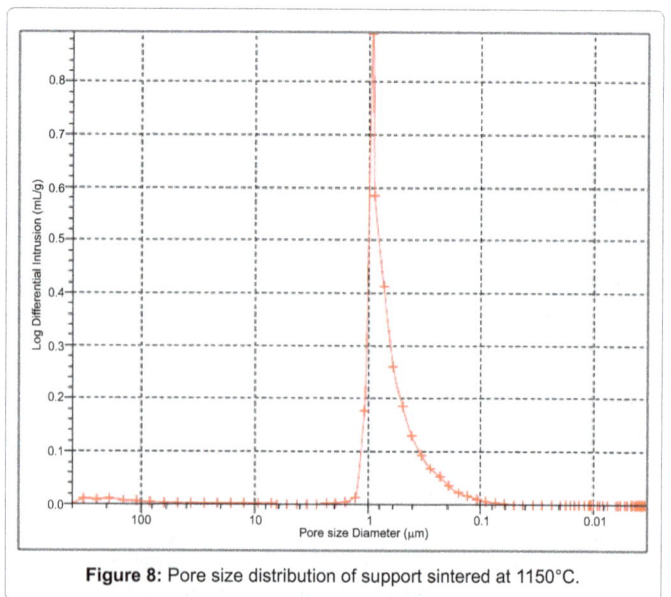

Figure 8: Pore size distribution of support sintered at 1150°C.

homogenous pore distribution. This is necessary for a good integrity of the membrane.

Mechanical resistance: Figure 9 shows the variation of flexural strength with sintering. In accordance with the SEM pictures and the porosity values, the increase of the sintering temperature is accompanied with a densification phenomenon and consequently an

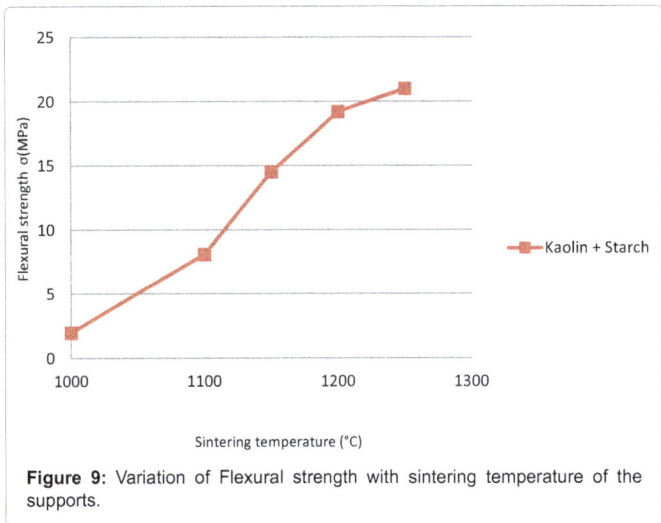

Figure 9: Variation of Flexural strength with sintering temperature of the supports.

increase in the tensile strength from 3 MPa at 1000°C to 21 MPa at 1250°C.

Chemical corrosion: The weight loss due to the corrosion of acids and alkali is shown in Figure 10. It can be seen that the membrane shows a better acid corrosion resistance, since its mass loss is much lower than those of supports after alkali corrosion. Therefore, the observed results in weight loss during corrosion tests suggest that the prepared support possesses a good chemical corrosion resistance and it is suitable for applications involving acidic and basic media.

Choice of the support: Finally the best condition to prepare the support are established for a firing temperature of 1150°C, at this condition the average pore diameter is 1 μm and the porosity is 44%. Moreover this support presents good mechanical resistance having a tensile strength of 15 MPa, and also good chemical properties towards the acid and basic solutions. Thus, membrane support sintered at 1150°C for 1 hour has been selected as substrates for filtration tests.

Rheological characterization of the suspension

For a good adhesion on the macroporous support, viscosity must be sufficient to facilitate the coating and to prevent that the support does not absorb solvent too quickly. Many parameters were studied in order to optimise the slip composition (percentage mass in PVA, percentage mass in kaolin, temperature, etc.).

Influence of temperature: The influence of the temperature on the suspension viscosity was studied using different concentrations of PVA (20%, 30% and 40%) and kaolin (2%-10%). The mineral study used for coating was tested at varying temperatures ranging from 25 to 80°C. As expected, a rise of temperature leads to a decrease of the suspension viscosity (Figures 11 and 12) according to an exponential relationship whose expression is given by Dufauda [56]:

$$\eta = A.e^{Ea/Rt}$$

Where A is the pre-exponential factor, Ea is the activation energy for viscous flow, R is the gas constant (8.314 J mol^{-1} K^{-1}), and T is the absolute temperature. The logarithm of the viscosity plotted against the inverse of temperature bits well aspect Arrhenius relation.

Figures 11a and 12a show that the viscosity increases considerably with increasing the concentration of kaolin powder and PVA but

decreases with the increase of the temperature in the range 25-80°C. The influence of temperature on the composition and viscosity can be described using Arrhenius Law. The determination of the activation energy Ea shows a pronounced sensitivity of the viscosity to temperature changes [57,58].

Figures 11b and 12b show an evolution accorded to Arrhenius law for the different prepared suspensions with an increase of the logarithmic viscosity with the percentage of loading of PVA and kaolin in the temperature range between 20°C and 70°C. A straight line was obtained using Arrhenius model which fitted reasonably well with the experimental data and gave high R2 equal to 0.99 with 99% confidence level.

Table 5 lists the corresponding flow activation energy ΔEa and the

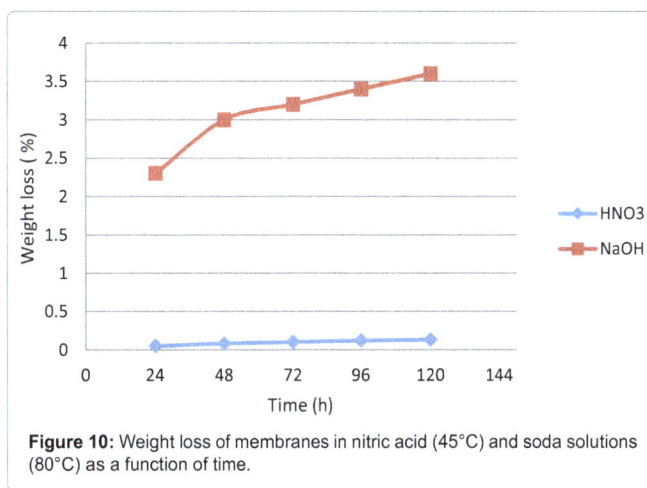

Figure 10: Weight loss of membranes in nitric acid (45°C) and soda solutions (80°C) as a function of time.

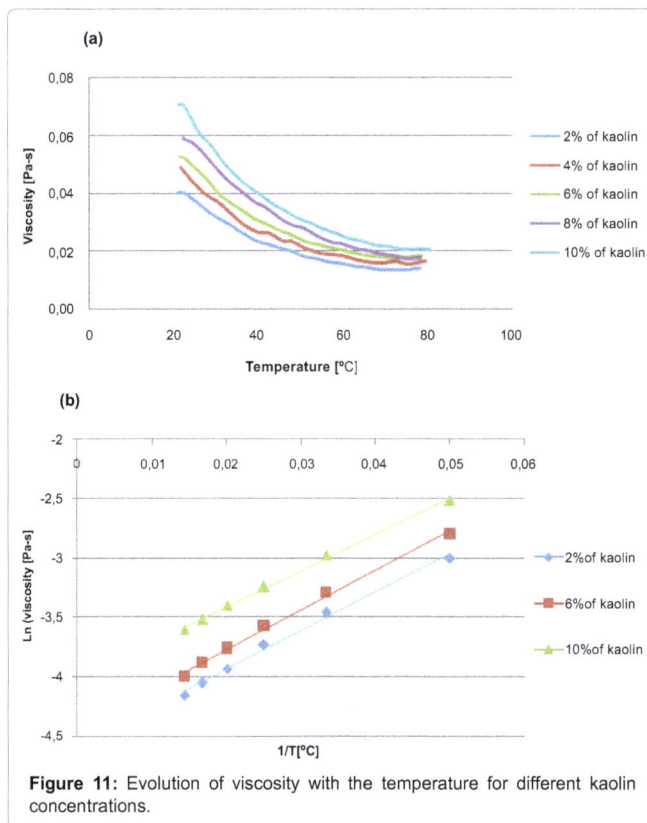

Figure 11: Evolution of viscosity with the temperature for different kaolin concentrations.

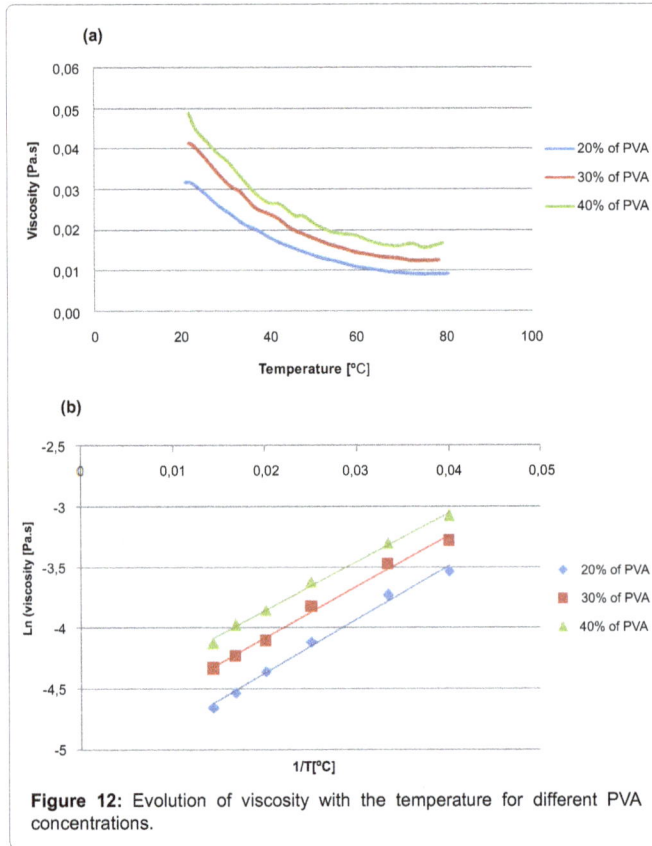

Figure 12: Evolution of viscosity with the temperature for different PVA concentrations.

related fit stability index R2 for different suspensions. With increasing concentration of kaolin powder and PVA, a pronounced increase of the flow activation energy can be observed. Such results are in agreement with the values given in the literature [57,58] and are due to the increasing interaction of the particles with the polymer.

In slip casting process, capillary suction is the driving force that causes the layer deposition on the inner side of the support. Then, to obtain a good and homogenous layer, the suspension does not have to be very viscous nor fluid. The effect of PVA and mineral powder percentages on the viscosity of the suspension is due to the fact that increasing concentration would have a direct effect on the fluid internal shear stress while the temperature effect is obviously due to a weakening temperature of inter-particle and inter-molecular adhesion forces [59].

The PVA and the mineral powder percentage, have a direct influence on the suspension viscosity. This influence is more important for low temperatures. A temperature of 25°C to 35°C will be retained since the layer deposition process is technically possible only at low temperatures. The PVA percentage has the stronger effect on the viscosity, so, it is of interest to study its effect on the rheological behavior of the suspension.

Influence of PVA percentage: Figure 13 shows that the viscosity of the suspension increases slightly depending on the stirring period. However, the increase of PVA percentage leads to a considerable increase of the viscosity. Such rheological behavior is explained by the different types of interactions susceptible to be established between PVA used as Binder and kaolin particles which depend on the concentration of both of them [57]. In conclusion, a stirring time of 90 minutes is a medium duration that can be used when preparing the suspension to be used.

From Figure 14, it can be noticed that dynamic viscosity decreases slightly according to the speed of shearing, but this variation remains insufficient for the determination of the nature of the behavior of such suspension. For identifying the rheological behavior, the variation of the shear stress (τ) according to the speed of shearing (D) was determined.

Figure 15 shows that the behavior of dip solution (B.1, B.2 and B.3) is of Newtonian type. The shear rate increases according to the speed of shearing D. These samples show a Newtonian behavior generated by an increase in inter-particle interactions due to increasing the concentration of PVA and formation of complex flocculated structures. Suspension with a PVA percentage comprised between 20 and 40 % can be apparently used to deposit the layer.

Finally, to obtain a suspension with a viscosity of 0.042 Pa.s and having a Newtonian behavior at room temperature of 25 to 30°C, the suspension should have a composition of 4% of kaolin and 40% of PVA and stirred for 90 minutes.

suspension	2% of kaolin[a]	6% of kaolin[a]	10% of kaolin[a]	20% of PVA[b]	30% of PVA[b]	40% of PVA[b]
ΔEa (KJ/mol)	17.44	19.43	20.28	16.85	17.50	18.64
R²	0.993	0.995	0.997	0.992	0.992	0.995

a: PVA aqueous solution was fixed at 40%
b: Kaolin powder % was fixed at 4%

Table 5: Estimation of the flow activation energy Ea for kaolin-PVA.

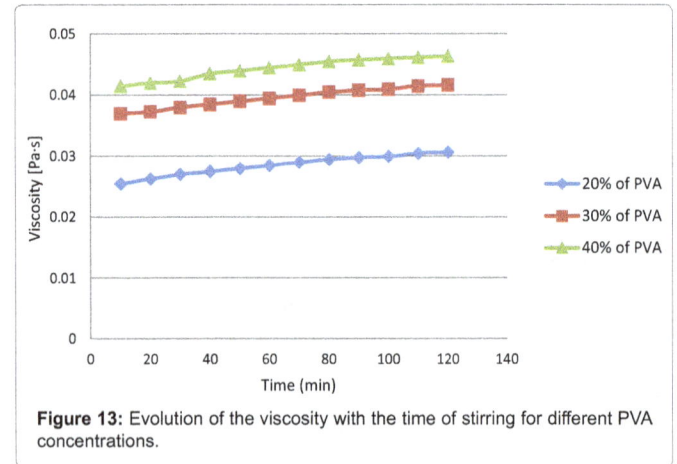

Figure 13: Evolution of the viscosity with the time of stirring for different PVA concentrations.

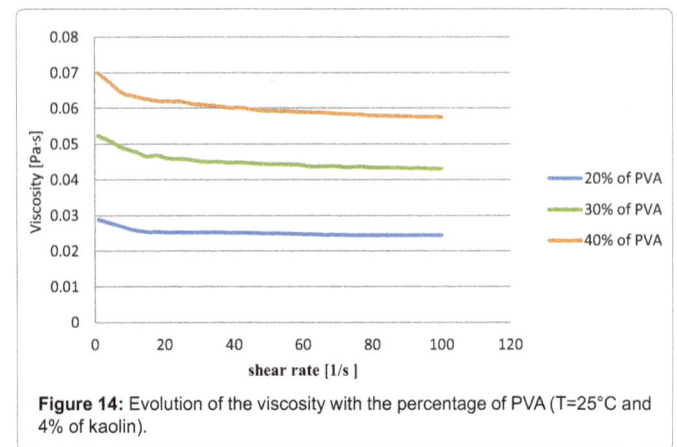

Figure 14: Evolution of the viscosity with the percentage of PVA (T=25°C and 4% of kaolin).

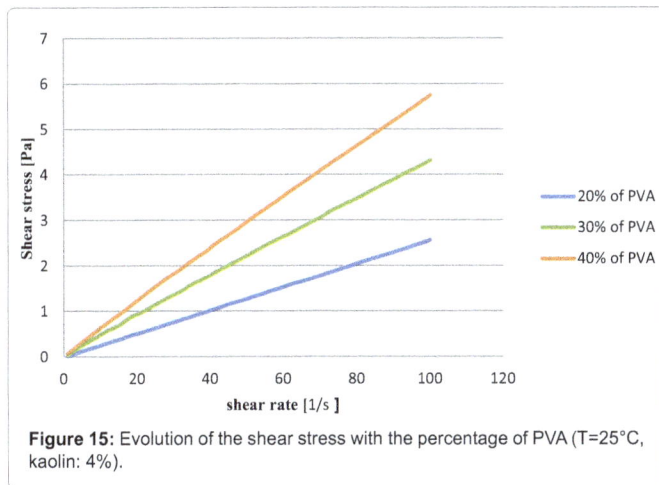

Figure 15: Evolution of the shear stress with the percentage of PVA (T=25°C, kaolin: 4%).

Membrane characterization

Scanning electron microscopy (SEM): Influence of PVA percentage: Three layers were cast on clay tubular supports using three different suspension compositions (B.1, B.2 and B.3). The same casting time was used during the slip casting operation (6 min). The sintering conditions, previously mentioned, were respected.

Figure 16, which shows SEM pictures for surface and cross-section of the obtained layers, give information about the thickness and texture. According to the SEM analysis, it can be noticed at first, that there is a good adhesion between the layer and the clay support. No detachment is observed on the different cross section photographs.

The optimal composition for coating suspension is obtained by using 40% PVA (B.3). This composition enables us to achieve a thickness of 8 μm for ultrafiltration and homogeneous surface free from defects (such as cracks, etc.). On the other hand, the composition corresponding to 20% of PVA (B.1) enables us to obtain a thickness of 3 μm. So, B.3 composition was retained for the preparation of an ultrafiltration membrane.

Influence of deposition time: The deposition of the slip B.3 on the clay support was performed by slip casting using a deposition time between 2 and 20 min. Different ultrafiltration membranes with different layers thickness (between 3 and 30 μm) were prepared. SEM (scanning electron microscopy) images of the resulting membranes are shown in Figure 16.

B.3 composition was retained for the preparation of an ultrafiltration membrane. The casting time was fixed to 6 min.

Mercury porosimetry: The average pore size of the optimized membrane was determined by mercury porosimetry. The pore diameter measured was centered near 11 nm (Figure 17), which confirms that an ultrafiltration layer was achieved.

Determination of membrane permeability: The membrane was initially characterized by the determination of water permeability. It can be seen that the pure water flux increases linearly with increasing the applied pressure (Figure 18). The membrane permeability was found to be equal to 78 l/h.m².bar.

Application to the treatment of the cuttlefish effluents

The elaborated ultrafiltration membranes have been applied to the cuttlefish effluents treatment. Figure 19 gives the variation of permeate flux with transmembrane pressure. Permeate flux increased linearly with transmembrane pressure until 6 bar and then became pressure independent. This behavior can be explained by the formation of a concentrated polarization layer. Beyond 6 bar, the flux value is about 64 l/h.m².

The variation of the permeate flux with time at different transmembrane pressures (TMP) from 2 to 6 bar and a temperature of 25°C, as shown in Figure 20. Permeate flux decreases with the time and stabilizes after the first 20 min whatever the pressure value. This behavior can be attributed to the fouling due to the interaction between membrane material and the waste water solution. It is important to notice that fouling is not very important since the flux decline did not exceed 15% for the different pressures used. This is a specific behavior of

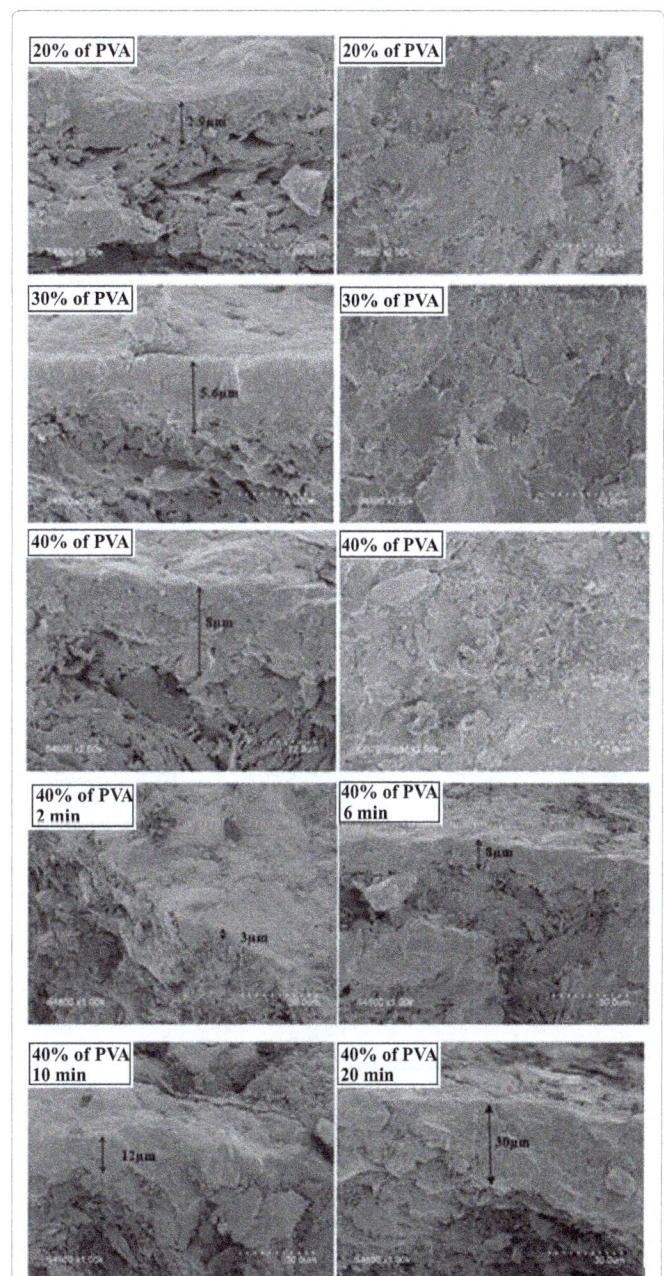

Figure 16: SEM micrographs (cross section and surface) of active layer obtained with different slip compositions and deposition time.

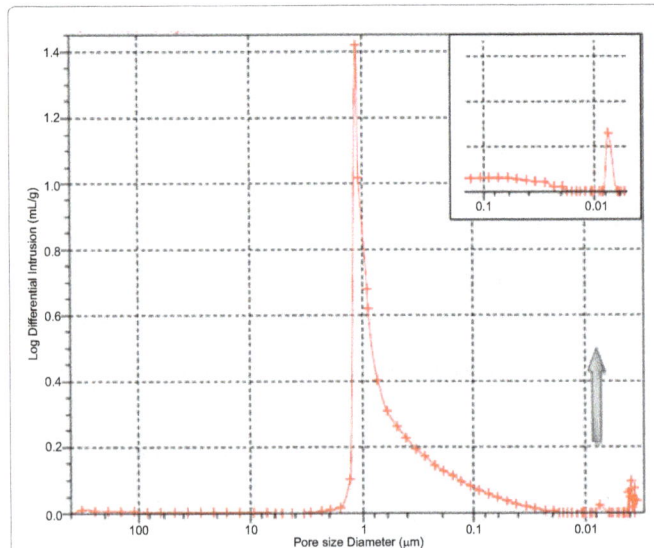

Figure 17: Pore diameters distribution of the kaolin ultrafiltration layer.

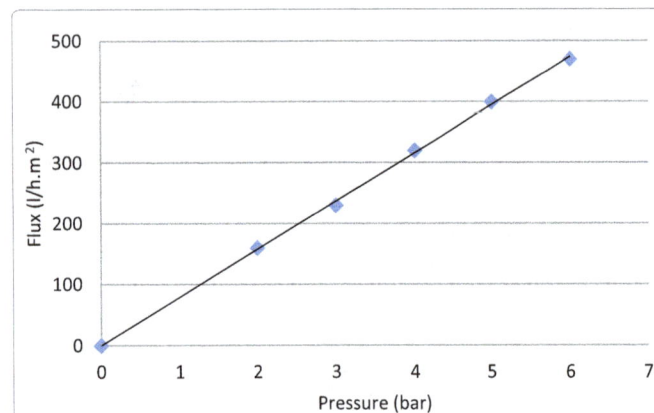

Figure 18: Water flux permeability versus working pressure.

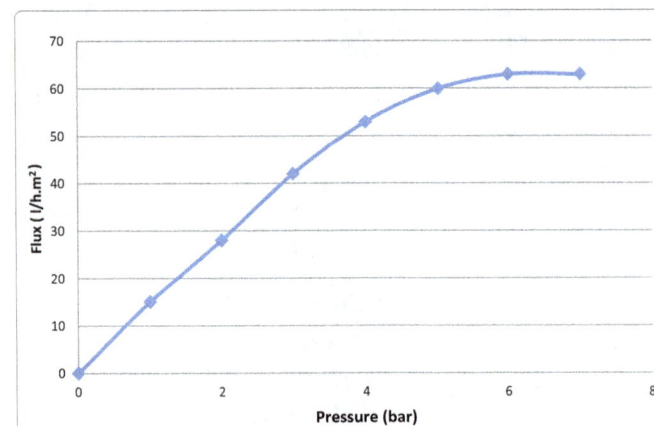

Figure 19: Variation of the permeate flux with pressure.

the UF membrane compared to MF membrane that showed generally an important decrease of flux due to pores blocking by retained particles [3,42].

Ultrafiltration performances: Table 6 gives the main physicochemical parameters analyzed for permeate obtained by UF kaolin membrane. These analyses show variability in the turbidity and COD values. The turbidity of the ultrafiltrated effluent was 0.62, 0.95 and 1.36 NTU respectively for 2, 4 and 6 bar. The values of COD (from 350 to 530 mg L^{-1} with TMP from 2 to 6 bar). The conductivity values were usually in the range of 140-148 ms.cm^{-1}.

Percentage reduction of turbidity and COD as a function of TMP has been shown in Figure 21. Both the turbidity and COD reduction have been found to increase with decrease in TMP, which could be attributed due to the higher rejection at lower TMP. As pressure increases, more melanin permeates through the membrane leaving most of solutes to through the pores of the membrane by increasing transmembrane pressure and subsequently decreased rejection.

Turbidity of permeate was found to get reduced by 99.8% when UF was carried out at a TMP of 2bar, whereas, COD was reduced by only 87%. However, the conductivity in the permeate decreases slightly and is pressure independent. In fact, at all the TMP level, turbidity

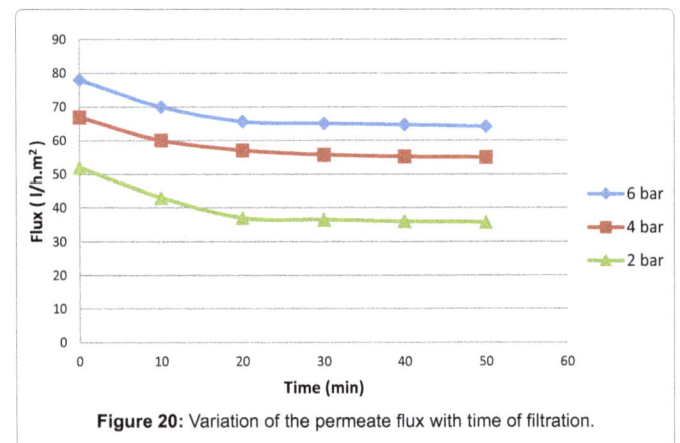

Figure 20: Variation of the permeate flux with time of filtration.

	Pressure (bar)	Turbidity (NTU)	COD (mg.L^{-1})	Conductivity (mS.cm^{-1})
Raw effluents		335	2615	204
Filtrate	2	0,62	350	140
	4	0,95	460	146
	6	1,36	530	148

Table 6: Characteristics of the effluent before and after filtration on kaolin membranes.

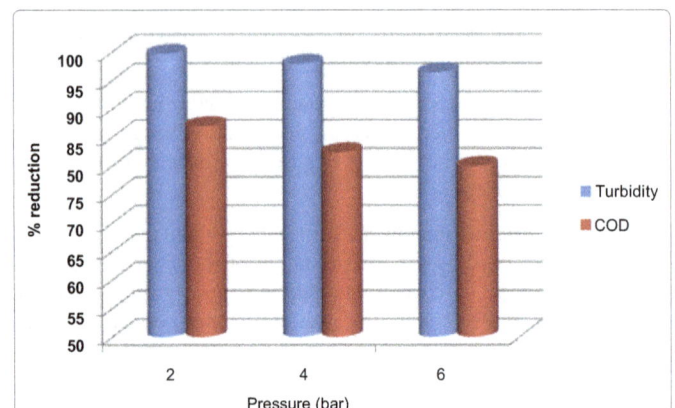

Figure 21: Effect of transmembrane pressures in ultrafiltration on turbidity and COD reduction.

reductions were found to be more than the corresponding COD reduction on percentage basis. As the COD is caused by the presence of low molecular inorganic chemicals also, which might pass through the membrane, may give less (%COD).

Evaluations of resistances: The volume flux in a pressure driven membrane process depends on the hydraulic resistance of the used membrane and the pressure drop over the membrane. This is generally expressed by the following formula:

$$J_w = \Delta P / \eta_w R_m J_w \qquad (1)$$

where, R_m is the intrinsic hydraulic resistance of the membrane and η_w is the viscosity of water.

The permeate flux (J_s) after filtration of the solution can be expressed by the resistance-in-series model.

$$J_s = \Delta P / \eta_s R_t J_s \qquad (2)$$

in which η_s is the permeate viscosity and R_t is the total resistance that can be defined as:

$$R_t = R_m + R_c \qquad (3)$$

where, R_c is the total resistance fouling.

Experimentally, the intrinsic membrane resistance R_m was calculated by measuring the pure water flux J_w and viscosity ηw.

Total resistance R_t was estimated from the solution flow rates under operating conditions using Equation (2):

$$R_t = \Delta P / \eta_s J_s \qquad (4)$$

The intrinsic membrane resistance and the total fouling resistance deduced from these Equations for kaolin membrane are presented in Table 7. The major part of the total resistance was due to fouling (($R_p + R_f)/R_t$) for different working pressure. Intrinsic membrane resistance (R_m/R_t) represented a small portion of the total resistance.

Membrane regeneration

After each experiment, the membrane must be regenerated. The efficiency of the using protocol is verified by the measurement of water flux. The regeneration of the membrane was carried out by firstly, the membrane was exposed to an acid-basic washing sequence (Table 8), and secondly, by leaving the membrane in distilled water. The used protocol appears sufficient because we obtained the value of the initial permeability of the membranes (Figure 22).

Conclusion

In this work, a comprehensive study on the fabrication and characterization of an asymmetric ultrafiltration membrane from naturally occurring- kaolin clays were performed. Ceramic supports have been obtained by extrusion using kaolin powder with corn starch as poreforming agent. It has been concluded that the supports sintered at 1150°C is the optimum supports allowed for depositing the membrane layers. The mechanical and structural properties of the supports are satisfying in terms of porosity and pore diameter.

Pressure (bar)	$R_t(10^{12})$ (m^{-1})	$R_m(10^{12})$ (m^{-1})	$R_c(10^{12})$ (m^{-1})	R_m/R_t (%)	R_c/Rt (%)
2	20.11	4.61	15.5	22.92	77.08
4	26.18	4.61	21.57	17.6	82.4
6	33.75	4.61	29.14	13.66	86.34

Table 7: Resistance values for kaolin membrane.

Sequence	Agent	Concentration	T (°C)	Pressure (Bar)	Time (min)
Rinsing then evacuation	water		RT	2	
Basic washing	NaOH	5-10 g L^{-1}	60-65	2	30
Rinsing until neutrality	water		RT	2	
Acidic washing	HNO$_3$	5 ml L^{-1}	40-50	2	30
Rinsing until neutrality	water		RT	2	

*RT: Room Temperature

Table 8: Acido-basic washing sequence.

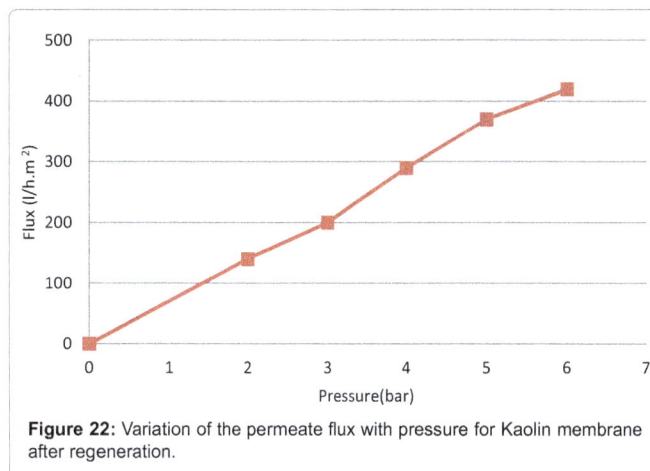

Figure 22: Variation of the permeate flux with pressure for Kaolin membrane after regeneration.

The UF layer, deposited on the supports, was obtained by the slip-casting technique. The composition of the slip was optimized using a rheological study. An excellent link between the support and the ultrafiltration layer was obtained. The thickness of the ultrafiltration layer is about 9 μm; it can be controlled by the percentage of the PVA powder added in the slip suspension and the duration of the deposition time. The obtained membrane has an average pore size diameter of about 11 nm and a water permeability of 78 l/h.m².bar. This result may allow the novel membrane to be used in water treatment in the ultrafiltration range.

The application of this membrane to the washing cuttlefish effluent treatment exhibited an acceptable stabilized permeate flux of almost 64 l/h.m² at 2 bar and 25°C with a very important decrease of different pollutants in terms of turbidity (99%) and COD (87%). These membranes can also be used as a support for nanofiltration layer.

Acknowledgments

Authors would like to thank IEM (European Institute of Membranes), UMR-CNRS 5635/ENSCM Montpellier/University Montpellier 2, for their help to carry out the analysis.

References

1. Xu L, Li W, Lu S, Wang Z, Zhu Q, et al. (2002) Treating dyeing waste water by ceramic membrane in cross flow microfiltration. Desalination 149: 199-203.

2. Ebrahimi M, Ashaghi KS, Engel L, Willershausen D, Mund P, et al. (2009) Characterization and application of different ceramic membranes for the oil-field produced water treatment. Desalination 245: 533-540.

3. Masmoudi S, Amar RB, Larbot A, El Feki A, Salah AB, et al. (2005) Elaboration of inorganic microfiltration membranes with hydroxyapatite applied to the treatment of waste water from sea product industry. J Membr Sci 247: 1-9.

4. Khemakhem S, Amar RB, Hassen RB, Larbot A, Medhioub M, et al. (2004) New ceramic membranes for tangential waste-water filtration. Desalination 167: 19-22.

5. Kumar SM, Madhu GM, Roy S (2007) Fouling behaviour, regeneration options

and on-line control of biomass-based power plant effluents using micro porous ceramic membranes. Sep Purif Technol 57: 25-36

6. Nandi BK, Das B, Uppaluri R, Purkait MK (2009) Microfiltration of mosambi juice using low cost ceramic membrane. J Food Eng 95: 597-605.

7. Wang BJ, Wei TC, Yu ZR (2005) Effect of operating temperature on component distribution of West Indian cherry juice in a microfiltration system. LWT 38: 683-689.

8. Zulewska J, Newbold M, Barbano DM (2009) Efficiency of serum protein removal from skim milk with ceramic and polymeric membranes at 50ºC. J Dairy Sci 92: 1361-1377.

9. Espina VS, Jaffrin MY, Frappart M, Ding LH (2008) Separation of case in micelles from whey proteins by high shear microfiltration of skim milk using rotating ceramic membranes and organic membranes in a rotating disk module. J Membr Sci 325: 872-879.

10. Saboya LV, Maubois JL (2000) Current developments of microfiltration technology in the dairy industry. Lait 80: 541-553.

11. Pouliot Y (2008) Membrane processes in dairy technology - from a simple idea to worldwide panacea. Int Dairy J 18: 735-740.

12. Kao PM, Huang SC, Chang YC, Liu YC (2007) Development of continuous chitin as e production process in a membrane bioreactor by Paenibacillus sp. CHE-N1. Process Bio chem 42: 606-611.

13. Larbot A (2003) Fundamentals on inorganic membranes: present and new developments. Pol J Chem Technol 6: 8-13.

14. Li K (2007) Ceramic Membranes for Separation and Reaction, John Wiley & Sons, Ltd, West Sussex, England.

15. Monash P, Pugazhenthi G (2011) Effect of TiO2 addition on the fabrication of ceramic membrane supports: a study on the separation of oil droplets and bovine serum albumin (BSA) from its solution. Desalination 279: 104-114.

16. Han JH, Oh E, Bae B, Song IH (2013) The effect of kaolin addition on the characteristics of a sintered diatomite composite support layer for potential microfiltration applications. Ceram Int 39: 8955-8962.

17. Han JH, Oh E, Bae B, Song IH (2013) The fabrication and characterization of sintered diatomite for potential microfiltration for applications. Ceram Int 39: 7641-7648.

18. Schillo MC, Park LS, Chiu WV, Verweij H (2010) Rapid thermal processing of inorganic membranes. J Membr Sci 362: 127-133.

19. Zhu J, Fan Y, Xu N (2010) Preparation and characterization of alumina membranes on capillary supports: effect of film-coating on crack-free membrane preparation Chin. J Chem Eng 3: 377-383.

20. Qin W, Peng C, Lv M, Wu J (2014) Preparation and properties of high-purity porous alumina support at low sintering temperature. Ceram Int 40: 13741-13746.

21. Li G, Qi H, Fan Y, Xu N (2009) Toughening macroporous alumina membrane supports with YSZ powders. Ceram Int 35: 1641-1646.

22. Liu C, Wang L, Ren W, Rong Z, Wang X (2007) Synthesis and characterization of a mesoporous silica (MCM-48) membrane on a large-pore α-Al2O3 ceramic tube. Microporous Mesoporous Mater 106: 35-39.

23. Wang YH, Tian TF, Liu XQ, Meng GY (2006) Titania membrane preparation with chemical stability for very hash environments applications. J Membr Sci 280: 261-269.

24. Yoshino Y, Suzuki T, Nair BN, Taguchi H, Itoh N (2005) Development of tubular substrates, silica based membranes and membrane modules for hydrogen separation at high temperature. J Membr Sci 267: 8-17.

25. Masmoudi S, Larbot A, Feki H, Amar RB (2007) Elaboration and characterization of apatite based mineral supports for microfiltration and ultrafiltration membranes. Ceram Int 33: 337-344.

26. Saffaj N, Persin M, Younsi SA, Albizane A, Cretin M, et al. (2006) Elaboration and characterization of microfiltration and ultrafiltration membranes deposited on raw support prepared from natural Moroccan clay: application to filtration of solution containing dyes and salts. Appl Clay Sci 31: 110-119.

27. Palacio L, Bouzerdi Y, Ouammou M, Albizane A, Bennazha J, et al. (2009) Ceramic membranes from Moroccan natural clay and phosphate for industrial water treatment. Desalination 245: 501-507.

28. Loukili H, Younssi SA, Albizane A, Bennazha J, Persin M, et al. (2008) The rejection of anionic dyes solutions using an ultrafiltration ceramic membrane. Phys Chem News 41: 98-1011.

29. Jana S, Purkait MK, Mohanty K (2010) Preparation and characterization of low-cost ceramic microfiltration membranes for the removal of chromate from aqueous solutions. Appl Clay Sci 47: 317-324.

30. Ayadi S, Jedidi I, Rivallin M, Gillot F, Lacour S, et al. (2013) Elaboration and characterization of new conductive porous graphite membranes for electrochemical advanced oxidation processes. J Membr Sci 446: 42-49.

31. Khemakhem M, Khemakhem S, Ayedi S, Amar RB (2011) Study of ceramic ultrafiltration membrane support based on phosphate industry subproduct: application for the cuttlefish conditioning effluents treatment. Ceram Int 37: 3617-3625.

32. Khemakhem M, Khemakhem S, Ayedi S, Cretin M, Amar RB (2015) Development of an asymmetric ultrafiltration membrane based on phosphates industry sub-products. Ceram Int 49: 10343-10348.

33. Bouzerara F, Harabi A, Achour S, Labrot A (2006) Porous ceramic supports for membranes prepared from kaolin and doloma mixtures. J Eur Ceram Soc 26: 1663-1671.

34. Boudaira B, Harabi A, Bouzerara F, Condom S (2009) Preparation and characterization of microfiltration membranes and their supports using kaolin (DD2) and CaCO3. Desalin Water Treat 9: 142-148.

35. Bouzerara F, Harabi A, Condom S (2009) Porous ceramic membranes prepared from kaolin. Desalin Water Treat 12: 415-419.

36. Harabi A, Guechi A, Condom S (2012) Production of supports and filtration membranes from Algerian kaolin and limestone. Procedia Eng 33: 220-224.

37. Zenikheri F, Boudaira B, Bouzerara F, Guechi A, Foughali L (2014) A new and economic approach to fabricate resistant porous membrane supports using kaolin and CaCO3. J Eur Ceram Soc 34: 1329-1340.

38. Tewari PK, Singh RK, Batra VS, Balakrishnan M (2010) Membrane bioreactor (MBR) for waste water treatment: filtration performance evaluation of low cost polymeric and ceramic membranes. Sep Pur Technol 71: 200-204.

39. Dong Y, Liu X, Ma Q, Meng G (2006) Preparation of cordierite-based porous ceramic micro-filtration membranes using waste fly ash as the main raw materials. J Membr Sci 285: 173-181.

40. Cao J, Dong X, Li L, Dong Y, Hampshire S (2014) Recycling of waste fly ash for production of porous mullite ceramic membrane supports with increased porosity. J Eur Ceram Soc 34: 3181-3194.

41. Jedidi I, Khemakhem S, Larbot A, Amar RB (2009) Elaboration and characterisation of fly ash based mineral supports for microfiltration and ultrafiltration membranes. Ceram Int 35: 2747-2753.

42. Jedidi I, Saïdi S, Khemakhem S, Larbot A, Elloumi-Ammar N, et al. (2009) Elaboration of new ceramic microfiltration membranes from mineral coal fly ash applied to waste water treatment. J Hazard Mater 172: 152-158.

43. Hanemann T (2006) Viscosity change of unsaturated polyester-alumina composites using polyethylene glycol alkyl ether based dispersants, Composites: Part A. Applied Science and Manufacturing 37: 2155-2163.

44. Tsetsekou A, Agrafiotis C, Milias A (2001) Optimization of the rheological properties of alumina slurries for ceramic processing applications Part I: Slip-casting. J Eur Cer Soc 21: 363-373.

45. Masmoudi S, Larbot A, Feki HE, Amar RB (2006) Elaboration and properties of new ceramic microfiltration membranes from natural and synthesised apatite. Desalination 190: 89-103.

46. Zhang Y, Yokogawa Y, Feng X, Tao Y, Li Y (2010) Preparation and properties of bimodal porous apatite ceramics through slip casting using different hydroxyapatite powders. Ceram Int 36: 107-113.

47. Chaari K, Bouaziz J (2005) Rheological behavior of organic suspensions of fluorapatite. J Colloid Interface Sci 285: 469-475.

48. Sahnoun RD, Bouaziz J (2012) Sintering characteristics of kaolin in the presence of phosphoric acid binder. Ceram Int 38: 1-7.

49. Boulmokh A, Berredjem Y, Guerfi K, Gheid A (2014) Kaolin from Djebel Debbagh Mine Guelma Algeria. J Appl Sci 2: 435-440.

50. Harrabi A, Bouzerara F, Condom S (2009) Preparation and characterization

Development of an Asymmetric Ultrafiltration Membrane from Naturally Occurring Kaolin Clays: Application...

33

of tubular membrane supports using centrifugal casting. Desal Water Treat 6: 222-226.

51. Bouzerara F, Harrabi A, Ghouil B, Medjemem N, Boudaira B, et al. (2012) Synthesis and characterization of multilayer ceramic membranes. Proc Eng 33: 278-284.

52. Tahri N, Jedidi I, Cerneaux S, Cretin M, Amar RB (2013) Development of an asymmetric carbon microfiltration membrane: application to the treatment of industrial textile wastewater. Sep Purif Technol 118: 179-187.

53. Majouli A, Younssi SA, Tahiri S, Albizane A, Loukili H (2011) Characterization of flat membrane support elaborated from local Moroccan Perlite. Desalination 277: 61-66.

54. Issaoui M, Bouaziz J (2015) Elaboration of membrane ceramic supports using aluminium powder, Desalin. Water Treat 53: 1037-1041.

55. Fakhfakh S, Baklouti S, Bouaziz J (2013) Elaboration and characterisation of low cost ceramic support membrane. Adv appl ceram 109: 31-38.

56. Dufauda O, Marchal P, Corbel S (2002) Rheological properties of PZT suspensions for stereolithography. J Eur Ceram Soc 22: 2081-2092.

57. Hanemann T (2008) Influence of particle properties on the viscosity of polymer-alumina composites. Ceram Int 34: 2099-2105.

58. Nasri H, Khemakhem S, Amar RB (2014) Physico-Chemical Study of Coating Formulation Based on Natural Apatite for the Elaboration of Microfiltration Membrane. Periodica Polytechnica 58: 171-178.

59. Nguyen AT, Desgranges F, Roy G, Galanis N, Mare T, et al. (2007) Temperature and particle-size dependent viscosity data for water-based nanofluids - Hysteresis phenomenon. International Journal of Heat and Fluid Flow 28: 1492-1506.

5

Adsorption of Heavy Metals Cations onto Zeolite Material from Aqueous Solution

Maria Visa[1]* and Nicoleta Popa[2]

[1]*Department of Renewable Energy Systems and Recycling, Transilvania University of Brasov, Romania*
[2]*Department of Forest District Teliu, National Administration of State Forests Romsilva, Romania*

Abstract

Increased urbanization and industrialization led to excessive release of wastes into the environment; one of them, causing many problems not only to the environment but also to human health, is fly ash resulted from coal combustion. One of the solutions for solving these problems would be fly ash reutilization as an adsorbent for wastewater treatment and for obtaining new and efficient materials such as zeolites. Class "F" fly ash collected from C.H.P. Craiova from Romania was used for obtaining new zeolite materials for advanced wastewater treatment. This material was characterized by AFM, XRD, FTIR, SEM to outline the crystalline and morphology modifications. Hydrothermal modified fly ash using NaOH was used for heavy metals removal (Cd^{2+}, Cu^{2+} and Ni^{2+}) from synthetic wastewaters containing one, two and three pollutants. For obtaining a maximum efficiency during the adsorption process the adsorption conditions (contact time, optimum amount of substrate) were optimized. These parameters were further used in thermodynamic and kinetic modeling of the adsorption processes. The uptake of Cd^{2+}, Cu^{2+} and Ni^{2+} was followed by a Langmuir adsorption isotherm and the maximum uptake capacity was estimated to be 95.24 mg/g Cd^{2+}, 107.52 mg/g Cu^{2+} from aqueous solution with two cations. Correlated with the surface structure, composition and morphology the adsorption kinetic mechanisms and the substrate capacities are further discussed.

Keywords: Heavy metals; Adsorption; Filtration; Zeolite; Wastewater treatment

Introduction

The pollution of water is one big problem in each country. More hundred pollutants are released every day into the environment from numerous branches of industries. The main industries responsible for large amount of wastewater with complex pollutant load are: the mining and extractive industry, electroplating and metal surface treatment processes, metallurgy, metal coating and plastics. Water interaction with chemical substances resulting wastewaters with a large quantities of heavy metals, different organic or inorganic pollutants and the last target are ecosystem and the health of humans.

Many compounds of heavy metals are easy soluble in water and can be adsorbed by living organisms part of the food chain. Some heavy metals (copper, cobalt, iron, manganese, vanadium, strontium and zinc) are accepted in small concentration for living organisms but excessive levels of essential metals can be detrimental to the organisms. Non–essential heavy metals (including cadmium, lead and zinc) are dangerous for living organisms [1,2].

Heavy metals are persistent pollutants, non-biodegradable and can be accumulated easily in organisms even at low concentrations, causing serious illness; common effects on humans are described as: increased salivation, severe stomach irritations leading to vomiting and diarrhea, abdominal pain, choking, high blood pressure, iron-poor blood, liver disease, pancreas and nerve or brain damage, poisoning by ingestion include vomiting, vomiting of blood, hypotension, coma, jaundice, and gastrointestinal pain [3,4]. For avoiding flora, fauna and health problems, the discharge limits are strict and require advanced wastewater treatment processes [5,6].

On the other hand the heavy metals can deactivate the active sludge (poisoning the bacteria) from the secondary treatment plants; therefore the chemical treatment must remove heavy metals before the biological step.

The conventional techniques used for heavy metal removal are expensive and with average efficiency. The most common and widely used methods for advanced heavy metal removal are: chemical precipitation, flotation, flocculation, sedimentation, solvent extraction, oxidation/reduction and dialysis/electro-dialysis, reverse osmosis, ultra filtration, electrochemical deposition, ion exchange and adsorption e.g.

Choosing one or more complex solutions is subject of efficiency and cost analysis reported at large quantity of wastewater. Among these, adsorption technologies have several advantages: easy operation and well known technology, inexpensive equipment, less sludge, adsorbents' reuse after desorption. In water the heavy metals cations are hydrated with different bipolar water molecules.

Heavy metals (cadmium, copper, zinc, nickel) removal by adsorption was reported on, sugar beet pulp [7], bituminous coal, peat [8], clay and diatomite [9] natural zeolite [10] or on zeolite obtained by conversion of fly ash [11].

This paper presents the synthesis, characterization and results of adsorption of zeolite ZCR40 obtained from fly ash by hydrothermal method, which were used as adsorbents in removal of Cd^{2+}, Cu^{2+} and Ni^{2+} from mono-, di-, and three-component systems.

Materials and Methods

Raw Fly Ash was collected from the electrostatic filters in the

*Corresponding author: Maria Visa, Department of Renewable Energy Systems and Recycling, Transilvania University of Brasov, Eroilor 29, 500036 Brasov, Romania, E-mail: maria.visa@unitbv.ro

combined heat and power plant S.E. Craiova II, Romania. The content of the main constituents are presented in Table 1.

The total percentage of SiO_2, Al_2O_3 and Fe_2O_3 is 78.37% (above 70%),thus, according the ASTM C618 classification, the fly ash –S.E.Craiova II., is of class F [12].

Fly ash was washed in ultra-pure water by stirring up (100 rpm), at room temperature, for 48h, to remove the soluble compounds, until constant pH was reached, at the value of 11.5. The water resulted after washing FA had an ionic conductivity of K=630 [μS/cm] and a TDS value of 640 mg/L, as result of the soluble compounds dissolution: CaO, MgO, Na_2O, K_2O. The dry fly ash was subject sieving to determine the size distribution Table 2.

The zeolite materials were obtained from washed fly ash (FACRw) with 40 μm granulometry by mixed with NaOH solution 3N in autoclave for 5h at 150°C. The fractions >40 μm can be used for obtaining geo-polimers, for cement, bricks, stone blocks. After running time the material was washed with ultra-pure water till constant pH, filtrated, dried at 105-120°C overnight and was analyzed. This new material denoted (ZCR40) was used in adsorption experiments.

The crystalline structures of the FAw and ZCR40 were evaluated by XRD (Bruker D8 Discover Diffractometer). The AFM images (Ntegra Spectra, MT-NDTmodel BL222RNTE) were used for surface morphology studies. The AFM was used to characterize the uniformity, grain size and pores size distribution of the zeolite material. Image analysis was carried out by means of WSxM software, to evaluate the pore size distribution [13]. By using scanning electron microscopy (SEM, S-3400N– Hitachi) further surface investigations were done at an accelerating voltage of 20 kV. Surface compositions of zeolite ZCR40 was measured before and after adsorption using energy dispersive X-ray spectroscopy (EDS Thermo Scientific Ultra Dry). By using porosity analysis and BET surface (Autosorb-IQ-MP, Quantachrome Instruments) characterization of the surface was completed.

The new substrates (ZCR40) were used in adsorption experiments for removing heavy metals from synthetic wastewater loaded with Cd^{2+}, Cu^{2+} and Ni^{2+} cations.

Adsorption experiments

Batch adsorption experiments were carried out under mechanical stirring, at room temperature 20-22°C considering that the vapors pressure of water at this temperature is low (0.023-0.024 bar), thermosetting was not necessary.

Three series of experimental test of adsorption were done:

Adsorption on ZCR40 substrate in solution containing one, two and three pollutants:

Compound	% in FA	Compound	% in FA	Compound	% in FA
SiO_2	46.13	MgO	5.24	MnO	-
Al_2O_3	21.39	K_2O	0.5-2	P_2O_5	-
Fe_2O_3	10.85	Na_2O	0.2-0.6	LOI*	-
CaO	10.65	TiO_2	<1%	SiAlFe**	78.37

Table 1: The composition of fly ash (FA) - C.H.P. Craiova [%].

Raw FA S.E. Craiova II	Particle size distribution [%]			
	20 μm	40 μm	100 μm	<100 μm
	5.2	53.5	31.8	9.5

Table 2: Particle size distribution [%].

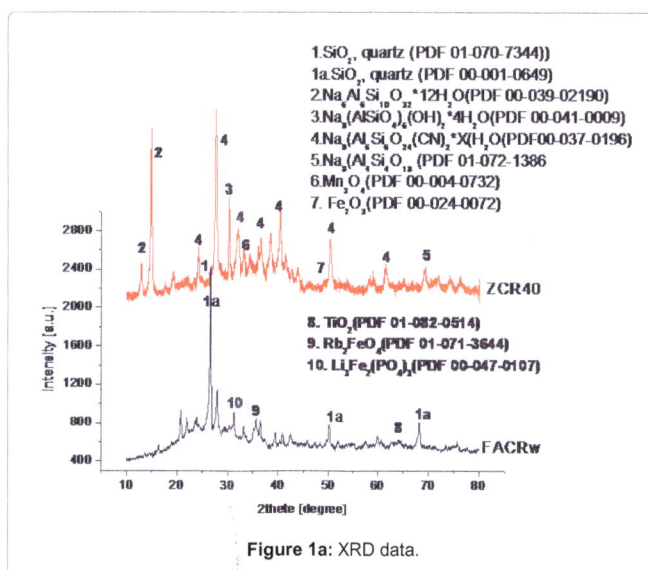

Figure 1a: XRD data.

- Cd^{2+}/Cu^{2+} from mono-cationic system, under mechanical stirring;

- Cd^{2+} + Cu^{2+} from di-cationic system, under mechanical stirring;

- Cd^{2+}, Cu^{2+} and Ni^{2+} from three-cationic system, under mechanical stirring.

In all experiments the pollutant systems were synthetically prepared using bidistilled water and $CdCl_2*2.5\ H_2O$ (Scharlau Chemie S.A., c<98%), $CuCl_2*2H_2O$ (Scharlau Chemie S.A., c<98%) and $NiCl_2*6H_2O$ (Scharlau Chemie S.A., c<98%). The experiments were done using heavy metals solutions in concentration of range of C_{Cd2+}= 0.560 mg/L, C_{Cu2+}=0.330 mg/L and C_{Ni2+}=0.300 mg/L.

The optimal contact time was evaluated in suspension, 0.1 g ZCR40 in 50 mL of equimolar multi-cation solutions of cadmium, copper and nickel; aliquots were taken at certain moments (10, 15, 30, 45, 60, 90, 120, 150 and 180 min), when stirring was briefly interrupted and after filtration the volumes of supernatant were analyzed. The residual metal concentration in the aqueous solution was analyzed by AAS (Analytic Jena, ZEEnit 700), at λ_{Cd}=228.8 nm, λ_{Cu}=324.75 nm and λ_{Ni}=232.00 nm. The working pH was in the range of 5.4-6.3. Preliminary experiments proved that heavy metal losses due to the adsorption to the container walls and to the filter unit 0.22 μm were negligible.

Results and Discussion

Characterization of the substrates

The composition of silicon-aluminous of the fly ash is confirmed by XRD spectra, Figure 1. The major crystalline components of raw fly ash are: αSiO_2 (quartz) identified by the sharp peaks in the range of 2θ =26° and hematite (Fe_2O_3).

The diffractogram (Figure 1) data show that some compounds: SiO_2 (quartz), $Li_3Fe_2(PO_4)_3$ (lithium iron phosphate), $Rb_2(FeO_4)$ (rubidium iron oxide) from FACRw are or not presented in low percentage, while the percent of sodium aluminum hydrate ($Na_6Al_6Si_{10}O_{32}*12H_2O$), sodium aluminum hydroxide hydrate ($Na_8(AlSiO_4)_6*(OH)_2*4H_2O$) sodium aluminum cyamide hydrare ($(Na_8Al_6Si_{10}O_{24}(CN)_2*xH_2O)$,) sodium aluminum silicate hydrate $Na_8(AlSiO_4)_6*4H_2O$ and other phases of the aluminosilicates typical the zeolites is increased in the zeolite material (ZCR40). The quartz syn, hematite and manganese oxide are in small proportion. The area picks of new alumino-silicates phases are higher. The hydrothermal process of FACRw further promotes the surface interactions, including dissolution, re-crystallization processes that confirm chemical restructuring. The crystalline phases increased from 38.5% at 51.1% for ZCR40.

An evaluation criterion of raw fly ash into zeolites can be the proportion of quartz from raw fly ash converted to aluminosilicates using equation (1).

$$R = \frac{\% \text{ quartz in ZCR40}}{\% \text{ quartz in rawFA}} \tag{1}$$

R value is low for ZCR40 materials showing high conversion of SiO_2 from FACRw to crystalline zeolite (NaP1) by hydrothermal treatment.

The content of quartz and of mullite was found to be mainly responsible which hinder formation of zeolite and their activity [14-20].

The composition of crystalline phases, crystalline degree, and morphological changes are presented in Table 3.

Other information related to the morphology and characteristics of the surface were obtained from the AFM and SEM micrographs (Figures 2 and 3).

These AFM images were used to characterize the surface morphology: the uniformity, grain size and pore size distribution [19] of the samples Figure 2. The FACRw and ZCR40 have a rough surface

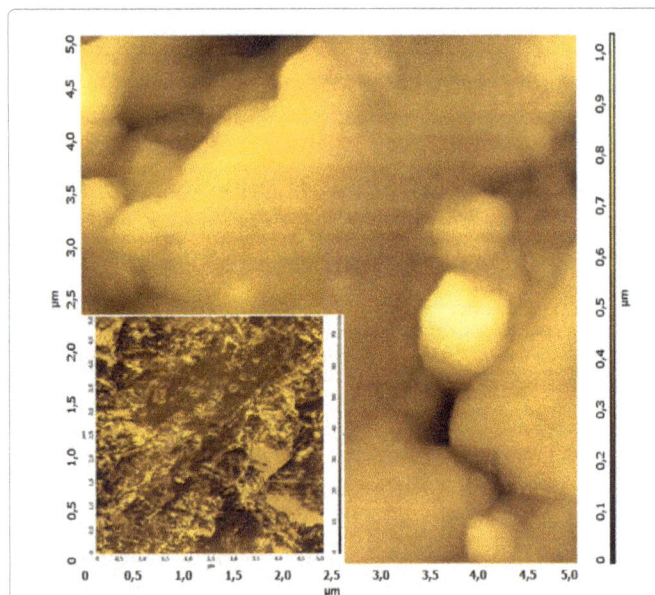

Maximum amplitude = 54.3 nm

Averageroughness = 74.1 nm

Mezo-pori = 265 nm

[Si/Al] = 2.49

Figure 1b: XRD data of FACRw.

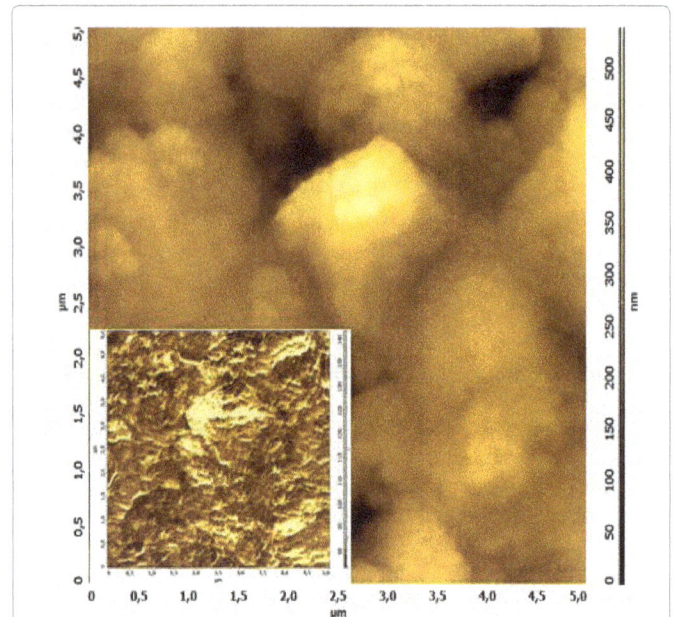

Maximum amplitude = 45.5 nm

Avreageroughnes = 69.4 nm

Mezo-pori = 150 nm

[Si/Al] = 1.37

Figure 1c: XRD data of ZCRs.

Crystalline degree				Morphology modifications		
Sample	Composition of crystalline phases	Crystalline degree [%]	R	Specific surface area (BET) [m²/g]	Micropores volume (t-plot) [cm³/g]	Average pores diameter [nm]
FACRw	Quartz, SiO_2, hexagonal (31.6%) Graphite, hexagonal (4.83%) Na/K Aluminosilicates /hydrate (57.45%) Hematite, Fe_2O_3, romboedric (5.65%), Mn_3O_4, tetragonal (0.46%)	38.5	-	4.09	0.004	27.2
ZCR40	Quartz, SiO_2, hexagonal (6,96%); Hematite, Fe_2O_3, romboedhric (2,41%) TiO_2, monoclinic (1.27%) Mn_3O_4, tetragonal (1.56%) Aluminosilicates simple/hydratated de Na/K (87.8%)	58.50.1	0.07	54.09	0,003	17.98

Table 3: The characteristics of the raw fly ashes and of the zeolite material.

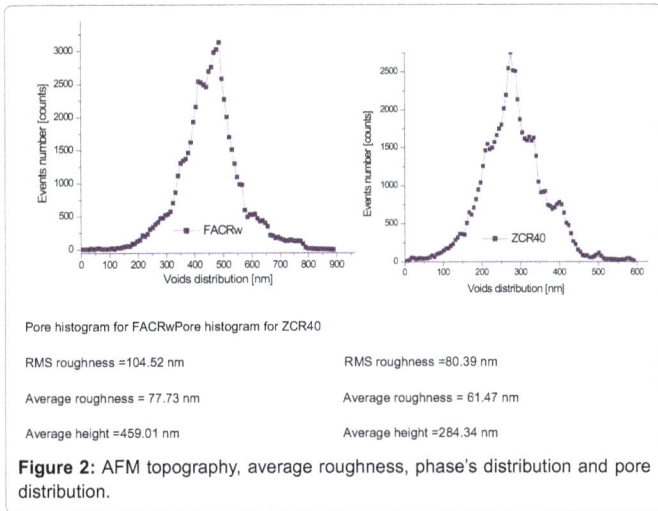

Figure 2: AFM topography, average roughness, phase's distribution and pore distribution.

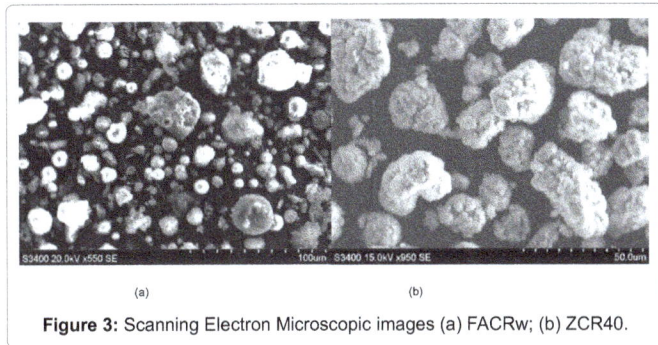

Figure 3: Scanning Electron Microscopic images (a) FACRw; (b) ZCR40.

Substrate	Surface Energy [mN/m]	Disperse contribution [mN/m]	Polar contribution [mN/m]
ZCR40	22.44	4.17	18.31

Table 4: Surface energy data for the substrates.

with larger pores/voids heterogeneously distributed, confirmed by the pores/voids distribution curves with two or three maxima.

Optimizing fly ash substrate by washing and hydrothermal treating the roughness, RMS roughness show a decrease from 74.1 nm to 45.5 nm, from 284.34 nm to 104.52 nm respectively. The roughness can give information knowing the level of investigation (the millimetric scale usually permits one to distinguish the main surface treatments) [15]. These chemical report (Si/Al) and structural changes are mirrored in morphology modifications, Figures 2a and 2b resulting large differences in the substrates' affinity for heavy metals. On phase's distributions images, it can be seen more agglomerates in the new material adsorbent so more macro-pores and meso-pores are ready to lodge the cations of heavy metals.

The SEM images Figure 3 show significant modifications during the fly ashes processing. Most FACRw particles are spherical and in zeolite materials large agglomerates are formed (Figure 3).

The polar and disperse contributions to the surface energy of ZCR40 was calculated according to the model developed by Owens, Wendt, Robel and Kaelble and are presented in Table 4.

The data show a large polar component, recommending the zeolite ZCR40 as a good adsorption substrate for heavy metals cations.

The zeolite material ZCR40 has composition to zeolites leading

to similar types of vibrations. The asymmetric stretch of the Si-O-Si or Si-O-Al bonds corresponds to the peak positioned around 1000 cm^{-1}. Moreover, the water molecules associated with the cations and hydrogen bonded to the oxygen ions of the framework; explain the peak recorded at 1640.38 cm^{-1}, characteristic to the bended mode in the water molecules [16,17]. The characteristic peaks for zeolite are known from the literature [18,19] as being positioned at 421.89 cm^{-1}; 666.03 cm^{-1} and 956.02 cm^{-1}, (Figure 4 and Table 5).

Adsorption of heavy metals on ZCR40 substrate

The adsorption experiments of Cd^{2+}, Cu^{2+} and Ni^{2+} on the zeolite substrate ZCR40 were carried out as a function of contact time, adsorbent dose and heavy metals concentration at optimum contact time. Maximum removal efficiency (η) of Cd^{2+}, Cu^{2+} respectively Ni^{2+} cations was evaluated using the following eqn (2).

$$\eta = \frac{(c_{HM}^i - c_{HM}^e) \times 100}{c_{HM}^i} \qquad (2)$$

where: c_{HM}^i - the initial concentration and c_{HM}^e - equilibrium momentary concentration of the heavy metals (HM).

Adsorption equilibrium is reached in most cases after 60 min with stabilization at 90 min (Figures 5a, 5b and 5c). These values are acceptable for technology process so each of these materials can be further used for technology transfer.

An important parameter governing the efficiency of the adsorption of heavy metals is the volume of the cation, copper cation (hydrated

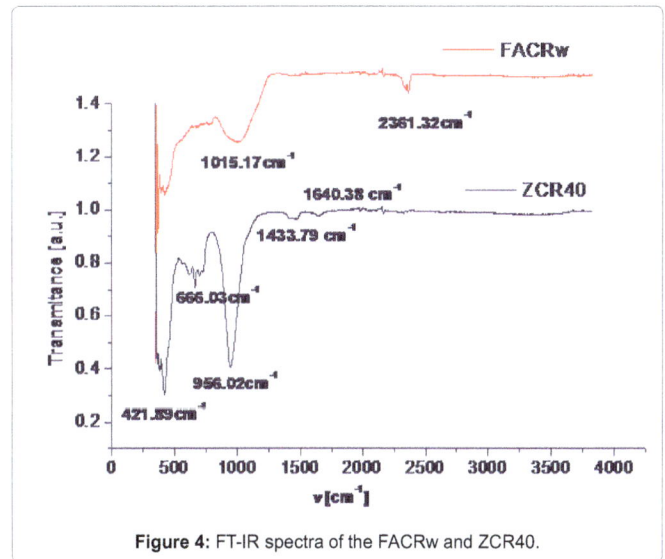

Figure 4: FT-IR spectra of the FACRw and ZCR40.

Compounds	Wavenumber	Allocation
Water	1630...1640	Vibration H - O or H - OH from reaction products hydroalumino-silicate
Kaolin/zeolites	1028...1032	Asymmetric stretching vibration of Si-O-Si bonds
Silicates of Ti/ zeolites	950...961	Asymmetrical stretching vibration of Si-O-Ti /asymmetrical stretching
Feldspars/zeolite	660...670	Symmetric stretching vibration ties Al-O and Si-O
Hematite/zeolite	570...590	Bending Vibration of the links Al-O, Si-O, Al-O-Si and Fe-O
Kaolin/zeolites	400...420	Vibration of the open pores, band O-Si-O

Table 5: IR bands characteristic of the zeolites material.

Figure 5a: Efficiency of heavy metals (Cd²⁺, Cu²⁺, Ni²⁺) removal from mono-cationic system on ZCR40.

Figure 5b: Efficiency of heavy metals (Cd²⁺, Cu²⁺) removal from bi-cationic system on ZCR40.

Figure 5c: Efficiency of heavy metals (Cd²⁺, Cu²⁺, Ni²⁺) removal from three-cationic system on ZCR40.

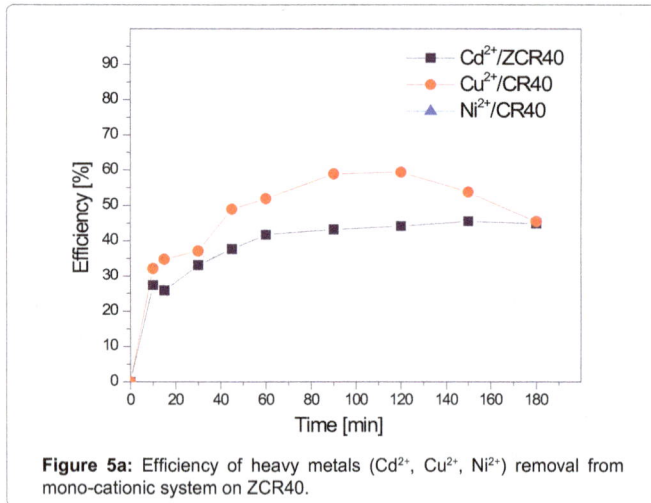

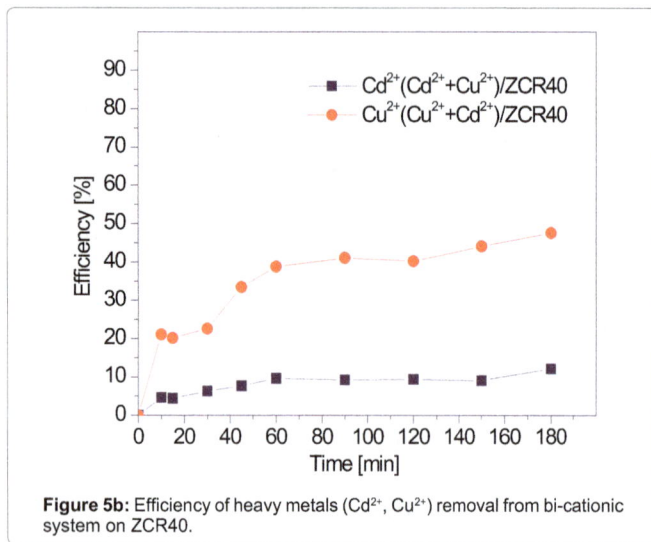

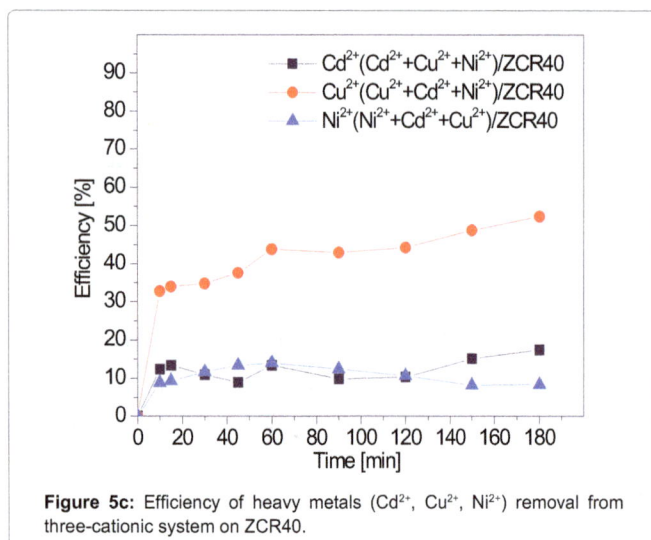

with water molecules 4 .. 6) having the smallest volume. The significant difference observed between single and binary metal adsorption is

in the initial metal removal rate. Slightly lower initial metal removal rate observed in the case of binary adsorption is probably due to of competing hydrated metal ions present in solution. The effect amount of ZCR40 vas analyzed and presented in Figure 6.

The maximum of efficiency is obtained after 90 min. in all cases of adsorption and was further used in next experiments. Increasing the amount of ZCR40 substrate will be an increase of amount of active site.

Heavy metals cations can establish bonds with active sites ($\equiv$ SiO⁻) and ($\equiv$ AlO⁻) forming complex structure accordind to reaction (3, 4):

$$2\,(\equiv\!SiO\text{-}) + M^{2+} \rightarrow (\equiv Si\text{-}O)_2\,M \tag{3}$$

$$2\,(\equiv AlO\text{-}) + M^{2+} \rightarrow (\equiv Al\text{-}O)_2\,M \tag{4}$$

The effect of concentration of Cd²⁺, Cu²⁺, Ni²⁺ solutions on the adsorption was studies at optimal time (90 min.) and optimal mass (0.35 g ZCR40) Figure 7.

High adsorption efficiency for (Cd²⁺, Cu²⁺, Ni²⁺) cations are registration at the concentration up to 0.0003125 val/L.

The adsorption capacity Z (Figure 8) is a function of the initial cation

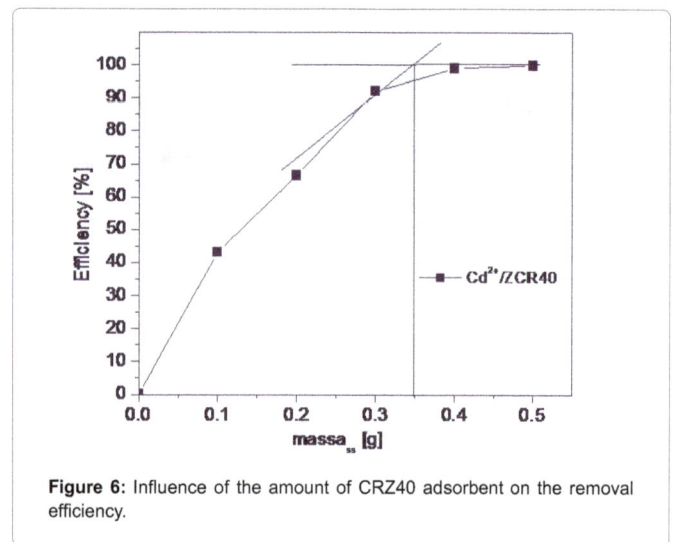

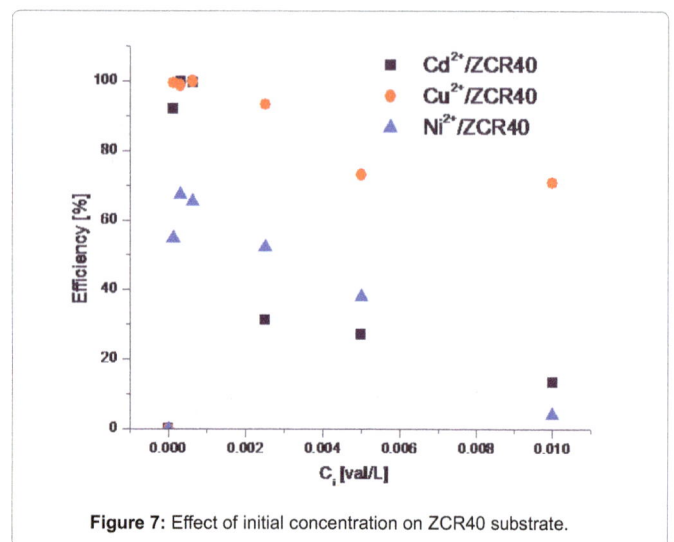

Figure 6: Influence of the amount of CRZ40 adsorbent on the removal efficiency.

Figure 7: Effect of initial concentration on ZCR40 substrate.

Substrate	Pseudo-first order kinetics		Pseudo-second order kinetics			Interparticle diffusion		
	K_L [min^{-1}]	R^2	k_2 [g/mg·min]	q_e [mg/g]	R^2	K_{id} (mg/g·min$^{1/2}$)	C	R^2
Cd^{2+}								
ZCRs 40	0.030	0.964	0.081	129.87	0.996	5.069	63.414	0.844
Cu^{2+}								
ZCR40	0.028	0.904	0.203	86.207	0.928	5.964	22.128	0.928
Cd^{2+}(Cd^{2+}+Cu^2)								
ZCR40	0.021	0.946	0.5156	27.701	0.988	1.717	7.702	0.848
Cu^{2+}(Cd^{2+}+Cu^2)								
ZCR40	0.015	0.907	0.309	81.967	0.983	2.155	9.502	0.906
Cd^{2+}(Cd^{2+}+Cu^{2+}+Ni^{2+})								
ZCR40	-	0.229	0.214	36.364	0.892	-	-	0.225
Cu^{2+}(Cd^{2+}+Cu^{2+}+Ni^{2+})								
ZCR40	0.012	0.853	0.194	86.957	0.981	3.232	38.904	0.829
Ni^{2+}(Cd^{2+}+Cu^{2+}+Ni^{2+})								
ZCR40	-	0.525	0.019	13.316	0.862	-	-	0.192

Table 6: Kinetic parameters of heavy metals adsorption.

concentration, the adsorbent dose, the energy surface of adsorbent (polar contribution) and the nature solute – sorbent interaction.

The adsorption kinetic

The metal uptake q_e (mg/g) was evaluated for the kinetic studies by using the initial and current, t, heavy metal concentrations (c^i_{cation} and c^t_{cation}) in a given solution volume (V=50 mL) for a given amount of ZCR40 (m_s = 0.1 g) as given in eq. (5):

$$q_e = \frac{(c^i_{cation} - c^t_{cation}) \cdot V}{m_s} \qquad (5)$$

By using the following equations (6) and (7) the kinetics of the heavy metals adsorption were modeled:

- The pseudo first-order eq. (6) [20]:

$$\log(q_e - q_t) = \log q_e - \frac{K_L}{2.303}t \qquad (6)$$

where K_L is the Lagergreen constant, q_e is the equilibrium uptake value and q_t the current metal uptake.

- The pseudo-second order kinetic eq. (7) [21,22]:

$$\frac{t}{q_t} = \frac{1}{k_2 q_e^2} + \frac{t}{q_e} \qquad (7)$$

where k_2 is rate constant.

The inter particle diffusion is another possible kinetic model that can be applied in adsorption processes eq. (8), [23].

$$q = k_{id}t^{1/2} + C \qquad (8)$$

where k_{id} is rate constant and C is the intercept at the ordinate. If the plot of qt vs. t1/2, according to eq.(8), gives a straight line, then intra particle diffusion was involved in the adsorption process; in addition, if the straight line passes through the origin, then the intra-particle diffusion was the rate-limiting step [24]. Generally, the intercept, C, gives an idea about boundary layer thickness: the larger value of intercept, the greater the boundary layer diffusion effect is shown in Table 6 [25].

The adsorption of Cu^{2+} cation on porous materials from mono-, di- and three-components system can by modeled with all kinetic model

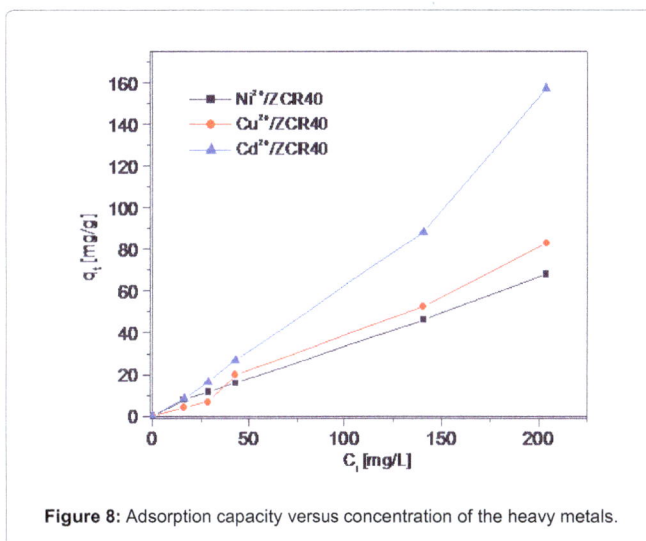

Figure 8: Adsorption capacity versus concentration of the heavy metals.

presented above, Table 6. These models is based on the assumption that the rate limiting step may be a chemical adsorption involving between the adsorbent (ZCR40) and the adsorbate [12,19].

Adsorption isotherm

The adsorption isotherm data were experimentally obtained based on: the optimized contact time (90 min), the substrate amount (0.35 g ZCR40 in 50 mL suspension). The adsorption parameters were calculated considering the Langmuir and Freundlich equations (9) and (10) [26].

the Langmuir isotherm – linearization:

$$\frac{C_{eq}}{q_{eq}} = \frac{1}{q_{max} \cdot a} + \frac{C_{eq}}{q_{max}} \qquad (9)$$

- where q_{max} (mg/g) represents the maximum monolayer adsorption capacity, a is a constant related to the adsorption free energy, q_{eq} is the amount of metal ions adsorbed from the solution with the equilibrium concentration, C_{eq}.

the Freundlich isotherm – linearization:

Substrate	Langmuir Parameters			Freundlich Parameters		
	q_{max} [mg/g]	a	R^2	n	K_F	R^2
Cd^{2+}						
ZCR40	95.238	0.0004	0.997	-	-	0.285
			Cu^{2+}			
ZCR40	107.527	0.0003	0.917	-	-	0.801
Cd^{2+}(Cd^{2+}+Cu^{2+})						
ZCR40	36.764	0.016	0.920	-	-	0.657
Cu^{2+}(Cd^{2+}+Cu^{2+})						
ZCR40	53.763	0.001	0.993	-	-	0.660
Cd^{2+}(Cd^{2+}+Cu^{2+}+Ni^{2+})						
ZCR40	18.657	0.04	0.978	-	-	0.652
Cu^{2+}(Cd^{2+}+Cu^{2+}+Ni^{2+})						
ZCR40	13.774	0.001	0.970	-	-	0.896
Ni^{2+}(Cd^{2+}+Cu^{2+}+Ni^{2+})						
ZCR40	26.525	0.066	0.999	-	-	0.665

Table 7: Adsorption isotherm parameters.

$$\ln q_{eq} = \ln k_f + \frac{1}{n} \ln C_{eq} \qquad (10)$$

- where k_f is Freundlich constant, an indicator of the adsorption capacity, and the $1/n$ dimensionless parameter is a measure of the adsorption density. The adsorption parameters are presented in Table 7.

The Langmuir model fits the equilibrium data much better than the Freundlich model for this adsorbent Table 4. Accordingly with table data the adsorption was rather homogeneous than heterogeneous and rather monolayer than multilayer for all cations studied.

Conclusions

Adsorption efficiency of cadmium, cooper and nickel cations is reduced compared to the adsorption efficiencies of mono-cationic systems, indicating competition on similar active centers (homogeneity substrates). Competition between the different metal ions for surface sites occurs and is dependent on the characteristics of the ions, substrate (crystalline phases, and morphological changes), respectively.

The adsorption isotherms and kinetic data fitted well the Langmuir and pseudo-second order kinetic models, respectively.

The affinities order of ZCR40 for the divalent metal ions reported is Cu^{2+}> Cd^{2+}>Ni^{2+} at optimized contact time 90 min., substrate dose 7g ZCR40/L from multicomponent solution.

This adsorbent material (ZCR40) presents several advantages such as high specific surface area, uniform pore distribution, thermal and mechanical stability, high adsorption capacity, and high fictionalizations ability.

The zeolitic material ZCR40 can be recommended for technologic wastewater treatment.

Acknowledgement

The paper was supported by a grant of the Romanian National Authority for Scientific Research, CNCS – UEFISCDI, project number PNII-RU-TE-2012-3-0177/2013

References

1. Shyam R, Puri JK, Kaur H, Amutha R, Kapila A (2013) Single and multi-binary adsorption of heavy metals on fly ash samples from aqueous solution. Journal of Molecular Liquids 178: 31-36.

2. Tomul F, Basoglu FT (2013) Adsorption of Copper and Zinc from Aqueous Solutions by Bentonite. Asian Journal of Chemistry 22: 615-628.

3. Mimura H, Yokota K, Akiba K (2001) Alkali hydrothermal synthesis of zeolites from coal fly ash and their uptake properties of cesium ion. Journal of Nuclear Science and Technology 38: 766-772.

4. Dieter H (2011) Drinking Water Toxicology in Its Regulatory Framework. Journal of Aquatic Chemistry and Biology 3: 377-415.

5. Wake H (2005) Oil refineries: A review of their ecological impacts on the aquatic environment, Estuarine. Journal of Coastal and Shelf Science 62: 131-140.

6. Visa M, Duta A (2008) Advanced Cd^{2+} Removal on Dispersed TiO$_2$ – Fly Ash. Journal of Environmental Engineering and Management 7: 373-378.

7. Mata YN, Blázquez ML, Ballester A, González F, Muñoz JA (2009) Sugar beet pulp pectin gels as biosorbent for heavy metals: Preparation and determination of biosorption and desorption characteristics. Chemical Engineering Journal 150: 289-301.

8. Ho YS, McKay G (2004) Sorption of Cooper (II) from aqueous solution by peat. Water Air Soil Pollution 158: 77-97.

9. Šljivić M, Smičiklas I, Pejanović S, Plećaš I (2009) Comparative study of Cu^{2+} adsorption on a zeolite, clay and a diatomite from Serbia. Journal of Applied Clay Science 43: 33-40.

10. Peri J, Trgo M, Medvidovi NV (2004) Removal of zinc, copper and lead by natural zeolite - a comparison of adsorption isotherms. Water Research 38: 1893-1899.

11. Lee MG, YiG, Ahn BJ, Roaddich F (2000) Conversion coal fly ash into zeolite and heavy metal removal characteristics of the products. Korean Journal of Chemical Engineering 3: 325-33.

12. Lee WKW, Deventer JSJ (2002) Structural reorganization of class F fly ash in alkaline silicate solutions. Colloids and Surfaces A: Physicochemical and Engineering Aspects 211: 49-66.

13. Horcas I, Fernandez R, Gomez-Rodriguez JM, Colchero J, Herrero GJ, et al. (2007) WSXM: A software for scanning probe microscopy and a tool for nanotechnology. Rev Sci Instrum 78: 013705.

14. Jha B, Singh DN (2014) A three step process for purification of fly ash zeolites by hydrothermal treatment. Journal of Applied Clay Science 90: 122-129.

15. Garbacz A, Courard L, Kostana K (2006) Characterization of concrete surface roughness and its relation to adhesion in repair systems. Journal of Materials Characterization 56: 281-289.

16. Visa M, Duta A (2013) TiO$_2$/fly ash as novel substrate for simultaneous removal of heavy metals and surfactants. Journal of Chemical Engineering 223: 860-868.

17. Visa M, Chelaru AM (2014) Hydrothermally modified fly ash for heavy metals and dyes removal in advanced wastewater treatment. Journal of Applied Surface Science 303: 14-22.

18. Martens JA, Jammaer J, Bajpe S, Aerts A, Lorgouilloux Y, et al. (2011) Simple synthesis recipes of porous materials. Journal of Microporous and Mesoporous Materials 140: 2-8.

19. Hui KS, Chao CYH, Kot SC (2005) Removal of mixed heavy metals ions in wastewater by zeolite 4A and residual products from recycled coal fly ash. Journal of Hazardous Materials 127: 89-101.

20. Lagergren S (1898) About the theory of so-called adsorption of soluble substances, Kungliga Svenska Vetenskapsakademiens. Handlingar 24: 39- 41.

21. Ho YS, McKay G (2006) Second-order kinetic model for the sorption of cadmium onto tree fern: a comparison of linear and non-linear methods. Journal of Water Research 40: 119-125.

22. Mohammad I, El- Khaiary, Gihan F, Malash, Ho YS (2010) On the use of linearized pseudo-second-order kinetic equations for modeling adsorption systems. Journal of Desalination 257: 93- 101.

23. Allen SJ, Mc Kay G, Khader ZH (1999) Interparticle diffusion of a basic dye during adsorption onto sphagnum peat. Journal of Environmental Pollution 56: 39-50.

24. Dinu MV, Dragan ES (2010) Evaluation of Cu^{2+}, Co^{2+} and Ni^{2+} Ions removal from aqueous solution using a novel chitosan/clinoptilolite composite kinetics and isotherms. Chemical Engineering Journal 160: 157-163.

25. Liu W, Zhang J, Zhang C, Wang Y, Li Y (2010) Adsorptive removal of Cr(VI) by Fe-modified activated carbon prepared from Trapa natans husk. Chemical Engineering Journal 162: 677-684.

26. Wang S, Li L, Zhu ZH (2007) Solid-state conversion of fly ash to effective adsorbents for Cu removal from wastewater. Journal of Hazardous Materials 130: 254 259.

6

Microfiltration, Nano-filtration and Reverse Osmosis for the Removal of Toxins (LPS Endotoxins) from Wastewater

Guizani Mokhtar[1,2]* and Funamizu Naoyuki[1]

[1]Graduate School of Engineering, Hokkaido University, Japan
[2]Center for Sustainability Science, Hokkaido University, Japan

Abstract

Lipopolysaccharide (LPS) endotoxin, a bacterial byproduct abundantly present in wastewater, is more and more representing a major concern in wastewater treatment sector for the potential health risk it represents. It is, therefore, more urgent than before to protect consumers from contaminating their fresh potable water reserves with LPS endotoxin through aquifer replenishment using reclaimed wastewater or by supplying reclaimed wastewater as potable water. Membrane treatment is an alternative to activated sludge process and is the most commonly used to treat wastewater. Moreover, nano-filtration and reverse osmosis are the most advanced technologies used to treat wastewater to a potable level. Removal efficiency of LPS endotoxin using Membrane bioreactors (MBRs) and Nano-filtration (NF) and Reverse Osmosis (RO) is subject of this paper. It revealed that these advanced technologies could remove a significant amount of endotoxin. However, levels of concentration in the product water are still much higher than the one found in tap water and it is not advisable to supply this water directly to consumers. Further investigations are required to determine the best management practices for a safe supply of potable water from reclaimed wastewater.

Keywords: Lipo-polysaccharide (LPS); Microfiltration, Nano-filtration and Reverse Osmosis; Waterwater management

Introduction

There is an increasing demand for better quality of effluents from wastewater treatment processes. As assumed and with some incertitude, Lipo-polysaccharide (LPS) endotoxin materials, a bacterial byproduct [1] found in treated wastewater, are a potential threat for human health [2-5].

Knowing that it is almost not possible to control these chemicals in activated sludge process [6,7], endotoxin removal from secondary treated water is crucial and a preliminary step towards safe potable reuse of reclaimed wastewater as well as for fresh water resources contamination prevention. Several treatment alternatives can be considered for achieving this goal ranging from high-tech processes to simple natural, but effective, processes. Alternative removal technologies include membrane filtration (microfiltration, ultra-filtration, nano-filtration and reverse osmosis), soil treatment and coagulation, flocculation and sedimentation. In this paper we investigate the removal of organic matter showing endotoxicity found in the secondary effluents of wastewater treatment plants using membrane filters, including micro-filters, nano-filters and reverse osmosis.

Membrane bioreactors (MBRs), operated, as an alternative of conventional activated sludge system, have become an attractive water treatment alternative. MBR systems offer several advantages over conventional treatment systems. For instance, permeates of membranes with pore size of about 0.1 μm to 0.4 μm is of higher quality as compared to conventional effluent [8]. As matter of fact, MBRs are of interest wherever high quality effluent is required, as they do not replace only secondary clarifier but also replaces treatment steps like sand filtration and ultra-violet (UV) disinfection as well. Large molecules of endotoxin, which represent up to 80% of the total organic matter as stated in previous chapter, could be removed using MBR. However, effect of sludge retention time (SRT) on the molecule aggregation and therefore endotoxin removal is crucial and has to be evaluated.

Nano-filtration (NF) and reverse osmosis (RO) membranes have been increasingly used in wastewater reclamation in recent years for the production of clean water from sewage, benefiting due to their improvement in terms of energy consumption [2]. For the time being, they are almost employed in all wastewater treatment plants producing potable water quality. Efficiency of NF and RO in removing endotoxin is discussed in this article. Samples from feed water (MBR permeate) and NF/RO permeate will be taken and LPS endotoxin and dissolved organic matter (DOC) quantified.

Organic matter, showing endotoxicity in reclaimed wastewater, aggregates into molecules larger than 100 KDa [8], which makes endotoxin removal from reclaimed water feasible using tight membranes. This can be enhanced by the domination of hydrophobic fractions of endotoxins [8]. In addition, ultra-filtration using membranes with about 10 KDa nominal-molecular weight cut-off are routinely employed to get ultrapure water in laboratory systems and it is found in operation wherever endotoxins are not allowed to enter sterile equipment [9]. Furthermore, dead end ultra-filtration showed a good removal of endotoxin (Rlog=1). Such a property has potential utility in endotoxin removal from treated wastewater using membrane technology. Removal of endotoxin from reclaimed wastewater using ultra-filter, micro-filter, nano-filter and/or reverse osmosis is of great importance. To understand efficiency of these membranes in removing endotoxin from wastewater, a lab scale experiment using a set of MBR-NF/RO was operated. A two-stage filtration was used in this study

*Corresponding author: Guizani Mokhtar, Graduate School of Engineering, Hokkaido University, Japan, E-mail: g_mokh@yahoo.fr

during which water contaminated with endotoxin pass through 0.4 μm micro-filter membrane. After that, the membrane permeate run through either nano-filter or reverse osmosis.

Materials and Methods

Water samples

MBR: Because secondary effluent showed toxicity, we focused on the treatment capacity of MBR as an alternative to the activated sludge process. Membranes at longer sludge retention time (SRTs) are expected to have better biodegradation of organic matter and removal of toxic compounds as compared to the activated sludge process. In this study, three types of MBRs with different SRTs were examined.

Influent of the primary sedimentation basin of a domestic wastewater treatment plant in Sapporo was used as a feed for three types of MBRs, operated in parallel in this study as an alternative to conventional activated sludge system (Figure 1). MBR18 (SRT = 18) is baffled submerged MBR where the reactor is divided into two zones by the baffles. The inner zone is kept aerobic due to continuous aeration. In the outer zone, aerobic and anoxic conditions are alternatively created at a constant interval. The MBR18 used in this study was equipped with 6.8 m^2 flat-sheet type of micro-filtration (MF) membranes (Toray, Tokyo, Japan). The membrane was made of polyvinylidene-fluoride (PVDF) and had a nominal pore size of 0.4 μm. On the other hand, MBR12 and MBR50 are normal submerged MBRs with, respectively, a solid retention time of 12 and 50 days. Hollow-fiber MF membranes (Mitsubishi Rayon Engineering, Tokyo, Japan) made from PVDF polymer were used in MBR12 and MBR50. Nominal pore size of the membranes was 0.4 μm. Table 1 summarizes the details and operational conditions of the three MBRs.

NF/RO: Permeate of MBR18 was treated using a parallel set of nano-filtration (NF) and reverse osmosis (RO) (Figure 2). The main pilot system equipped with two separate membrane filtration modules with two 2-inch spiral-wound configurations: one NF membrane module (LES 90, nittouDenkou, Tokyo, Japan) and one RO membrane (ES10, NittouDenkou, Tokyo, Japan). Membrane characteristics of NF/RO membranes are given in Table 2. Membrane flux and recovery, in the continuous operation, were fixed at 460 Lm^{-2} day^{-1} and 70% respectively. Permeate samples were collected after one day of operation.

The unit was operated for 6 weeks. No chemical cleaning or back washing was performed. During operation, changes in permeates endotoxin activity and trans-membrane pressure (TMP) were monitored.

Assays

Endotoxin presence in the samples was assessed using the *Limulus* amoebocyte lysate (LAL) test. The LAL assay is exquisitely sensitive to endotoxin, and the use of endotoxin-free glassware and plastic ware (including tubes and pipette tips), diluent is compulsory. Results were obtained in endotoxin units (EU), where 10 EU corresponds to 1 ng endotoxin. Commercially available endotoxin standards were used as positive controls and LAL reagent water, water free from endotoxin, was used as negative control.

The dissolved organic carbon (DOC) was measured using a Shimadzu TOC-5000 Total Organic Carbon Analyzer according to analytical standards [10]. A Shimadzu TOC-5000 analyzer is able to perform reproducible accurate and sensitive measurements. Detection limits are as low as 4 ppb and standard deviation is less than 1% of full scale for the range less than 2000 ppm.

Water samples from membrane reactors (mixed liquor) were filtered using 0.45 μm filter prior to the measurement.

Results and Discussion

Continuous operation of MBR18

MBR18 unit was operated continuously for 6 six weeks. No chemical cleaning or back washing was performed. During operation, changes in feed water and permeates' endotoxin activity, DOC and trans-membrane pressure (TMP) were monitored. DOC and endotoxin values give an idea about the level of organic matter showing endotoxicity, while TMP is a translation of the fouling level. For instance high TMP associated with high organic matter showing endotoxicity, would mean that fouling is occurring due to microbial by-products.

Figure 3 shows that the dissolved organic carbon in the feed water is oscillating. The DOC of permeate is slightly increasing. The TMP data (Figure 5) reveal that fouling is slowly and gradually occurring on the membrane. This indicates that bio-film is slowly and gradually formed on the surface of membrane. Endotoxin concentration of permeate was also measured. The value of endotoxin concentration in both the feed water and permeate is illustrated in Figure 4. While, this concentration is fluctuating in the feed water, it increased slightly in permeate water from about 600 EU/ml to about 1000 EU/ml. After twenty days of operation, the endotoxin concentration in permeate is around 1000

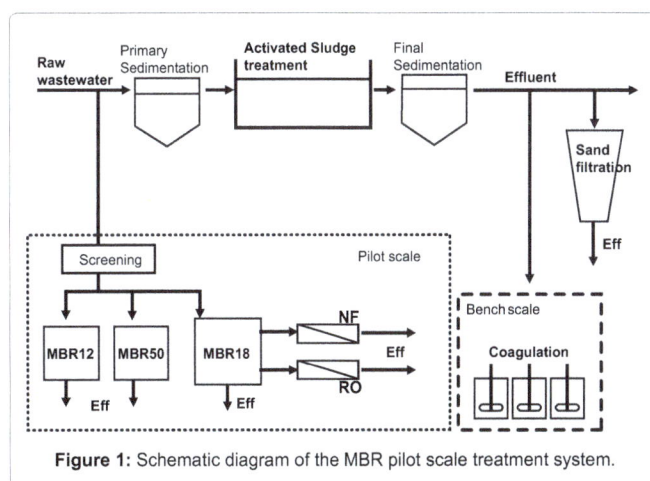

Figure 1: Schematic diagram of the MBR pilot scale treatment system.

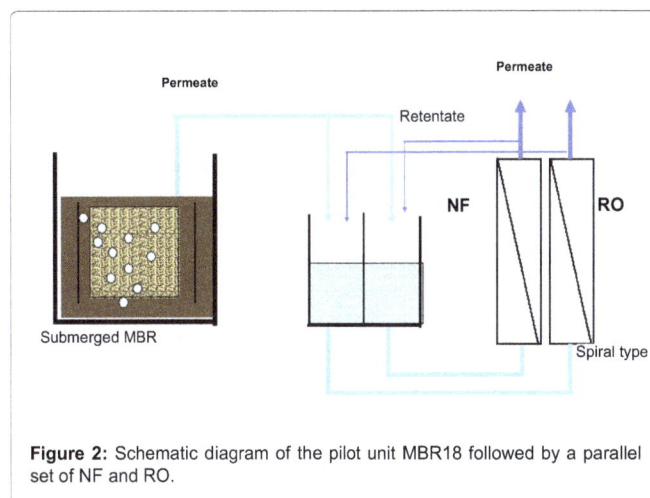

Figure 2: Schematic diagram of the pilot unit MBR18 followed by a parallel set of NF and RO.

EU/ml. The increase is associated to the increase of TMP. Two reasons might explain this increase in endotoxin concentration in the MBR permeates. First, Bio-film can be formed on the surface of MBR and it contributes to the release of some endotoxin material (especially low molecular weight MW) that passes through MBR. Second, endotoxic active material is mainly composed of large molecules as reported by Guizani et al. [8]. These large molecules are being biodegraded on the surface of MBR and are converted into smaller molecules that could pass through the membrane.

SRT effect

The pilot plant, subject of this study, is equipped with three different MBRs, endotoxin was measured in membrane permeates and in membrane reactors of the three MBRS. Figure 6 illustrates endotoxin level in both permeates and reactors. Comparing SRT12, SRT18 and SRT50, at longer SRT the endotoxin and DOC levels in the reactors are the highest (Figure 6 and Table 3). Holakoo et al. [11] reported that longer SRT might lead to the accumulation of higher MW fractions of biomass decay associated soluble microbial products (SMP). Bin and Shuangying [12], using sequencing batch MBR, reported that high MW components become more evident at long SRT. Holakoo et al. [11] concluded that longer SRT might lead to the accumulation of higher MW fractions (>100 kDa) of biomass-decay-associated SMP. Hence, smaller endotoxin molecules at shorter SRT pass through the membrane. It suggests that in shorter SRT (SRT12), permeates show higher endotoxicity and DOC concentration as compared to other MBRS (Table 3 and Figure 6). At longer SRT, larger molecules and aggregates are formed and cannot pass through the MBR pores, thus showing less endotoxicity and DOC concentration in their permeates. These aggregates increase the endotoxicity and DOC concentration in mixed liquor as shown in Figure 6 and Table 3.

Similarly, Guizani et al. [2] applied heat shock protein (HSP) assays to the membrane permeates. The HSP is a very sensitive measurement (bio-assay test) of stress response in cells exposed to pollutants. They found that no significant stress response was detected in the MBR permeates except MBR12, which was operated under a shorter SRT (12 days). It means that at SRT18 and SRT50, toxic molecules could not pass through the membrane. Hence, it is thought that biological organic matter and endotoxin rejected by membrane are being accumulated in the reactors.

To study the effect of SRT on stress response, HSP assays were applied to the supernatant of mixed liquor of the two MBRs with SRT50 and SRT12, respectively [2]. The supernatants showed significant stress response, and in both cases, the stress was significant in samples including higher MW fractions (> 0.1 um). The study reports that in the SRT12 case, stress response was significant at all MW ranges. In comparison, the DOC concentration (< 0.45 mm) of MBR12 was higher than MBR50. Holakoo et al. [11] concluded that longer SRT might lead to the accumulation of higher MW fractions (> 100 kDa) of biomass-decay-associated SMP. Using sequencing batch MBR, Bin et al. [12] reported that high MW components become more evident at long SRT. MBR operation at shorter SRT induced inadequate biodegradation of the toxic organic matter and resulted in toxicity of all MW fractions. Therefore, in the case of MBR12 the MBR permeate showed a stress response because smaller MW fractions carrying toxicants pass through the membrane. The toxic compounds in a supernatant of mixed liquor cannot be removed at longer SRT, but, toxicity was not detected in permeates. This is probably caused by the fact that small molecules aggregated together into larger molecules and were then removed by MBR. Knowing that the detected toxicity was somehow correlated to the existence of microbial by-products and precisely LPS endotoxin [2], therefore we can conclude that LPS endotoxin can be removed by MBR and the removal is enhanced by their aggregation into larger molecules. However, selection of the pore size of the membrane will be important because the size (MW) of organic matter and the aggregates, showing endotoxicity, are strongly dependent on SRT.

Endotoxin removal using NF and RO

In MBR18 and MBR 50, large aggregates were formed and this has led to a lower endotoxicity in permeate. In general, at longer SRT, better efficiency of MBR is expected. However longer SRT will induce more cost for treatment. Therefore for a better optimization, NF/RO unit is tested with a permeate water from MBR operated at a moderate SRT (SRT18). It might be interesting that further studies investigate the optimization of MBR-NF/RO treatment, to study at which SRT we can get better treatment at low cost and with less fouling.

The MBR18 permeate was treated using NF and RO. The DOC, TMP and endotoxin were measured. Dissolved organic carbon was below detection limits in NF and RO permeates. The NF/RO system removed a significant amount of the remaining endotoxin

ID	MBR Type		Volume	Nominal pore size	SRT
			m³	μm	Days
MBR50	Submerged	Hollow fiber	0.00255	0.4	50
MBR18	Baffled-submerged	Flat sheet	0.712	0.4	18
MBR12	Submerged	Hollow fiber	0.00255	0.4	12

Table 1: Characteristics of the three MBRs.

ID	Membrane material	Zeta potential	Water permeability	Salt rejection	Recovery rate
		mV	Lm⁻²day⁻¹Kpa⁻¹	(%)	(%)
NF(LES90)	Polyamide	-8.6	1.6	90	80
RO(ES10)	Polyamide	-15.3	1.2	95.5	80

Table 2: Characteristics of NF and RO membranes.

	DOC (mg/L)		
	MBR50	MBR18	MBR12
Feed water	21	21	21
Permeate	4	3.8	4.5
Mixed liquor (supernatant)	15.1	15	8.5

Table 3: DOC concentration of feed water, permeate and of the mixed liquor of the three MBRs.

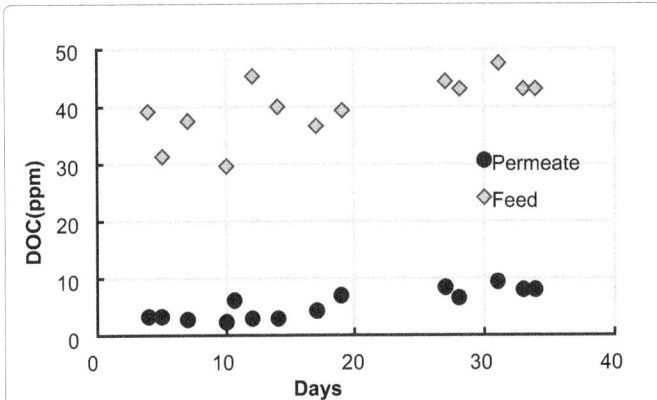

N.B.: There is no preference in choosing the membrane material. We used the polyamide thin film composites, simply because it is a Common *membrane material. It is recommended that further studies focus on the effect of membrane material on endotoxin removal.*

Figure 3: Dissolved organic carbon (DOC) concentration in MBR18.

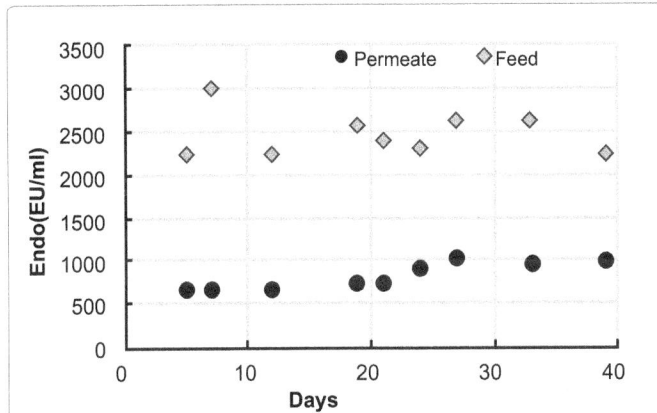

Figure 4: Endotoxin concentration in MBR18.

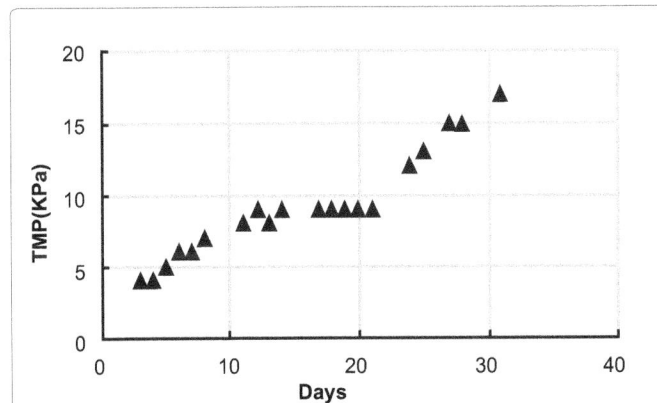

Figure 5: Trans-Membrane Pressure (TMP) in MBR18.

concentration in the MBR permeates. The endotoxin concentration, as illustrated in Figure 7, is lower than 100 EU.mL^{-1} but still higher than tap water level. Endotoxin concentration in tap water ranges from 1 EU.mL^{-1} to 50 EU.mL^{-1} [13]. The endotoxin concentration in NF/RO permeates increases as a function of time. Even though, TMP of

NF/RO is increasing (Figure 8), it is not well understood whether this increase in endotoxin concentration is due to biofilm formation or because the endotoxin has increased in the feed water. Feed water in MBRs permeate experiences an increase of endotoxin concentration as discussed above. In addition, a study on stress response to cells exposed to NF and RO permeates reported that endotoxin removal by these membrane units, led to a decrease in the stress response level [2]. We should notice that DOC value in NF and RO membranes were below detection limits.

Discussion

Investigations on the fate of endotoxin in advanced reclamation processes revealed that the MBR-NF/RO set could lead to a substantial removal of endotoxin. Within the observed period of time, efficiency of removal of MBR-NF/RO has decreased. However, based on the available data on endotoxin in tap water [13], the endotoxin concentration in product water (membrane permeates) was 1.5 to 3 fold of endotoxin

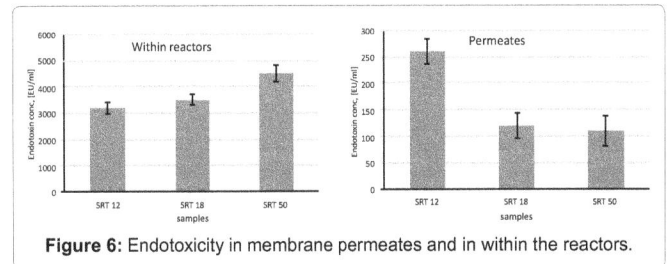

Figure 6: Endotoxicity in membrane permeates and in within the reactors.

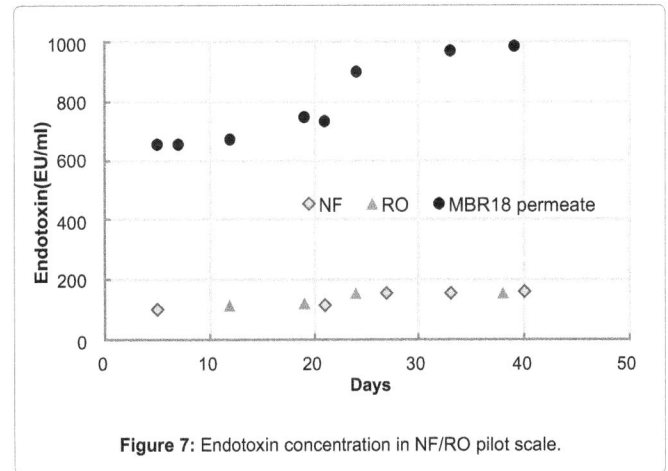

Figure 7: Endotoxin concentration in NF/RO pilot scale.

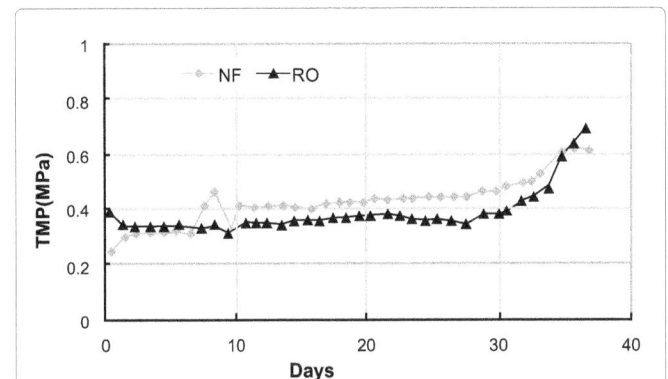

Figure 8: Trans-Membrane Pressure TMP in NF/RO pilot scale.

present in drinking water.

In the absence of health standards related to allowable endotoxin consumption, and with a final concentration of about 100 EU.ml^{-1} for NF and RO permeates, subsequent of MBR, it not advisable to reuse these reclaimed waters for direct potable practices as they show a concentration much higher than that present in drinking water. In terms of this study, it becomes obvious that in some cases treated water can be diluted with drinking water at an appropriate proportion. Dilution of recycled water in the environmental buffer also minimizes any potential risk by decreasing the concentration of endotoxins that may be present. In some other cases we need to investigate the effect of storage on the fate of endotoxin. In some extreme cases, further treatments are needed.

Conclusion

MBR-NF/RO treatment system is effective for endotoxin removal, and this system can be multi-barrier to achieve the stable and high quality of reclaimed water. However, it remains possible that endotoxin removal patterns change depending on combination of treatment or membrane fouling due to long-term operation. Further investigations are needed.

Within the observed period of time, endotoxin concentration in permeates of MB-NF/RO is higher than that of tap water. For a successful reuse practices, two ways can be approached: Dilution of MB-NF/RO permeates with drinking water to reach acceptable concentration, or further treatments are required. Effect of storage of water contaminated with endotoxin is one of the issues to be further investigated.

Acknowledgments

The authors wish to thank Kimura Katsuki (Assoc. Prof.) as well as wastewater treatment plant operators for cooperation with regard to the pilot unit and assistance. Special thanks are also addressed to Sapporo Sewer Bureau for financial support of this research.

References

1. Williams LK (2007) Endotoxins: Pyrogens, lal testing and depyrogenation (chap.4: Endotoxin structure, function and activity), CRC Press.

2. Guizani M, Nogoshi Y, Ben Fredj F, Han J, Isoda H, et al. (2012) Heat shock protein 47 stress responses in Chinese hamster ovary cells exposed to raw and reclaimed wastewater. J Environ Monit 14: 492-498.

3. Hindman SH, Favero MS, Carson LA, Peterson NJ, Schonberger LB, et al. (1975) Pyrogenic reactions during haemodialysis caused by extramural endotoxin. Lancet 2: 732-734.

4. Muittari A, Kuusisto P, Virtanen P, Sovija.rvi A, Gro.nroos P (1980) An epidemic of intrinsic allergic alveolitus caused by tap water. Clin Allergy 10: 77-90.

5. Rylander R, Haglind P, Lundholm M, Mattsby I, Stenqvist K (1978) Humidifier fever and endotoxin exposure. Clin Exp Allergy 8: 511-516.

6. Guizani M, Dhahbi M, Funamizu N (2009) Assessment of endotoxin activity in wastewater treatment plants. J Environ Monit 11: 1421-1427.

7. Guizani M, Dhahbi M, Funamizu N (2009) Survey on LPS endotoxin in rejected water from sludge treatment facility. J Environ Monit 11: 1935-1941.

8. Guizani M, Yusuke N, Dhahbi M, Funamizu N (2011) Characterization of endotoxic indicative organic matter (2-keto-3deoxyoctulosonic acid) in raw and biologically treated domestic wastewater. Water Res 45: 155-162.

9. Li L, Luo RG (1999) Quantitative determination of ca^{2+} effects on endotoxin removal and protein yield in a two-stage ultrafiltration process. Sep Sci Technol 34: 1729 -1741.

10. American Public Health Association (1985) Standard Methods for the Examination of Water and Wastewater, 16th edn. Washington, DC.

11. Holakoo L, Nakhla G, Yanful EK, Bassi AS (2006) Chelating properties and molecular weight distribution of soluble microbial products from an aerobic membrane bioreactor. Water Res 40: 1531-1538.

12. Dong B, Jiang S (2009) Characteristics and behaviors of soluble microbial products in sequencing batch membrane bioreactors at various sludge retention times. Desalination 243: 240-250.

13. Anderson WB, Slawson RM, Mayfield CI (2002) A review of drinking-water-associated endotoxin, including potential routes of human exposure. Can J Microbiol 48: 567- 587.

7

Turning Commercial Ceramic Membranes into a First Stage of Membranes for Post-Combustion CO$_2$ Separation

Sophie Cerneaux[1], Vincent Germain[1], Gil Francisco[1], David Cornu[1], Cédric Loubat[2], Eric Louradour[3], André Larbot[1] and Eric Prouzet[4]*

[1]Institut Européen des Membranes de Montpellier (IEMM) (UMR 5635), Université Montpellier II (UM2) /Centre National de la Recherche Scientifique (CNRS)/ Ecole Nationale Supérieure de Chimie de Montpellier (ENSCM), Site Halle de Technologie, 276 rue de la Galéra, F-34000 Montpellier, France

[2]Specific Polymers, ZAC Via Domitia, 150 Av. Des Cocardières, 34160 Castries, France

[3]CTI-Céramiques Techniques Industrielles, 382 Avenue du Moulinas, F-30340 Salindres, France

[4]University of Waterloo, Department of Chemistry, 200 University Avenue West, Waterloo N2L 3G1, Ontario, Canada

Abstract

This report describes how commercial tubular ceramic membranes, initially designed for liquid filtration, can be modified to provide the core separation components of a first stage of flue gas treatment and enrichment in post-combustion CO$_2$ separation. Commercially available tubular NanoFiltration (NF) ceramic membranes were turned into a membrane for CO$_2$ separation by a two-step process including additional ceramic coating and chemical grafting. The combination of ceramic coating and chemical grafting drastically modify the membrane properties and turn the membrane initially designed for liquid filtration into a membrane that displays CO$_2$ vs N$_2$ selectivity at the opposite of Knudsen-based selectivity, with a CO$_2$:N$_2$ ideal selectivity of 2.3. A second step of this study addressed the reduction of membrane cost, by starting with a low ultrafiltration (UF) 200 nm ceramic support specifically manufactured for this application in place of a NF membrane. After successful coating of a 5 nm and a 1 nm ceramic membranes, this membrane, grafted with a commercial fluorosilane molecule was tested in pure gas permeation of CO$_2$ and N$_2$, with an ideal selectivity CO$_2$:N$_2$=3. Finally, the same membrane, grafted with glymo, was tested against separation of a CO$_2$ (20%):N$_2$ (80%) mixture, and as a function of the permeation stage-cut. A CO$_2$:N$_2$ selectivity of 4 was obtained for a stage-cut of 0.5, and even higher (CO$_2$:N$_2$ selectivity=14) for low stage-cuts usually used for testing dense polymer membranes. These results demonstrate that commercial ceramic porous membranes can be used as starting elements for a first stage of CO$_2$ post-combustion gas cleaning and CO$_2$ enrichment.

Introduction

Today, anthropogenic emission of greenhouse gases embraces CO$_2$ emissions peaking now at 2 billion tons per year [1]. Among them, capture of post-combustion CO$_2$ released by the combustion of fossil fuels in gas turbines or coal-fired power plants, is a major challenge because flue gas are emitted at atmospheric pressure, with rather low concentration in CO$_2$ (5-25%) compared with other sources, and contain a high proportion of water (~ 25%) as well as acid aerosols, flying ashes and corrosive elements like SO$_2$ and NO [2]. The problem is clearly defined by the scale of emissions to deal with: a 600 MW coal-fired power plant emits 500 STP m^3/s flue gas, that is, 460 tons CO$_2$/h [3]. This means that among the elegant scientific responses like optimized chemical scrubbing [4], zeolites and Metal Oxide Frameworks (MOF) used as absorbent [5,6], or others [7,8], only technical pertinent solutions that can afford to treat such a huge amount of CO$_2$ will be worth being farther explored. For example, among the best results reported until now for the adsorption of CO$_2$ by MOFs (MOF-177: 1.54g CO$_2$/g @ 4 bar [9], MOF-200: 2.4 g CO$_2$/g @ 50 bar [10], Ni-MOF-74: 0.19g CO$_2$/g @ 1 bar [11]), and without noticing the unacceptable energy cost required for pressurizing gas at 4 bars, the amount of MOF-177 required for continuously capturing CO$_2$ emission with a one hour adsorption/desorption cycle, would be close to 600 tons (300 for the adsorption stage, 300 for the desorption), which can be hardly considered. Combining the outstanding properties of MOFs for molecular selection with membranes is also on progress, but many problems have still to be solved [12]. So, as pointed recently, "... *no current technologies for removing CO$_2$ from large sources like coal-based power plants exist which satisfy the needs of safety, efficiency, and economy; further enhancement and innovation are much needed*" [13].

Separation processes such as cryogenic, absorption, adsorption and membranes are usually compared in technical-economical studies, but it has been reported that membranes used to be discarded in the early stage of these evaluations, as being considered as inappropriate [14]. However, with the development of both theoretical and experimental works, the solutions provided by membranes have been highlighted, and they happen now to be promising candidates for post-combustion CO$_2$ capture if some problems are correctly addressed and solved by involving both material science and system design engineering [2,3,14-16]. The challenges listed by these reports are the following: (i) membranes must display high selectivity toward CO$_2$ –at least 50-, (ii) it would be suitable to overcome the low CO$_2$ concentration in the feed flow (~ 12%), (iii) the large amount of gaseous or aerosol water (25%) present in flue gas would foul the membranes stages if not removed or drastically reduced beforehand, (iv) the low pressure of the feed gas (~ 1 bar) reduces the CO$_2$ recovery, which is a major drawback for low permeance membranes, and (v) the raw flue gas, even after the gas cleaning and desulfurization through the FGD stage, still contain corrosive components that will destroy polymer membranes if directly exposed.

*Corresponding author: Eric Prouzet, University of Waterloo, Department of Chemistry, 200 University Avenue West, Waterloo N2L 3G1, Ontario, Canada
E-mail: eprouzet@uwaterloo.ca

It seems therefore that a double stage membrane system, designed according to the "enricher" double stage membrane concept [17], with a first one made of high permeance and resistant membranes, devoted to clean the raw flue gas and concentrate CO_2 in the permeate side - the feed side of the second stage -, and a second one with high selectivity but fragile membranes, is a solution worth being studied, as far as all components can be produced and implemented for treating very large gaseous volumes. We report in the present study how commercial ceramic membranes can be modified by well-known procedures to achieve the requirements of this first treatment stage. As these membranes can be chemically modified to become hydrophobic, the solution offered in the present study can solve also the need for water removal, which will be demonstrated in another report.

Ceramic membranes, with pore size ranging from 2 to 50 nm, have proven to be promising in liquid separation technology but not really in gas separation. NanoFiltration (NF) membranes with pore sizes smaller or equal to 2 nm, having molecular weight cut-offs (MWCO) ranging from 200 to 1000 Dalton (Da) can be the good starting elements as their pore is small enough to allow surface interaction [18,19]. Indeed, separation mechanisms with NF membranes generally involve both size and Donnan exclusion effects in contrast to UF membranes, giving high salts rejection rates in liquid separation [20-22]. For example, Larbot et al. successfully prepared γ-alumina NF membranes using a sol-gel approach, with a MWCO of 375 Da (pore size of 1 nm), corresponding to 95% rejection of polyethylene glycol (375 Da) solutions. These NF membranes were successfully tested for mineral salts separation using Donnan exclusion [23,24].

Compared with polymer membranes that offer high selectivity but rather low permeability and poor chemical durability [25], developing PIMs (Porous Inorganic Membrane) for gas separation, especially CO_2 recovery, was pursued by research groups and companies, for many years now, to use their surface property to enhance molecular sieving. Indeed, when considering flue gas treatment, porous ceramic membranes offer the suitable chemical and temperature resistance, but their porosity, which enhances permeability, is a drawback for selectivity, if it is too large, or if surface defects exist. Decreasing the membrane pore size appears as a key factor to enhance selectivity without altering the gas flux and permeance, and initial reports demonstrated that silica microporous membranes with pore size of 1 nm could provide successful gas separation [26]. Since then, the domain of PIMs has evolved and includes now different types of materials like silica, zeolites and MOFs [25]. Surprisingly, non-siliceous metal oxides are not broadly studied.

Since the actual recovery of CO_2 will require very large membrane surface area, which can only be obtained by extrapolating from current industrial manufacturing, we attached our research to explore how current commercial non-siliceous PIMs could be adapted. More specifically, we studied how the suitable porosity could be achieved by starting from commercial PIMs, and how surface modification could improve gas selectivity to hit the requirements for a first membrane stage of post-combustion CO_2 capture (high permeance, selectivity in the 5-10 range, chemical durability). This study was based on preliminary works that demonstrated that surface modification induced by fluorosilane grafting could improve CO_2 separation [27].

The present work reports the combination of PIM improvement by adding first a low ultrafiltration (UF) or nanofiltration (NF) membrane layer onto the internal surface of commercially ceramic supports, with the suitable chemical grafting to provide a hydrophobic surface property in order to enhance CO_2 permeance. The hydrophobic

treatment was obtained using the fluorinated molecule since CO_2 was reported to have a higher affinity than N_2 [10-13] with fluorinated molecules, which is by far beyond the single Knudsen mechanism [28], frequently observed with PIMs.

A first part of this work was devoted to explore step-by-step the influence of each component on the membrane structure and property. Therefore, we developed a zirconia membrane (Membrane n°1) onto commercial 5 nm low UF alumina tubular membrane, by following the sol-gel route without plasticizer using $Zr(OPr)_4$. This membrane was fully characterized, and the surface of this zirconia membrane was chemically modified in a second step with perfluorophosphonic acid $C_6F_{13}C_2H_4PO(OH)_2$. Its performance was evaluated in pure gas separation with CO_2 and N_2, to determine ideal the gas selectivity. We used these results as a starting point to optimize both the performances and the manufacturing cost of membranes.

As the membrane used in this first part, is rather expensive, with the need for coating several ceramic layers up to the 5 nm alumina film, a second part of this work explored if similar performances could be obtained by starting from a 200 nm ceramic support directly prepared by a single step extrusion process (Membrane n°2). These supports were directly coated with a 5 nm layer, then a 1 nm one, followed by chemical grafting with either a commercial fluorosilane, or glymo. Ideal selectivity of the membrane grafted with the hydrophobic fluorosilane was in the same range as for membrane n°1. As for the membrane grafted with glymo, it was tested against gas mixture (CO_2:20%; N_2:80%), as a function of the membrane stage-cut. An exceptional CO_2:N_2 selectivity of 14 for a porous ceramic membrane prepared from industrial components, was obtained at low stage-cut –where dense polymer membranes are usually tested-, and a selectivity of 4 was obtained for stage-cut of 0.5, which is demonstrated to be sufficient for using these membranes in a first membrane stage of cleaning and CO_2 enrichment of post-combustion CO_2 separation.

Experimental Section

Materials

For the membrane n°1, ultrafiltration γ-alumina membranes with pore size of 5 nm (T1-70) were purchased from Pall-Exekia, France. Zr(IV) tetraisopropoxide at 70 weight percent (wt.%) in propanol was purchased from Fluka and HNO_3 (70wt.%) from SDS. For the membrane n°2, the 5 nm alumina layer was made by peptisation of a Plural SB sol (Sasol, Germany). The second coating of the 1 nm alumina layer was made with boehmite (Sasol, Germany). Ethanol, polyethylene glycols (PEGs) with molecular weights of 400, 600, 1000, 1500, 2000 and 3000 Dalton (Da) were purchased from Sigma. (3,3,4,4,5,5,6,6,7,7,8,8,8)-tridecafluorooctyl perfluorophosphonic acid ($C_6F_{13}C_2H_4PO(OH)_2$) denoted CFP hereafter was synthesized by Specific Polymers. (3-glycidoxypropyl)-trimethoxysilane –glymo- was purchased from Sigma-Aldrich. All the chemicals were used as received without any purification. DI water (Elga UHQ apparatus with 18 MΩ quality) was used for the solutions preparation.

Synthesis

Membrane n°1: Prior to its use, the commercial 5 nm γ-alumina membrane was thoroughly rinsed with 18 MΩ water and ethanol successively, and stored in an oven at 120°C for 1 hour. The ZrO_2 polymeric sol was prepared at room temperature by adding 0.7 mL of Zr(IV) tetraisopropoxide (in propanol) ($2.25.10^{-3}$ mol) to 10 mL of a 0.45 M HNO_3 solution. The sol was stirred for 24 hours and kept for another 48 hours at room temperature to allow the reduction by 15wt.

% of the sol concentration. The transparent resulting sol was then slip-casted for 5 minutes on a γ-alumina membrane. The casted support was finally air-dried at room temperature for 24 hours then flash-fired at 450°C for 1h to give the zirconia membrane. To evidence the influence of the layer thickness, a second coating was realized following the same procedure. Unless specified in the text, all the membranes described in the here below manuscript were prepared with two coatings. Grafting a 10^{-2} M solution of CFP in ethanol turned this hydrophilic zirconia membrane into a hydrophobic one. The membrane was maintained for 4 hours at reflux at 65°C in the grafting solution. After completion of the reaction, the membrane was removed from the grafting solution, rinsed with ethanol and acetone successively using an ultrasonication bath to remove the unreacted molecules and placed in an oven at 60°C for 2 hours then at 150°C for 12 hours. Durability of this grafted membrane was tested in real environment by exposing it directly for four months to flue gas from a coal-fired power plant, thanks to the kind contribution of a partner of the FP6 European Program NanoGLOWA (www.nanoglowa.com).

Membrane n°2: For the membrane n°2, a 10 mm single channel 0.250 μm titania porous tubular support was prepared by direct extrusion according to usual industrial processes by CTI (France). The preparation of the 5 nm alumina layer was derived from a method previously developed in our group [23]. The ceramic sol was made by peptisation of a Plural SB solution by nitric acid and then the addition of polyvinyl alcohol (PVA) to enhance viscosity. The prepared solution was then tap-casted into the support. After the solution removal, the supports were left to dry for a day a room temperature and fired at 540°C for 2 hours. The second coating of the 1 nm layer alumina was performed on the top of the previous alumina layer. The alumina layer was made from boehmite obtained by the precipitation of an aluminum alkoxide in a water solution at 85°C. The slurry was then tap-casted into the tube and left drying for 24 hours at room temperature. The support was then flash fired at 450°C for 1 hour. Membrane surface modification was made by grafting these membranes with either a octylfluorosilane (C8FSi) or (3-glycidoxypropyl)trimethoxysilane (glymo).

Techniques

X-ray Diffraction patterns were recorded with a X'pert Pro diffractometer (PanAnalytical, Netherlands) between 10-70 degrees in the 2θ range, using the Cu Kα radiation (λ= 1.54Å). Attenuated Total Reflectance Infrared *(ATR-IR)* spectra were recorded on a FTIR Nicolet 510 instrument (Thermo Scientific). To facilitate the spectra acquisition, the measurements were realized on powder grafted or not. Surface morphology and thickness of the zirconia layer deposited onto the γ-alumina membrane were observed by SEM with a Hitachi S4800 SEM. A fraction of the membrane was broken and flushed with compressed air to remove the dust and then coated with platinum for electron conduction prior to being imaged. The porous volume Vp and specific surface area S of ZrO_2 were evaluated by N_2 adsorption-desorption (B.E.T. method) at 77K using a Micromeritics ASAP 2010 equipment. The ceramic powder was outgassed at 523K for 12 hours prior to the analysis.

Membrane tests

Pure water and PEG solutions permeations were performed using a home-made tangential filtration pilot using a 15 cm long membrane [23]. As the driving force of the system is a pressure difference across the membrane, a nitrogen pressure (5 and 10 bar) was applied to the liquid circulating inside the membrane (1L of solution in the feed tank). The temperature was maintained at 20°C during the experiment

using an external cooling system (Thermo scientific, Germany). The velocity of the circulating water in the filtration loop was set at 2.5 m.s^{-1}. The rejection rate of PEG solutions was followed by chromatography analysis of permeate aliquots collected on a 30 minutes time period. Before the measurement, the membrane was immersed in 18 MΩ water for a minimum of 2 hours to reach rapidly a stable flux at the beginning of the filtration experiment.

Gas permeation measurements were conducted using a typical gas permeation pilot. A 20 cm long tubular stainless steel housing was used to give a membrane permeation area of $3.4.10^{-3}$ m^2. The module was linked to pressure gauges and gas flow systems that delivered either pure CO_2 or N_2, or a 20:80 CO_2:N_2 mixture. The feed pressure was maintained in the range of 1.5 to 4.5 bar, while the downstream permeate was vented to the atmosphere. The permeate flow rate of each gas was measured by volumetric displacement method using a soap bubble flow meter and the permeance J (m^3 (STP)m^2.h^{-1}.bar^{-1}) was calculated using Equation 1:

$$J = \frac{273.15}{T} \cdot \frac{v}{t} \cdot \frac{1}{A} \cdot \frac{1}{\left(P^0 - P^L\right)} \qquad (1)$$

Where T (K) is the room temperature, v (m^3) the volume of permeate collected over a period of time t (h) through the membrane characterized by an effective area A (m^2). P^0 and P^L (bar) correspond to the upstream and downstream pressures, respectively, and the term P^0-P^L to the corresponding pressure differential. The feed and permeate gases were analyzed using gas chromatography analysis (Varian, MicroGC 4800) directly connected to the pilot. The permeation data were recorded thirty minutes after that the feed composition was similar to the canister one.

The stage cut is defined as the permeate gas flow over retentate gas flow [29]. For the tests with pure CO_2 and N_2, the retentate line was closed and permeate was kept opened, which corresponds to a stage-cut of 1. The gas permeance was measured every 0.5 bar between 1.5 and 4.5 bar. All data presented are the average of five measurements in the same conditions. The CO_2/N_2 permeance ratio was calculated to evaluate the permselectivity of the zirconia membrane and thus the ideal separation factor. For tests with a gas mixture, the stage-cut was set at values lower than 1 by adjusting the outlet valve of the retentate side.

Results and Discussion

Validation of membrane n°1

Characterization of the additional zirconia coating: The nature of the zirconia crystalline phase present in the zirconia layer, was checked by following the same procedure for the sol. preparation, as the one used for the support coating, with the powder being collected and fired it at 450°C. The analyses of the material after flash sintering, confirm that a well-crystallized tetragonal zirconia is obtained even after a very short calcination time. The resulting fine powder was grinded and analyzed by X-Ray Diffraction (XRD) (Figure 1). The XRD pattern of the as-synthesized powder (dashed line) displays only broad peaks. After calcination, diffraction peaks appeared at 30.11, 34.59, 35.14, 50.29, 59.46, 60.08 and 62.84 degrees for the lattice family plans (101), (002), (110), (112), (103), (211), (202), respectively. This is in good agreement with the reported peak positions for tetragonal zirconia [30,31]. Sherrer analysis of the (101) diffraction peak broadening gives an average crystal size of 15 nm. It is worth noting that even the (202) family plan, with a reported relative intensity of 5% is clearly visible in Figure 1, demonstrating therefore the high crystallinity of the obtained

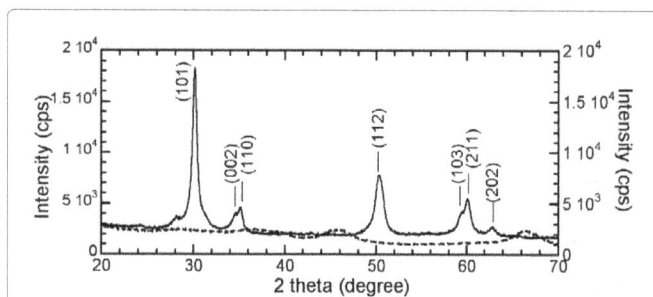

Figure 1: X-ray diffraction pattern of the zirconia powder flash-fired at 450°C for 1 h.

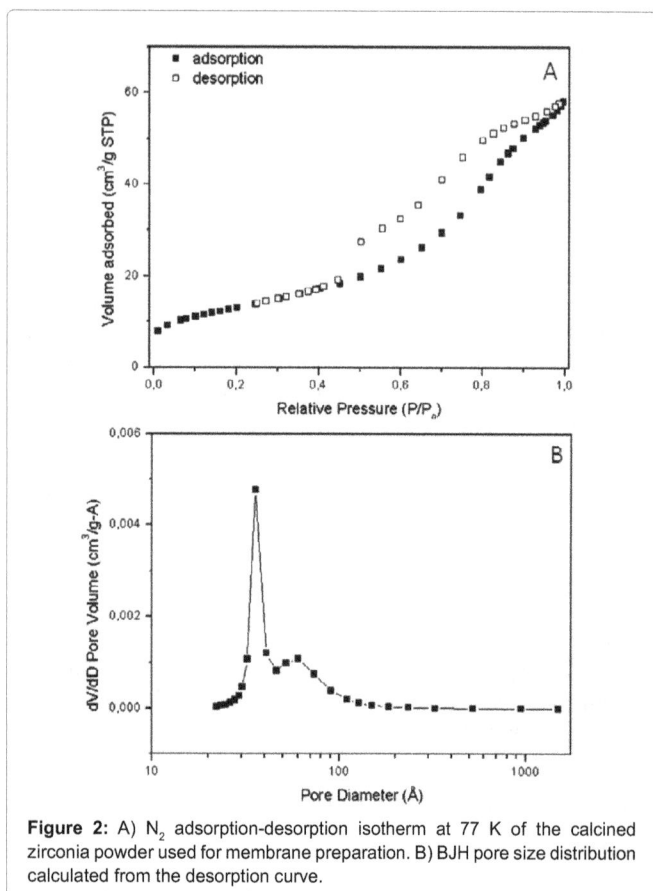

Figure 2: A) N_2 adsorption-desorption isotherm at 77 K of the calcined zirconia powder used for membrane preparation. B) BJH pore size distribution calculated from the desorption curve.

zirconia sintered for only 1h at rather low temperature.

We characterized also the porous structure of this ZrO_2 powder after thermal treatment by N_2 adsorption-desorption (Figure 2). Both the porous volume (Vp=0.09 $cm^3.g^{-1}$) and the BET specific surface (ss=47 $m^2.g^{-1}$) are small, compared to alumina or zirconia coatings previously prepared following a similar procedure [23,32]. We demonstrate in the following how this limited porous volume enhances performances in gas permeation. The N_2 isotherm (Figure 2A) is assigned to a Type IV curve, with a well-defined textural porosity between particles as a result of their small size (no adsorption at high partial pressure is observed). Figure 2B displays the pore size distribution (PSD) calculated by the BJH method. The PSD is defined within the 3 – 10 nm range, with a sharp peak at 3.5 nm, probably better explained as a result of a curve artifact resulting from the selection of data points. This zirconia powder appears rather dense compared to similar alumina or zirconia coatings

prepared following a similar procedure, which explains well by the lack of organic porogens used in the current method.

We confirmed by SEM that the morphology of the zirconia coating is better compared with a thin ceramic skin instead of a membrane (Figure 3). The zirconia layer (Figure 3C and D) deposited on the alumina support (Figure 3A and B) shows indeed a very thin layer of ZrO_2 nanoparticles above the 5 nm alumina top layer. The 15 nm ZrO_2 nanoparticles could actually not enter the 5 nm porous alumina top layer, and remain on the surface (Figure 3C), with this ceramic porous skin being formed.

The influence of this additional ZrO_2 porous skin was tested with pure water permeation and compared to the γ-alumina 5 nm support alone. The flux is drastically reduced with this zirconia film compared to the bare support, decreasing from 20 $L/m^2.h$ for the alumina support to 8 $L/m^2.h$ at 5 bars for the zirconia membrane and from 40 to 16 $L/m^2.h$, respectively at 10 bars. The water permeability deduced from these values is 1.6 $L/m^2.h.bar$ for the zirconia membrane against 4 $L/m^2.h.bar$ for the bare alumina support.

The influence of this zirconia porous skin was confirmed by measuring the Molecular Weight Cut-Off (MWCO) of the modified membrane characterized with the rejection rate of 1mM PEG solutions with different molecular weights. The rejection rate was measured under different nitrogen pressure without any noticeable difference in the rejection rate of the different PEG solutions. The results obtained with a 5 bar pressure as a function of the Molecular weight is displayed in Figure 4. The MWCO defined as the 90% rejection rate is set at 2.2 kDa, which is far below the 20 kDa MWCO delivered by a 5 nm alumina membrane.

We conclude from this first series of test that the addition of a zirconia coating modifies drastically the performances of the commercial alumina membrane, by reducing its water permeability from 4.0 $L/m^2.h.bar$ down to 1.6 $L/m^2.h.bar$, and the MWCO from 20 kDa to 2.2 kDa.

Characterization of the hydrophobic chemical grafting: Low UF membranes are promising candidates for large volume gas separation because of their higher permeability than dense polymer membranes, but their poor selectivity prevents them from being used as such. We confirm in this study that a true covalent grafting of perfluorophosphonic acid improves gas selectivity toward CO_2 without hampering their durability.

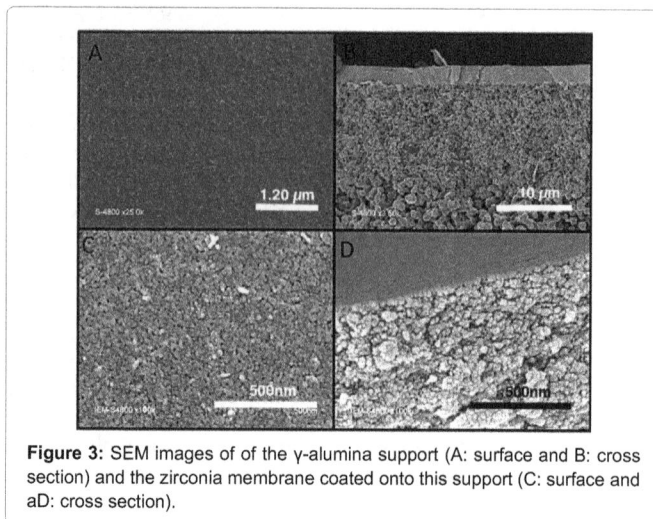

Figure 3: SEM images of of the γ-alumina support (A: surface and B: cross section) and the zirconia membrane coated onto this support (C: surface and aD: cross section).

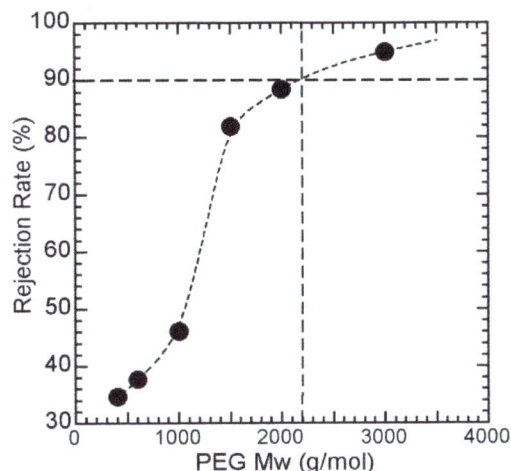

Figure 4: Rejection rate of the zirconia membrane as a function of the PEG molecular weight.

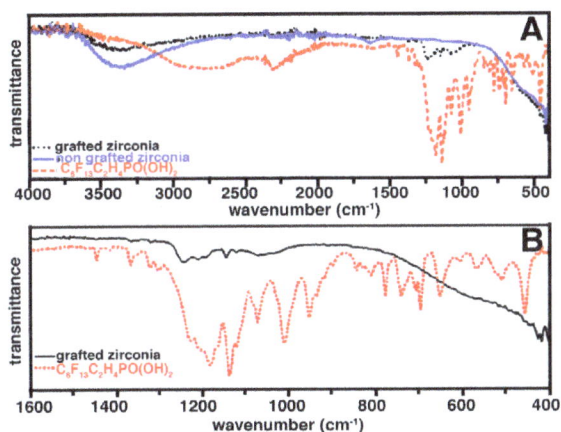

Figure 5: ATR FTIR spectra of the grafting agent (red dash), the grafted zirconia (black dots) and non grafted zirconia (blue line), A) in the 4000- 400 cm^{-1} range and B) an enlarged section in the 1600-400 cm^{-1} range.

As it had been reported that the modification of the surface of a ceramic membrane by fluorinated molecule modifies the membrane gas selectivity [27], we modified the surface of the zirconia coating by grafting (3,3,4,4,5,5,6,6,7,7,8,8,8)-tridecafluorooctyl perfluorophosphonic acid (CFP) through its reaction between the acid function and the oxide surface.

The FT-IR spectra of bare and chemically modified zirconia as well as the CFP molecule are reported in Figure 5. Zirconia bare powder is characterized by a broad signal centered at around 3400 cm^{-1} and a small vibration band at 1650 cm^{-1}, corresponding to hydroxyl groups or adsorbed water. The CFP FT-IR spectrum is characterized by several vibration modes in the 1,500-800 cm^{-1} domain. Upon surface modification resulting from the chemical reaction between the acid functionality of CFP and the metal oxide surface, the OH groups react with the phosphonic acid groups (3,000-2,000 cm^{-1} range) to give Zr-O-P bonds. This is evidenced by the reduction in the OH stretching domain (3,000-2,000 cm^{-1}) disappearance of the broad signal between 2700 and 2550 cm^{-1} and the appearance of several vibration bands (1205,1130, 1070, 1020 cm^{-1} for ν CFx, ν P-C at 1145 cm^{-1}, ν P=O

around 1350-1250) at low frequency, ranging from 1,400 to 800 cm^{-1}, characteristic of the perfluoroalkyl chains [33-35].

We tested the surface property of the grafted zirconia by measuring the water contact angle on a flat zirconia support made with the same conditions. The value obtained with a homemade apparatus, is equal to 149° (Figure 6A), which ranges the hydrophobic support close to superhydrophobic materials [36]. Before going further in the study of gas separation, we validated that this membrane can be exposed for a long time (4 months) to actual flue gas emitted by a coal-fired power plant (Figure 6B and C) without any macroscopic degradation of structure or property. Unlike polymeric materials tested in parallel, which were destroyed within several hours, our membranes did not display any obvious deterioration, and the water angle contact, measured after this 4 month exposure, remained the same and even higher (153°) as a result of soot deposited onto. This stability is the result, not only of the high thermal and chemical resistance of ceramic membranes, but also of the CFP polymer, which is very stable up to 250°C, as displayed in Figure 7.

Gas permeation tests and ideal selectivity: The influence of each addition to the initial 5 nm alumina membrane was studied against gas permeance with pure CO_2 and N_2. We compared the bare support, a membrane with a single zirconia layer (ZrO_2/1/), a membrane with two zirconia layers (ZrO_2/2/), and this double-coated membrane after grafting (ZrO_2/2/-CFP). We observed a significant reduction in gas permeance with the double-coated zirconia membrane, and grafting of this double-coated membrane (ZrO_2/2/-CFP), provided an increase in the CO_2/N_2 selectivity up to 2.5.

For the bare support and the ZrO_2/1/membrane (Figure 8A), the single coating of the ZrO_2 skin does not modify drastically the trans-membrane gas flow, and the values of permeance, which varies linearly with the applied pressure, are in the same range (~ 45 m^3/m^2.bar).

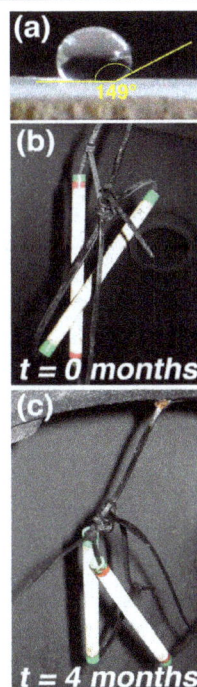

Figure 6: (a) contact angle for the superhydrophobic zirconia surface obtained after grafting; grafted tubular membranes exposed to a power plant flue gas at (b) t = 0, and (c) t = 4 months.

Applying a second coating ($ZrO_2/2/$) divides by 2 the permeance (Figure 8A), but the difference in gas flow does not demonstrate a high selectivity.

Figure 8B displays the results of permeance measurements made with a double-coated zirconia membrane modified by grafting ($ZrO_2/2/$-CFP). The permeance was measured for pure N_2 and CO_2 within a broader pressure difference range (0.4 to 3.5 bars) than in Figure 8A. The gas permeance varies linearly with the applied pressure, and it is markedly reduced from 25 m^3/m^2.bar for the initial non-grafted membrane ($ZrO_2/2/$), down to less than 0.15 m^3/m^2.bar after

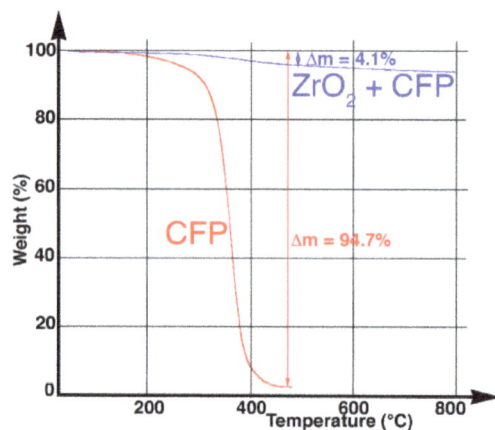

Figure 7: TGA analysis of the CFP polymer alone (red) and the ZrO_2/CFP grafted ceramic.

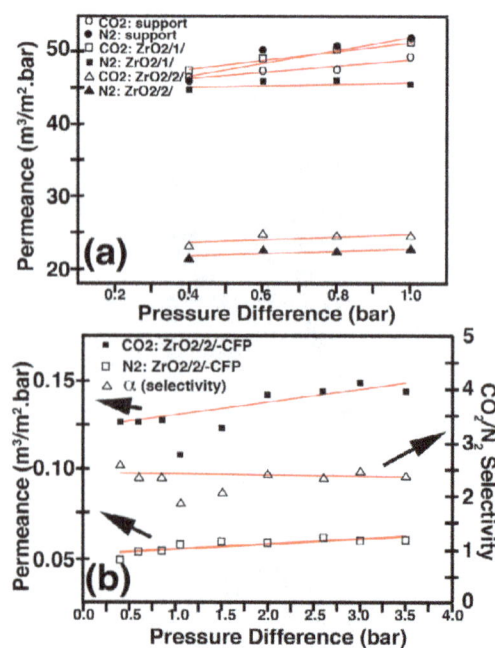

Figure 8: (a) N_2 and CO_2 permeance for the bare support (circle), the zirconia membrane $ZrO_2/1/$ with one coating (square), and the zirconia membrane $ZrO_2/2/$ with 2 coatings (triangle) as a function of the applied gas pressure; (b) N_2 and CO_2 permeance and selectivity of the double coated zirconia CFP-modified membrane ($ZrO_2/2/$-CFP).

Figure 9: SEM micrographies of the surface (a) and cross section (b) of the 250 nm ceramic support, and (c) of the successive 5 nm (bottom) and 1 nm alumina layers coated on the previous support.

grafting ($ZrO_2/2/$-CFP). The ideal selectivity calculated from the CO_2 over N_2 permeance (Figure 8B) is equal to 2.3. This ideal selectivity of 2.3 obtained with a tubular commercial ceramic support modified by a well-known slip casting method and the chemical grafting of a suitable chemical, compares well with the most recent values obtained with hybrid materials like ZIF-9 metal oxide framework (MOFs) membranes prepared onto limited alumina supports, which offered an ideal selectivity of 6.9 but with a method that can hardly be extended to large membrane surface area [37], or ZIF-8 MOFs prepared on stainless steel porous tubes, which presented an opposite pure Knudsen-based CO_2/N_2 selectivity of 0.75 [38].

Validation of membrane n°2

Membrane preparation: Figure 9 displays SEM micrographies of the CTI 250 nm support (a,b). Hg porosimetry gives 47% porosity and a median pore diameter of 0.235 μm. Observation of the additional membranes (Figure 9C) show homogeneous defect-free layers.

This membrane was grafted with either the C8FSi molecule, or a silane bearing an epoxy function (glymo), which is known for displaying specific interaction with CO_2.

Membrane properties: This membrane was tested against pure gas and the evolution of gas flow (STP conditions) for pure CO_2 and N_2 as a function of the pressure difference, is given in Figure 10. Permeance calculated from the slope of the linear fit for each gas, gives a value of 0.104 STP m^3/m^2.h and of 0.03 STP m^3/m^2.h for CO_2 and N_2, respectively, with an ideal selectivity evaluated at 3.4. This value confirms that the membrane obtained by a direct coating of a 5 nm and 1 nm.

Membrane testing: selectivity *vs* stage-cut

As defined before, the stage-cut (SC) is defined as the permeate gas flow over retentate gas flow [29]. Without this parameter, the actual evaluation of membranes performances can hardly be conducted. Indeed, the measurement of pure gas permeance, which corresponds to SC=1, is principally an evaluation of the material affinity, and the ideal selectivity defined as the ratio of pure gas permeance corresponds to a parameter that quantifies the relative material affinity toward different gases. At the opposite, for a SC=0, there would not be any driving force for the cross membrane flow. We show in the following that the actual membrane selectivity must be evaluated under the light of stage-cut, which defines how much of the feed gas is actually treated.

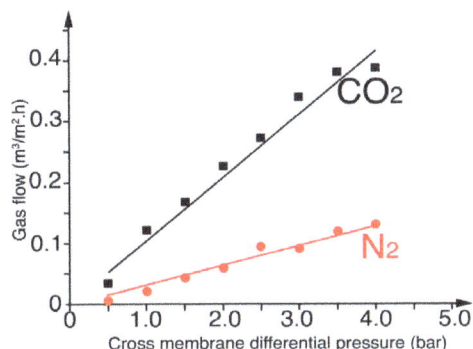

Figure 10: Flow of pure CO_2 and N_2 through the membrane n°2 grafted with a C8 fluorosilane (C8FSi).

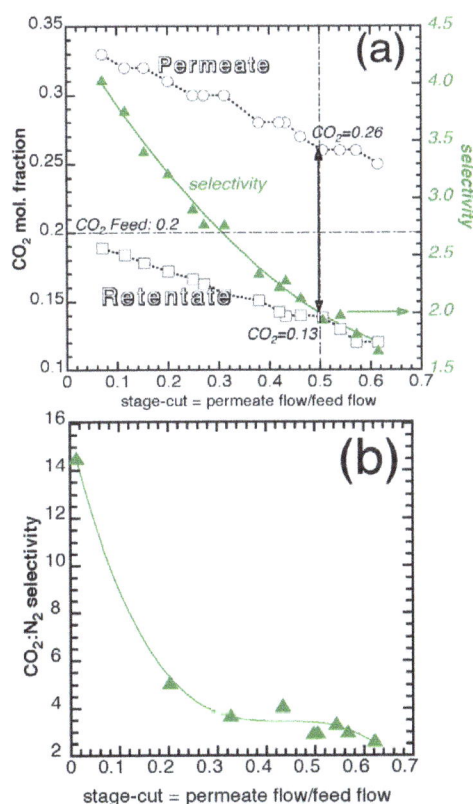

Figure 11: (a) Evolution of (i) the molar fraction of CO_2 in both the permeate and retentate, and (ii) CO_2:N_2 selectivity, as a function of the stage-cut, for a (20% CO_2: 80% N_2) feed mixture, for the membrane n°2 grafted with fluorosilane (C8FSi). The horizontal line corresponds to the initial concentration in CO_2 (CO_2 Feed: 0.2); (b) Evolution of the CO_2:N_2 selectivity, as a function of the stage-cut for a (20% CO_2: 80% N_2) feed mixture, for the membrane n°2 grafted with glymo.

Testing the actual membrane properties requires a stage-cut lower than one, which will create a trade-off between the actual membrane selectivity, hence the gas purity, and the gas flow in the permeate, hence the global gas recovery. This is illustrated in Figure 11A with the membrane n°2 grafted with C8FSi where the actual CO_2 molar fractions in both permeate and retentate sides, of a (20% CO_2; 80% N_2) feed mixture, are reported as a function of the stage-cut (SC). For a

SC of 0.5 (50% feed flow across the membrane), the permeate gas is 30% enriched in CO_2 (0.2 to 0.26), and the retentate is impoverished in parallel (0.2 to 1.3).

The CO_2:N_2 selectivity is also reported. This selectivity decreases from 4.0 at SC=0.1, where only 10% of feed flow goes across the membrane, down to 1.7 at SC=0.6, with 60% of the feed flow going across the membrane (the line plotted in Figure 11A is the result of a second degree polynomial fit of the experimental values). These results illustrate well the difference between the ideal CO_2:N_2 selectivity of the material equal to 3.4, and the actual membrane selectivity, which can be evaluated for different stage-cuts (0.1; 0.3; 0.5; 0.6) and equal to 3.8, 2.7, 2.0, and 1.8, respectively.

Despite their promises for blocking the water aerosol present in the flue gas, which makes the membranes grafted with perfluoro chemical groups, good candidates for a first stage of postcombustion CO_2 separation, the CO_2:N_2 selectivity remains too small to achieve the performances in gas enrichment expected for such a first separation stage. We confirmed that our method of PIM surface functionalization could work with other chemical groups. (3-glycidoxypropyl)-trimethoxysilane is mostly used as a precursor for dense hybrid membranes for proton exchange in fuel cells [39], but the epoxy group was demonstrated to exhibit affinity toward CO_2 [40,41]. Tests with a membrane n°2 grafted with (3-glycidoxypropyl)-trimethoxysilane (glymo) demonstrate a significant improvement with a CO_2:N_2 selectivity doubled for a stage-cut of 0.5, and equal to 10 for a stage-cut of 0.1 (Figure 11B). These values do not reach selectivity of 50 observed with dense hybrid membranes [41], but a selectivity of 4 for a stage-cut of 0.5, would allow the feed gas in the second membrane stage to be 50% enriched in CO_2, compared with the initial flue gas (see Figure 11A), which could significantly increase the CO_2 purity and recovery in a second high selectivity membrane stage, as demonstrated by modeling studies [2,15].

Conclusion

We have successfully modified commercial gamma alumina 5 nm membranes by applying an additional coating of tetragonal zirconia, using a sol-gel process, in order to achieve a final pore size reduction down to 3.5 nm, compatible with gas permeation. This pore size reduction was confirmed by the drastic reduction in water flow. Surface modification of this modified membrane, with the chemical grafting of (3,3,4,4,5,5,6,6,7,7,8,8,8)-tridecafluorooctyl perfluorophosphonic acid, using a phosphonic acid function as grafting head, led the membrane surface to become superhydrophobic. This new surface property allowed us to achieve a significant CO2 over N2 selectivity in gas permeation, especially by tuning the membrane stage-cut, to allow a significant flow going to the permeate. These properties, along with the chemical and abrasion resistance of ceramic membranes, make them the perfect candidates for a first membrane stage dealing with gas cleaning and CO_2 enrichment in the treatment of power plant raw flue gas, before sending this cleaned flue gas to a second high selectivity stage of polymer membranes.

Acknowledgment

The authors thank the European Union through the NanoGLOWA FP6 program, for financial support.

References

1. Ciferno JP, Fout TE, Jones AP, Murphy JT (2009) Capturing Carbon from Existing Coal-Fired Power Plants. Chem Eng Prog 105: 33-41.

2. Brunetti A, Scura F, Barbieri G, Drioli E (2010) Membrane technologies for CO2 separation. J Memb Sci 359: 115-125.

3. Merkel TC, Lin H, Wei X, Baker R (2010) Power plant post-combustion carbon dioxide capture: An opportunity for membranes. J Memb Sci 359: 126-139.

4. Sharma SD, Azzi M (2014) A critical review of existing strategies for emission control in the monoethanolamine-based carbon capture process and some recommendations for improved strategies. Fuel 121: 178-188.

5. Sabouni R, Kazemian H, Rohani S (2014) Carbon dioxide capturing technologies: a review focusing on metal organic framework materials (MOFs). Environ Sci Pollut Res 21: 5427-5449.

6. Krishna R, van Baten JM (2012) A comparison of the CO2 capture characteristics of zeolites and metal-organic frameworks. Sep Purif Technol 87: 120-126.

7. D'Alessandro DM, Smit B, Long JR (2010) Carbon Dioxide Capture: Prospects for New Materials. Angew Chem-Int Edit 49: 6058-6082.

8. Choi S, Drese JH, Jones CW (2009) Adsorbent Materials for Carbon Dioxide Capture from Large Anthropogenic Point Sources. Chem Sus Chem 2: 796-854.

9. Férey G (2008) Hybrid porous solids: past, present, future. Chemical Society Review 37: 191-214.

10. Furukawa H, Yaghi OM (2009) Storage of Hydrogen, Methane, and Carbon Dioxide in Highly Porous Covalent Organic Frameworks for Clean Energy Applications. J Am Chem Soc 131: 8875-8883.

11. Bae Y-S, Snurr RQ (2011) Development and Evaluation of Porous Materials for Carbon Dioxide Separation and Capture. Angew Chem-Int Edit 50: 11586-11596.

12. Shah M, McCarthy MC, Sachdeva S, Lee AK, Jeong HK (2012) Current Status of Metal-Organic Framework Membranes for Gas Separations: Promises and Challenges. I&EC res 51: 2179-2199.

13. Li B, Duan Y, Luebke D, Morreale B (2013) Advances in CO_2 capture technology: A patent review. Appl Energy 102: 1439-1447.

14. Favre E (2007) Carbon dioxide recovery from post-combustion processes: Can gas permeation membranes compete with absorption? J Memb Sci 294: 50-59.

15. Brunetti A, Drioli E, Lee YM, Barbieri G (2014) Engineering evaluation of CO2 separation by membrane gas separation systems. J Memb Sci 454: 305-315.

16. Luis P, Van der Bruggen B (2013) The role of membranes in post-combustion CO2 capture. Greenhouse Gases-Science and Technology 3: 318-337.

17. Zhao L, Riensche E, Blum L (2010) Stolten D Multi-stage gas separation membrane processes used in post-combustion capture: Energetic and economic analyses. J Memb Sci 359: 160-172.

18. Sarbolouki MN (1982) A general diagram for estimating pore-size of ultrafiltration and reverse-osmosis membranes. Sep Sci Technol 17: 381-386.

19. Sharma RR, Agrawal R, Chellam S (2003) Temperature effects on sieving characteristics of thin-film composite nanofiltration membranes: pore size distributions and transport parameters. J Memb Sci 223: 69-87.

20. Bowen WR, Mukhtar H (1996) Characterisation and prediction of separation performance of nanofiltration membranes. J Memb Sci 112: 263-274.

21. Cadotte J, Forester R, Kim M, Petersen R, Stocker T (1988) Nanofiltration membranes broaden the use of membrane separation technology. Desalination 70: 77-88.

22. Rautenbach R, Groschl A (1990) Separation potential of nanofiltration membranes. Desalination 77: 73-84.

23. Larbot A, Alami-Younssi S, Persin M, Sarrazin J, Cot L (1994) Preparation of a g-Alumina Nanofiltration Membrane. J Memb Sci 97: 167.

24. Levenstein R, Hasson D, Semiat R (1996) Utilization of the Donnan effect for improving electrolyte separation with nanofiltration membranes. J Memb Sci 116: 77-92.

25. Pera-Titus M (2014) Porous Inorganic Membranes for CO_2 Capture: Present and Prospects. Chem Rev 114: 1413-1492.

26. De Vos RA, Maier WF, Verweij H (1999) Hydrophobic silica membranes for gas separation. J Memb Sci 158: 277-288.

27. Abidi N, Sivade A, Bourret D, Larbot A, Boutevin B, et al. (2006) Surface modification of mesoporous membranes by fluoro-silane coupling reagent for CO2 separation. J Memb Sci 270: 101-107.

28. Villet RH, Wilhelm RH (1961) Knudsen flow-diffusion in porous pellets. Industrial and Engineering Chemistry 53: 837-840.

29. Koros WJ, Ma YH, Shimidzu T (1996) Terminology for membranes and membrane processes. J Memb Sci 120: 149-159.

30. Kikkawa S, Kijima A, Hirota K, Yamamoto O (2002) Crystal structure of zirconia prepared with alumina by coprecipitation. J Am Ceram Soc 85: 721-723.

31. Srivastava A, Dongare MK (1987) Low-temperature preparation of tetragonal zirconia. Mater Lett 5: 111-115.

32. Etienne J, Larbot A, Julbe A, Guizard C, Cot L (1994) A microporous zirconia membrane prepared by the sol-gel process from zirconyl oxalate. J Memb Sci 86: 95-102.

33. Kim JD, Mori T, Honma I (2007) Anhydrous proton conductivity of a larnella-structured inorganic-organic zirconium-monododecyl phosphate crystalline hybrid. J Power Sources 172: 694-697.

34. Wang X, Ma X, Chen T, Qin X, Tang Q (2011) The homogenization of zirconium hydroxide phosphonate-supported ruthenium catalyst in asymmetric hydrogenation. Catal Commu 12: 583-588.

35. Vermeulen LA, Burgmeyer SJ (1999) Reactivity of a new zirconium phosphonate phase, Zr-2(O3P-CH2CH2-bipyridinium-CH2CH2-PO3) X-6 center dot 2H(2)O, toward organic and inorganic monophosphonates. J Solid State Chem 147: 520-526.

36. Erbil HY, Demirel AL, Avci Y, Mert O (2003) Transformation of a Simple Plastic into a Superhydrophobic Surface. Science 299: 1377-1380.

37. Liu Y, Zeng G, Pan Y, Lai Z (2011) Synthesis of highly c-oriented ZIF-69 membranes by secondary growth and their gas permeation properties. J Memb Sci 379: 46-51.

38. Venna SR, Zhu MQ, Li SG, Carreon MA (2014) Knudsen diffusion through ZIF-8 membranes synthesized by secondary seeded growth. J Porous Mater 21: 235-240.

39. Inoue, Uma T, Nogami M (2008) Performance of H-2/O-2 fuel cell using membrane electrolyte of phosphotungstic acid-modified 3-glycidoxypropyl-trimethoxysilanes. J Memb Sci 323: 148-152.

40. Sforca ML, Yoshida IVP, Nunes SP (1999) Organic-inorganic membranes prepared from polyether diamine and epoxy silane. J Memb Sci 159: 197-207.

41. Nistor C, Shishatskiy S, Popa M, Nunes SP (2009) CO_2 Selective membranes based on epoxy silane. Rev Roum Chim 54: 603-610.

8

Study of Polysulfone and Polyacrylic Acid (PSF/PAA) Membranes Morphology by Kinetic Method and Scanning Electronic Microscopy

Chamekh Mbareck[1]* and Quang Trong Nguyen[2]

[1]Université des Sciences, de Technologie et de Médecine; Faculté des Sciences et Techniques, B.P. 5026, Nouakchott, Mauritanie
[2]P.B.S. UMR 6270 CNRS - Université de Rouen, 76821 Mont-Saint-Aignan, France

Abstract

This work focuses on the study of the morphology forming of Polysulfone (PSf) and polyacrylic acid (PAA) membranes which are prepared by mixing both of the polymers in DMF and the obtained blend is precipitated in water (non-solvent). The precipitation kinetic and the effects of polysulfone concentration, drying-time in free-air and proportions of both polymers are investigated.

Based on the SEM technique, the kinetic and viscosity measurements, and the visual observations, this study brings to the fore the different steps of the morphology forming of PSf/PAA membranes: formation of finger-like structures, sponge-like-structures, inner pores, superficial pores and the craters. Appearance of these structures is governed by the exchange process, between casting polymer-solution and the precipitation-bath (coagulation), which is controlled by operating conditions. This work constitutes a great tool in understanding the mechanisms of the morphology forming of PSf/PAA membranes and to improve their performances: preparation of tailored membrane.

Keywords: Polysulfone; Poly (acrylic acid); Membrane morphology; SEM; Blends; Macrovoids

Introduction

Polysulfone is a hydrophobic polymer with excellent chemical and mechanical properties. However, the hydrophobic behaviour prevents its use in several areas such as in water treatment. In order to overcome this drawback, some researchers have proposed the incorporation of hydrophilic groups, to increase the polymer hydrophilicity which improves the membrane water- permeability and reduces membrane-fouling [1,2].

In literature, the synthesis of carboxylated polysulfone membranes was realized through different ways. Membranes were prepared by (i) substitution with functionalized groups, (ii) by polymer graft modification or (iii) by surface modification of prefabricated membranes [2-9].

In a previous, work we proposed the preparation of carboxylated polysulfone membranes by mixing PSf and PAA in dimethy formamide (DMF) solvent and, the obtained blend, was precipitated in water-bath (non-solvent) [10]. Incorporation of carboxylic groups in PSf matrix imparted the obtained membrane high ionic-exchange and complexation capacities. It also increased the membrane hydrophilicity which enhanced the resistance to organic fouling and made membrane suitable for many applications. As a result, it was found that, the membrane properties depended on: (i) uniform distribution of PAA chains in the network formed by the PSf matrix, (ii) rate of immobilization of PAA, and (iii) morphology of the membrane. The PSf/PAA membrane has shown good performances for the ultrafiltration of different heavy metal solutions and dye molecules [10,11]. Here, our interest will be focused on the PSf/PAA membrane morphology in order to learn much about the mechanisms of the morphology forming and to know how to improve the membrane performances in view to prepare tailored membranes.

SEM technique, visual observation, precipitation speed, viscosity measurement are used to characterize the morphology of the PSf/PAA membranes.

Experiment

Polymer properties

Polysulfone (PSf) and poly (acrylic acid) (PAA) were supplied by Aldrich and dimethyl formamide (DMF) by Prolabo. All polymers and chemical products were used as supplied, without any further purification. Molecular weights of PAA and PSf were 450 000 and 26 000 g/mol, respectively.

Membrane preparation

Polysulfone (PSf) and Poly (acrylic acid) (PAA) were dissolved, separately, in dimethyl formamide (DMF) in a glass reactor equipped with a mechanical stirrer and thermostated at 90°C for over 3 hours. The PSf concentration was 17 (or 19) wt. % and that of PAA was 5 wt. %. Afterwards, both solutions were mixed together in known proportions, stirred for 30 min and de-bubbled. Such de-bubbled casting-solution was casted on a glass plate with a lab made Gardner knife, dried in free-air for a known time (generally 20 seconds) and finally immersed in a coagulation-bath containing a sufficient volume of MilliQ water (18.2 MΩcm) at 18°C. Membranes were thoroughly washed with water, and stored in a dilute sodium azide solution till their use.

Kinetic measurement

Five polymer-solutions with the same mass (2 g) and different composition (PSf/PAA 100/0, 96/04, 92/08, 89/11 and 83/17) were poured on a glass plate to get exactly a same surface area. Afterwards,

*Corresponding author: Chamekh MBareck, Université des Sciences, de Technologie et de Médecine; Faculté des Sciences et Techniques, B.P. 5026, Nouakchott, Mauritanie, E-mail: chamec1@yahoo.fr

polymer casting-solution is immersed in equal volume of water coagulation-bath (2 liters).

Kinetic measurements were used to determine the time and the speed of PSf/PAA precipitation. The time of precipitation corresponds to the period separating the moment of immersion of casting-solution in the water-bath and the moment at which the formed membrane peels spontaneously off from the glass plate. The speed of precipitation was calculated according to the following formula: $S_p = m / t_p$

Where, S_p is the precipitation speed of the polymer solution (in gr/sec);m: is the mass of casting-solution (in gram); t_p: is the time of precipitation (in second).

Viscosity measurement

Rheological experiments were carried out using a controlled shear stress (advanced rheometer: AR 2000 from TA instruments). A double

PAA%	0	0.04	0.08	0.11	0.17
S_p (gr./Sec)	0.066	0.055	0.029	0.027	0.025

Table 1: Precipitation speed of polymer casting-dopes as function of PAA percentage.

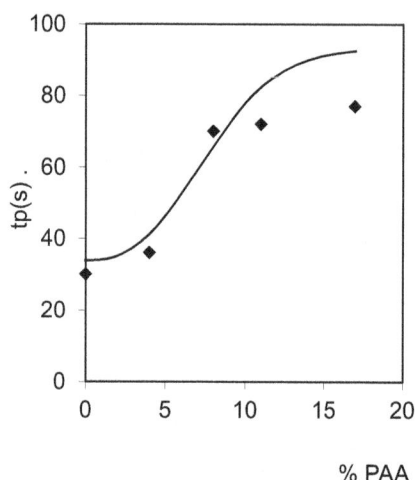

Figure 1: Precipitation time for PSf/PAA blends at different PAA fraction.

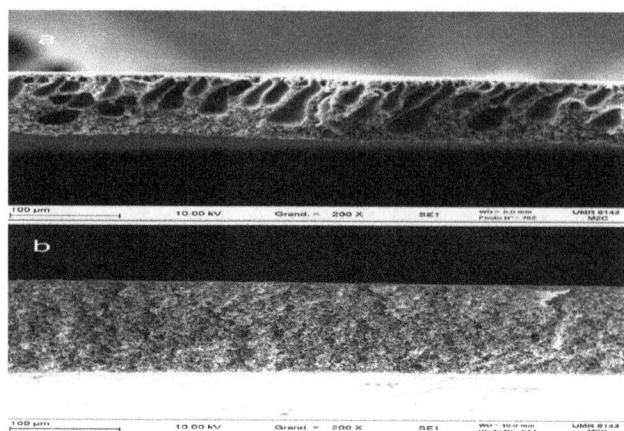

Figure 2: SEM micrographs of PSf membranes for (a) PSf/DMF 17/83 and (b) PSf/DMF 19/81.

gap concentric cylinder device was used. Measurements were done at $25 \pm 0.1°C$ with the shear rates ranging from 0.1 to 660 s^{-1} and shear stress from 0.01 to 622 Pa.

Scanning Electronic Microscopy

Membrane surfaces and sections were analyzed by a Zeiss EVO 40 EP microscope with the Secondary Electron Backscattered Detector (CE-BSD). Samples were frozen in liquid nitrogen at -190°C, fractured and vacuum coated with a thin gold film. The SEM observations were carried out at various magnifications while the electron beam energy was fixed at 10 keV.

Results and Discussion

Kinetic measurement

Kinetic measurements aim at characterizing the behaviour of PSf/PAA membranes during precipitation step. Therefore, precipitations time and speed were determined. It is interesting to note that precipitation time may be also regarded as time of total exchange of solvent and non-solvent between casting-solution and precipitation-bath.

As shown in Table 1, the high precipitation speed of PSf solution-film, compared with those of PSf/PAA solution-films, is the consequence of larger difference in the solubility parameter values of PSf (12.9 $cal^{1/2}/cm^{3/2}$) and water (23.4 $cal^{1/2}/cm^{3/2}$) [12]. The addition of PAA (14.6 $cal^{1/2}/cm^{3/2}$) on the PSf solution - to form a blend - decreases the difference in solubility between polymer-solution and water; and increases the precipitation time of PSf/PAA blends as shown in Figure 1 [13].

It's well known that PAA is a hydrophilic polymer which can develop strong interactions with water. Then, the increase of PAA content in the casting-solution reduces chemical driving forces of exchange with non-solvent and such a thing delays the precipitation of the casting-solution [14,15].

As a result, the precipitation of PSf/PAA blends depends on the rate of the solvent expulsion from casting-solutions. This expulsion is governed by the difference in the solubility parameter or the affinity (interactions) between polymer and coagulation-bath.

Effect of the PSf concentration in the casting- solution

Figure 2a and b presents the SEM micrographs of PSf membranes obtained from the casting-solution containing, respectively, 17% and 19 wt. % of PSf in DMF. These micrographs show that with the increase in the PSf concentration, the membrane morphology changes from finger-like structures to sponge- like structures. Some researches assign the appearance of finger-like structures (macrovoids) to instantaneous precipitation of PSf-film [15-17]. Here, suppression of macrovoids, within membrane with higher percent (19 wt % PSf), is mainly due to the increase of the viscosity of casting-solution as both of the casting-solutions have practically the same precipitation time [15,18].

At higher polymer content, growth of the nucleus of poor-polymer phase is disrupted by the high viscosity of the liquid medium; the vitrification of the rich-polymer matrix surrounding the nucleus would occur more easily. However, the macrovoid suppression is accompanied by a change in the porosity, pore size and pore distribution in the membrane [19-21].

Effect of drying time in free air (pre-coagulation) (t_d)

Three PSf/PAA 96/04 membranes are used to study the effect of drying time on the membrane morphology. Two samples are,

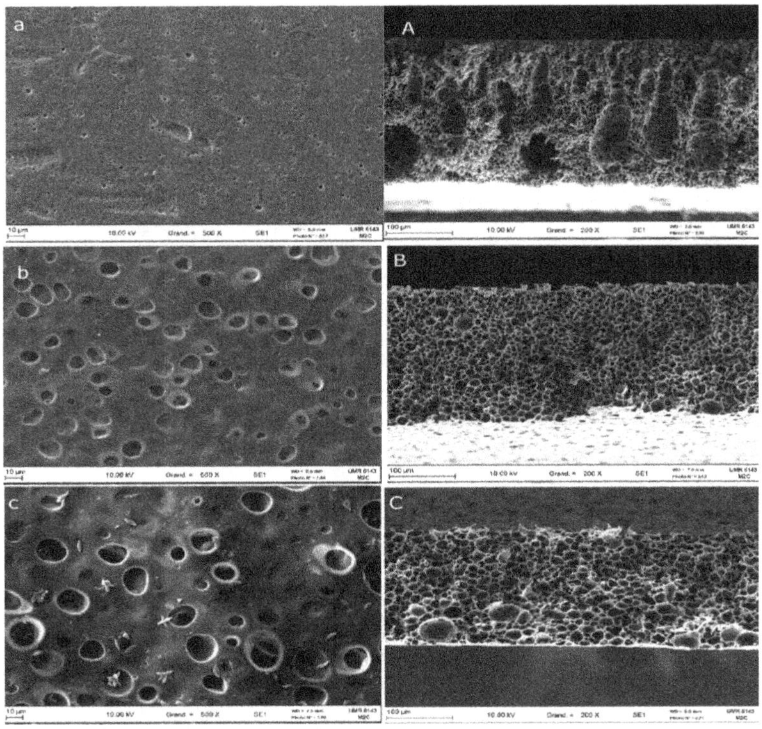

Figure 3: Surface and section SEM micrographs of PSf/PAA 96/04 at different precoagulation drying time: (a, A) 10 sec, (b, B) 300 sec and (c, C) + ∞, respectively. The magnification of surface images is 500X whereas that of section images was 200X. The concentration of initial PSf/DMF solution was 19% (wt).

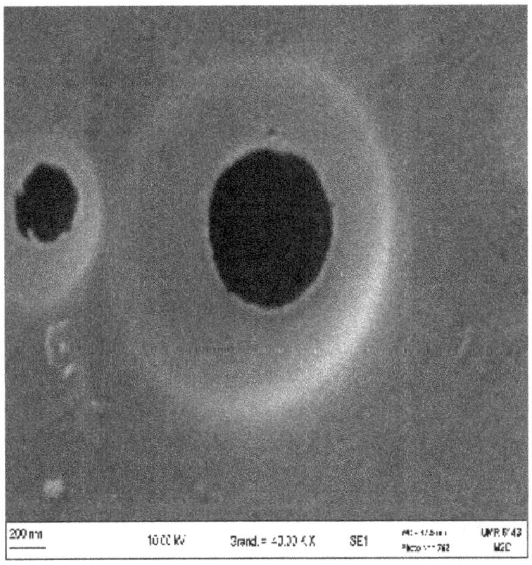

Figure 4: SEM image of PSf/PAA 92/08 membrane with different levels of pore structure (40 000X, 10KV).

respectively, dried during 10 and 300 seconds before immersing in water-bath whereas the third is completely dried in free air (water vapour 40%).

This experiment shows that the size of superficial pore increases with the increase in drying time (Figure 3), and macrovoids in the section are completely suppressed for t_d higher than 300 seconds (Figure 3C). The size of superficial pores increases from 100 nm at a short time to 10 µm at a long drying time.

In fact, it is difficult to give a single value to the pore size as the majority is composed of different porous structures as shown in Figure 4. The SEM image of membrane surface shows that each superficial pore contains smaller inner-pores at different levels. For instance, the average size of the superficial pore in the PSf/PAA 92/08 membrane was ca. 680nm, while that of the first inner-pores level was 130nm and the second inner-pores level was 25 nm. It appears that the pore sizes in the membrane active-layer decrease progressively from the upper surface towards the inner section. Progressive decrease of these pore sizes and good performances of these membranes for ultrafiltration of heavy metal ions, published previously [11], suggest the high membrane tortuosity from one face to another.

The exposure of PSf/PAA/DMF film in free-air provokes an exchange between the solvent in casting-film (DMF, b.p. 152.8°C) and the atmospheric water vapour (at 40% relative humidity) [22]. Due to the low solvent volatility and low water content in air, the exchange rate of solvent and nonsolvent (water) is lower in free-air than in water-bath.

A short drying time (a rapid quenching) of the casting liquid-film in the coagulation bath induces a rapid film vitrification (phase separation) due to further solvent outflow. This phenomenon provokes high growth of rich-polymer phase in the sub-layer and growth of poor-polymer phase in the bottom layer. These circumstances favor the formation of finger-like structure forms. For long drying time, the exchange of solvent and non-solvent is slow, as well as, the solution-film vitrification. This behavior leads to a more homogeneous structure (more sponge- like structure).

The precipitation of casting-solution in free-air is very slow in comparison with the precipitation in water-bath which is fast. In fact, the high boiling-point of solvent and low inflow rate of water-vapour

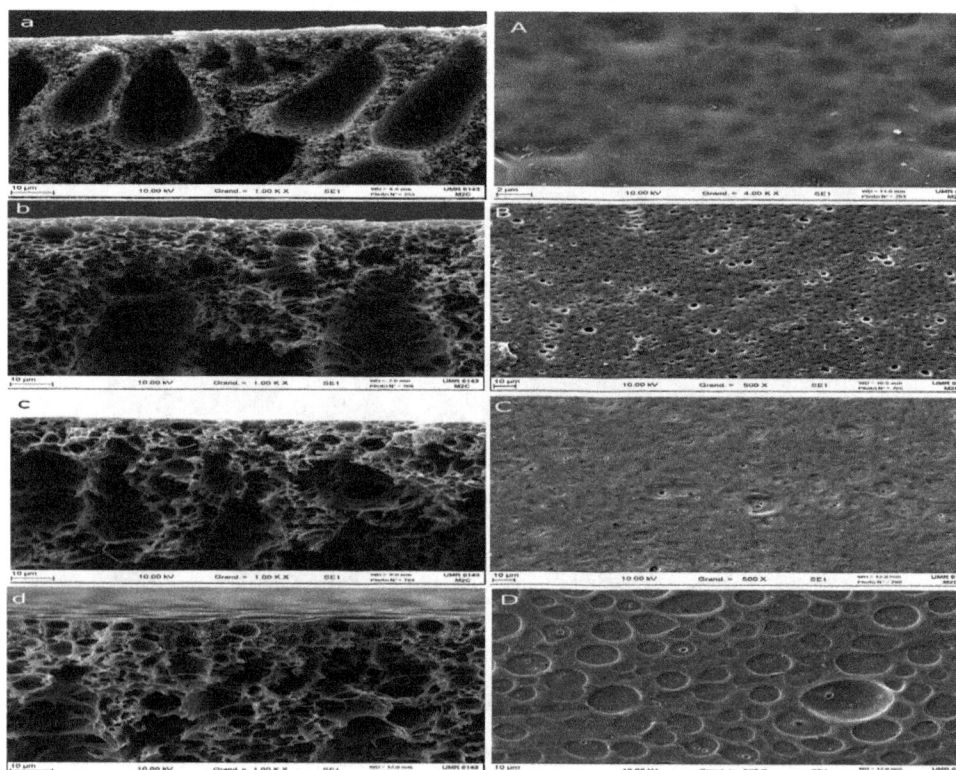

Figure 5: Cross- section SEM micrographs of PSf/PAA membranes (a) PSf/PAA 100/0, (b) 96/04, (c) 92/08 and (d) 89/11. The concentration of initial PSf/DMF solution was 17% (wt).

in casting-solution reduced sensibly the exchange driving forces between casting-solution and water. Thus, the vitrification of solution-film progresses slowly leading to homogeneous matrix with inner and superficial pores.

These results show again the role of the driving forces arising from solvent and non-solvent exchange: rapid exchange of solvent and non-solvent favors the formation of finger-like structures, whereas the slow exchange enhances the formation of sponge-like structures.

Effect of the blend composition

Figure 5 shows the cross- section and surface morphologies of PSf/PAA 100%, 96/04, 92/08 and 89/11, respectively. The SEM micrographs show that the morphology of PSf 100% is mainly composed of finger-like structures which progressively disappear in favor to sponge-like structures with the increase in PAA content.

The progressive suppression of finger-like structure is in agreement with the previous results (section 3.1). The addition of PAA increases the intermolecular interactions between casting-solution and the non-solvent (water). Such a thing delays the precipitation of casting-solution and enhances the homogenous structures forming. These results are in agreement with those of Wu et al. [23] who suggest that weak interactions between polyvinylidene fluoride and poly (ethersulfone) chains tend to create a large distance between the polymers chains which induces the formation of macrovoid-structures.

Visual observation of casting-solution precipitation

Visual observation of the precipitation of PSf/PAA/DMF casting-solution shows a strong film-contraction and thin turbulence-liquid

streams which escape rapidly from the top-face of the nascent-membrane. These liquid-streams appear at numerous superficial points. Afterwards, the film peels off from the glass plate pushed by a large amount of solvent which emerges from the membrane bottom-face. This phenomenon strength decreases with the increase of PAA percentage.

The simple analysis of this phenomenon leads to admit that, the immersion of PSf/PDF casting-solution in non-solvent-bath creates strong repulsive forces between PSf chains (hydrophobic) and water molecules (hydrophilic). These forces provoke the rapid retraction of polymers' chains and the formation of rich-polymer domains and rich-liquid domains. The progressive retraction of polymers' chains pushes the rich-liquid domains to escape through the polymer matrix inducing the formation of inner and superficial pores, on the one hand, and the formation finger-like structures and sponge-like structures, on the other hand. The increase of PAA percentage reduces the repulsive forces and encourages the homogeneity of membrane morphology.

Interpretation of PSf/PAA morphology Mechanism

Under the driving forces of exchange between casting-solution and non-solvent (water), large fluctuations in concentration appear at the beginning of the phase separation, leading to large stresses between the poor-polymer and the rich-polymer phases, as illustrated in Figure 6a. Transient gel-clusters would be formed due to the dynamic asymmetry in stresses between the system components. Afterwards, the clusters aggregate into large gel domains [24]. Chemical potential changes induce an osmotic pressure which causes contraction of the elastic gel domains and liquid expulsion as illustrated in Figure 6b.

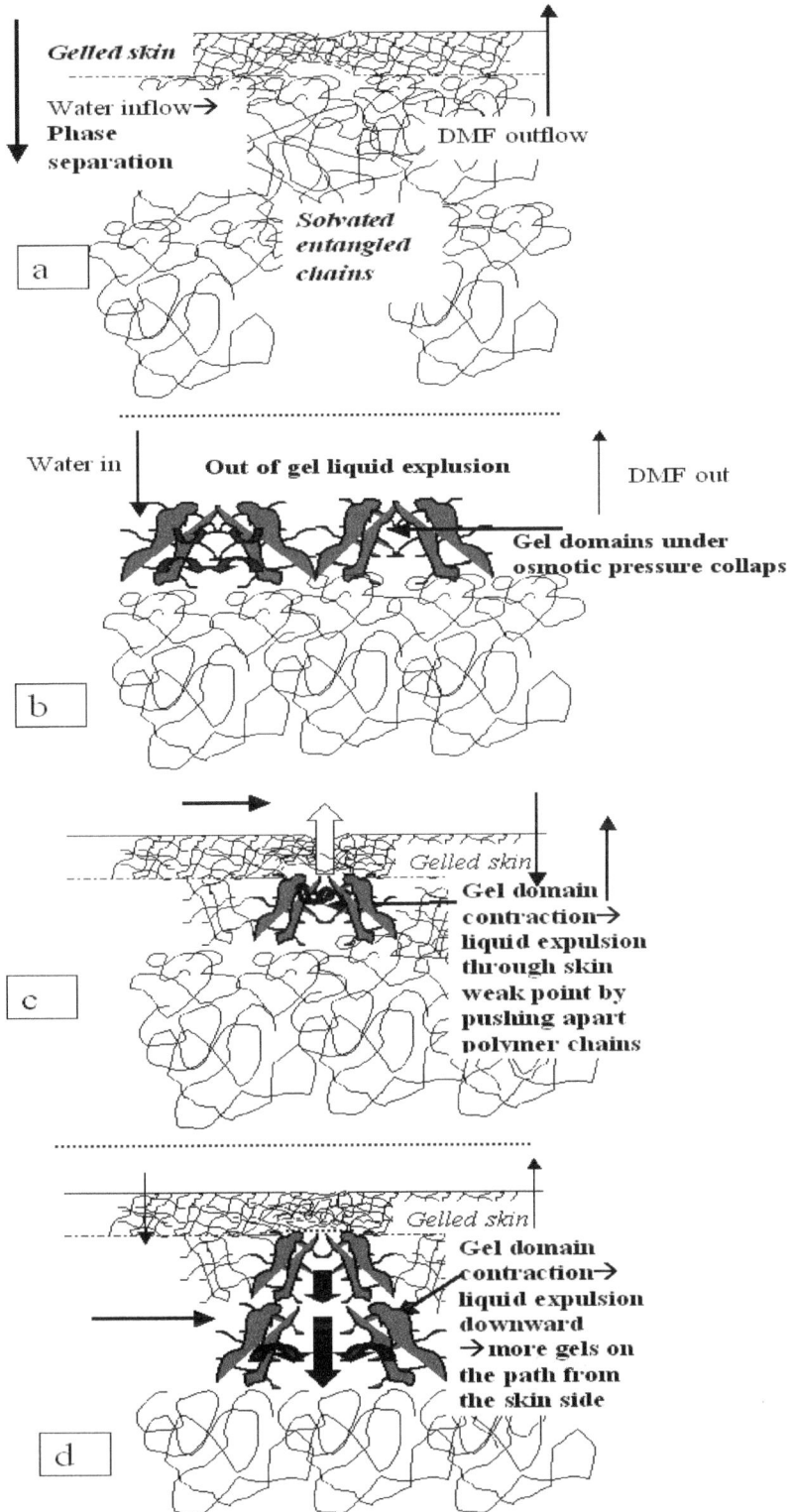

Figure 6: The mechanism of the formation of macrovoids structure and surface craters. Evolution in time of the structure of the solution of polymer film cast on a glass plate:

a) in the first step of polymer gelling into a surface gel skin, just after immersion in the water coagulation bath ;

b) seconds later, in the next step of gel domain formation in the sub – gel skin layer ;

c) in the step of formation of a surface crater by solvent ejection under gel collapse pressure through weak points in the gel skin, followed by a self- sealing of gel skin at the break point ; d) in the step of formation of a macrovoid by water- rich solvent flow towards the bottom face under gel collapse pressure, and subsequent preferential phase separation and gel domain formation on the flow path.

PAA%	0	0.04	0.08	0.11	0.15
η_0(Pa S)	0.6	1.3	3.6	5.5	14.6
Average superficial pore size (µm)	0	2.36	0.49	0.42	--

Table 2: Zero-shear viscosity values of polymer casting-dopes and Average superficial pore size of PSf/PAA membranes as a function of PAA percentage.

Gel-collapse is physically described by Tanaka et al. [24] who propose that the water penetration induces stresses (osmotic stresses) throughout the different parts of gel structures and the liquid squeezing out of gel. As a result, rich polymer-domains and rich liquid-domains are formed across the gel. If the gel domain is closer to the surface, it will be easier for the squeezed liquid to escape through the weak spots on the nascent gel layer [25]. The ejected DMF-water mixture causes turbulence streams in the water-bath that we observed in the regions surrounding the weak points on the surface, as illustrated in Figure 6c. As early as, the stress is released and the solvent escaped at the weak points on the surface of the nascent gel layer, small pores would be formed at low layer viscosity and craters at high layer viscosity. This behavior is confirmed by the decrease of average pore size and the increase of zero-shear viscosity, with the increase of PAA fraction, as shown in Table 2. The round crater observed on the upper layer is due to a local collapse of the elastic skin after the escape of the liquid from the casting-film.

When there are no breakable weak points in the gel skin, the rich-water liquid would be pushed downwards to the bottom face by the collapsing gel domains. As the water transport, from the liquid-sub-layer towards the bottom face, is much faster than that of molecular diffusion, the separation of the polymer-poor phase occurs faster on the path of the squeezed water-rich liquid leading to finger-like structures (macrovoids) in the membrane section. Bottom parts of these structures are generally larger than the top parts. Such morphology can also be explained on the basis of viscoelastic properties: larger the distance from the skin layer, lower the medium viscosity. Thus, the poor-polymer phase tends to expand more in the lower part of the nascent macrovoids, as illustrated in Figure 6d.

The addition of PAA increases the viscosity of polymer-solutions (as shown by Table 2) and decreases the intensity of turbulence streams observed above the nascent membrane surface. It also induces the decrease of precipitation time. All these results show that with the reduction of solvent expulsion forces or the delay of precipitation, the morphology tends to be more homogeneous and the sponge like-structures dominate as shown by Figure 5.

Higher affinity of the casting-solution to non-solvent due to the presence of hydrophilic PAA makes the polymer-gel domains more swollen and weaker. As a result, the breaking of weak spots in the gelled skin is still possible, and the deformation of the softer-gel domains is larger at higher PAA contents, leading to larger craters on the surface.

Exploring the morphology of the membrane prepared from casting solutions of polysulfone (PSf), polyvinylpyrrolidone (PVP) and N-methylpyrrolidone (NMP), Han and Nam [26] report that a membrane with a low PVP content (5%) shows larger macro pores than that with a high PVP content (20%). They suggest that the introduction of a low PVP content enhanced sufficiently the demixing ability of the casting solution to provoke a rapid collapse of polymer chains and to induce the formation of macrovoids. On the contrary, at high PVP content, the increase in the viscosity of the polymer solution due to the enhanced entanglement of polymer chains hinders the solvent-non-solvent exchange, delays the demixing of casting-solution and induces

a suppression of macrovoids. These results may be interpreted by the mechanisms that we propose to explain the morphology of PSf/PAA membrane.

Conclusion

This work has elucidated many points of the morphology mechanisms forming of PSf/PAA membrane morphology. Generally, these membranes present two morphologies: sponge-like structure and finger-like structure. The appearance of these morphologies is governed by the driving forces of solvent and non-solvent exchange. These forces provoke the retraction of polymers' chains which push the solvent to escape through the polymer matrix inducing the formation of different morphology forms: inner and superficial pores, finger-like structures and sponge-like structures. The decrease of these driving forces favors the appearance of sponge-like structure whereas their increase favors the appearance of finger-like structure.

References

1. M Ulbricht, Belfort G (1996) Surface modification of ultrafiltration membranes by low temperature plasma II. Graft polymerization onto polyacrylonitrile and polysulfone. J Memb Sci 111: 193-215.

2. Mockel D, Staude E, Guiver MD (1999) Static protein adsorption, ultrafiltration behavior and cleanability of hydrophilized polysulfone membranes. J Memb Sci 158: 63-75.

3. Lu Z, Liu G, Duncan S (2005) Morphology and permeability of membranes of polysulfone-graft-poly(tert-butyl acrylate) and derivatives. J Memb Sci 250: 17-28.

4. Fang JK, Chiu HC, Wu JY, Suen SY Preparation of polysulfone based cation - exchange membranes and their application in protein separation with a plate and frame nodule, React Funct Polym 59: 171-183.

5. Gancarz I, Pozniak G, Bryjak M, Frankiewiez A (1999) Modification of polysulfone membranes. 2. Plasma grafting and plasma polymerization of acrylic acid. Acta Polym 50: 317-326.

6. Kato K, Uchida E, Kang ET, Uyama Y, Ikada Y (2003) Polymer surface with graft chains. Prog Polym Sci 28: 209-259.

7. Nabe A, Staude E, Belfort G (1997) Surface modification of polysulfone ultrafiltration membranes and fouling by BSA solutions. J Memb Sci 133: 57-72.

8. Matsumoto Y, Sudoh M, Suzuki Y, (1999) Preparation of composite UF membranes of sulfonated polysulfone coated on ceramics. J Memb Sci 158: 55-62.

9. Béquet S, Remigy JC, Rouch JC, Espenan JM, Clifton M, et al. (2002) From ultrafiltration to nanofiltration hollow fiber membranes: a continuous UV-photografting process. Desalination 144: 9-14.

10. Mbareck C, Nguyen QT, Alexandre S, Zimmerlin I (2006) Fabrication of ion -exchange ultrafiltration membranes for water treatment. I. Semi-interpenetrating polymer networks of polysulfone and poly (acrylic acid). J Memb Sci 278: 10-18.

11. Mbareck C, Nguyen QT, Alaoui OT, Barillier D (2009) Elaboration, characterization and application of polysulfone and poly acrylic acid blends as ultrafiltration membranes for removal of some heavy metals from water. J Hazard Mater 171: 93-101.

12. Polymer handbooks (1999) Brandrup pages VII/701, 705, 711.

13. Guan Y, Jiang W, Zhang W, Wan G, Peng Y (2001) Polytetrahydrofuran amphiphilic networks. IV. Swelling behaviour of poly (acrylic acid)-l-polytetrahydrofuran and poly(methacrylic acid)-l-polytetrahydrofuran networks, J Polym Sci A- Polym Chem 39: 1784-1790.

14. Ruaan RC, Chang T, Wang DM (1999) Selection criteria for solvent and coagulation medium in view of macrovoid formation in the wet phase inversion process. J Polym Sci: Part B: Polymer Physics 37: 1495-1502.

15. Albrecht W, Weigel Th, Tiedemann MS, Kneifel K, Peinemann PK, et al. (2001) Formation of hollow fiber membranes from poly(ether imide) at wet phase inversion using binary mixtures of solvents for the preparation of the dope. J Membr Sci 192: 217-230.

16. Smolders CA, Reuvers AJ, Boom RM (1992) Microstructures in phase-inversion membranes Part 1. Formation of macrovoids. J Memb Sci 73: 259-275.

17. Reuvers AJ, Van der Berg JWA, Smolders CA (1987) Formation of membranes by means of immersion precipitation. I. A model to discribe mass transfer during immersion precipitation. J Memb Sci 34: 45-67.

18. Ismail AF, Lai PY (2003) Effects of phase inversion and rheological factors on formation of defect-free and ultrathin-skinned asymmetric polysulfone membranes for gas separation. Separ Purif Tech 33: 127-143.

19. Pekny MR, Greenberg AR, Khare V, Zartman J, Krantz WB, et al. (2002) Macrovoid pore formation in dry-cast cellulose acetate membranes: buoyancy studies. J Membr Sci 205: 11-21.

20. Machado PST, Habert AC, Borges CP (1999) Membrane formation mechanism based on precipitation kinetics and membrane morphology: at and hollow fiber polysulfone membranes. J Memb Sci 155: 171-183.

21. Mckelvey SA, Koros WJ (1996) Phase separation, vitrification, and the manifestation of macrovoids in polymeric asymmetric membranes. J Membr Sci 112: 29-39.

22. Han MJ, Bhattacharyya D (1995) Changes in morphology and transport characteristics of polysulfone membranes prepared by different demixing conditions. J Membr Sci 98: 191-200.

23. Wu L, Sun J, Wang Q (2006) Poly (vinylidene fluoride)/polyethersulfone blend membranes: Effects of solvent sort, polyethersulfone and polyvinylpyrrolidone concentration on their properties and morphology. J Membr Sci 285: 290-298.

24. Tanaka H (1999) Viscoelastic model of phase separation in colloidal suspensions and emulsions. Phys Rev E 59: 6842-6852.

25. Kim JH, Lee KH (1998) Effect of PEG additive on membrane formation by phase inversion. J Memb Sci 138: 153-163.

26. Han MJ, Nam ST (2002) Thermodynamic and rheological variation in polysulfone solution by PVP and its effect in the preparation of phase inversion membrane. J Memb Sci 202: 55-61.

9

Nonequilibrium Dissolution-diffusion Model for PDMS Membrane Pervaporation of ABE Water Binary System

Xia Yang[1], Zhen Wu[2], Fang Manquan[1] and Li Jiding[1]*

[1]State Key Laboratory of Chemical Engineering, Department of Chemical Engineering, Tsinghua University, Beijing 100084, China
[2]Ordos Redbud Innovation Institute, Ordos 017000, China

Abstract

Previous models of equilibrium dissolution-diffusion, pore flow and virtual phase change cannot describe the mass transfer process of pervaporation precisely. The fact that dissolution process on the surface of the membrane does not reach equilibrium is seldom emphasized in the literature. The aim of the present work is to develop the nonequilibrium dissolution-diffusion model (nonequilibrium model) for membrane pervaporation process. In this research, the steps of dissolution and desorption were treated as the pseudo surface reaction processes on the surface based on the hypothesis of nonequilibrium dissolution at the interface of the feed liquid and membrane. The semi-experimental model was set based on steady state mass transfer, ignoring the concentration polarization and adsorption at the permeation side. Through linear fitting of the flux with different thickness of the membrane, the diffusion coefficients and adsorption kinetic rate constants of the model were achieved with equilibrium partition coefficient estimated by UNIFAC-ZM model. The calculated values of the model were well in consistent with experimental flux in the vacuum pervaporation of acetone, butanol and ethanol with polydimethylsiloxane membrane. The nonequilibrium model and its parameters will be further applied for prediction of separation performance and selection of operation conditions.

Keywords: Pervaporation; Nonequilibrium; Dissolution; Diffusion; Polydimethylsiloxane

Introduction

Rapid growths in population and economy have resulted in energy and water shortage on a global scale [1]. Among separation techniques for organics recovery from aqueous solutions, membrane based processes are very promising ones [2]. Membrane technology is witnessing an era of rapid growth due to the great demand of renewable energy production and water purification. Membrane pervaporation is first and mainly applied for continuous production of renewable biofuel from bio-fermentation of acetone, butanol and ethanol (ABE) aqueous solution. As an efficient technique to separate oil/water mixture, pervaporation is a permeation process through the membrane with the thermodynamic phase change. Feed liquid is passing over on one side of the membrane while the permeable component is changed to gas on the other side. Selective separation is realized via chemical potential difference between the solvents with the membrane [3]. Compared with traditional methods like distillation, adsorption, freeze crystallization, gas stripping and liquid–liquid extraction for ABE fermentation products recovery, pervaporation has the advantages of high selectivity, low energy consumption, moderate cost to performance ratio and compact and modular design [4]. There are several pervaporation modes, such as sweeping gas pervaporation thermos-pervaporation and vacuum pervaporation. Vacuum pervaporation is the most commonly used and investigated pervaporation configuration [5].

Many researches on polymer membrane pervaporation for ABE solution have been reported [6,7]. Polydimethylsilicone (PDMS) of moderate selectivity and high permeability to many organics [8,9], is one of the most widely used polymer material in ABE fermentation pervaporation [10-12]. However, the separation performance of pervaporation is not high enough especially for organic permselective removal for industrial application [7]. Moreover, in spite of fact that very thin polymeric membranes are employed in several miniaturized devices, the dependence of permeability from membrane thickness

is not known enough. So it is extremely important to study the mechanism of mass transfer for pervaporation, which involves various interactions between membrane and components. It is difficult to describe very precisely via present theoretical models for dense homogeneous membrane [13], including models of equilibrium dissolution diffusion [14-16], pore flow [17], virtual phase change [18-20], evaporation-permeation [21] and irreversible thermodynamics [22]. Dissolution-diffusion model, the most widely accepted model, was first found by Lonsdale et al. [15], who divided pervaporation into dissolution (adsorption), diffusion and desorption three steps. From the mathematical model based on the hypothesis of equilibrium dissolution, the flux was in inverse proportion to membrane thickness and separation factor was independent of membrane thickness [23]. The conclusion was not in consistent with recent experiment results [24-27]. Pore flow model defined the dense layer as "pore" like nanofiltration which was not very reasonable, since the "pore" free volume formed by random movement of polymer chain was not fixed. Virtual phase change model was the combination of dissolution-diffusion model and pore flow model which was of some self-contradiction. Evaporation-permeation model treated the pervaporation as two separate processes, liquid evaporation and vapor permeation. The total separation factor was not equaled to the product of that two separation factors in the real operation. Irreversible thermodynamics model was set up on the chemical potential considering the coupling interaction of the

*Corresponding author: Li Jiding, State Key Laboratory of Chemical Engineering, Department of Chemical Engineering, Tsinghua University, China
E-mail: lijiding@mail.tsinghua.edu.cn

components, but the phenomenological parameters needed to be determined by experiments which could not be deduced from the present theoretical models.

Pervaporation process, additional new chemical engineering operation, has many similarities to other equilibrium unit operations. However, it is not an equilibrium process in fact. Different from the main trend of equilibrium dissolution, there were few reports based on the non-equilibrium dissolution [28]. Yu et al. [29] presented nonequilibrium dissolution-diffusion model from dynamic analysis of mass transfer. The semi-experimental model, ignoring the desorption resistance, defined "apparent" mass transfer coefficient K_s. K_s was an experimental parameter related with follow-up diffusion. Nonequilibrium model proposed by Islam [30] optimized the analysis of mass transfer in dissolution and desorption process, in which a dimensionless parameter (similar to second Damköler number) was proposed to measure for the deviation of the nonequilibrium surface reaction from equilibrium. Islam's model was further applied in recent studies of gas separation [26,31].

Above nonequilibrium hypothesis agreed with our opinion. It was considered that the dissolution equilibrium of the feed in the membrane could not be really reached in this research. Dissolution and desorption step was treated as pseudo surface reaction with the consideration of desorption resistance. The adsorption at the permeation side could be ignored with the high vacuum pervaporation method. Adsorption and desorption rate constant was applied to help for the kinetic analysis of surface reaction. The effect of thermodynamic partition and diffusion kinetic on permeation flux of acetone-water, butanol-water and ethanol-water binary system in PDMS membrane was analyzed, which could help the optimization of operation conditions, chosen of membrane material and development of theoretical model for pervaporation.

Experiments

Materials

PDMS with viscosity of 20 kg·m^{-1}·s^{-1} and average molecular weight of 80,000 was achieved by Beijing Second Chemistry Company of China. Acetone, butanol, ethanol, hexane and triethyl phosphate (TEP) of analytical grade were purchased from China Medicine Group (Shanghai Chemical Reagent Corporation). Crosslinking agent tetraethylorthosilicate (TEOS) of analytical grade was obtained from Beijing Beihua Fine Chemicals Company of China. Catalyst di-n-butyltin dilaurate (DBTL) was achieved from Beijing Jingyi Chemical Reagents Corporation. All the reagents were used without further purification.

Preparation and characterization of PDMS membrane

Polyvinylidene fluoride (PVDF) support layer was prepared by the dissolution of PVDF in TEP solvent to form 10 wt % solution, which was then casted on the non-woven fiber by spin coating method and immersed into water to induce polymer precipitation. The effect of support layer [32,33] on pervaporation flux was eliminated by reduction of the thickness of the support layer. The residual solvent was exchanged with alcohol for 5 minutes and dried at room temperature.

PDMS membrane was prepared as the way proposed by Zhan et al. [34]. Different mass of PDMS was dissolved in n-hexane and TEOS and DBTL was subsequently added. Homogeneous PDMS solution after stirring was coated on the PVDF. The thickness of PDMS layer δ_m was controlled by the mass ratio of PDMS-solvent and characterized by SEM monitor.

Evaluation of membrane pervaporation performance

Pervaporation experiments were performed by pervaporation laboratory rig the same as reported by Han et al. [35] in our laboratory. The pressure on the back of the membrane was controlled at 100-200pa, which was low enough to remove the effect of vacuum degree and adsorption at the permeation side. The experiments were carried out with different membrane thickness, feed concentration of acetone-water (A-W), butanol-water (B-W), ethanol-water (E-W) and feed temperature. The composition of permeate liquid was analyzed by the gas chromatography GC-14C (Shimadzu Co. Ltd, Japan) equipped with a thermal conductivity detector.

Membrane performance of pervaporation was evaluated via permeate flux J and separation factor α. J was defined by equation (1)

$$J = \frac{N_{mol}}{At} \tag{1}$$

where N_{mol} was permeate mole amount, A membrane area and t time over which the permeate sample was collected. α was calculated by equation (2)

$$\alpha = \frac{c_{p,1}}{c_{p,2}} / \frac{c_{f,1}}{c_{f,2}} = \frac{J_1}{J_2} / \frac{c_{f,1}}{c_{f,2}} \tag{2}$$

where c_p and c_f were concentrations in the permeate stream and in the bulk feed stream respectively, and subscript 1, 2 permselective component and the other component.

Nonequilibrium dissolution-diffusion model for vacuum pervaporation

Model hypothesis, proposition and discussion

Nonequilibrium model for pervaporation was proposed first. Concentration polarization was neglected in high flow rate of feed. So the concentration near the surface of membrane was identified the same as the main body of the liquid. Coupling effect of the components was also neglected for convenient analysis [36].

Through previous hypothesis of equilibrium dissolution [15], the mathematical model was got as equation (3),

$$J = \frac{D_m}{\delta_m}\left(c_f - c_p\right) \tag{3}$$

where D_m was diffusion coefficient in the membrane. However, for steady pervaporation process, the dissolution of liquid in membrane could not be a thermodynamic equilibrium process, or the mass transfer would not be carried on with equal chemical potential on both side of the membrane. In this research, mass transfer of dissolution and desorption was treated as pseudo surface reaction on the surface of the membrane based on the nonequilibrium hypothesis. The flux was the net amount of the reaction per time and surface of unit. Non-equilibrium model was as equation (4),

$$J = k_s c_f - k_d c_m^f = \frac{D_m}{\delta_m}\left(c_m^f - c_m^p\right) = k_d c_m^p - k_s c_p \tag{4}$$

where c_m^f and c_m^p were mole volume concentration contacted with feed and production in the membrane, and k_s, k_d adsorption and desorption rate constants (m·s^{-1}). Rate of capture and departure from the surface depended non-linearly on the solution composition, geometrical dimension of the membrane surface and interaction between the components and membrane [37]. Partition coefficient of the components parted between the membrane and liquid, K, could be defined as equation (5),

$$K = \frac{c_m^{f,\mathrm{e}}}{c_f} = \frac{c_m^{p,\mathrm{e}}}{c_p} = \frac{k_s}{k_d} \tag{5}$$

where $c_m^{f,\mathrm{e}}$ and $c_m^{p,\mathrm{e}}$ were the equilibrium dissolution and desorption concentration of component for feed solution and production respectively. For high vacuum pervaporation, the adsorption at the permeation side with high vacuum and high flux could be ignored. The modified nonequilibrium model was as equation (6).

$$J = k_s c_f - k_d c_m^f = \frac{D_m}{\delta_m}\left(c_m^f - c_m^p\right) = k_d c_m^p \tag{6}$$

J and α could be calculated by equation (7) and equation (8) according to the definition of equation (3) and equation (4).

$$J = \frac{c_f}{\dfrac{\delta_m}{K D_m} + \dfrac{2}{k_s}} \tag{7}$$

$$\alpha = \frac{\dfrac{\delta_m}{K_2 D_{m,2}} + \dfrac{2}{k_{s,2}}}{\dfrac{\delta_m}{K_1 D_{m,1}} + \dfrac{2}{k_{s,1}}}\frac{c_{f,2}}{c_{f,1}} \tag{8}$$

Then J and α of the different systems and operation conditions could be predicted with known δ_m, K, D and k_s of both components. It could be deduced that J was not in verse ratio to δ_m and α was related with δ_m. The relative size of $\dfrac{\delta_m}{K D_m}$ to $\dfrac{2}{k_s}$ characterized the nonequilibrium degree of dissolution, which was caused by the kinetic adsorption and desorption at the membrane interface. When $\dfrac{\delta_m}{K D_m} \gg \dfrac{2}{k_s}$, $J = \dfrac{K D_m}{\delta_m}c_f$, $\alpha = \dfrac{K_1 D_{m,1}}{K_2 D_{m,2}}\dfrac{c_{f,2}}{c_{f,1}}$, in which the equilibrium solubility selectivity and diffusion selectivity could be characterized by $\beta = \dfrac{K_1}{K_2}$ and $\gamma = \dfrac{D_{m,1}}{D_{m,2}}$, respectively. The conclusion was in accordance with equilibrium dissolution-diffusion model. When $\dfrac{\delta_m}{K D_m} \ll \dfrac{2}{k_s}$, or the membrane was very thin, $J = \dfrac{k_s}{2}c_f$, $\alpha = \dfrac{k_{s,1}}{k_{s,2}}\dfrac{c_{f,2}}{c_{f,1}}$, which meant that flux was independent of membrane thickness and separation factor was only determined by the kinetic adsorption and desorption. The kinetic adsorption rate selectivity could also be defined as $\varepsilon = \dfrac{k_{s,1}}{k_{s,2}}$.

Model parameter deduction

The model parameter K was calculated by group contribution method of UNIFAC-ZM model, while D and k_s were achieved by regression analysis of pervaporation experiments data.

Partition coefficient: It was not precise enough to calculate the partition coefficient as the function of solubility parameters independent of solution concentration [38]. Swelling experiment was very complicated and of high requirement of accuracy. Partition equilibrium of ABE-water binary system with PDMS was calculated via the method proposed by Huang et al. [13] in this research. UNIFAC model [39] based on the conception of group contribution was introduced to calculate the activity of penetrants in the polymer. UNIFAC model consisted of a combinatorial and a residual part. Zhong et al. [40] added a universal constant in the volume fraction expression for correction of the combinatorial part with polymer solution. Through equal of the activity in the solution and membrane, equilibrium concentration in the membrane ($c_m^{f,\mathrm{e}}$) was calculated by UNIFAC-ZM model and partition coefficients under different concentration and

temperature could be deduced by Matlab program of our group. Then K was calculated by equation (5). K and defined solubility selectivity β under different feed concentration and temperature was showed in Figures 1 and 2. Boiling point of acetone was near 329.4K at standard atmospheric pressure, so the pervaporation experiment of acetone-water was not carried on at higher temperature than 323.2K.

Figures 1 and 2 showed first that butanol had the largest partition coefficient K and solubility selectivity β, while the water had the smallest at the same condition. Figure 1a, 1b and Figure 1c also revealed that all the partition coefficients of ABE and water decreased with increasing of ABE concentration at 313.2K. At the same time, β of acetone and ethanol decreased while butanol was on the contrary. Augment of butanol content promoted much better compatibility butanol with PDMS, along with larger increment of butanol activity in water solution than acetone and ethanol. Figure 2 illustrated that the partition coefficients of ABE decreased with enlargement of temperature while water was on the contrary at 5 wt% of feed solution. All the solubility selectivity decreased with temperature increasing. The reason might be that the increasing temperature diminished the solubility difference of ABE and water with PDMS besides expanding the polymer chain distance of PDMS. Separation performance at different membrane thickness, feed concentration and temperature was then analyzed according the nonequilibrium model in the following.

Diffusion coefficient and adsorption rate constant: In order to apply the nonequilibrium model for mass transfer analysis of pervaporation and diffusion coefficient and adsorption rate constant

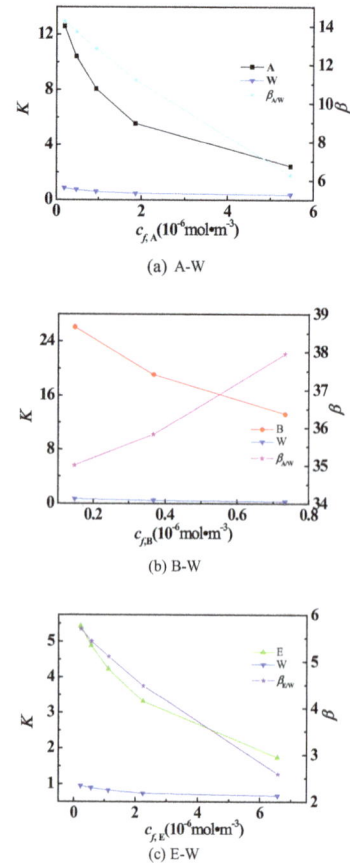

(a) A-W

(b) B-W

(c) E-W

Figure 1: Partition coefficient and equilibrium solubility selectivity of different concentration of ABE solution with PDMS membrane pervaporation at 313.2K.

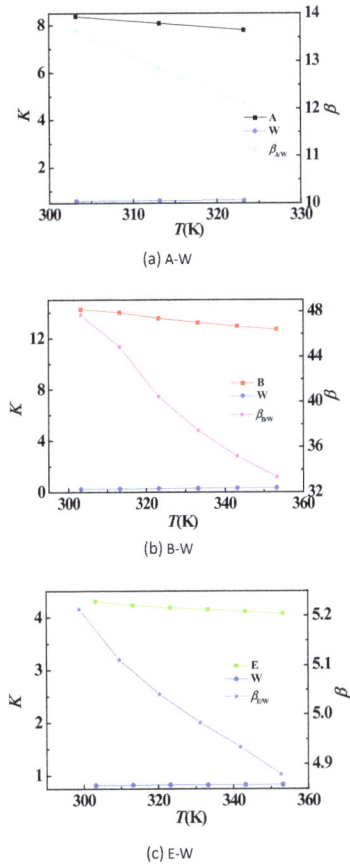

Figure 2: Partition coefficient and equilibrium solubility selectivity of 5 wt% ABE solution with PDMS membrane pervaporation at different temperature.

calculation, the equation (7) was transformed to equation (9).

$$\frac{1}{J} = \frac{\delta_m}{KD_m c_f} + \frac{2}{k_s c_f} \qquad (9)$$

Equation (9) demonstrated that J^{-1} versus δ_m gave a linear plot. According to the linear fitting of J to different membrane thickness at the same concentration, D_m and k_s which were difficult to be determined directly were achieved with known K. Experimental data of pervaporation for ABE binary aqueous solution at different membrane thickness was displayed in Figure 3. The thickness of membrane was controlled in the range of 2-50 μm.

Figure 3 showed that all the flux of components decreased with increasing membrane thickness, while the separation factor increased. Trade off effect between flux and separation factor occurred as in other membrane separation techniques. J and α were both in the order, acetone > butanol > ethanol. When the membrane thickness was over 10μm, α changed very little. That meant the dissolution process might be approximated with equilibrium process. The flux could be treated by Origin software according to Equation (9) to achieve linear fitting result as Figure 4.

Figure 4 showed J^{-1} had relative good linear relationship with δ_m as predicted by the model. When K was known from Huang's method, D_m and k_s of components could be evaluated from the slope and intercept of the line, which could be used in the following prediction for the separation performance.

From Equation (9) and Figure 4, the slope of fitting line was $\frac{1}{KD_m c_f}$ and the intercept was $\frac{2}{k_s c_f}$. Then the average value of D_m and k_s of the components at 313.2K determined by the nonequilibrium model with known K and c_f were listed in the Table 1.

From Table 1, acetone had the largest D_m and k_s in the PDMS membrane, while water had the smallest D_m and k_s. D_m and k_s were not only related with relative size of component and polymer molecule, but also determined by the affinity among them, which might be accused for the difference.

Pervaporation performance evaluation and predicted values via nonequilibrium model

Nonequilibrium model was then applied for the prediction of pervaporation performance with various operation conditions.

Feed concentration effect

Feed concentration was an important factor for pervaporation flux and separation factor. The concentration of ABE feed solution was adjusted to carry on the pervaporation experiments. Pervaporation experimental results and predicted values via nonequilibrium model with different feed concentration were shown in Figure 5. Mass concentration of butanol was controlled lower than 5% because phase separation would occur over 7 wt% at room temperature.

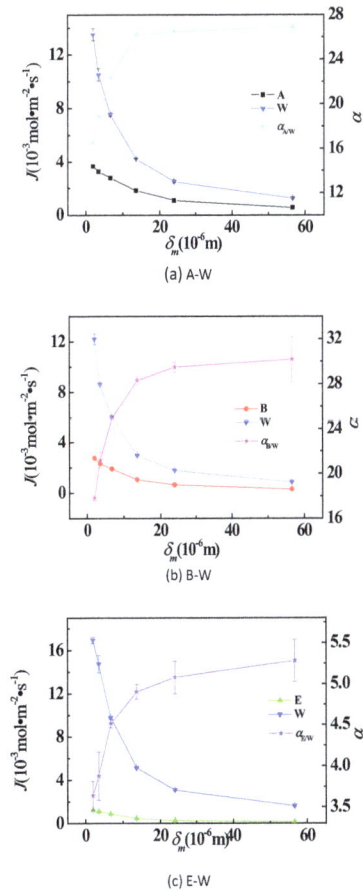

Figure 3: Flux and separation factor of 5 wt% ABE solution with different thickness of PDMS membrane at 313.2K.

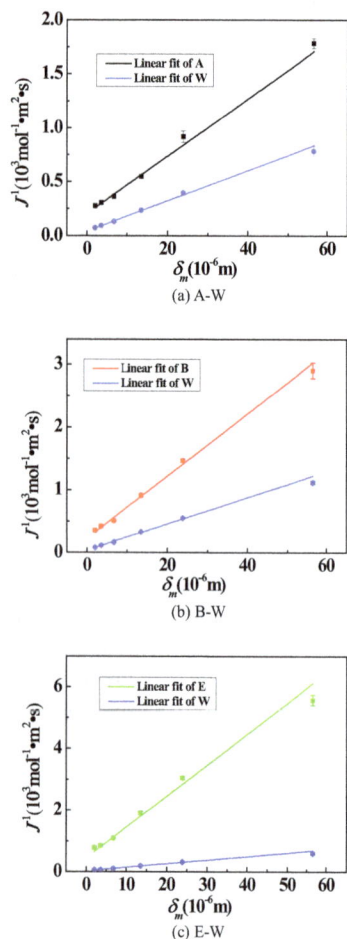

(a) A-W

(b) B-W

(c) E-W

Figure 4: Linear fitting of J^{-1} versus δ_m for 5 wt% ABE solution at 313.2K.

Component	D_m (10^{-12}m·s^{-2})	k_s (10^{-6}m·s^{-1})
Acetone	6.106	7.722
Butanol	2.589	6.221
Ethanol	2.050	3.208
Water	1.867	0.7988

Table 1: Calculated D_m and k_s of acetone, butanol, ethanol and water at 313.2K.

From Figure 5, the pervaporation flux of ABE increased with increasing concentration of ABE, while the water flux declined. Separation factors of acetone and ethanol decreased with the increasing concentration, while butanol was on the verse. It was attributed to the increasement of K with high concentration illustrated in Figure 1b. The calculated values according to equation (7) were in moderate accordance with experimental results in low concentration (<5 wt%), which explained that k_s was independent of concentration. Huge deviation of calculated and experimental results at higher concentration might be caused by the intensified swelling of organics with PDMS polymer chain.

Temperature effect

Temperature effect on the pervaporation performance was further examined. The operation temperature for pervaporation was in the range 303-353K for 5 wt% ABE solution. The results were shown in Figure 6.

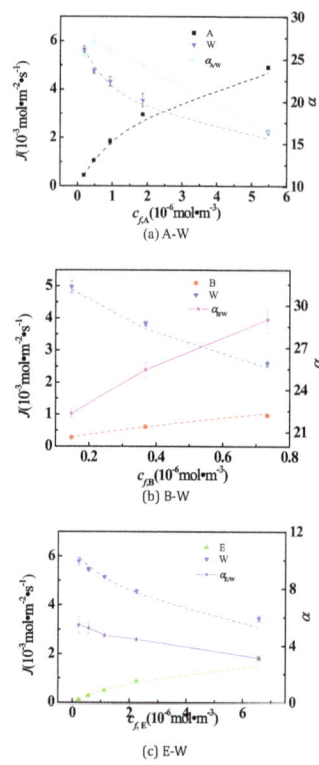

(a) A-W

(b) B-W

(c) E-W

Figure 5: Experimental flux, separation factor and predicted values via non-equilibrium model with different feed concentration at 313.2K (Dash line represented the model calculated results).

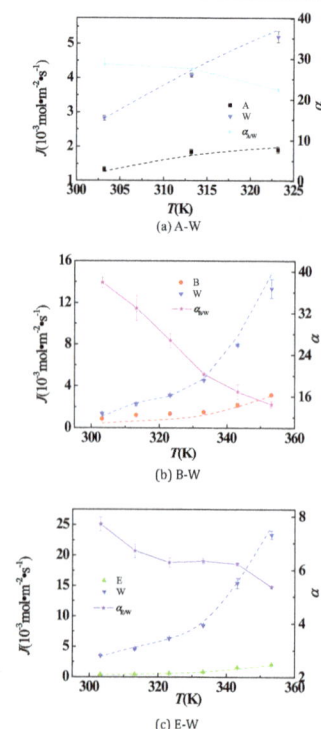

(a) A-W

(b) B-W

(c) E-W

Figure 6: Experimental flux, separation factor and predicted values via non-equilibrium model for 5 wt% ABE solution at different temperature(Dash line represented the model calculated results).

From Figure 6, all the flux increased with temperature elevating, while the separation factor decreased. The predicted curve of nonequilibrium model agreed with the experimental results well. D_m and k_s at different temperature were determined the same as the method presented above. The change of D_m and k_s with temperature was showed in Figure 7.

As shown in Figure 7, D_m and k_s of all the components increased with higher temperature. Based on the pseudo reaction hypothesis, the kinetic adsorption process could be treated by Arrhenius fitting like diffusion as equation (11) and equation (12),

$$D_m = D_{m,0} e^{-\frac{E_D}{RT}} \tag{11}$$

$$k_s = k_{s,0} e^{-\frac{E_s}{RT}} \tag{12}$$

where E_D and E_s were the diffusion and kinetic adsorption activation energy respectively, $D_{s,0}$ and $k_{s,0}$ the pre-exponential factors. Through linear fitting of $\ln D_m$ versus $1/T$ and $\ln k_s$ versus $1/T$ according to equation (11) and equation (12) in Figure 7, E_D and E_s were characterized by the slope of the fitting line, while the intercept characterized the $D_{s,0}$ and $k_{s,0}$. The value of were collected in Table 2.

From Table 2, water had the largest E_s and E_D, which meant that k_s and D_s of water changed the most with temperature. Butanol had the smallest $k_{s,0}$ and $D_{s,0}$, which might be accused to the large volume of the butanol molecule. Those values would be applied for the D_m and k_s determination and flux calculation at the other temperature.

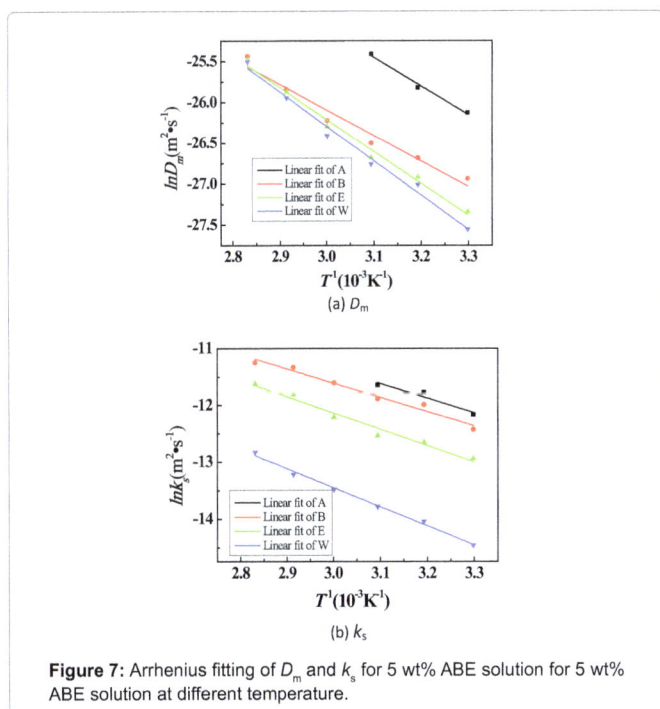

Figure 7: Arrhenius fitting of D_m and k_s for 5 wt% ABE solution for 5 wt% ABE solution at different temperature.

Component	E_s(10^3kg·m·s^{-2})	E_D(10^3kg·m·s^{-2})	$k_{s,0}$ (10^{-2}m·s^{-1})	$D_{m,0}$ (10^{-7}m·s^{-2})
Acetone	-2.588	-3.497	2.753	4.504
Butanol	-2.516	-3.128	1.729	0.5533
Ethanol	-2.881	-3.901	3.046	4.994
Water	-3.344	-4.204	3.285	11.41

Table 2: Calculated E_s, E_D, $k_{s,0}$ and $D_{m,0}$ of acetone, butanol, ethanol and water.

Conclusions

In our pervaporation research, the dissolution and desorption steps were not equilibrium processes, which was different from traditional opinions and proved by the separation factor changing with membrane thickness. So the dissolution and desorption was treated as pseudo surface reaction in this work and nonequilibium dissolution-diffusion model for pervaporation membrane separation was built and analyzed. Membrane thickness, partition coefficient, diffusion and adsorption velocity were all related with flux from the semi-experimental model. Only the thickness of the membrane was very thick or the dissolution-desorption was extremely fast, the mass transfer resistance of dissolution and desorption could be ignored. After partition coefficient achieved by UNIFAC-ZM model and diffusion coefficient and adsorption rate constant regressed from the experiments, the model was applied in prediction of pervaporation separation for low organics concentration of acetone-water, butanol-water and ethanol-water with PDMS membrane. The operation conditions like membrane thickness, feed concentration and temperature were all examined. The experimental results were in good accordance with model calculation. Kinetic solubility process, as well as diffusivity, affected strongly the overall permeation and separation behavior of a PDMS membrane, which could thus be a strong function of the penetrant. Though the nonequilibrium dissolution - diffusion model is not very popular at present, this discussion might introduce the interest of research with membrane surface and dissolution kinetics in future. Those thermodynamics and kinetic parameters could not only be applied for the further performance prediction and material selection, but also offer the data to the further theoretical models for gas separation, pervaporation and reverse osmosis based on the same nonequilibrium mechanism.

Acknowledgement

We highly appreciated the financial supports of Jiangsu National Synergetic Innovation Center for Advanced Materials (SICAM), National Natural Science Foundation of China (21576150), Science Foundation of Tsinghua University (20131089399) and the Special funds for technological development research of Research Institutes from National Ministry of Science and Technology (2013EG111129).

References

1. Shannon MA, Bohn PW, Elimelech M, Georgiadis JG, Mariñas BJ (2008) Science and technology for water purification in the coming decades. Nature 452: 301-310.

2. Kujawska A, Kujawski J, Bryjak M, Kujawoki W (2015) Removal of volatile organic compounds from aqueous solutions applying thermally driven membrane processes. 1. Thermopervaporation. Chem Eng Process 94: 62-71.

3. Padaki M, Surya Murali R, Abdullah MS, Misdan N, Moslehyani A, et al. (2015) Membrane technology enhancement in oil–water separation. A review, Desalination 357: 197-207.

4. Chapman PD, Oliveira T, Livingston AG, Li K (2008) Membranes for the dehydration of solvents by pervaporation. J Memb Sci 318: 5-37.

5. Mortaheb HR, Ghaemmaghami F, Mokhtarani B (2012) A review on removal of sulfur components from gasoline by pervaporation. Chem Eng Res Des 90: 409-432.

6. Rozicka A, Niemistö J, Keiski RL, Kujawski W (2014) Apparent and intrinsic properties of commercial PDMS based membranes in pervaporative removal of acetone, butanol and ethanol from binary aqueous mixtures. J Memb Sci 453: 108-118.

7. Kujawski J, Rozicka A, Bryjak M, Kujawski W (2014) Pervaporative removal of acetone, butanol and ethanol from binary and multicomponent aqueous mixtures. Sep Purif Technol 132: 422-429.

8. Cocchi G, De Angelis MG, Doghieri F (2015) Solubility and diffusivity of liquids for food and pharmaceutical applications in crosslinked polydimethylsiloxane

(PDMS) films: II. Experimental data on mixtures. J Memb Sci 492: 612-619.

9. Cocchi G, De Angelis MG, Doghieri F (2015) Solubility and diffusivity of liquids for food and pharmaceutical applications in crosslinked polydimethylsiloxane (PDMS) films: I. Experimental data on pure organic components and vegetable oil. J Memb Sci 492: 600-611.

10. Romdhane IH, Danner RP (1993) Polymer-Solvent Diffusion and Equilibrium Parameters By Inverse Gas-Liquid-Chromatography. AIChE J 39: 625-635.

11. Zhao CW, Li JD, Jiang Z, Chen CX (2006) Measurement of the infinite dilution diffusion coefficients of small molecule solvents in silicone rubber by inverse gas chromatography. Eur Polym J 42: 615-624.

12. Zhao S, Zhang WW, Zhang F, Li B (2008) Determination of Hansen solubility parameters for cellulose acrylate by inverse gas chromatography. Polym Bull 61: 189-196.

13. Huang J, Li J, Chen J, Zhan X, Chen C (2009) Pervaporation Separation of N-Heptane/Organosulfur Mixtures With PDMS Membrane: Experimental and Modelling. Can J Chem Eng 87: 547-553

14. J.G.Œ.B. Wijmans, Baker RW (1995) The solution-diffusion model: a review. J Memb Sci 107: 1-21.

15. Lonsdale HK, Merten U, Riley RL (1965) Transport properties of cellulose acetate osmotic membranes. J Appl Polym Sci 9: 1341-1362.

16. Valentínyi N, Cséfalvay E, Mizsey P (2013) Modelling of pervaporation: Parameter estimation and model development. Chem Eng Res Des 91: 174-183.

17. Okada T, Matsuura T (1991) A new transport model for pervaporation. J Memb Sci 59: 133-149.

18. Chang CL, Hsuan C, Chang YC (2007) Pervaporation performance analysis and prediction - using a hybrid solution-diffusion and pore-flow model. Journal of the Chinese Institute of Chemical Engineers 38: 43-51.

19. Shieh JJ, Huang R (1998) A pseudophase-change solution-diffusion model for pervaporation. II. Binary mixture permeation, Sep Sci Technol 33: 933-957.

20. Shieh J, Huang RYM (1998) A Pseudophase-Change Solution-Diffusion Model for Pervaporation. I. Single Component Permeation, Sep Sci Technol 33: 767-785.

21. Wijmans JG, Baker RW (1993) A simple predictive treatment of the permeation process in pervaporation, J Memb Sci 79: 101-113.

22. Kedem O (1989) The role of coupling in pervaporation. J Memb Sci 47: 277-284.

23. Brun JP, Larchet C, Melet R, Bulvestre G (1985) Modelling of the pervaporation of binary mixtures through moderately swelling, non-reacting membranes. J Memb Sci 23: 257-283.

24. Chen GQ, Scholes CA, Doherty CM, Hill AJ, Qiao GG, et al. (2012) The thickness dependence of Matrimid films in water vapor permeation. Chem Eng J 209: 301-312.

25. Vicinanza N, Svenum I, Næss LN, Peters TA, Bredesen R, Borg A, et al. (2015) Thickness dependent effects of solubility and surface phenomena on the hydrogen transport properties of sputtered Pd77%Ag23% thin film membranes. J Memb Sci 476: 602-608.

26. Firpo G, Angeli E, Repetto L, Valbusa U (2015) Permeability thickness dependence of polydimethylsiloxane (PDMS) membranes. J Memb Sci 481: 1-8.

27. Flynn EJ, Keane D, Holmes JD, Morris MA (2012) Unusual trend of increasing selectivity and decreasing flux with decreasing thickness in pervaporation separation of ethanol/water mixtures using sodium alginate blend membranes. J Colloid Interface Sci 370: 176-182.

28. Hwang S (2011) Fundamentals of membrane transport. Korean J Chem Eng 28: 1-15.

29. Yu L, Jiang W (1994) Non Equilibrium Dissolution Diffution Model of Pevaporation 45: 153-510.

30. Islam MA, Buschatz H, Paul D (2002) Non-equilibrium surface reactions—a factor in determining steady state diffusion flux. J Memb Sci 204: 379-384.

31. Islam MA, Buschatz H (2005) Assessment of thickness-dependent gas permeability of polymer membranes. Indian Journal of Chemical Technology 12: 88-92.

32. Sukitpaneenit P, Chung T, Jiang LY (2010) Modified pore-flow model for pervaporation mass transport in PVDF hollow fiber membranes for ethanol–water separation. J Memb Sci 362: 393-406.

33. Trifunović O, Trägårdh G (2005) The influence of support layer on mass transport of homologous series of alcohols and esters through composite pervaporation membranes. J Memb Sci 259: 122-134.

34. Zhan X, Li JD, Huang JQ, Chen CX (2010) Enhanced Pervaporation Performance of Multi-layer PDMS/PVDF Composite Membrane for Ethanol Recovery from Aqueous Solution. Appl Biochem Biotechnol 160: 632-642.

35. Han XL, Wang L, Li JD, Zhan X, Chen JA, Yang JC (2011) Separation of Ethanol from Ethanol/Water Mixtures by Pervaporation with Silicone Rubber Membranes: Effect of Silicone Rubbers. J Appl Polym Sci 119: 3413-3421.

36. She M, Hwang S (2006) Effects of concentration, temperature, and coupling on pervaporation of dilute flavor organics. J Memb Sci 271: 16-28.

37. Wang D, Gou SY, Axelrod D (1992) Reaction rate enhancement by surface diffusion of adsorbates. Biophys Chem 43: 117-137.

38. Brookes PR, Livingston AG (1995) Aqueous-aqueous extraction of organic pollutants through tubular silicone rubber membranes. J Memb Sci 104: 119-137.

39. Gmehling J, Constantinescu D, Schmid B (2015) Group Contribution Methods for Phase Equilibrium Calculations. Annu Rev Chem Biomol Eng 6: 267-292.

40. Zhong C, Sato Y, Masuoka H, Chen X (1996) Improvement of predictive accuracy of the UNIFAC model for vapor-liquid equilibria of polymer solutions. Fluid Phase Equilibria 123: 97-106.

10

Development of High Water Selective Sodium Alginate-Silica Hybrid Membranes via Sol-gel Technique for Pervaporation Dehydration of Ethanol

Shang Han, Shasha Na, Weixing Li* and Weihong Xing

College of Chemistry and Chemical Engineering, Nanjing Tech University, Nanjing 210009, China

Abstract

In order to improve the performance of sodium alginate (SA) membrane for dehydration of ethanol, a new method for preparing SA-silica membranes using sol-gel was proposed. The hybrid pervaporation membranes were prepared by hydrolysis and condensation of tetraethyl orthosilicate (TEOS) within SA aqueous solution. The obtained membranes were characterized by scanning electron microscope (SEM), energy-dispersive X-ray spectroscopy (EDX), Fourier Transform Infrared Spectroscopy (FTIR), X-ray diffraction (XRD), Atomic Force Microscope (AFM), thermogravimetry (TG) and differential scanning calorimetry (DSC). Then the membranes were tested by pervaporation dehydration of ethanol. FTIR indicated that -Si-O-C bonds were obtained. XRD showed the SiO_2 particles were generated in the SA matrix. The thermal stability of hybrid membranes was enhanced after incorporating TEOS into SA from the DSC results. The tensile strength of SA-40 membrane was improved after incorporating TEOS into SA. The effect of mass ratio of TEOS to SA on the separation performance was investigated. The permeate flux was improved with increasing mass ratio of TEOS to SA. And the result showed that the permeate flux reached 274 $g \cdot m^{-2} \cdot h^{-1}$ with a high separation factor 17990 when the amount of water in feed was 10 wt% at 50°C. The Arrhenius apparent activation energy for permeation has been estimated from the temperature dependence of permeation values. The activation energy for permeation was 15.1 kJ/mol.

Keywords: Sodium alginate; Tetraethyl orthosilicate; Pervaporation; Ethanol; Dehydration

Introduction

Pervaporation (PV) has attracted growing interests in the separation process because of its energy-saving characteristics and high separation performance compared to traditional techniques [1,2]. It's widely used in dehydration of ethanol, methanol and isopropanol. Sodium alginate (SA), a kind of polysaccharides extracted from seaweed, is considered as a prospective dehydration membrane material for its intrinsic properties such as good hydrophilicity, outstanding water solubility and good film-formation characteristics [3,4]. The development of SA membrane has been focused on for the decades. The performance of pure SA membrane is not so good to dehydrate organics. In order to improve the separation performance of SA membrane, modification methods such as blending, cross-linking and adding fillers have been applied. Dong et al. [5] blended SA with poly-(vinyl alcohol) (PVA) for separating ethanol aqueous solutions. The prepared membranes showed better pervaporation performance for ethanol aqueous solution with a permeate flux of 384 $g \cdot m^{-2} \cdot h^{-1}$ and a separation factor of 384 for 90 wt% ethanol aqueous solution at 45°C. Pan et al. [6] incorporated reduced grapheme oxide into SA matrix. The hybrid membranes exhibited optimum separation performance with a separation factor of 1566 and a permeate flux of 1699 $g \cdot m^{-2} \cdot h^{-1}$. Generally, metal oxides were added into the SA membrane for higher mechanical property and permeability. Organic-inorganic hybrid materials may be a proper candidate for having the advantages of organic moiety and inorganic moiety which have been recognized in various fields [7-9]. Inorganic particles have a good thermal stability as well as high mechanical strength. The separation performance of the membranes can be improved by incorporating inorganic particles into PV membranes [10-15]. Blending inorganic particles into polymer matrix is a simple way to prepare organic-inorganic hybrid membranes.

However, the inorganic particles often behave serious aggregation. In order to improve the dispersion performance of inorganic particles

in casting membrane solution, in situ generation of inorganic particles via sol-gel method in polymer matrix is focused on. The generated inorganic particles can be dispersed in the organic membranes homogeneously. Kariduraganavar et al. [16] prepared chitosan based hybrid membranes by incorporating 2-(3, 4-epoxycyclohexyl) ethyltrimethoxysilane into chitosan matrix using a sol-gel technique. And the developed hybrid membranes could be effectively used to break the azeotropic point of water-isopropanol mixtures with separation selectivity of 17990 and a flux of 29.2 $g \cdot m^{-2} \cdot h^{-1}$ at 30°C for 10 mass% of water. Jiang et al. [17] incorporated TiCl4 into CS membrane, the membrane exhibited the optimal pervaporation performance with a permeate flux of 1403 $g \cdot m^{-2} \cdot h^{-1}$ and a separation factor of 730 for 90 wt% ethanol aqueous solution at 77°C. The sol-gel reaction is helpful to the hybridization of organic and inorganic components which can form covalent bonds and hydrogen bonds between the polymeric phase and inorganic phase [18-20]. Clearly, it is efficient to hybridize the organic and inorganic components homogeneously.

The aim of this work is to attempt preparation of a SA-silica hybrid membrane with high water selectivity via sol-gel technique for dehydration of ethanol. The proposed preparation method of the hybrid SA-silica membrane via sol-gel was rarely reported, and the

---

*Corresponding author: Weixing Li, College of Chemistry and Chemical Engineering, Nanjing Tech University, Nanjing 210009, China
E-mail: wxli@njtech.edu.cn

latest report about hybrid membranes prepared by incorporating silica precursors into alginate matrix was study by Choudhari [21]. Here, we try to investigate the membrane preparation parameters and characterization of the separation characteristics for PV of ethanol solution.

Experimental

Materials

Sodium alginate, tetraethyl orthosilicate, ethanol (99.7%), glutaraldehyde solution (25%) and sulfuric acid were supplied by Sinopharm Chemical Reagent Co., Ltd. Deionized water was produced by a Milli-Q system (Millipore, US).

Membrane fabrication

The 3wt% SA solution was prepared by dissolving SA in deionized water with stirring for 3 h at 60°C. Then a known amount of TEOS was added to the SA solution. Subsequently, quantitative of glutaraldehyde and sulfuric acid were added into the mixtures which were used as cross-linking agent and catalyst, respectively. Then the solution was stirred for 24 h. After that the solution was cast onto an organic glass plate with the aid of automatic film blowing machine. Dried membranes were peeled off from the glass plate. The mass ratio of TEOS to SA was varied as 0, 10, 20, 30, 40 and 60%, and the resulting hybrid membranes were designated as SA, SA-10, SA-20, SA-30, SA-40 and SA-60.

Membrane characterization

The interaction properties among different chemical groups of the hybrid membranes were characterized by Fourier Transform Infrared Spectroscopy (FTIR) spectrometer (AVATAR360, Thermo Nicolet, USA). The morphologies of membranes were conducted with a field-emission scanning electron microscopy (FESEM) (S-4800, Hitachi, Japan). The silicon element was recorded by energy-dispersive X-ray spectroscopy (EDX) equipped on (FESEM). The crystalline structure of membrane was investigated using an X-ray diffraction (XRD) (Miniflex 600, Rigaku, Japan) in the range of 6-80° at the scan rate of 15° min⁻¹. Tensile strength of the SA, SA-10, SA-20, SA-30, SA-40 and SA-60 matrix membranes were measured using the universal testing machine (CMT-6203, MTS SANS, China). Thermogravimetric (TG) and differential scanning calorimetry (DSC) analysis were conducted by a thermoanalyzer (Sta 449 F3, Netzsch, Germany) at a heating rate of 10°C/min under nitrogen atmosphere to analysis the thermal stability of all membranes. The membrane thickness was determined by a field-emission scanning electron microscopy (FESEM) (S-4800, Hitachi, Japan). The increase in surface area and surface roughness was calculated by Atomic Force Microscope (AFM) (XE100, Park systems, Korea).

Swelling experiments

The dry membrane was weighed as W_d, and then it was immersed into 10wt% water-ethanol solutions for 24 h at room temperature to achieve equilibrium. The swollen membranes were taken out carefully and the solution on the membranes surface was wiped off by tissue paper, and then weighed as quickly as possible. The mass of the swollen membranes was measured as Ws (supplementary Table 1). The degree of swelling (DS) was calculated by the following Eq. (1).

$$DS(\%) = \frac{W_s - W_d}{W_d} \times 100 \tag{1}$$

Where W_d and W_s are the mass of the dry and swollen membranes, respectively.

Pervaporation experiments

PV experiments were carried out using an indigenously designed apparatus. The feed temperature was controlled by constant temperature oil bath. The effective membrane area was 0.0011 m². The permeated solution was condensed downstream by liquid nitrogen. A vacuum pump in the downstream maintained the pressure at about 300 Pa. The permeate flux was defined by Eq. (2).

$$J = \frac{W}{A \times t} \tag{2}$$

Where W represents the mass of permeate over a certain time interval t, A represents the effective membrane area.

The compositions of feed and permeate were determined by gas chromatograph (GC-2014, Shimadu, Japan) which was equipped with a thermal conductivity detector (TCD). The length of PORAPAK⁰ Q (mesh 50-80) column was 2 m. Helium (99.9999%) was used as the carrier gas. Both the injector and detector temperatures were 200°C and the column temperature was 180°C. The bridge current was 90 mA. The separation factor (α) was defined by Eq. (3).

$$\alpha = \frac{Y_W / Y_E}{X_W / X_E} \tag{3}$$

Where X_W, X_E, Y_W and Y_E are the weight fractions of water and alcohol in the feed and permeate, respectively.

Results and Discussion

Membrane characterization

FTIR spectra analysis: Silanol groups were obtained by hydrolyzing TEOS. The silanol groups yielded siloxane bonds due to the dehydration or dealcoholysis reaction with other silanol or SA during the membrane drying [22]. Figure 1 shows the FT-IR spectra of SA membrane and hybrid membranes. A characteristic band at around 3243 cm⁻¹ in SA pristine membrane spectra corresponds to -OH stretching vibrations. The peaks at 1591 cm⁻¹ and 1406 cm⁻¹ correspond to asymmetric and symmetric stretching of carboxyl group of SA, respectively. Peaks appeared at around 1010-1030 cm⁻¹ are assigned to -C-O-C stretching of SA membrane. However, the intensity of these bands for the hybrid membranes increased which suggested the formation of -Si-O-C bonds [23]. That's because -Si-O stretching also appears at the same wave numbers of -C-O stretching.

XRD analysis: The XRD patterns of SA membrane and hybrid membranes are shown in Figure 2. The SA membrane exhibited a broad peak around 13° [24], which is attributed to the presence of amorphous region in the polymer. After incorporating TEOS into SA matrix, broad peaks of SiO₂ at around 21° appeared in hybrid membranes [25]. It's clear that the generated SiO₂ particle was in an amorphous form. The intensity of the SiO₂ peaks increased gradually from SA-10 to SA-60 membrane with increasing mass ratio of TEOS to SA. This is because more SiO₂ particles were generated upon increasing TEOS content.

Membrane	Tensile strength (MPa)
SA	26.8
SA-10	26.9
SA-20	16.8
SA-30	16.1
SA-40	28.8
SA-60	31.9

Table 1: Mechanical strength data of SA and hybrid membranes.

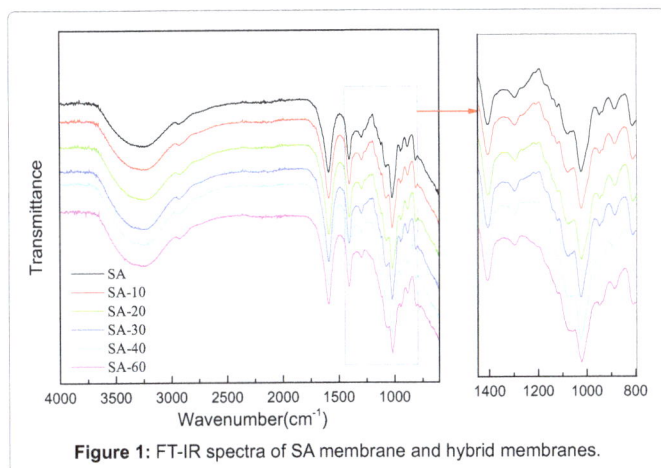

Figure 1: FT-IR spectra of SA membrane and hybrid membranes.

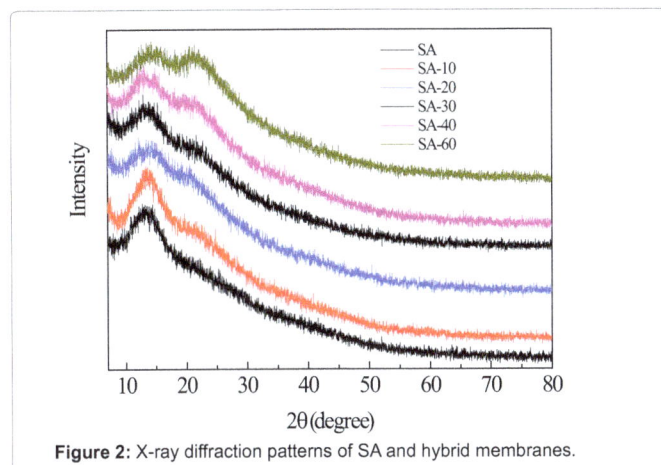

Figure 2: X-ray diffraction patterns of SA and hybrid membranes.

TG and DSC analysis: The thermal stability of the hybrid membranes was evaluated by TG and DSC under nitrogen flow. The results are shown in Figure 3. The membranes had three stages of weight loss from Figure 3a. The first weight loss of 20% occurred between ambient temperature and 200°C corresponds to the physically absorbed water molecules. The second thermal event occurring in the range of 200-250°C, which is attributed to the decomposition of SA matrix and the weight loss was about 25%. The third stage, for temperature higher than 250°C, corresponds to the residual decomposition reactions. What's more, it can be clearly seen that the decomposition temperature of hybrid membranes were higher than the SA membrane in the DSC patterns (Figure 3b). It indicated that the thermal stability of membranes was virtually improved after hybridization, which is mainly due to a hindrance of chain mobility of SA by the generation of SiO_2 and the hydrogen bonding between SA and SiO_2.

SEM and EDX characterizations: Figure 4 illustrates the SEM and EDX photographs of hybrid membranes. The surface view of SA membrane was smooth. But the surface of hybrid membranes became rougher with increasing mass ratio of TEOS to SA. From the EDX pictures, it can be seen that the SA membrane had no silicon element. However, the silicon element was obtained from the hybrid membranes, suggesting that the generation of SiO_2 particles. This is agreement with the XRD results, because the formation of SiO_2 particles was also verified by XRD analysis. The EDX Si-mapping of the SA-40 membrane in Figure 4e-3 showed that silicon element was distributed uniformly in the SA matrix which indicated the homogeneous hybrid structure.

Tensile strength test: The maximum tensile strengths of all membranes are given in Table 1. The tensile strengths of SA, SA-10, SA-20, SA-30, SA-40 and SA-60 membrane were 26.8 MPa, 26.9 MPa, 16.8 MPa, 16.1 MPa, 28.8 MPa and 31.9 MPa, respectively. These data indicated higher mechanical strengths for the hybrid membranes expect SA-20 and SA-30 membrane as compared to SA membrane. The decrease of tensile strengths of SA-20 and SA-30 membranes may be because of the uniform dispersion in SA matrix as can be observed in the SEM images. The tensile strengths of the other membranes were improved by incorporating TEOS into SA matrix. It is due to the hydrogen and covalent bonds between SiO_2 and SA.

Swelling behaviors

Membrane swelling plays a key role in separation property of the membrane which depends on the membrane structure. Figure 5 shows the effect of the mass ratio of TEOS to SA on the degree of swelling for hybrid membranes in 10wt% water-ethanol mixtures at 25°C. It can be seen that the degree of swelling was enhanced with increasing mass ratio of TEOS to SA. This is attributed to increased hydrophilic nature of the hybrid membranes owing to the presence of SiO_2 particles. And thereby adsorption of water and ethanol molecules increased resulting to degree of swelling.

Pervaporation performance

TG and DSC analysis: In general, the diffusion of pervaporation process plays an important role in permeate flux and the diffusion is influenced by membrane thickness. Table 2 illustrates that the permeate flux decreased with increasing membrane thickness. That's because the

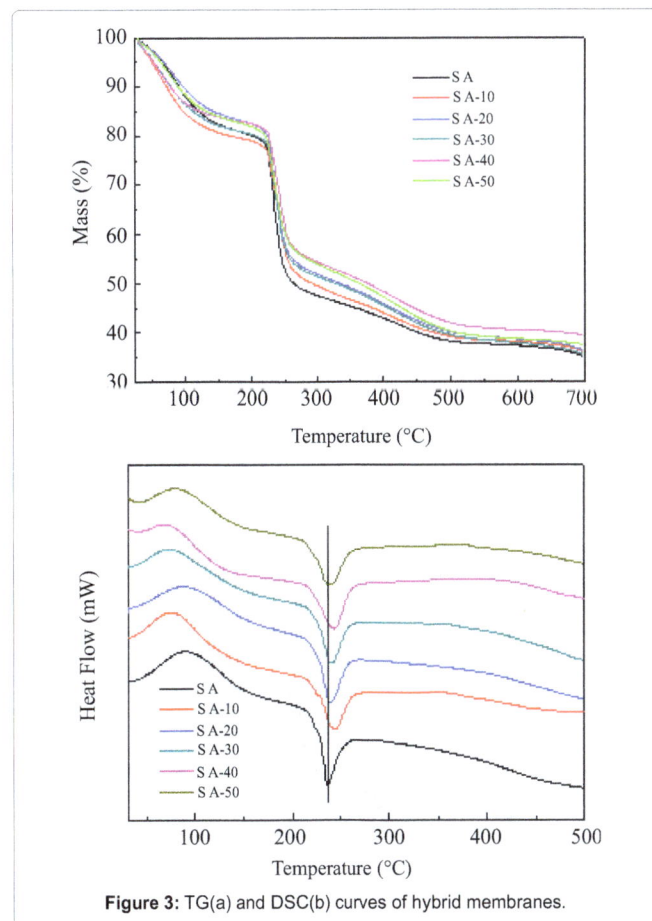

Figure 3: TG(a) and DSC(b) curves of hybrid membranes.

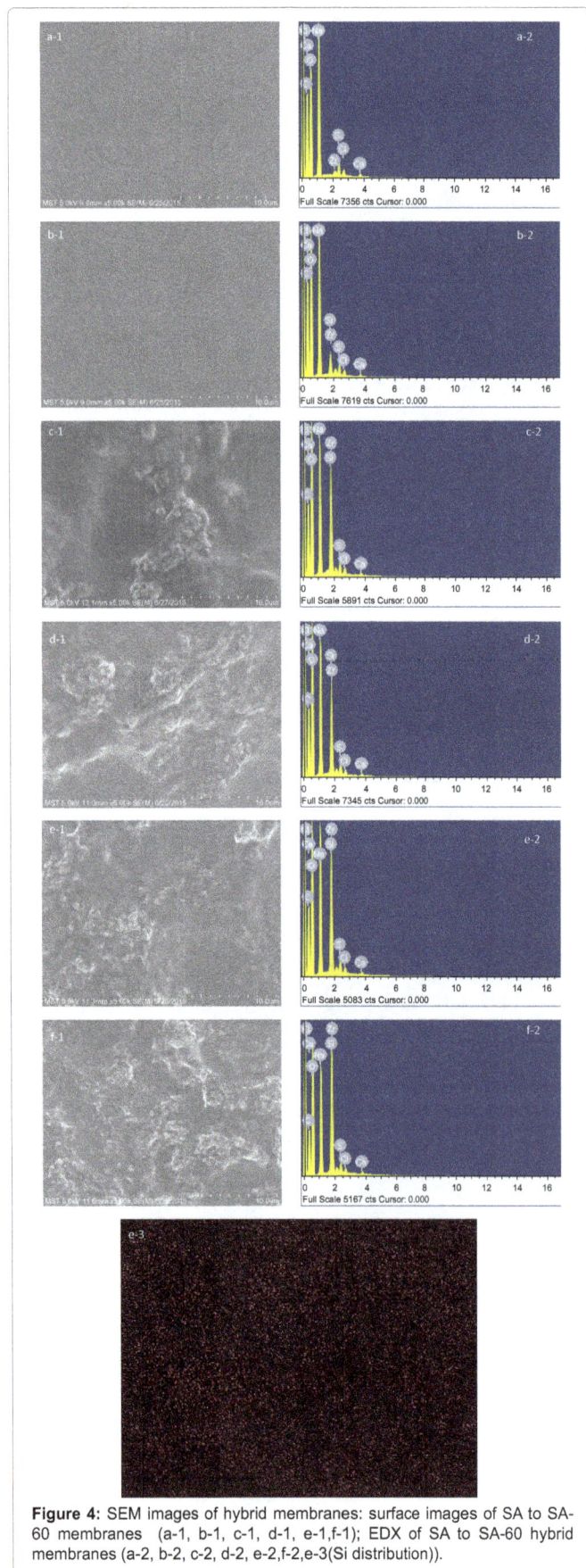

Figure 4: SEM images of hybrid membranes: surface images of SA to SA-60 membranes (a-1, b-1, c-1, d-1, e-1,f-1); EDX of SA to SA-60 hybrid membranes (a-2, b-2, c-2, d-2, e-2,f-2,e-3(Si distribution)).

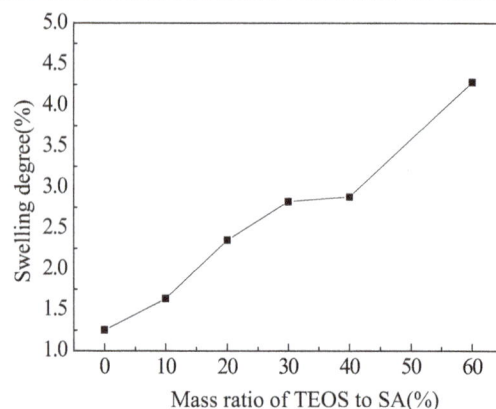

Figure 5: Effect of mass ratio of TEOS to SA on the degree of swelling of hybrid membranes (12 μm).

Membrane Thickness(μm)	Flux(g·m⁻²·h⁻¹)	Water content in permeate
10	274	99.95
15	248	99.96
20	169	99.96
26	137	99.94

Table 2: Influence of membrane thickness on pervaporation performance (SA-40 membrane, 10 wt% water in the feed, 50°C).

increased mass transfer resistance led to reduced diffusion rate. The upstream side of the membrane was swollen and plasticized due to adsorbed liquid molecules and allowed unrestricted transport of feed components [26]. However, the downstream side of the membrane was dry in vacuum condition, which allowed smaller sized molecules to pass through. It's notable that the water content of permeate side was nearly 100wt%.

Effect of mass ratio of TEOS to SA: As shown in Figure 6, Tables 3 and 4, the generated SiO_2 particles can improve the roughness of membrane surface, and the permeability is influenced by membrane roughness. Pervaporation experiment was carried out to study the effect of mass ratio of TEOS to SA on the dehydration performance of 10 wt% water in the feed at 50°C. The membrane thickness was 12 μm. Figure 7 demonstrates that the permeate flux increased with increasing mass ratio of TEOS to SA. The permeate flux increased from 225 to 282 g·m⁻²·h⁻¹ when the mass ratio increased from 0% to 60%. This is mainly due to increased interaction between water molecules and membrane. The free-volume in membrane matrix increased because of generated SiO_2 particles [27]. The water content of permeate side was nearly 100wt% when the mass ratio of TEOS to SA was lower than 30% and decreased slightly over 30%. It may be that the interaction between water molecules and membrane matrix becomes weak because of slight agglomeration of SiO_2 particles.

Effect of feed temperature: Effect of feed temperature ranging from 30°C to 50°C on the PV performance for water-ethanol mixtures was studied using SA-40 membrane at 10wt% water in the feed, and the results are presented in Figure 8. It's observed that the permeate flux increased with increasing feed temperature which can be explained by solution-diffusion mechanism. With the increase of temperature, the vapor pressure difference increased between the upstream and downstream side of the membranes, which resulted in the enhancement of the driving force of transport. Higher temperature leads to higher molecular diffusivity [28]. Permeation of diffusing

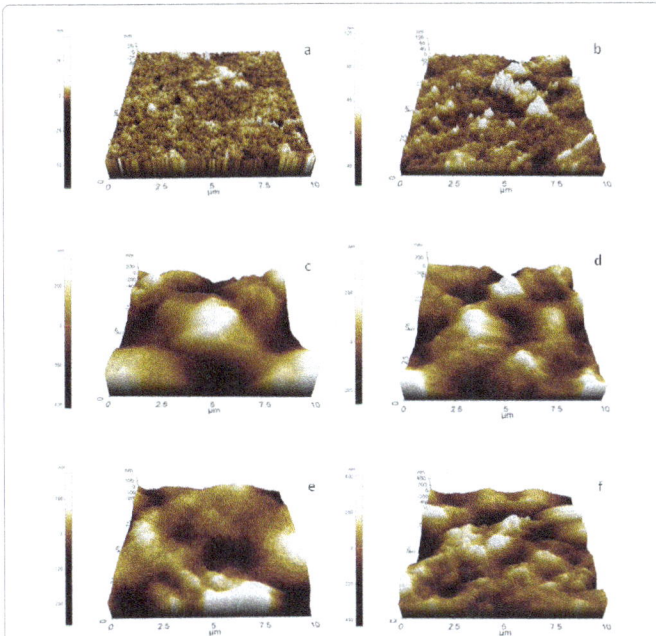

Figure 6: AFM surface images of hybrid membranes: SA(a), SA-10(b), SA-20(c), SA-30(d), SA-40(e), SA-60(f).

Membrane	T(°C)	Flux(g·m⁻²·h⁻¹)	Separation factor	Reference
PVA-SA/PS	45	384	384	[5]
NaAlg-zeolite beta	50	166	465	[8]
NaAlg-zeolite beta	30	132	1598	[8]
CS-TEOS	80	284	460	[18]
NaAlg-mPTA10	50	393	2423	[29]
NaAlg-mPTA10	30	316	8991	[29]
SA- heteropolyacid	30	570	14991	[11]
SA-40	30	189	7013	This study
SA-40	50	274	17990	This study
SA-60	50	282	4496	This study

Table 3: Comparisons of pervaporation performance for dehydration of ethanol/water mixtures with 10 wt% water in the feed.

Membrane	Increase in surface area (%)	Surface roughness(Ra) (nm)
SA	0.59	3.63
SA-10	1.19	11.33
SA-20	2.88	112.82
SA-30	3.50	63.34
SA-40	0.76	31.68
SA-60	9.36	90.81

Table 4: The increase in surface area and surface roughness of SA and hybrid membranes calculated by AFM.

molecules pass through the membrane becomes easier, therefore, the mass transport is faster and the total flux increases. What's more, as the temperature increased, the thermal mobility and the free volume of polymer were elevated, which led to the increase of the solubility of solution on the surface [29,30]. Meanwhile, the activation energy for permeation through the membrane can be described by Arrhenius relationship in Eq. (4).

$$J = J_0 \exp(\frac{-E_P}{RT}) \qquad (4)$$

Where J is the permeate flux, J0 is the pre-exponential factor, EP refers to the activation energy for permeation, R and T are the gas constant and the operating temperature, respectively. Figure 9 was the normalized Arrhenius plot. From Figure 9, the activation energy for permeation through the SA-40 membrane was calculated from the slope of the fit liner and its value was 15.1 kJ/mol [30].

Effect of feed water composition: Feed water composition exhibits a considerable effect on membrane performance. Figure 10 shows the permeate flux against the feed water composition of ethanol/water system, and the SA-40 membrane was used. The total flux increased from 161 to 985 g·m⁻².h⁻¹ with increasing water content in the feed. This can explain that an increase of feed water concentration may lead to enhancement of membrane swelling. The surface of the membrane became more compact, which led to a positive impact on flux.

Comparisons with literature data: The present PV data compared with former results provided by other researchers are listed in Table 3. It shows that the SA-40 membrane had higher separation factor with good permeate flux for separation of ethanol-water mixtures compared to similar data published in the literature.

Conclusions

The SA-Silica hybrid membranes were fabricated using sol-gel method for pervaporation dehydration of ethanol aqueous solution. The XRD characterization and SEM images indicated that SiO2 inorganic particles were generated by in situ hydrolysis and condensation of tetraethyl orthosilicate (TEOS) within SA aqueous

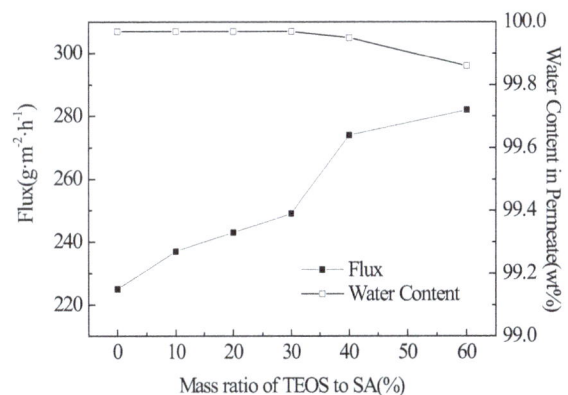

Figure 7: Effect of mass ratio of TEOS to SA on the pervaporation properties (10 wt% water in the feed, 50°C, 12 μm).

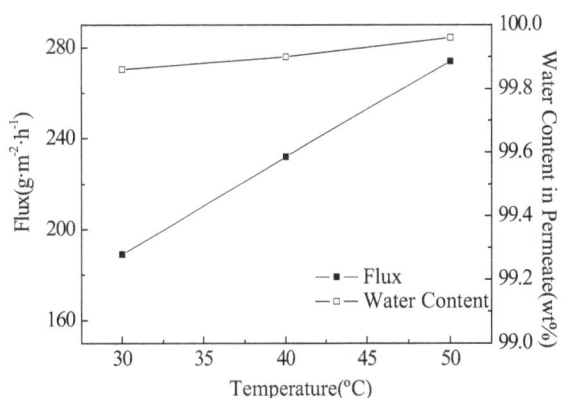

Figure 8: Effect of feed temperature on pervaporation performance (SA-40 membrane, 10 wt% water in the feed, 12 μm).

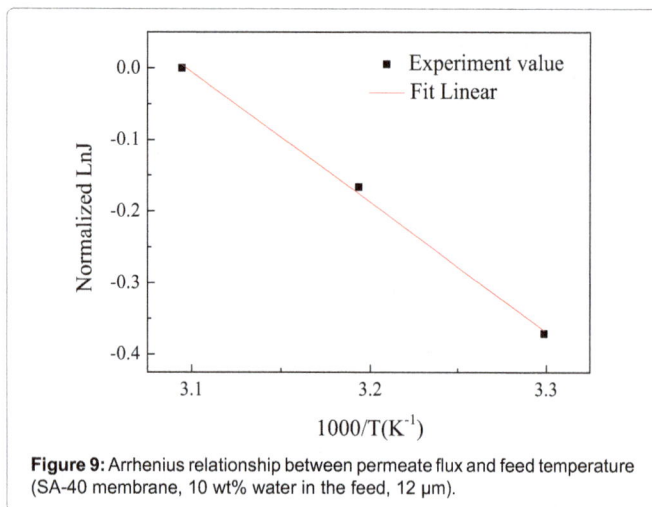

Figure 9: Arrhenius relationship between permeate flux and feed temperature (SA-40 membrane, 10 wt% water in the feed, 12 μm).

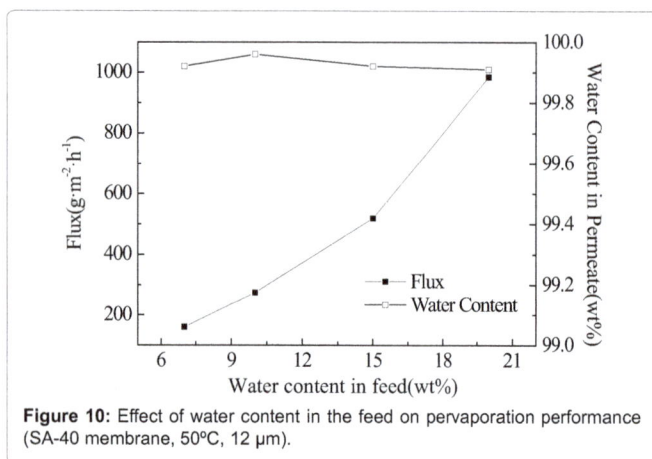

Figure 10: Effect of water content in the feed on pervaporation performance (SA-40 membrane, 50ºC, 12 μm).

solution. The thermal stability of hybrid membranes was improved after incorporating TEOS into SA. The tensile strength of SA-40 membrane was enhanced compared to SA membrane. The permeate flux was improved after incorporating TEOS into SA matrix. When the mass ratio of TEOS to SA was 40%, permeate flux reached 274 g·m^{-2}·h^{-1} while the water content of permeate side was nearly 100wt% with separation factor 17990.

Acknowledgements

This work was financially supported by National Natural Science Foundation of China (No. 21576132), National Key Science and Technology Program of China (No. 2013BAE11B01), Jiangsu Province Foundation of China (No. 2013-XCL-027).

References

1. Chapman PD, Oliveira T, Livingston AG, Li K (2008) Membranes for the dehydration of solvents by pervaporation. J Membr Sci 318: 5-37.

2. Jullok N, Martinez R, Wouters C, Luis P, Sanz MT, et al. (2013) A biologically inspired hydrophobic membrane for application in pervaporation. Langmuir 29: 1510-1516.

3. Saraswathi M, Rao KM, Prabhakar MN, Prasad CV, Subha MCS (2010) Pervaporation studies of sodium alginate (SA)/dextrin blend membranes for separation of water and isopropanol mixture. Desalination 269: 177-183.

4. Gao CY, Zhang MH, Ding JW, Pan FS, Jiang ZY, et al. (2014) Pervaporation dehydration of ethanol by hyaluronic acid/sodium alginate two-active-layer hybrid membranes. Carbohydr Polym 99: 158-165.

5. Dong YQ, Zhang L, Shen JN, Song MY, Chen HL (2006) Preparation of poly(vinyl alcohol)-sodium alginate hollow-fiber hybrid membranes and pervaporation dehydration characterization of aqueous alcohol mixtures. Desalination 193: 202-210.

6. Cao K, Jiang ZY, Zhao J, Zhao CH, Gao CY, et al. (2014) Enhanced water permeation through sodium alginate membranes by incorporating graphene oxides. J Membr Sci 469: 272-283.

7. Amirilargani M, Sadatnia B (2014) Poly(vinyl alcohol)/zeolitic imidazolate frameworks (ZIF-8) mixed matrix membranes for pervaporation dehydration of isopropanol. J Membr Sci 469: 1-10.

8. Adoor SG, Manjeshwar LS, Bhat SD, Aminabhavi TM (2008) Aluminum-rich zeolite beta incorporated sodium alginate mixed matrix membranes for pervaporation dehydration and esterification of ethanol and acetic acid. J Membr Sci 318: 233-246.

9. Kulkarni SS, Tambe SM, Kittur AA, Kariduraganavar MY (2006) Modification of tetraethylorthosilicate crosslinked poly(vinyl alcohol) membrane using chitosan and its application to the pervaporation separation of water-isopropanol mixtures. J Appl Polym Sci 99: 1380-1389.

10. Premakshia HG, Ramesh K, Kariduraganavar MY, Aminabhavi TM (2015) Modification of crosslinked chitosan membrane using NaY zeolite for pervaporation separation of water-isopropanol mixtures. Chem Eng Res Des 94: 32-43.

11. Magalad VT, Supale AR, Maradur SP, Gokavi GS, Aminabhavi TM (2010) Preyssler type heteropolyacid-incorporated highly water-selective sodium alginate-based inorganic-organic hybrid membranes for pervaporation dehydration of ethanol. Chem Eng J 159: 75-83.

12. Flynn EJ, Keane DA, Tabari PM, Morris MA (2013) Pervaporation performance enhancement through the incorporation of mesoporous silica spheres into PVA membranes. Sep Purif Technol 118: 73-80.

13. Kariduraganavar MY, Kittur AA, Kulkarni SS, K Ramesh (2004) Development of novel pervaporation membranes for the separation of water-isopropanol mixtures using sodium alginate and NaY zeolite. J Membr Sci 238: 165-175.

14. Nigiz FU, Dogan H, Hilmioglu ND (2012) Pervaporation of ethanol/water mixtures using clinoptilolite and 4A filled sodium alginate membranes. Desalination 300: 24-31.

15. Hua DU, Ong YK, Wang Y, Yang TX, Chung TS (2014) ZIF-90/P84 mixed matrix membranes for pervaporation dehydration. J Membr Sci 453: 155-167.

16. Rachipudi PS, Kittur AA, Sajjan AM, Kariduraganavar MY (2013) Synthesis and characterization of hybrid membranes using chitosan and 2-(3,4-epoxycyclohexyl) ethyltrimethoxysilane for pervaporation dehydration of isopropanol. J Membr Sci 441: 83-92.

17. Zhao J, Wang F, Pan F, Zhang M, Yang X, et al. (2013) Enhanced pervaporation dehydration performance of ultrathin hybrid membrane by incorporating bioinspired multifunctional modifier and TiCl$_4$ into chitosan. J Membr Sci 446: 395-404.

18. Ma J, Zhang MH, Lu LY, Yin X, Chen J, et al. (2009) Intensifying esterification reaction between lactic acid and ethanol by pervaporation dehydration using chitosan-TEOS hybrid membranes. Chem Eng J 155: 800-809.

19. Xie ZL, Hoang M, Duong T, Ng D, Dao B, et al. (2011) Sol-gel derived poly(vinyl alcohol)/maleic acid/silica hybrid membrane for desalination by pervaporation. J Membr Sci 383: 96-103.

20. Mosa J, Durán M, Aparicio M (2010) Epoxy-polystyrene-silica sol-gel membranes with high proton conductivity by combination of sulfonation and tungstophosphoric acid doping. J Membr Sci 361: 135-142.

21. Choudhari SK, Premakshi HG, Kariduraganavar MY (2016) Development of novel alginate-silica hybrid membranes for pervaporation dehydration of isopropanol. Polym Bull 73: 743-762.

22. Kariduraganavar MY, Kulkarni SS, Kittur AA (2005) Pervaporation separation of water-acetic acid mixtures through poly(vinyl alcohol)-silicone based hybrid membranes. J Membr Sci 246: 83-93.

23. Robertson MAF, Mauritz KA (1998) Infrared investigation of the silicon oxide phase in [perfluoro-carboxylate/sulfonate (bilayer)]/[silicon oxide] nanohybrid membranes. J Polym Sci 36: 595-606.

24. Li YF, Jia HP, Pan FS, Jiang ZY, Cheng QL (2012) Enhanced anti-swelling property and dehumidification performance by sodium alginate-poly(vinyl alcohol)/polysulfone composite hollow fiber membranes. J Membr Sci 407: 211-220.

25. Obaid M, Tolba GMK, Barakat NAM (2015) Effective polysulfone-amorphous SiO2 NPs electrospun nanofiber membrane for high flux oil/water separation. Chem Eng J 279: 631-638.

26. Kalyani S, Smitha B, Sridhar S, Krishnaiah A (2008) Pervaporation separation of ethanol-water mixtures through sodium alginate membranes. Desalination 229: 68-81.

27. Sajjan M, Kumar BKJ, Kittur AA, Kariduraganavar MY (2013) Novel approach for the development of pervaporation membranes using sodium alginate and chitosan-wrapped multiwalled carbon nanotubes for the dehydration of isopropanol. J Membr Sci 425: 77-88.

28. Svang-Ariyaskul A, Huang RYM, Douglas PL, Liu LJ (2006) Blended chitosan and polyvinyl alcohol membranes for the pervaporation dehydration of isopropanol. J Membr Sci 280: 815-823.

29. Adoor SG, Rajineekanth V, Nadagouda MN, Chowdoji RC, Dionysiou DD, et al. (2013) Exploration of nanohybrid membranes composed of phosphotungstic acid in sodium alginate for separation of aqueous-organic mixtures by pervaporation. Sep Purif Technol 113: 64-74.

30. Wang J, Zhang WY, Li WX, Xing WH (2015) Preparation and characterization of chitosan-poly (vinyl alcohol)/polyvinylidene fluoride hollow fiber composite membranes for pervaporation dehydration of isopropanol. Korean J Chem Eng 32: 1369-1376.

11

Studies of Cellulose Acetate Supported Ergosterol Liquid Membrane

Vandana Upadhyay*

Department of Chemistry, Marwar Business School, Gorakhpur, Uttar Pradesh, India

Abstract

Supported liquid membrane on cellulose acetate matrix has been formed by using ergosterol, a fungal product as a surfactant. The behaviour of this membrane has been simulated with biological membranes by its electrochemical characterization. For this purpose membrane potential has been measured using NaCl and $CaCl_2$ solutions of different concentrations. Ionic transport numbers have been used to estimate fixed charge density and permselectivity. The variation of membrane potential has also been examined as a function of concentration and pH values of NaCl & $CaCl_2$ solutions.

Keywords: Ergosterol; Liquid membrane

Introduction

Liquid membrane technology has attracted attention recently because of its numerous applications which include gas transport, metal ion recovery, waste water treatment, reverse osmosis, desalination, biotechnology and biomedical engineering [1-6]. Liquid membranes are usually formed at every interface encountered by a surfactant solution. Permeation through liquids is orders of magnitude faster than that through solid polymers of comparable thickness. Among various studies on surfactant generated liquid membranes [7], the studies with surfactants of fungal origin are rare. Ergosterol is a fungal product and a biologically important material known as provitamin D which can be converted to vitamin D2 on exposure to ultraviolet radiation. The epidermal cells of the skin of man and animal contain large amount of ergosterol, which is converted to vit D2 by exposure of the skin to sunlight. Vitamin D2 is a powerful antirachitic compound for man and rat. It helps in maintaining the levels of calcium and phosphorus in the body and is involved in resorption of calcium from calcified bones by osteolytic cells. Vitamin D2 is necessary for proper growth of bone and teeth.

Ergosterol has therefore been used in the present investigation as a surfactant for the formation of liquid membrane supported on the cellulose acetate matrix. Ergosterol is a white crystalline compound. There is a hydroxyl group present in ergosterol. Three double bonds are also present in its nucleus. Ergosterol is non ionic as such. The liquid membrane has been electrochemically characterized by measuring membrane potentials by the use of NaCl, and $CaCl_2$ solutions of various concentrations. Structure of ergosterol is given in Figure 1.

Materials and Methods

The film of cellulose acetate was prepared as already reported [8]. The thickness of cellulose acetate film used was 7.3×10^{-5} m and its cross sectional area was 1.78×10^{-4} m^2 · Liquid membrane generation was accomplished [9,10] following the method suggested by Gershfeld and Pagano [11] . A large amount of water was added to a known amount of ergosterol (ACROS ORGANICS) dissolved in ethanol. The mixture was constantly stirred. Dispersions of ergosterol of different known concentrations were thus prepared. Generation of liquid membrane was carried out by first equilibrating the film with water and then keeping it in contact with ergosterol solution of lowest concentration. To ascertain the minimum concentration needed for liquid membrane formation, Ergosterol suspension of different concentrations were prepared and used. The minimum ergosterol concentration needed

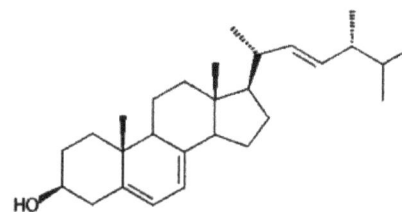

Figure 1: Ergosterol.

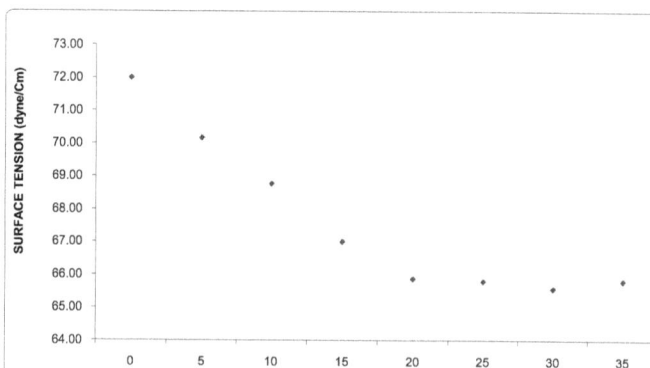

Figure 2: Critical Micelle Concentration Determination by Surface Tension Measurements.

for complete liquid membrane formation was found to be 20 µM. It is believed that liquid membrane formation occurs when the critical micelle concentration (CMC) is exceeded. The CMC was determined by measuring surface tension and it was found to be 20 µM. Membrane resistance was measured. Ergosterol dispersions of higher concentration (35 µM) were used in the present study.

***Corresponding author:** Vandana Upadhyay, Department of Chemistry, Marwar Business School, Gorakhpur, Uttar Pradesh, India, E-mail: vandana_gkp@yahoo.com

Membrane potential was measured as usual [12] with the help of digital multimeter (MAS 830) using calomel electrodes. Sodium chloride and calcium chloride solutions of unequal concentrations with and without ergosterol were kept on either side of the supported matrix to allow equilibration and liquid membrane formation. Membrane potentials were also measured in the same manner using sodium chloride and calcium chloride solutions of different pH values.

Results and Discussion

The formation of liquid membrane was ascertained by measuring membrane resistance using ergosterol suspensions of different concentrations. Due to accumulation of ergosterol in the interfacial region, the resistivity changed and approached a constant value after 3-4 h. This was attributable to the accumulation of ergosterol in the interfacial region. The plot of surface tension against concentrations of ergosterol solutions is given in Figure 2. The trend shows that upto the CMC (20 μM) of ergosterol and thereafter the values of surface tension become constant. This is in consonance with Kesting's hypothesis. With increasing concentration of ergosterol, the surface tension of water is lowered showing that the interface is being progressively covered. The critical micelle concentration must be exceeded for the occurrence of complete liquid membrane formation. Surface tension was also measured when the surfactant solution contained different concentrations of electrolytes (Nacl and Cacl2). The value of CMC in presence of all the electrolytes remains almost constant (Figure 3,4). A curve almost parallel to concentration axis ensures that the interface is completely covered. Membrane potential was measured using the following type of experimental cell:

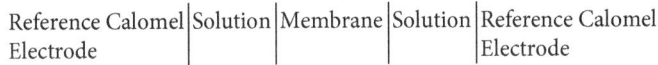

| Reference Calomel Electrode | Solution | Membrane | Solution | Reference Calomel Electrode |

The condition I=0 is fulfilled by short circuiting electrodes and then taking measurements.

The formation of liquid membrane is indicated by lowering of membrane potential and it is clear from the membrane potential values, given in Table 1,2.

Since cellulose acetate is practically uncharged its selectively is considerably low. Permselectivity was estimated using the equation [13]

$$P_s = \frac{\overline{t_+} - t_+}{t_+ - (2t_+ - 1)\overline{t_+}} \tag{1}$$

The transport number on the basis of TMS theory may be given as [14], for 1:1 electrolyte i.e. for NaCl and KCl.

$$\overline{t_+} = \frac{E}{2E_{max}} + 0.5 \tag{2}$$

Where the symbols have their usual meaning. A simplification of TMS theory gives:

$$E = \frac{RT}{F}\left(2\overline{t_+} - 1\right)\ln\frac{a_1}{a_2} \tag{3}$$

and if the membrane is ideally selective,

$$E_{max} = \frac{RT}{F}\ln\frac{a_1}{a_2} \tag{4}$$

Permselectivity is related to fixed charge density as [15]

$$\overline{\varnothing}x = \frac{2\overline{c}P_s}{\sqrt{1 - P_s^2}} \tag{5}$$

Where $\overline{c}$ is the mean concentration.

For CaCl$_2$, which is 2:1 electrolyte, the membrane potential which is infact the liquid junction potential is given as [16]

$$E_L = \left[1 - \frac{3}{2}t_+\right]\frac{RT}{F}\ln\frac{a_1}{a_2} \tag{6}$$

Where the activities are the geometric mean activities. For an ideally selective membrane this potential is given as

$$E_{max} = \frac{-RT}{2F}\ln\frac{a_1}{a_2} \tag{7}$$

Teorell-Meyer-Sievers theory gives the following expression for membrane potential for 2:1 electrolyte [17]

$$E = \left[1 - \frac{3}{2}\overline{t_+}\right]\frac{RT}{F}\ln\frac{a_1}{a_2} \tag{8}$$

Solving Eqs (6) and (8) the transport number $\overline{t_+}$, may be given as :

$$\overline{t_+} = \frac{E}{E_L}\left[\overline{t_+} - \frac{2}{3}\right] + \frac{2}{3} \tag{9}$$

For calculating $\overline{t_+}$ the experimentally measured values of membrane potential E, have been used.

In Table 3 the values of transport numbers for CaCl$_2$ are given which are calculated using eqn. 9. E$_L$ and E$_{max}$ values, calculated by using eqs (6) and (7) are also recorded in this table.

C (mol-dm⁻³)	C₁ (mol-dm⁻³)	f1	f2	C₂ (mol-dm⁻³)	E (mV) Without ergosterol	E (mV) With ergosterol
0.15	0.2	0.6949	0.7730	0.1	14.9	12.5
0.30	0.4	0.5976	0.6949	0.2	13.6	11.1
0.45	0.6	0.5263	0.6351	0.3	12.3	10.3
0.60	0.8	0.4828	0.5976	0.4	11.6	9.7
0.75	1.0	0.4430	0.5623	0.5	10.3	9.1

Table 1: Membrane potential data at different mean concentration of NaCl with and without ergosterol.

C (mol-dm⁻³)	C₁ (mol-dm⁻³)	f1	C₂ (mol-dm⁻³)	f2	E (mV) Without ergosterol	E (mV) With ergosterol
0.075	0.1	0.5322	0.05	0.6402	14.5	13.6
0.15	0.2	0.4828	0.1	0.5322	13.1	12.3
0.30	0.4	0.2833	0.2	0.4828	12.8	11.4
0.60	0.8	0.1680	0.4	0.2833	11.9	10.1
0.75	1.0	0.1361	0.5	0.2441	7.5	6.4

Table 2: Membrane potential data at different mean concentration of CaCl2 with and without ergosterol at 25° ± 0.1°c.

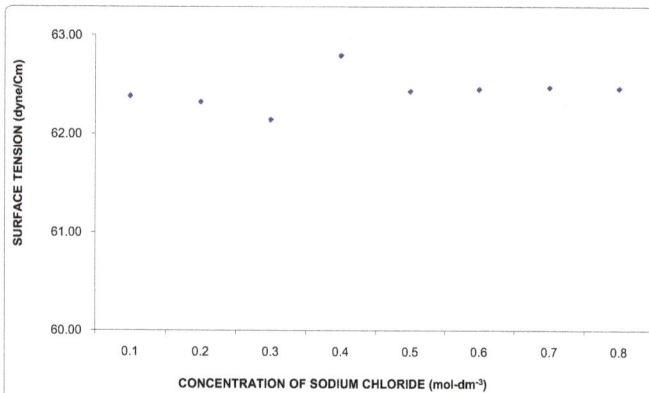

Figure 3: Variation of Surface Tension with Different Concentration of Sodium Chloride.

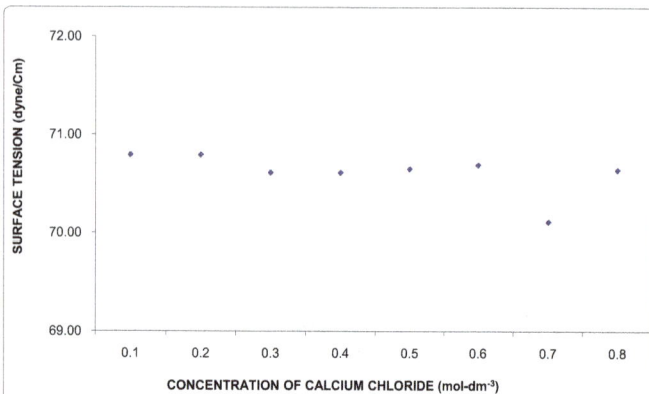

Figure 4: Variation of Surface Tension with Different Concentration of Calcium Chloride.

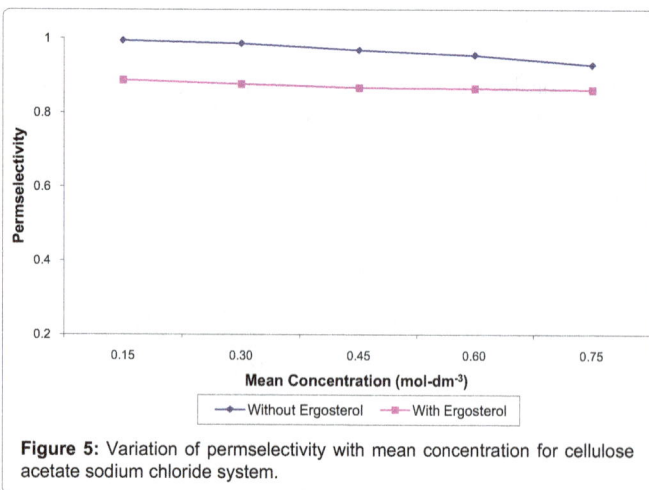

Figure 5: Variation of permselectivity with mean concentration for cellulose acetate sodium chloride system.

Activity coefficient values needed for the estimation of activity were calculated using the equation [18]

$$\text{Log } f = - AZ_+ Z_- \sqrt{I}$$

Where f = activity coefficient of the electrolyte Z_+ and Z_- are ionic charges of the cation and anion respectively and A is a constant.

In Table 3, the transport number values for calcium chloride system are given. The results show that cellulose acetate film exhibits

change in selectivity i.e. at lower mean concentration, E is greater than E_L even when cellulose acetate supports ergosterol liquid membrane, it indicates towards anion acceleration, but as the mean concentration increases, E_L becomes greater than E either with or without ergosterol liquid membrane. Thus at higher mean concentration cation is accelerated, since $t_+ > t_-$ transport number is found to increase with increasing mean concentration.

In case of NaCl, the transport number increases with decreasing pH values showing that the anion selectivity increases with lowering of pH however a decrease in transport number with increasing concentration indicates the decrease of selectivity of the film with increasing mean concentration.

For NaCl , the permselectivity and fixed charge density, both increase with decreasing pH (Table 4) but a decrease in permselectivity and increase in fixed charge density with increasing mean concentration is observed in the case of sodium chloride (Figure 5,6) as well as potassium chloride systems. It is known that cellulose acetate undergoes swelling when kept in aqueous solutions and the swelling decreases with increase in concentration. The deswelling of the cellulose acetate membrane is likely to enhance openness of its matrix

C (mol-dm⁻³)	t_+		t_+
	Without ergosterol	With ergosterol	
0.075	0.0142	0.0251	0.4041
0.15	0.1050	0.1392	0.3980
0.30	0.4601	0.4830	0.3842
0.60	0.4861	0.5060	0.3650
0.75	0.4900	0.5162	0.3520

Table 3: Transport number values at different mean concentration of $CaCl_2$ derived using eqn. (9).

pH	P_s		$\varphi \overline{X}$ (mol dm⁻³)	
	Without ergosterol	With ergosterol	Without ergosterol	With ergosterol
6.5	0.4683	0.3879	0.3580	0.2525
5.5	0.5572	0.4439	0.4458	0.2972
4.5	0.6477	0.5339	0.5474	0.3788
3.5	0.6806	0.5802	0.6223	0.4275
2.5	0.7342	0.6532	0.6789	0.5177
1.5	0.7499	0.7022	0.6803	0.5918

Table 4: Permselectivity and fixed charge density values at different pH for NaCl : C = 0.30, C_1 = 0.40, C_2 = 0.20 (mol dm⁻³).

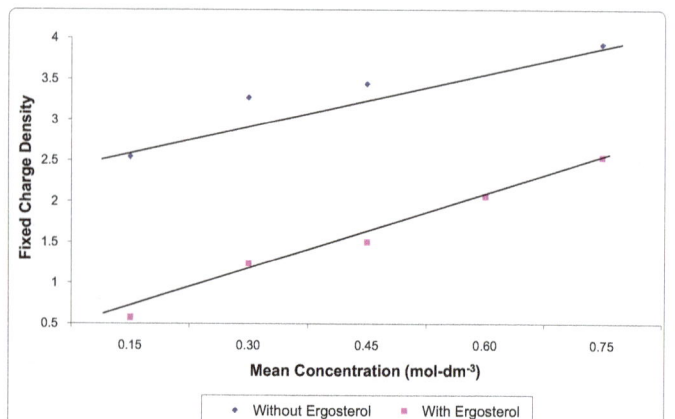

Figure 6: Variation of fixed charge density for cellulose acetate- sodium chloride system at different mean concentrations.

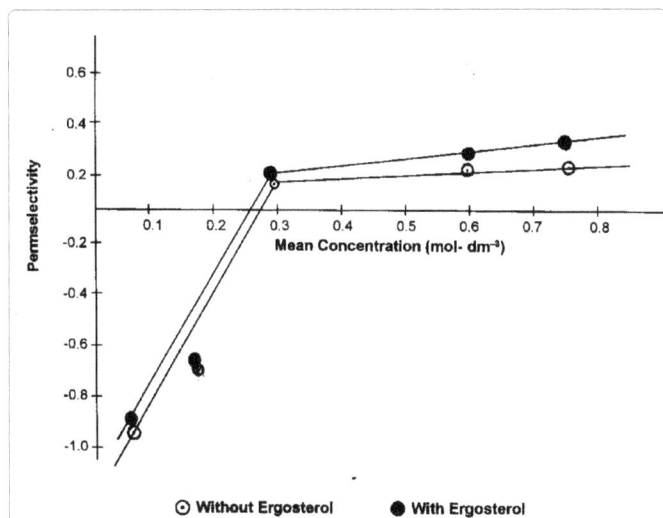

Figure 7: Variation of permselectivity with mean concentration for cellulose acetate-calcium chloride system.

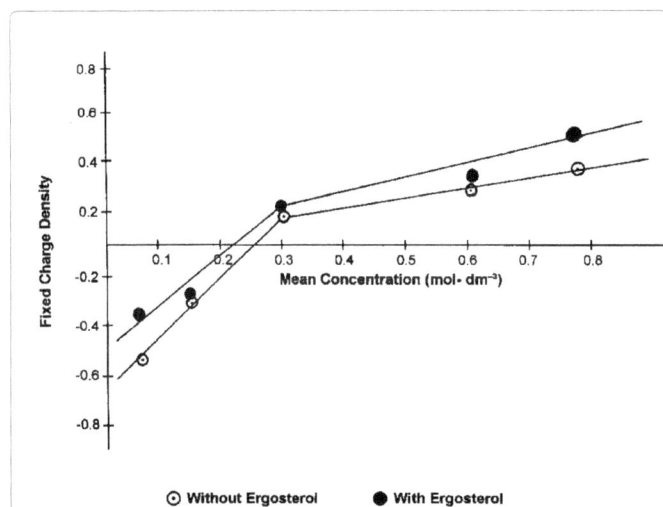

Figure 8: Variation of fixed charge density with mean concentration for cellulose acetate-calcium chloride system.

leading to reduction in permselectivity inspite of increased fixed charge density with concentrations of NaCl.

For $CaCl_2$ system, at lower mean concentration, the permselectivity values are negative and with increasing concentration the values become positive (Figure 7). These values indicate that the selectivity of the cellulose acetate membrane is changing from anion to cation. The same trend is obtained for fixed charge density (Figure 8). An increasing tendency of both the parameters ensures an increase in selectivity of cellulose acetate membrane. An increase in permselectivity and fixed charge density with increasing pH values justifies the increase in selectivity.

When a surfactant is added to a membrane, the transmission characteristics of membrane is modified due to formation of immobilized structure [19,20]. Surfactants are amphipathic compounds. They consist of both hydrophilic and hydrophobic portions in the same molecule. In the case of ergosterol, only –OH group is hydrophilic and the rest part is hydrophobic. When dispersed in aqueous solution at

low concentration, the surfactant migrates towards any available hydrophobic interface and orients its hydrophobic segment towards the interface and its hydrophilic portion in the aqueous phase [21-23]. During studies, it has been found that immobilization of the surfactant depends on many factors like pH, ionic strength, temperature and surfactant structure.

The values for various parameters in the case of ergosterol liquid membrane supported on cellulose acetate film are larger than those obtained earlier [24]. This may be attributed to the structural feature of ergosterol which contains three double bonds in its structure resulting in the presence of electron rich environment. It is also responsible for enhanced surface activity.

Acknowledgement

The author is thankful to the Head, Department of Chemistry, D.D.U. Gorakhpur University, Gorakhpur for providing laboratory facilities. Special thanks are given to Dr.M.L.Srivastava, Professor, Deptt. of Chemistry,D.D.U. Gorakhpur University, Gorakhpur.

References

1. Mohapatra PK, Manchanda VK (2003) Indian J Chem A 42: 2925-2938.

2. Reits EAJ, Neefjes JJ (2001) Nat Cell Biol 3: 145-147.

3. Kenworthy AK, Nichols BJ, Remmert CL, Hendrix GM (2004) Dynamics of putative raft-associated proteins at the cell surface. J Cell Biol 165: 735-746.

4. Gambin Y, Massiera G, Ramos L, Ligoure C, Urbach W (2005) Bounded step superdiffusion in an oriented hexagonal phase. Phys Rev Lett 94: 110602

5. Girard P, Pecreaux J, Lenoir G, Falson P, Rigaud JL, et al. (2004) A new method for the reconstitution of membrane proteins into giant unilamellar vesicles. Biophys J 87: 419-429.

6. Lee V, Petersen NO, Revington M, Dunn SD (2003) The lateral diffusion of selectively aggregated peptides in giant unilamellar vesicles. Biophys J 84: 1756 -1764.

7. Srivastava ML, Shukla NK (2003) On the electrochemical characterization of ion-selective cellulose acetate membranes with and without stigmasterol liquid membranes. J Colloid Interface Sci 267: 132-135.

8. Singh K, Tiwari AK (1987) J Membrane Sci 3: 155-163.

9. Srivastava RC, Jakhar RPS (1981) J Phys Chem 85: 1457.

10. Singh K, Tiwari AK (1987) J Colloid Interface Sci 116: 42.

11. Gershfeld NL, Pagano RE (1972) Physical chemistry of lipid films at the air-water interface. 3. The condensing effect of cholesterol. A critical examination of mixed-film studies. J Phys Chem 76: 1244-1249.

12. Singh K, Tiwari AK, Rai JP (1985) Indian J Chem 24: 825-827.

13. Kobatke Y, Kamo N (1973) Pog Polym Sci Jpn 5: 257.

14. Lakshminarayanaiah N (1969) Transport Phenomena in Membranes, 199.

15. Singh K, Shabd R, Srivastava VN (1981) Indian J Chem 20: 391.

16. Glasstone S (1974) An Introduction to Electrochemistry 122.

17. Lakshminarayanaiah N (1976) Membrane Electrodes, Academic Press 64.

18. Singh K, Tiwari AK, Dwivedi CS (2002) Indian J Chem A 41: 921-927.

19. Singh K, Shahi VK (1991) J Memb Sci 59: 27.

20. Srivastava RC, Jakhar RPS (1982) J Phys Chem 86: 1441.

21. Hiemenz PC (1986) Principles of colloid and surface chemistry (2nd Edn). Marcel Dekker Inc.

22. Hoeft CE, Zollars RL (1996) Adsorption of Single Anionic Surfactants on Hydrophobic Surfaces. J Colloid Interface Sci 177: 171-178.

23. Rosen MJ (1986) Surfactant and interfacial phenomena. Wiley 43.

24. Srivastava ML, Shukla NK (1991) Bull Chem Soc Jpn.

12

Performance Modeling and Analysis of a Hollow Fiber Membrane System

Shuang Liang[1], Haifeng Zhang[2], Yubo Zhao[1] and Lianfa Song[3]*

[1]*School of Environmental Science and Engineering, Shandong University, Jinan 250100, Shandong, PR China*
[2]*Department of Chemical Engineering, Northeast Dianli University, Jilin 132012, Jilin, PR China*
[3]*Department of Civil and Environmental Engineering, Texas Tech University, Lubbock TX 79409-41023, USA*

Abstract

A mathematical model was developed for performance of hollow fiber membrane filtration system by considering both frictional and kinetic pressure losses along the fiber. The model was solved numerically as a whole and analytically with the kinetic term omitted with constant driving pressure as the primary control parameter. Numerical simulations first demonstrated that the kinetic pressure loss in the hollow fiber was negligible compared to the frictional pressure loss for the current hollow fiber membranes. It was further demonstrated that the productivity (exit velocity) of a hollow fiber was greatly affected by the radius of the fiber. The axial velocity reached a plateau value rapidly along a fiber of a small diameter while increased linearly along a fiber of sufficiently large radius. For given membrane materials (resistance) and fiber length, an optimal diameter of the fiber can be determined to maximize exit velocity.

Keywords: Hollow fiber membrane; Mathematical modeling; System simulation; Transmembrane pressure; Exit axial velocity; Frictional pressure loss

Nomenclature:

dP_f: differential frictional pressure loss (Pa)

dP_k: differential kinetic pressure loss (Pa)

f: Darcy friction factor

L: fiber length (m)

m: mass in the differential volume in Figure 1 (m)

r: inner radius of fiber (m/s)

R_e: Reynolds number

R_m: membrane resistance (1/m)

u: average axial lumen velocity (m/s)

u_0: exit axial velocity (m/s)

u_{lim}: limiting exit axial velocity (m/s)

v: permeate flux or velocity (m/s)

v_0: permeate velocity or flux at the exit (m/s)

x: axial coordinate (m)

Introduction

Microfiltration and ultrafiltration membranes in the form of hollow fibers have many desirable advantages over other forms of membranes [1-3], such as high surface/volume ratio, lower cost in fabrication, and diversified applications, etc. In the last couple of decades, hollow fiber membranes have been growingly used in water and wastewater treatments process [4-8], especially in membrane bioreactors (MBRs) for the removal of organic contaminants from wastewater [9-11]. Obviously, there is an urgent need for a better understanding of the performance of hollow fiber membrane system and the affecting factors in order to use this promising technology more effectively and efficiently [12-15].

A hollow fiber with a dead-end operated at either constant transmembrane pressure or constant average flux mode is basically a heterogeneous filtration system, in which the key parameters vary significantly along the fiber length [14-16]. It is well established that the flow pattern and pressure field in a hollow fiber are fully governed by the Navier-Stokes and continuity equations. However, direct solution of Navier-Stokes and continuity equations remains a challenging task [13,17,18] nowadays for the performance modeling and simulation of hollow fiber membrane system. Instead, it would be more practical to develop models that determine the locally varying parameters at constant transmembrane pressure operation mode with the existing theories for tube flows combined with membrane filtration theories. Frictional pressure loss was first considered for the transmembrane pressure variation along the hollow fibers [19-24]. Lately it was realized that the axial flow in the hollow fiber was different from the common tube flow in that the flow accelerates as it flows from the dead-end to the open-end due to the addition of permeate along the fiber. Therefore, the pressure loss as a result of momentum changes might contribute to the change of local transmembrane pressure [3,25]. This impact has not been rigorously quantified and assessed in the performance modeling of hollow fiber membrane systems.

When friction of axial flow with the inner fiber wall is the main reason for the variation of local transmembrane pressure, the distribution of transmembrane pressure along the fiber was found to be governed by a second order ordinary differential equation [2,19]. An analytical solution of the governing equation was obtained by Chang and Fane [19] and Chang et al. [20,21] with the average initial permeate flux as a control variable. Although the concept of average permeate flux is practically useful in the operation of membrane process, it is unsuitable for exploration of the fundamental principles of the system. One of the

*Corresponding author: Lianfa Song, Department of Civil and Environmental Engineering, Texas Tech University, Lubbock TX 79409-41023, USA
E-mail: lianfa.song@ttu.edu

serious drawbacks of the model based on the average permeate flux is that the impact of various parameters on the performance of hollow fiber membrane system cannot be simulated with the model because the average permeate flux that should be used as an indicator of the system performance is already given as the control parameter.

As a matter of fact, the primary control parameter of a membrane process is the transmembrane pressure and the (average) permeate flux should be a natural performance indicator of a membrane system. Even in the membrane processes operated at constant permeate flux operation mode, the constant flux is obtained by adjusting the transmembrane pressure through a feed-back mechanism. Furthermore, the valid range of the average permeate flux for a hollow fiber membrane cannot be determined within the theoretical framework of the model based on average permeate flux. Therefore, there is a chance to pick up a value for the averages permeate flux that is unachievable in the membrane system. On the contrary, when the transmembrane pressure is used as the primary control parameter, the possible maximum permeate flux of a membrane system can be rigorously determined.

This present work presented an effort to address the three fundamental issues mentioned above. First, a mathematical model was developed for a hollow fiber membrane system with the transmembrane pressure as the control parameter. In addition to the frictional pressure loss, the kinetic pressure loss due to fluid acceleration inside of the fiber would be rigorously formulated and assessed. Instead the flow rate or average permeate flux of a fiber membrane, the exit axial flow velocity at the open-end of the fiber was used as the performance indicator of the fiber membrane system. Second, a numerical solution procedure was developed for the complete model with both frictional and kinetic pressure losses and an analytical solution was derived for the reduced model without the kinetic pressure term. The analytical solutions were obtained based on the constant driving pressure operation mode that is different from those based on the average flux as often found in the literature. Third, the performance of hollow fiber system was simulated with the new analytical solution under various conditions. The impacts of fiber length and radius on the performance of hollow fiber membrane system were particularly reported with many interesting findings.

Model Development

A schematic of a vertical placed hollow fiber is presented in Figure 1a. The end of the fiber on the bottom is sealed (dead-end). Permeate comes out of the open-end of the fiber on the top as a result of either applying a negative suction pressure to the lumen side or a positive pressure from outside dependent on the configuration of the membrane filtration system. In either case, the transmembrane pressure always ascribes a positive value measured from the outside to the inside of the hollow fiber. Because the fiber is always filled with water, the hydraulic head of water depth outside of the fiber has no role to play in the filtration process. The only driving force for permeate is the transmembrane pressure $\Delta P(x)$. In this sense, the orientation of the hollow fiber in either vertical direction or horizontal direction or any arbitrary direction does not affect the driving force for permeate flux.

In the hollow fiber membrane system, the fibers are commonly characterized with inner radius, fiber length, and membrane resistance. The operating parameter is the transmembrane pressure at the open-end of the fiber ΔP. The overall permeate production rate is a more important concern for hollow fiber membrane system. Therefore, the exit axial velocity at the open-end u0 is selected over the average membrane flux as the performance indicator of the fiber membrane system. Because these important parameters are defined at the open-

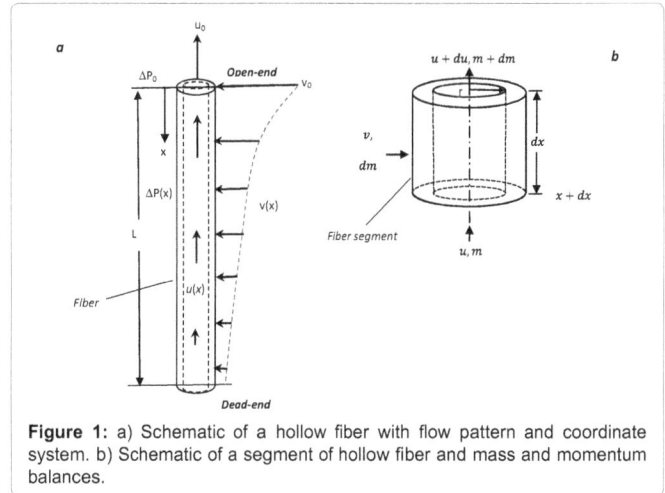

Figure 1: a) Schematic of a hollow fiber with flow pattern and coordinate system. b) Schematic of a segment of hollow fiber and mass and momentum balances.

end of the fiber, it is natural to select the open-end of fiber as the origin of coordinate x. As shown in Figure 1a, the section of the fiber that contributes more to permeate production has smaller x value in such a coordinate system. A hollow fiber membrane system is completely described by the distributions of transmembrane pressure, permeate flux, and average axial flow velocity.

Average axial velocity and permeate flux

The average axial velocity inside a hollow fiber is determined by the mass balance relationship with the permeate flux

$$u(x) = \frac{2}{r}\int_x^L v(x)dx \tag{1}$$

where x is the coordinate along the fiber starting from the open-end, $u(x)$ is the average axial velocity with the positive direction defined pointing to the open-end of the fiber, r is the fiber inner radius, $v(x)$ is the permeate flux, and L is the fiber length. At any location x, the permeate flux of the membrane is calculated with the membrane filtration equation

$$v(x) = \frac{\Delta P(x)}{\mu R_m} \tag{2}$$

where $\Delta P(x)$ is the transmembrane pressure, μ is the viscosity, and Rm is the membrane resistance. The membrane resistance is a constant along a clean fiber. It needs to point out here that both permeate velocity $v(x)$ and membrane resistance Rm is defined with respect to the inner radius of the fiber. In case that the membrane is actually on the outer shell of the fiber, the ratio of outer radius to inner radius is needed to convert these parameters into the right form.

The transmembrane pressure is usually only known at the open-end of a fiber as a given control parameter. The transmembrane pressure decreases along the hollow fiber as a function of location x and needs to be determined with fluid dynamics. In general, the pressure loss in the hollow fiber can be divided into frictional and kinetic components. The calculations of each component are detailed below separately.

Frictional pressure loss in hollow fiber

The frictional pressure loss in a small segment of the fiber as shown in Figure 1b is related to the average axial velocity with the Darcy–Weisbach equation [26,27].

$$dP_f = f\frac{\rho}{4r}u^2dx \tag{3}$$

where dP_f is the frictional pressure loss, f is the Darcy friction factor, ρ is the density of water, dx is the length of the hollow fiber segment. Darcy friction factor of laminar flow in a circular tube is given as [27,28].

$$f = \frac{64}{R_e} = \frac{32\mu}{\rho u r} \tag{4}$$

where Re is the Reynolds number, which is defined $R_e = 2\rho u r/\mu$ for the fiber, and μ is the viscosity of water.

Substituting Eq. (4) into Eq. (3) and integrating from 0 to x gives the pressure loss as

$$\Delta P_f(x) = \frac{8\mu}{r^2}\int_0^x u\,dx \tag{5}$$

where ΔP_f is the frictional pressure loss from the open-end to location x.

Kinetic pressure loss in hollow fiber

Water flow in the fiber lumen accelerates as it moves from the dead-end to the open-end because of the addition of permeate along the fiber. Energy is consumed for water movement acceleration. The pressure loss as a result of water acceleration is termed as kinetic pressure loss, which can be calculated from the principle of Newton's second law

$$\left(\pi r^2\right)dP_k = -\frac{d}{dx}\left(\frac{mu^2}{2}\right) \tag{6}$$

where dP_k is the pressure loss along the length of dx. The term $\pi r2$ is the cross sectional area of the fiber lumen and the term $mu2/2$ is the kinetic energy carried by the volume of water in the segment of fiber. Using the mass and momentum flows presented in Figure 1b, it can be found that

$$dP_k = \frac{2\rho}{r}uv\,dx \tag{7}$$

Then the total kinetic pressure loss from the open-end to location x is

$$\Delta P_k(x) = \frac{2\rho}{r}\int_0^x uv\,dx \tag{8}$$

Therefore the transmembrane pressure at location x of the fiber is determined as

$$\Delta P(x) = \Delta P - \frac{8\mu}{r^2}\int_0^x u\,dx - \frac{2\rho}{r}\int_0^x uv\,dx \tag{9}$$

where ΔP is the transmembrane pressure at the open-end of the hollow fiber membrane.

Up to now, the equations that fully govern the hollow fiber membrane system have been derived, which are Eqs. (1),(2), and (9). These three equations are highly coupled and the analytical solutions are usually difficult to find for the general case. Even for some special cases where analytical solution is available for clean membrane, they may not be valid for fouled membranes because the total hydraulic resistance cannot remain constant for the whole fiber length. Numerical solutions are much more versatile that can be used for both clean and fouled membranes.

Numerical Solution

In order to construct the numerical solution, the fiber is discretized into n equal segments with the length of $\Delta x = L/n$ each. The open-end of the fiber is designated step as $i = 0$, and grids are continuously numbered downward with $i = 1, 2, 3, \ldots$, and n for the dead-end of the fiber. The axial velocity at step n is known to be zero because it is the dead-end. In principle, the axial velocity in the previous steps can

be determined step by step backward from the step n. However, the method has a difficulty to execute because the pressure at the step n is unknown. In this study, an iteration scheme was developed to find the transmembrane pressure for the last step n. When this transmembrane pressure is known, the whole problem has been solved at the same time. The method of bisection was used to find the transmembrane pressure for the step n. The possible range of transmembrane pressure is initially the ΔP_{max} and ΔP_{min}, which are given as

$$\Delta P_{max} = \Delta P \tag{10}$$

$$\Delta P_{min} = 0 \tag{11}$$

With the method of bisection, the middle value of the pressure range is assumed the transmembrane pressure, i.e.

$$\Delta P_n = \frac{\Delta P_{max} + \Delta P_{min}}{2} \tag{12}$$

and the permeate flux is calculated as

$$v_n = \frac{\Delta P_n}{\mu R_m}$$

For $i < n$, the axial velocity, transmembrane pressure, and permeate flux are determined by

$$u_i = \frac{2\Delta x}{r}v_{i+1}$$

$$\Delta P_i = \Delta P_{i+1} + \frac{8\mu}{r^2}u\Delta x + \frac{2\rho}{r}uv\Delta x$$

$$v_i = \frac{\Delta P_i}{\mu R_m}$$

The process is repeated until $i = 0$. During the repetition, whenever $\Delta P_i > \Delta P$, the upper pressure bound ΔP_{max} is set to the newly determined ΔP_n and goes back to Eq. (12) to start over again. By the end of repetition, if ΔP_0 is equal to the driving pressure ΔP of the fiber filtration systems, the solution is reached and the process stops. Otherwise, the lower pressure bound ΔP_{min} is set to the newly determined ΔP_n and goes back to Eq. (12) to start over again. In each iteration, the domain length of the transmembrane pressure reduces by half. The calculation procedure is presented in the flow diagram shown in Figure 2. The same numerical procedure can be used directly for the fouled fibers where the total hydraulic resistance is a variable instead of a constant.

The numerical solution procedure was coded in C++ language that worked smoothly for all cases of simulations. In most case studies carried out in this study, numerical solutions were obtained in a fraction of second on a common personal computer. The detailed simulation results will be discussed later in the section of Simulations and Discussions. However, one important finding should be pointed out from the numerical solutions: the kinetic pressure loss is less than one percent of the frictional pressure loss for the typical parameters of the current hollow fibers. This finding stimulated the derivation of the analytical solution below.

Analytical Solution

When the kinetic pressure loss term is omitted from Eq. (9), the remained equation is analytically solvable. Taking derivative of both sides of Eq. (9) without the kinetic pressure loss term results in

$$\frac{d\Delta P(x)}{dx} = -\frac{8\mu}{r^2}u \tag{13}$$

Combining Eqs. (1), (2) and (13) gives

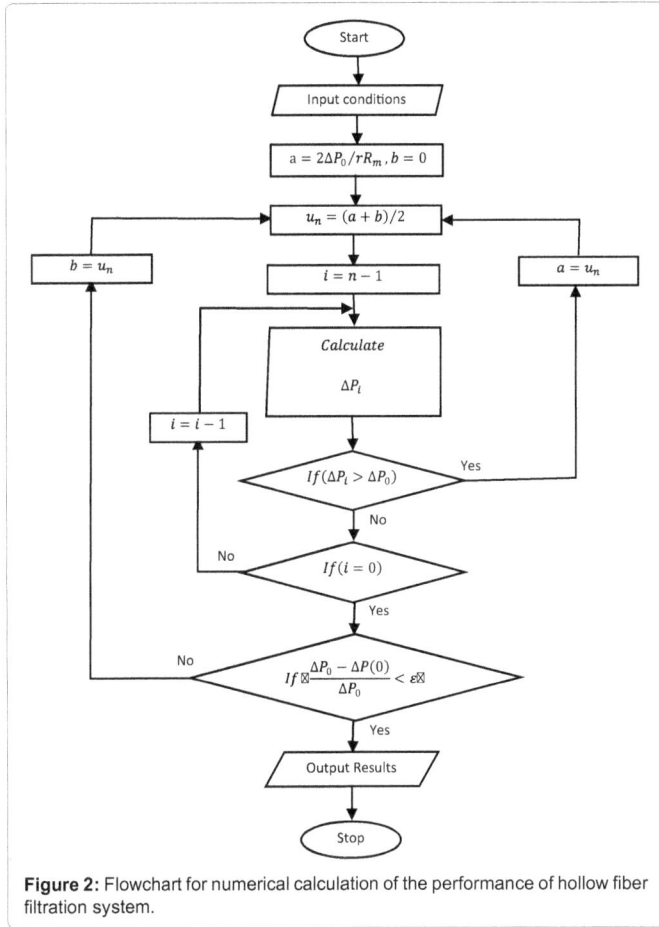

Figure 2: Flowchart for numerical calculation of the performance of hollow fiber filtration system.

$$\frac{d\Delta P(x)}{dx} = -\frac{16}{r^3 R_m} \int_x^L \Delta P(x) dx \tag{14}$$

Taking derivative of both side of Eq. (14) results in

$$\frac{d^2 \Delta P(x)}{dx^2} = \frac{16}{r^3 R_m} \Delta P(x) \tag{15}$$

The boundary conditions for the problem under consideration are

$$\Delta P(0) = \Delta P \tag{16}$$

and

$$\left. \frac{d\Delta P(x)}{dx} \right|_{x=L} = 0 \tag{17}$$

Eq. (16) means that the pressure at the open-end of the fiber is given as an operational parameter. On the other hand, Eq. (17) reflects the fact that the axial velocity in the hollow fiber is zero at the dead-end.

The solution of Eq. (15) with boundary conditions (16) and (17) is

$$\Delta P(x) = \Delta P \frac{e^{\lambda(L-x)} + e^{-\lambda(L-x)}}{e^{\lambda L} + e^{-\lambda L}} \tag{18}$$

where λ is given as

$$\lambda = \frac{4}{\sqrt{r^3 R_m}} \tag{19}$$

With the known transmembrane pressure distribution, the exit axial velocity of the fiber, which is the performance indicator of the membrane system, is readily determined. Substituting Eq. (1) into Eq. (2) and setting the lower integration limit to 0, one has the exit axial

velocity of the fiber

$$u_0 = \frac{2}{r} \int_0^L \frac{\Delta P(x)}{\mu R_m} dx \tag{20}$$

Substituting Eq. (18) for $\Delta P(x)$ in Eq. (20), the exit axial velocity is obtained

$$u_0 = v_0 \frac{\sqrt{rR_m}}{2} \frac{e^{\lambda L} - e^{-\lambda L}}{e^{\lambda L} + e^{-\lambda L}} = u_{lim} \frac{e^{\lambda L} - e^{-\lambda L}}{e^{\lambda L} + e^{-\lambda L}} \tag{21}$$

where v_0 is the initial permeate flux at $x=0$, and ulim is the limiting exit axial velocity of the fiber. The initial permeate flux and the limiting exit axial velocity are, respectively, given by

$$v_0 = \frac{\Delta P}{i R_m} \tag{22}$$

and

$$u_{lim} = \frac{\Delta P \sqrt{r}}{2 i \sqrt{R_m}} \tag{23}$$

Eq. (21) explicitly shows the impacts of various parameters on system performance. It can be directly used to analyze system performance in many cases. Eq. (23) shows that the maximum possible exit velocity is proportional to the transmembrane pressure and the square root of fiber inner radius, while inversely proportional to the water viscosity and the square root of membrane resistance.

Simulations and Discussions

In this section, the performance of the hollow fiber membrane system was simulated for various conditions. Unless other stated, the parameter values in Table 1 were used in the subsequent simulations. All the simulations were obtained with the numerical procedure. As a comparison and double check, the analytical solutions were also tried in some cases and almost exactly the same results were obtained. The choice of numerical procedure in the simulations was mainly for its easy expansion to the planned later study of the fouled fibers, for which the analytical solution will no longer be available.

Frictional and kinetic pressure losses

A numerical solution for the hollow fiber membrane system was presented in Figures 3a and 3b. The local transmembrane pressure was shown in Figure 3a with frictional and kinetic pressure losses. It can be seen that the kinetic pressure loss was very small that was practically zero in this case. The decrease on the transmembrane pressure was almost entirely caused by the frictional pressure loss. The corresponding local permeate velocity and the axial velocity were presented in Figure 3b. Similar to the transmembrane pressure, the local permeate velocity decreased from the open-end to the dead-end of the fiber. The average axial velocity always started from zero at the dead-end and increased to the maximum value at the open-end of the fiber.

Membrane resistance is reasonably anticipated to decrease with the advance in membrane materials and fabrication technology. It is of great interest to know how the decreased membrane resistance affects the frictional and kinetic pressure losses in the hollow fibers. The local transmembrane pressure and the frictional and kinetic pressure losses

Parameter	Unit	Value
Inner Radius	m	0.2×10^{-3}
Fiber length	m	3.0
Membrane resistance	1/m	1.12×10^{12}
Waterviscosity	Pa·s	0.89×10^{-3}
Pressure	Pa	0.5×10^5

Table 1: Default parameter values used in numerical simulations.

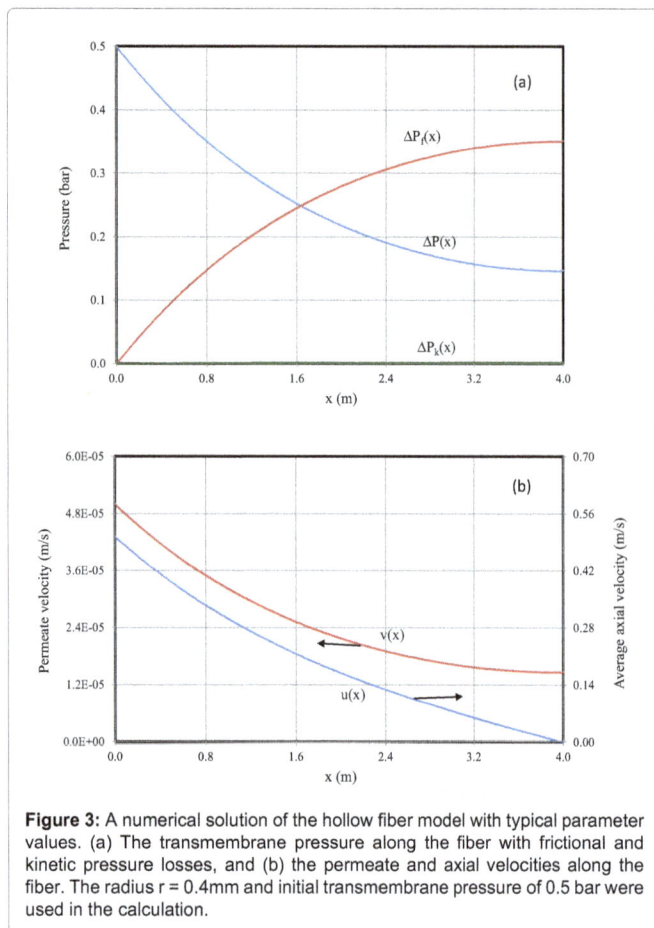

Figure 3: A numerical solution of the hollow fiber model with typical parameter values. (a) The transmembrane pressure along the fiber with frictional and kinetic pressure losses, and (b) the permeate and axial velocities along the fiber. The radius r = 0.4mm and initial transmembrane pressure of 0.5 bar were used in the calculation.

were simulated and shown in Figures 4a-4c for membrane resistance of 1.12×10^{11} 1/m, 5.62×10^{10} 1/m, and 1.12×10^{10} 1/m, respectively. It can be seen that the kinetic pressure loss is no longer negligible for the lower membrane resistances. Figure 4c shows that the frictional and kinetic pressure losses are comparable in magnitude when the membrane resistance is 1.12×10^{10} 1/m. For all the three membrane resistances used in the simulations, the kinetic pressure losses have to be considered for the sake of higher accuracy.

Performance of hollow fiber filtration system

For given membrane materials and fabrication technology of hollow fibers, the fiber length and radius are two important parameters of hollow fibers. The knowledge of their impacts on system performance is of paramount importance to the design of membrane system. In this section, the impacts of the fiber length and radius on the system performance (as represented by axial exit velocity) were studied under various conditions.

The impact of fiber length on the exit velocity for different membrane resistances was presented in the Figure 5. The resistance decreases with the increasing number on the graph. It can be seen that the exit velocity generally increases with fiber length. However, the increasing rate decreases as fiber length increases. The curves 1 to 3 in the figure represent the cases for the membrane resistance range of current hollow fibers. Curves 4 and 5 in the figure were produced with the membrane resistances of one third or one fifth of the typical resistance for current fibers as used to produce curve 2. It can be seen from curves 4 and 5 that

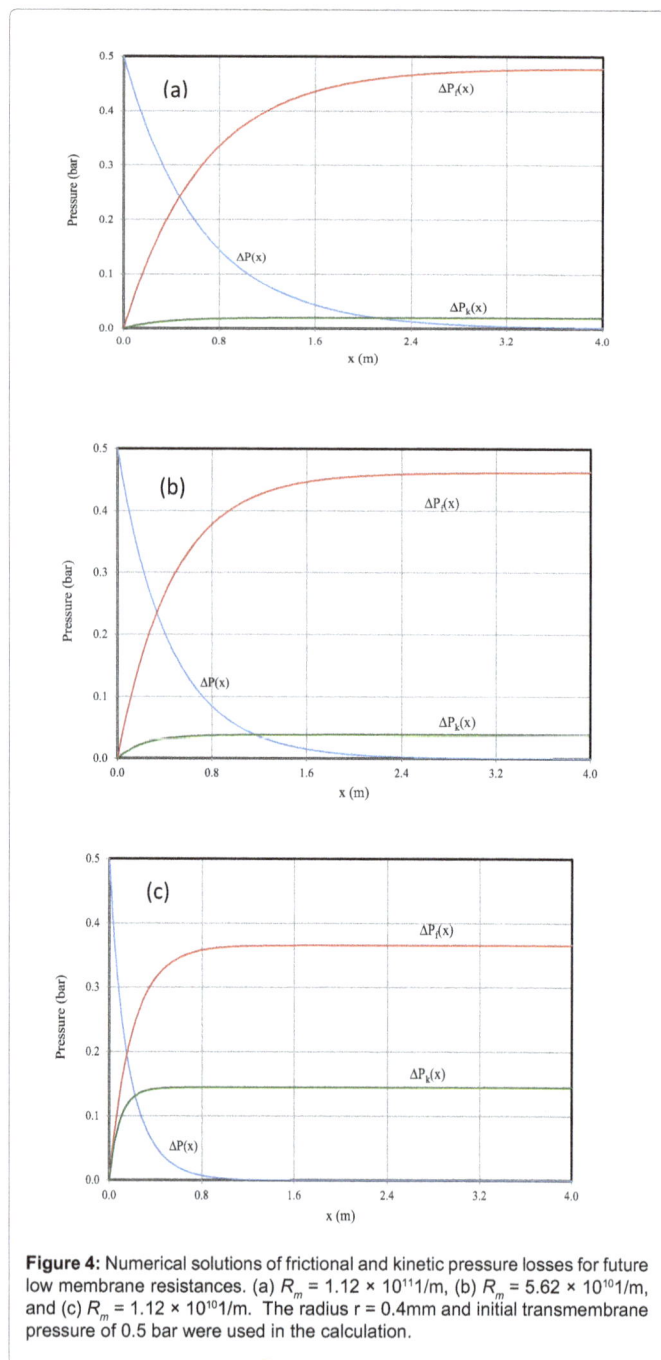

Figure 4: Numerical solutions of frictional and kinetic pressure losses for future low membrane resistances. (a) $R_m = 1.12 \times 10^{11}$ 1/m, (b) $R_m = 5.62 \times 10^{10}$ 1/m, and (c) $R_m = 1.12 \times 10^{10}$ 1/m. The radius r = 0.4mm and initial transmembrane pressure of 0.5 bar were used in the calculation.

the exit velocities of the fiber no longer increase with fiber length when the length exceeds certain values. The limiting exit velocities calculated with Eq. (23) were presented in Table 2. The analytical solution can produce exactly the same exit velocities for all the membrane resistances. However, the numerical simulations showed the value of fiber length at which the limiting exit velocities could be obtained. Similar impact of fiber length on the exit velocity of the fiber can be seen for different initial transmembrane pressures as presented in Figure 6. The numbers on the curves indicated increasing transmembrane pressure used in the simulations. For all the transmembrane pressures simulated, the exit velocity keeps increasing with the fiber length.

Figure 7 presented profiles of the exit velocity as functions of fiber

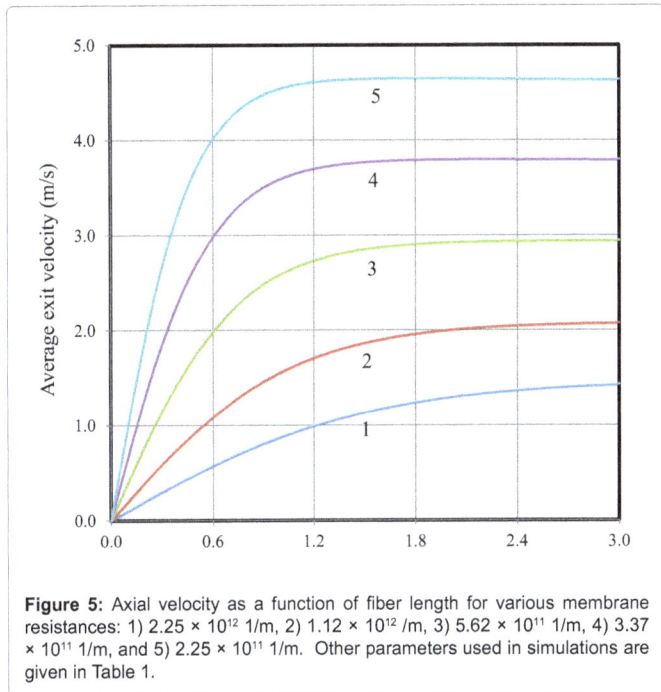

Figure 5: Axial velocity as a function of fiber length for various membrane resistances: 1) 2.25 × 10^{12} 1/m, 2) 1.12 × 10^{12} /m, 3) 5.62 × 10^{11} 1/m, 4) 3.37 × 10^{11} 1/m, and 5) 2.25 × 10^{11} 1/m. Other parameters used in simulations are given in Table 1.

Resistance , 1/m	Limiting velocity, m/s	Note
2.25 × 10^{12}	1.48	Curve 1 in Figure 5
1.12 × 10^{12}	2.10	Curve 2 in Figure 5
5.62 × 10^{11}	2.96	Curve 3 in Figure 5
3.37 × 10^{11}	3.82	Curve 4 in Figure 5
2.25 × 10^{11}	4.68	Curve 5 in Figure 5

Table 2: Limiting exit velocity of hollow fiber.

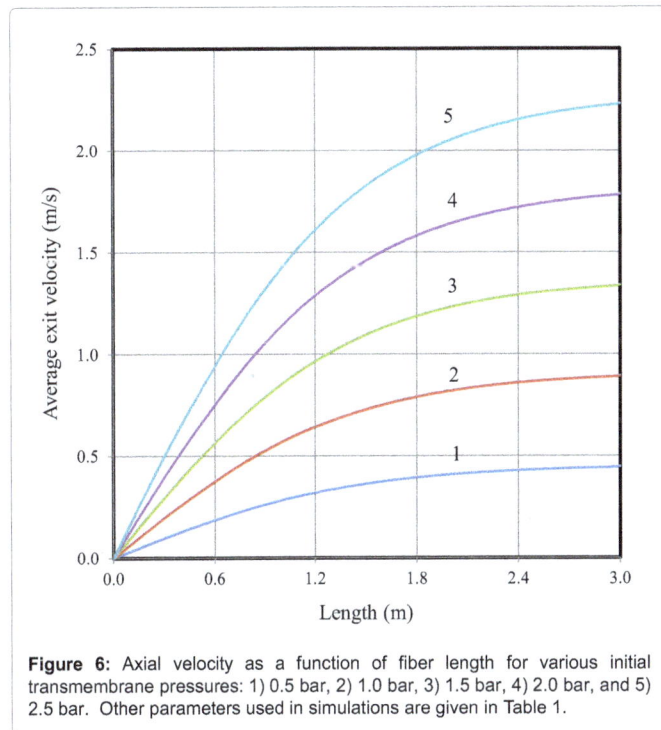

Figure 6: Axial velocity as a function of fiber length for various initial transmembrane pressures: 1) 0.5 bar, 2) 1.0 bar, 3) 1.5 bar, 4) 2.0 bar, and 5) 2.5 bar. Other parameters used in simulations are given in Table 1.

on the performance of the fiber membrane. The numbers on the curves in the figure indicate simulation results for increasing fiber radius. Instead of monotonic trend as for resistance and pressure, the impact of fiber radius on performance is more complicated and interesting. The hollow fiber of smaller radius produces more permeates (higher exit velocity) for shorter fiber but reaches a plateau value sooner as fiber length increases. It can be seen on Figure 7 that the hollow fiber with the radius of 0.15 mm (curve 1) produces the most permeate for fiber length up to 0.85 m. However, the maximum axial velocity (ca. 1.6 m/s)

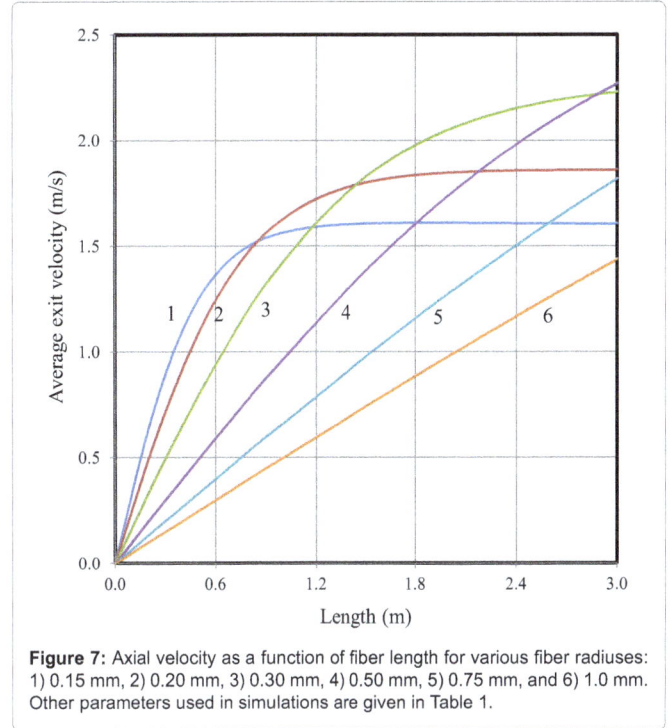

Figure 7: Axial velocity as a function of fiber length for various fiber radiuses: 1) 0.15 mm, 2) 0.20 mm, 3) 0.30 mm, 4) 0.50 mm, 5) 0.75 mm, and 6) 1.0 mm. Other parameters used in simulations are given in Table 1.

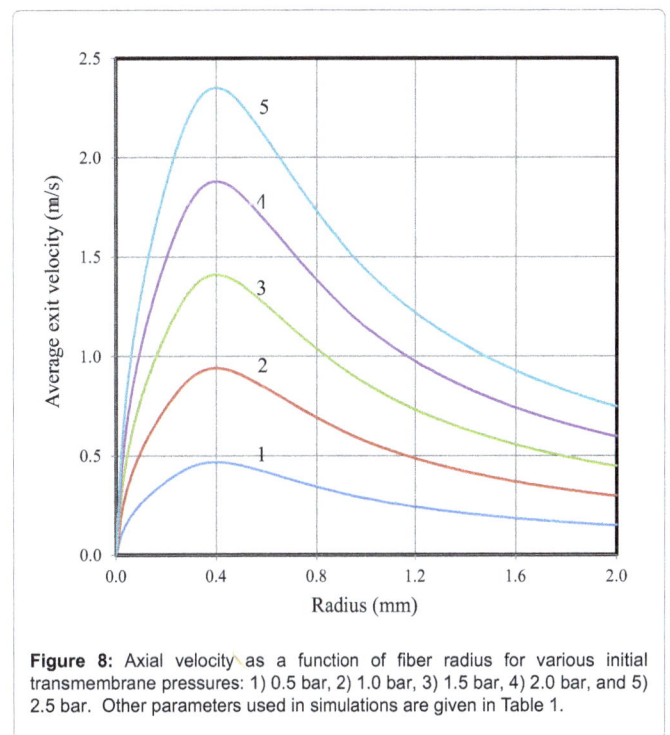

Figure 8: Axial velocity as a function of fiber radius for various initial transmembrane pressures: 1) 0.5 bar, 2) 1.0 bar, 3) 1.5 bar, 4) 2.0 bar, and 5) 2.5 bar. Other parameters used in simulations are given in Table 1.

length for different fiber radii. Comparing to resistance and pressure, the radius of hollow fiber is by far a much more influential parameter

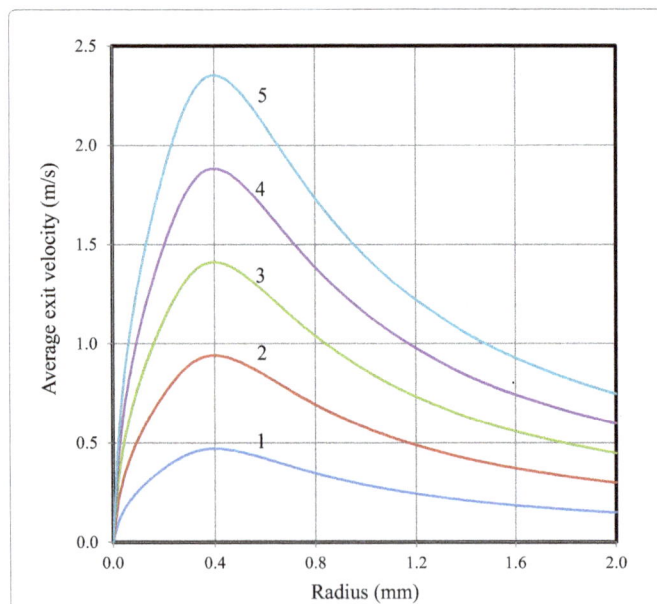

Figure 9: Axial velocity as a function of fiber radius for various membrane resistances: 1) 2.25×10^{12} 1/m, 2) 1.12×10^{12} /m, 3) 5.62×10^{11} 1/m, 4) 3.37×10^{11} 1/m, and 5) 2.25×10^{11} 1/m. Other parameters used in simulations are given in Table 1.

for the fiber is reached at the length of 1.8 m. The fiber with radius of 0.20 mm (curve 2) produces the most permeate for the fiber length between about 0.85 m to 1.40 m, where it is overtaken by the fiber of radius 0.30 mm (curve 3). Curves 5 and 6 represented radius of 0.50 mm and 1.0 mm and the radii are too large for the conditions used in the simulations. They cannot produce the highest exit velocity in the whole range of fiber length simulated.

The impact of fiber radius on exit velocities was presented in Figures 8 and 9 for groups of pressures and resistances, respectively. It can be seen that there is a peak value of exit velocity at certain radius for each curve, which do not appear in the previous figures. Figure 8 shows that transmembrane pressure does not affect the radius value at which the peak exit velocity occurs although the peak velocity is strongly affected by pressure. Figure 9, on the contrary, shows that the decrease in the membrane resistance postpones the occurrence of peak exit velocity to a larger radius. It suggests that fibers with larger radius should be made to accommodate the membranes with lower resistance. It is possible to mathematically find the optimal radius for given sets of parameters by taking derivative of Eq. (13) and setting it to zero. However, the simulations of the performance of the hollow fiber as a function of radius can provide a direct graphical and vivid presentation of the result that cannot be matched by the single value from mathematical determination.

Conclusions

A model for the performance of a hollow fiber membrane system was developed by rigorously considering both frictional and kinetic pressure losses along the fiber. By numerical simulations of the complete model, it was demonstrated that the kinetic pressure loss is much smaller than the frictional pressure loss for the current hollow fiber membranes. Therefore, the performance of the hollow fiber membrane systems can be reasonably modeled by considering frictional pressure loss only. The analytical solution for the reduced model was derived

with the driving pressure as the primary control parameter. Simulations showed that there exists a maximum exit axial velocity (correspondingly a maximum average permeate flux) with respect to driving pressure for given set of parameters. For given membrane materials (resistance) and fiber length, an optimal radius of the fiber can be determined to maximize axial velocity. The interesting simulations findings cannot be done with the model derived based on average permeate flux as a control parameter.

References

1. Baker RW (2004) Membrane Technology and Applications. (2ndedn), John Wiley & Sons Ltd., Chichester.

2. Carroll T (2001) The effect of cake and fibre properties on flux declines in hollow-fibre microfiltration membranes. J Membr Sci 189: 167-178.

3. Yeh HM (2009) Exponential model analysis of permeate flux for ultrafiltration in hollow-fiber modules by momentum balance. Chem Eng J 147: 202-209.

4. Keller AA, Bierwagen BG (2001) Hydrophobic hollow fiber membranes for treating MTBE-contaminated water. Environ Sci Technol 35: 1875-1879.

5. Guo H, Wyart Y, Perot J, Nauleau F, Moulin P (2010) Low-pressure membrane integrity tests for drinking water treatment: A review. Water Res 44: 41- 57.

6. Sun SP, Hatton TA, Chung TS (2011) Hyperbranched polyethyleneimine induced cross-linking of polyamide-imide nanofiltration hollow fiber membranes for effective removal of ciprofloxacin. Environ Sci Technol 45: 4003-4009.

7. Xiao P, Xiao F, Wang DS, Qin T, He SP (2012) Investigation of organic foulants behavior on hollow-fiber UF membranes in a drinking water treatment plant. Sep Purif Technol 95: 109-117.

8. Liu TY, Zhang RX, Li Q, Van der Bruggen B, Wang XL (2014) Fabrication of a novel dual-layer (PES/PVDF) hollow fiber ultrafiltration membrane for wastewater treatment. J Membr Sci 472: 119-132.

9. Aziz CE, Fitch MW, Linquist LK, Pressman JG, Georgiou G, et al. (1995) Methanotrophic biodegradation of trichloroethylene in a hollow-fiber membrane bioreactor. Environ Sci Technol 29: 2574-2583.

10. Fenu A, De Wilde W, Gaertner M, Weemaesde M, Gueldre G, et al. (2012) Elaborating the membrane life concept in a full scale hollow-fibers MBR. J Membr Sci 421: 349-354.

11. Robles A, Ruano MV, Ribes J, Ferrer J (2013) Factors that affect the permeability of commercial hollow-fibre membranes in a submerged anaerobic MBR (HF-SAnMBR) system. Water Res 47: 1277-1288.

12. Chellam S, Jacangelo JG, Bonacquisti TP (1998) Modeling and experimental verification of pilot-scale hollow fiber, direct flow microfiltration with periodic backwashing. Environ Sci Technol 32: 75-81.

13. Kim AS, Lee YT (2011) Laminar flow with injection through a long dead-end cylindrical porous tube: Application to a hollow fiber membrane. AIChE J 57: 1997-2006.

14. Serra C, Clifton MJ, Moulin P, Rouch JC, Aptel P. (1998) Dead-end ultrafiltration in hollow fiber modules: Module design and process simulation. J Membr Sci 145: 159-172.

15. Yoon SH, Lee S, Yeom IT (2008) Experimental verification of pressure drop models in hollow fiber membrane. J Membr Sci 310: 7-12.

16. Li X, Li J, Wang J, Wang H, He B, et al. (2014) Experimental investigation of local flux distribution and fouling behavior in double-end and dead-end submerged hollow fiber membrane modules. J Membr Sci 453: 18-26.

17. Ghidossi R, Daurelle J V, Veyret D, Moulin P (2006) Simplified CFD approach of a hollow fiber ultrafiltration system. Chem Eng J 123: 117-125.

18. Ghidossi R, Veyret D, Moulin P (2006) Computational fluid dynamics applied to membranes: State of the art and opportunities. Chem Eng Process 45: 437-454.

19. Chang S, Fane AG (2001) The effect of fibre diameter on filtration and flux distribution – relevance to submerged hollow fibre modules. J Membr Sci 184: 221-231.

20. Chang S, Fane AG, Vigneswaran S (2002) Modeling and optimizing submerged hollow fiber membrane modules. AIChE J 48: 2203-2212.

21. Chang S, Fane AG, Waite TD (2006) Analysis of constant permeate flow filtration using dead-end hollow fiber membranes. J Membr Sci 268: 132-141.

22. Lee S, Park PK, Kim JH, Yeon KM, Lee CH (2008) Analysis of filtration characteristics in submerged microfiltration for drinking water treatment. Water Res 42: 3109-3121.

23. Liu L, Ding Z, Lu Y, Ma R (2010) Modeling the bubbling enhanced microfiltration for submerged hollow fiber membrane module. Desalination 256: 77-83.

24. Yoon SH, Kim HS, Yeom IT (2004) Optimization model of submerged hollow fiber membrane modules. J Membr Sci 234: 147-156.

25. Lee HM, Lee CH, Chung KY, Lee S (2004) Prediction model for transmembrane pressure in a submerged hollow-fiber microfiltration membrane. Separ Sci Technol 39: 1833-1856.

26. Azarbadegan A, Eames I, Sharma S, Cass A (2011) Computational study of parallel valveless micropumps. Sensor Actuator B 158: 432-440.

27. Romeo E, Royo C, Monzón A (2002) Improved explicit equations for estimation of the friction factor in rough and smooth pipes. Chem Eng J 86: 369-374.

28. Chen NH (1979) An explicit equation for friction factor in pipe. Ind Eng Chem Fundam 18: 296-297.

13

Evaluation of Step Resistance in Multilayered Ceramic-Supported Pd-Based Membranes for Hydrogen Purification in the Presence of Concentration Polarisation

Caravella A[1]* and Sun Y[2]

[1]*Department of Environmental and Chemical Engineering (DIATIC), University of Calabria, Via P. Bucci, Rende (CS), Italy*

[2]*Department of Materials Engineering, Hanyang Univeristy, Ansan, Gyeonggi-do, South Korea*

Abstract

In this work, a systematic approach is used to quantify the single-step influence in composite Pd-based membranes on hydrogen permeation in the presence of concentration polarisation. To perform this study, an already developed permeation model is applied to a membrane supported on a five-layered asymmetric porous support. The results are presented in terms of both single-layer influence (calculated using an expression involving the permeation limiting fluxes) and the here introduced Support Resistance Coefficient, SRC, which is a coefficient measuring quantitatively the extent of the driving force in the entire support, analogously to what done for the definition of the Concentration Polarisation Coefficient, CPC. Analysing the membrane behaviour in different conditions of temperature, total feed pressure and Pd-layer thickness, it is eventually shows that, the presence of polarization determine a decreasing effect of the porous support in the considered configuration, i.e., with the selective layer placed on the high-pressure side and the support placed on the permeation one. This conclusion indicates that, for sufficiently thin metal layers, the hydrogen permeation is mostly influenced by concentration polarisation and, thus, the fluid dynamic conditions in the upstream side become a crucial parameter to optimise.

Keywords: Porous layers; Concentration polarisation; Palladium; Membranes; Hydrogen

Introduction

The permeation properties of composite membranes and thin films deposited on appropriate substrates and supports nowadays induce such materials to be used for a number of different industrial applications. Furthermore, multilayered structures are useful to prevent inter-diffusion between selective layers and support [1-4]. Such an interest has been boosting the development of enhanced fabrication and characterisation techniques (see, for example, refs [5-15]) adopted to optimise the membrane structure maximising the permeating flux and, dually, minimising the overall mass transfer resistance.

Zhang et al. [7] prepared a 5 micron-thick membrane with a selectivity higher than 3000, obtaining slightly lower performance than Itoh et al. [6], who deposited a thin PdAg layer of 2-4 mm with a good measured H_2/N_2 selectivity exceeding 5000, and Dittmar et al. [12], who prepared a 10-13 micron-thick membrane with variable selectivity (700-10000). Tong et al. [3] deposited a metal layer of 5 micron with virtually infinite selectivity, similarly to what done by Li et al. [5] and the research groups in SINTEF, who developed a method to sputter 2-3 mm-thick Pd-based layers on porous substrates [8,9].

Examples of ultra-thin membranes were provided by Lim et al. [10], with a 0.16 micron-thick membrane with a selectivity of around 710, and Yun et al. [11], with a membrane as thick as 1 micron within a selectivity range of 3000-9000. Furthermore, a research group of TECNALIA has recently developed a systematic methodology to prepare stable and 4-5 mm-thick selective supported Pd-based membranes [15,16]. A more exhaustive state-of-the-art on preparation and characterization of thin Pd-based membranes can be found in the recent review of Gallucci et al. [17].

Because of the high flux allowed by such thin membranes, the evaluation of the influence in the single permeation steps has been becoming progressively more important. These steps commonly include both kinetic phenomena and mass transfer, which are in general paired to each other eventually determining the overall membrane performance [18-22].

Therefore, for design purposes, a deep knowledge of the transport mechanisms involved in membranes and thin layers is required; this implying an appropriate mathematical model of the mass transfer involved in the single permeation steps.

In the particular case of hydrogen purification using membranes composed of Pd-alloy thin layers deposited on ceramic supports – which can be both symmetric and asymmetric multilayered ones – the effect of the meso-porous structure of the intermediate and top-layers can be significant, as intermediate layers and top-layer usually have a meso-structure characterised by a relatively low mean pore diameter (within [5-50] nm ca.23-29].

Furthermore, an additional permeation resistance affecting the actual membrane separation systems – and in particular the Pd-based membrane devices – is the external mass transfer resistance in the upstream mixture side, commonly named as concentration polarisation.

In the past, this phenomenon was considered not to be important

*Corresponding author: Alessio Caravella, Department of Environmental and Chemical Engineering (DIATIC), University of Calabria, Italy
E-mail: alessio.caravella@unical.it

in gas separation systems, as the diffusion coefficients of gases are around four orders of magnitude higher than those of liquids [30-32]. However, this hypothesis is strictly valid just for sufficiently thick selective layers providing relatively low permeating flux and, thus, with the above mentioned improvements in membrane fabrication, the state-of-the-art selective layers are sufficiently thin to provide a relatively high flux, causing the mass transport in the selective layer not to be the only permeation-determining step. Hence, a certain effect of concentration polarisation is expected in these conditions. For this reason, more complex approaches are needed to identify the permeation limiting steps, as the analysis of the support influence on permeation cannot leave aside the effect of the external mass transfer resistance, which has a direct influence on the extent of the support effect as well.

Pioneers in dividing the hydrogen permeation into several elementary steps were Ward and Dao in their modelling work [18], which involves external mass transfer, adsorption/desorption, absorption/de-absorption and internal diffusion. Later, several authors started from the Ward and Dao approach to develop more complex permeation models involving additional steps, like transport in the porous supports [20,22], concentration polarization [33-38] and external mass transfer based on a multicomponent film theory [20,21], and inhibition effects [39-42]. However, there is a lack in the existing modelling approaches to systematically relate to each other the number of mass transport phenomena playing a role in a metal membrane.

In this context, the aim of this paper is to provide a systematic way to evaluate the effect of each elementary step on permeation, showing also the mutual relationship between concentration polarisation, transport in the support and diffusion through the selective metal layer.

Description of the System

The asymmetric multilayered Pd-based membrane considered in this study (Figure 1) is similar to that considered in Caravella et al. [22]. Beside the geometrical properties of the support (Table 1), the main difference from that work is that in this paper the effect of concentration polarisation is also taken into account. For this purpose, the mixture side, which is supposed to be placed on the Pd-based layer side, is considered to be composed of four species, i.e., H_2, N_2, Ar and H_2O, whereas the pure-hydrogen side is on the support side.

With these hypotheses, the hydrogen permeation is composed of three steps: mass transport in the feed film, diffusion through the Pd-based layer and transport through the layers of the porous supports.

In the present investigation, the hydrogen content is varied independently keeping the composition ratio among the other species

Layer	Thickness, mm	Mean Pore Diameter, nm	Porosity, -	Tortuosity, -
1 (Top-Layer, Silica)	10	5	0.25	5
2 (g-alumina)	40	50	0.35	5
3 (a-alumina)	20	250	0.35	5
4 (a-alumina)	700	500	0.50	3
5 (a-alumina)	720	500	0.50	3

Table 1: Geometrical characteristics of the considered alumina porous support.

Side	Composition, %				Total Pressure	T$_{Inlet}$
	H$_2$	CO$_2$	H$_2$O	CO	kPa	°C
Feed	48.5	4.9	36.9	9.7	[400, 600, 800]	[300:20:400]
Sweep	100	-	-	-	[10, 20, 50]	Same as feed

ID^{Shell} = 1.2 cm, OD^{Mem} = 1 cm, d^{Shell} = 1 mm, d^{Mem} = 5 mm

Table 2: Operating conditions considered for simulation.

constant and equal to the unity. The other operating conditions are reported in Table 2.

The multicomponent-based permeation model already introduced elsewhere is used for calculation, using the permeation properties (permeability and solubility) of a membrane characterised in a previous work accounting for the non-ideal internal hydrogen diffusion in the Pd-based layer [21,22].

Mathematical Approach

The objective of the present investigation is to provide a systematic way to evaluate the influence of the mass transfer in the porous support on the permeation process. For this purpose, two different ways of measuring the support layer effect on permeation are used.

First, the influence of the generic jth permeation step (a_j) is evaluated by using the concept of limiting fluxes [22,43] (Section 4.1), reminding that the limiting flux of a particular permeation step is calculated at certain operating conditions by considering that step as the only rate determining one and all the others as they were at the equilibrium. The calculation of a_j is based on the following expression (Equation 1):

$$\alpha_j \equiv \frac{\dfrac{1}{J_{H_2,Lim}^j}}{\displaystyle\sum_{k=1}^{m}\dfrac{1}{J_{H_2,Lim}^k}}, \quad \sum_{j=1}^{m}\alpha_j = 1 \tag{1}$$

Where m is the number of permeation steps considered, which in this work is equal to 3 (i.e., related to external mass transfer in the feed, transport in the Pd-based layer and transport in the porous support). Equation 1 practically states that the influence of each step is given by the ratio of the limiting flux inverse of the jth step divided by the sum of all the limiting flux inverses. It is remarked here that, in the particular case where all driving forces were of the same type, a_j would coincide with the more conventional resistance evaluation.

Such an approach is necessary because, in general, each permeation step is characterised by a different type of driving force (Sieverts' one, linear, quadratic and so on) and, thus, the single step resistance cannot be simply evaluated by dividing the single driving forces by flux (= inverse of permeance), as the resulting resistances would have different units and, thus, would not be comparable to each other.

In parallel to the step influence, and analogously to what done in defining an appropriate Concentration Polarisation Coefficient, CPC [35], a convenient Support Resistance Coefficient, SRC, is here

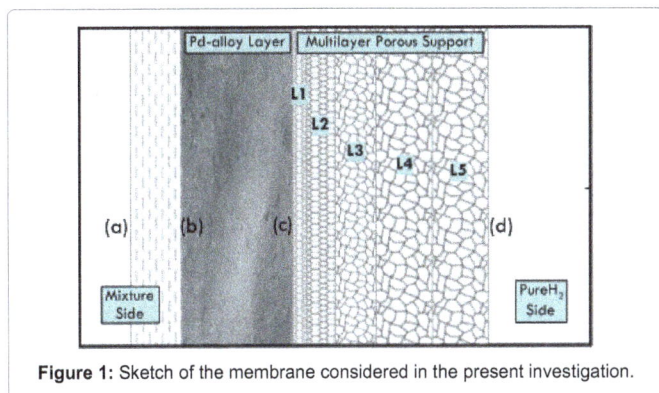

Figure 1: Sketch of the membrane considered in the present investigation.

introduced to evaluate the extent of the mass transfer resistance in the support with respect to the total resistance offered by the whole supported membrane (Section 4.2). The general definition of this coefficient for each ith species is provided as follows (Equation 2):

$$SRC_i = \frac{DrivingForce_i\big|_{Support}}{DrivingForce_i\big|_{Total}} \quad (2)$$

Where the subscripts "Support" and "Total" indicate the driving force within the support and within the whole supported membrane, respectively. In the case of Pd-based membranes, only hydrogen can basically pass through and, thus, just a single SRC is needed (Equation 3):

$$SRC = \frac{DrivingForce\big|_{Support}}{DrivingForce\big|_{Total}} \quad (3)$$

The form of the characteristic driving force can be conveniently chosen based on the operating conditions, even though in most cases it is convenient to choose the Sieverts one [35], as done in the present investigation. Therefore, with reference to the notation indicated in Figure 1, Equation 3 is made explicit as reported in Equation 4:

$$SRC = \frac{\left[\sqrt{P_{H_2}^{(c)}} - \sqrt{P_{H_2}^{(d)}}\right]}{\left[\sqrt{P_{H_2}^{(a)}} - \sqrt{P_{H_2}^{(d)}}\right]} \quad (4)$$

However, as also remarked elsewhere [38], it could be convenient sometimes to choose a different driving force, such as, for example, the pressure difference or another difference of pressure functionality taking into account the non-ideality of internal diffusion in the metal lattice in terms of a pressure exponent different from 0.5 [22, 44-46].

Regarding its definition, SRC is defined in such a way that it is close to zero for negligible influence of the support, whereas it is close to the unity for support-controlled permeation. The analogy of this coefficient to CPC is clear, although there is a mathematical difference between them arising from physical reasons. In fact, in case of pure hydrogen, there is no external mass transfer resistance in the film side and, thus, there is no concentration polarisation, i.e., CPC = 0.

Differently, in the porous support there is a concentration drop owing to the effect of at least Knudsen and viscous flow, which are effective in decreasing the permeation flux even under pure hydrogen conditions. As a result, the SRC value cannot be completely zero. In this particular case, where pure hydrogen is considered in the permeation side, just Knudsen and viscous flow take place.

Before going through results and discussion of the present paper, it is useful to remark the difference between the pieces of information coming from the values of the support influence and SRC. In fact, the former provides the actual contribution of permeation steps to the overall permeation process, regardless of the particular governing driving force.

On the contrary, the latter provides the driving force loss owing to the support, which implies a choice of a characteristic driving force, as already pointed out above. In doing that, it must be considered that the farther the chosen characteristic driving force is from the actual governing one, the more sensitive SRC is to such a choice.

Results and Discussion

The subsequent sub-sections report the analysis concerning the influence of permeation steps and the support resistance coefficient under different working conditions. Although the reported quantitative

results are specific for the considered conditions, the qualitative trends are general and can be studied to eventually maximise the permeating flux, which is the final objective of a membrane designer.

Permeation step influence

Figure 2 shows the calculated hydrogen profiles through membrane for different values of hydrogen mole fraction in the feed considering a total feed pressure of 1000 kPa and a temperature of 300°C. Two different cases are investigated, i.e., (a) in absence and (b) in the presence of external mass transfer resistance on the feed side (concentration polarisation).

In the former case, two permeation steps are active, i.e., Pd-alloy layer and porous support, whereas in the latter one, the external mass transfer on the feed side is additionally considered.

Considering the case without polarisation (Figure 2a), it can be observed that the majority of partial pressure drop is in the Pd-based layer, although there is a non-negligible drop in the porous support, mostly concentrated in the first two porous layers. This means that, under the considered conditions, the internal diffusion in the metal selective layer is not the only rate-determining step of the overall permeation.

Furthermore, the hydrogen profile through the metal layer is generally more than linear due to the "non-ideality" of the diffusion coefficient in the Pd-based layer, as diffusivity is an increasing function of hydrogen concentration in the conditions of interest of the present paper [44-48]. In light of such functionality, the profile non-linearity is progressively more pronounced as the hydrogen concentration increases.

This trend is shown by the dashed lines reported in Figure 2, each of which have the slope equal to that of the true profile calculated at the permeation side (low hydrogen pressure).

As for the values of the hydrogen mole fraction shown in Figure 2,

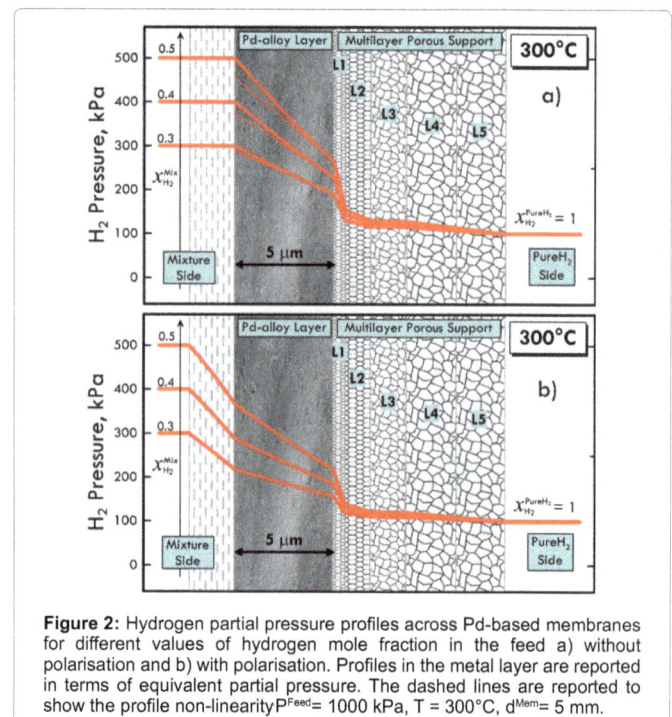

Figure 2: Hydrogen partial pressure profiles across Pd-based membranes for different values of hydrogen mole fraction in the feed a) without polarisation and b) with polarisation. Profiles in the metal layer are reported in terms of equivalent partial pressure. The dashed lines are reported to show the profile non-linearity P^{Feed}= 1000 kPa, T = 300°C, d^{Mem}= 5 mm.

the profile non-linearity is more evident at a hydrogen mole fraction of 0.5, whilst the profiles corresponding to 0.4 and 0.3 show trends that are progressively more linear. This is due to the fact that a higher hydrogen concentration in the lattice is shown to experimentally favour the hydrogen internal diffusion, at least within a concentration range where the lattice-hydrogen interactions are dominant on the hydrogen-hydrogen ones [47].

As for the profiles in the support layer, in pure hydrogen conditions at a fixed permeate pressure, a second order functionality of the profiles with the hydrogen partial pressure is theoretically found because of the presence of both Knudsen diffusion (linear along the support) and viscous flow (quadratic trend). However, since the contribution of the viscous flow is really small in the considered conditions, the profiles in the support layers can be considered practically linear.

The situation depicted in Figure 2b is slightly different. In fact, the presence of external mass transfer resistance on the feed side causes a hydrogen partial pressure drop. As a consequence, the hydrogen concentration in the metal lattice is lower with respect to the case in absence of mass transfer resistance, this implying more linear profiles in the Pd-based selective layer. Therefore, the mass transfer resistance acts in decreasing the permeating flux in two ways: the first one, which is more direct, acts by offering an additional resistance to permeation. The second one acts to decrease the hydrogen concentration in the metal layer, causing the non-ideality effect to be weaker and, thus, a consequent lower flux.

Concerning the effect of the support, it offers a non-negligible pressure drop. However, it must be remarked that the characteristic driving forces of each permeation steps are different, i.e., approximately linear with DP_{H2} for the external mass transfer in the feed and in the support layers, and approximately linear with $\Delta P_{H_2}^{0.5}$ (Sieverts' law) in the Pd-based layer.

Therefore, the single-step influence should not be evaluated by considering the pressure drop only, as this quantity is not representative of the permeation in each step. This is the reason why the quantitative influence based on the permeating limiting flux is required [43]. To this regard, Figure 3 shows the step influence as a function of temperature at certain operating conditions, calculated in the way mentioned in the previous section.

Considering the temperature of 300°C, it is possible to notice that the influence of external mass transfer, Pd based layer and support is around 40%, 35% and 25%, respectively. Therefore, it is not possible

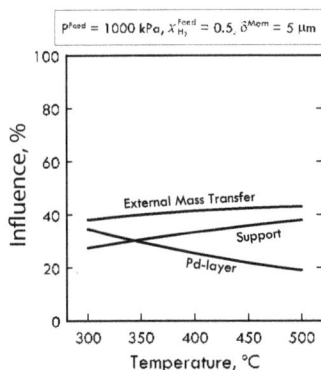

Figure 4: Step influence as a function of the hydrogen mole fraction in the feed for the three permeation steps considered in this work. $P^{Feed}=1000$ kPa, T=400°C, $d^{Mem}=5$ mm.

to recognise a single permeation-determining step. As temperature increases, the ratios of the steps rate changes, causing the relative influence to change as well.

This behaviour is related to the fact that temperature favours the activated processes, which, thus, becomes progressively faster with increasing temperature. Among the three permeation steps considered in the present work, the only activated one is the transport through the Pd-based layer, whereas external mass transfer and transport in the porous support have a weaker dependency on temperature. More specifically, the external mass transfer has a functionality that is slightly lower than the linear one [43], whilst the transport in the support is even slower with increasing temperature owing to the presence of both Knudsen and viscous diffusion mechanism.

The overall results of these considerations is that a higher temperature causes the external mass transfer and the transport in the support to be relatively slower than the hydrogen transport in the metal layer, whose contribution, therefore, becomes gradually smaller and less important. The rate by which its contribution decreases with increasing temperature is quite high because the other two mechanisms increase their respective contributions at the same time.

Figure 4 shows the functionality of the step influence with another key-working condition: the hydrogen composition in mixture. It can be observed that the influence of the Pd-layer increases with increasing hydrogen feed pressure, whereas that of the external mass transfer decreases. As for the support, a non-monotone trend is found, a minimum being present at a hydrogen composition of around 0.3.

As for the external mass transfer, we can consider that the hydrogen permeating flux increases approximately linearly with hydrogen feed pressure [21]. As for the transport in the Pd-layer, we can consider that the hydrogen flux is approximately proportional to the square root of the hydrogen feed pressure. Such a functionality becomes stronger as the hydrogen feed pressure increases due to the effect of the non-ideal contribution of hydrogen diffusion in the metal lattice, which is favoured by the hydrogen pressure in a wide range of pressure conditions [21,22,44-47,49]. Therefore, as the hydrogen feed pressure increases, the external mass transfer becomes gradually faster than the transport in the Pd-layer and in the support, this resulting in the behaviour shown in Figure 4.

As for the support influence, the presence of the minimum can be understood by taking into account that, as external mass transfer

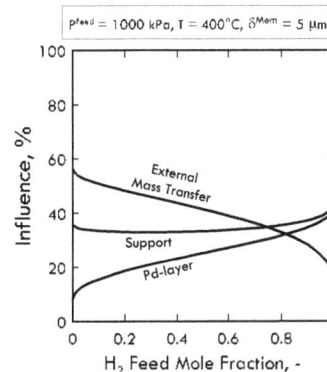

Figure 3: Step influence as a function of temperature for the three permeation steps considered in this work. $P^{Feed}=1000$ kPa, $x_{H_2}^{Feed}=0.5$, $d^{Mem}=5$ mm.

influence progressively decreases with increasing hydrogen content in the feed, the residual influence is redistributed between the Pd-layer and the support. In this redistribution, most of this residual influence gets assigned to the Pd-layer at lower hydrogen feed concentration, as the profiles in the support are less sensitive to the hydrogen content (Figure 2). Therefore, the Pd-based layer influence increases and that of the support slightly decrease.

Towards higher values of hydrogen feed content, the transport in the Pd-based layer becomes faster and, thus, the influence of the support increases, this resulting in a non-monotone trend.

Concerning the optimal membrane design, it is necessary to remark the obvious/non-obvious fact that the optimal operating conditions for membranes are those maximising the permeating flux, independently of the limiting steps controlling the permeation. From this point of view, the best case would be working under adsorption-controlled permeation [43], for which the permeating flux is just function of temperature and hydrogen partial pressure and not a function of membrane thickness.

However, working under these conditions is very difficult, if not impossible, with the actual technology, as the external mass transfer, internal diffusion and transport in the porous support are actually much slower that the hydrogen adsorption rate [43], with the second one mostly being the slowest step.

Therefore, the key-target of membrane designers is to find conditions minimising the influence of the steps external to the selective layer, whose influence is then maximised. In this particular case, the influence of the selective layer is higher towards higher hydrogen content, which actually reflects the real physical behaviour.

Figure 5 shows the trend of the step influence with membrane thickness. In this case, the effect of increasing membrane thickness is to increase the resistance of the Pd-layer, as the permeating flux under diffusion-controlled permeation is inversely proportional to membrane thickness.

As a result, the trend related to the Pd-layer increases, whereas those of the other two decrease. In this case, therefore, the minimisation of the external steps does not correspond to the flux maximisation, as membrane thickness is not kept constant in this analysis.

Furthermore, it can be observed that, going towards gradually lower values of membrane thickness, the influence of internal diffusion

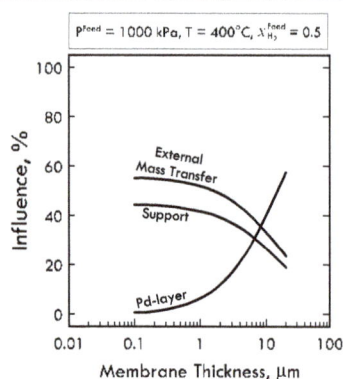

Figure 6: Influence of the support porous layers as a function of temperature in both the presence and absence of external mass transfer resistance (concentration polarisation). P^{Feed}=1000 kPa, x_{H2}^{Feed}=0.5.

tends to zero, whereas those of external mass transfer and transport in the support tend to respective constant values.

In a context where the general efforts of membrane researchers is to continuously decrease the selective layer to increase the permeating flux, this represents a useful indication to identify the lowest limit of the membrane thickness in dependence on the external operating conditions.

To analyse in a more detail the effect of the support, Figure 6 shows the influence of the different single porous layers along with that of the overall support as a function of temperature in the presence and absence of concentration polarisation for a total feed pressure of 1000 kPa and a hydrogen mole fraction of 0.5.

This type of analysis is useful to understand the bottleneck porous layers providing the largest mass transfer resistance in the support, allowing a more systematic membrane design.

From this figure, it can be observed that, in both the presence and absence of polarisation, the majority of influence is offered by the layer 1 of the support (L_1), which actually is in contact with the Pd-based selective layer. Differently, it can be considered with a good approximation that the other layers do not provide any appreciable contribution, their sum being around 4% and 8% with and without polarisation, respectively.

Moreover, by comparing Figure 6a and 6b, it is shown that polarisation reduces the influence of the support on permeation. Although the details of the physical explanation of this important fact are reported in the next section, it is here anticipated that the relative position of the mixture with respect to the support based on the flux direction plays an important role in creating an additional transport resistance causing the pressure drop in the support to be reduced.

Support resistance coefficient (SRC)

As already mentioned, it is also useful to have a single parameter

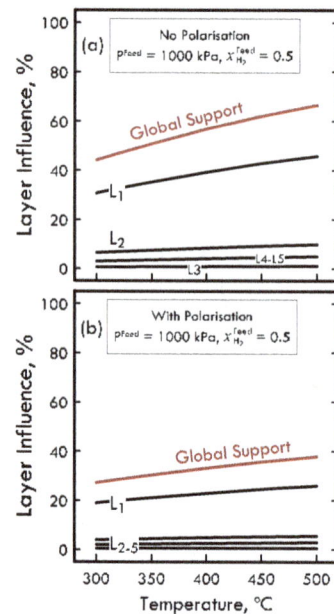

Figure 5: Step influence as a function of membrane thickness for the three permeation steps considered in this work. P^{Feed}=1000 kPa, T=400°C, x_{H2}^{Feed}=0.5.

indicating the driving force loss owing to the porous support (SRC, Equations 3,4), in analogy to what done with the introduction of concentration polarisation coefficient (CPC35], which is a measure of the driving force loss in the external film on the mixture side. Let us remind that the Sieverts driving force is chosen as the characteristic one.

Figure 7 represents a map depicting the behaviour of SRC with the hydrogen mole fraction for different values of temperature and membrane thickness at a total feed pressure of 1000 kPa.

Considering a temperature of 300°C and a membrane thickness of 10 mm, starting from pure hydrogen conditions (i.e., hydrogen mole fraction equal to the unity) and going towards a gradually less hydrogen content, a decreasing SRC is observed, which means that the loss of driving force in the support is progressively smaller. Such trend can be understood by considering that, as the hydrogen feed content decreases, the influence of the external mass transfer on the feed side (i.e., concentration polarisation) increases more than that of the support, as the former is a resistance acting at first based on the flux direction. This eventually causes a driving force re-distribution in the three permeation steps considered such that there is more and more pressure drop in the external feed film rather than in the support and in a sufficiently thin Pd-based layer.

For increasing temperature, the influence of the Pd-based layer decreases, as the internal diffusion becomes faster and, at the same time, the external mass transfer on the feed side is just slightly influenced by temperature, as the gas diffusivity is not an activated process.

As well, for progressively thinner Pd-based layers, the transport through the metal lattice becomes faster and, thus, the most of pressure drop is found in the gas phase on the film side and in the support, this holding for 1 mm with a quite good approximation. This is the reason why SRC decreases faster with decreasing hydrogen content.

Concerning the information provided by the maps depicted in Figure 7, it can be observed that, although such maps provide a direct measure of the pressure drop lost in the support, the effect of external mass transfer is hidden in SRC trend. Therefore, in order to provide also such information, SRC is plotted versus CPC (Figure 8).

Considering a Pd-layer thickness of 10 mm and 300°C, SRC is observed to decrease with increasing CPC. This means that, for increasing driving force in the feed film that in the support decreases. The physical meaning of such a trend is that the effect of concentration polarisation is to decrease the effect of the porous support. This can be

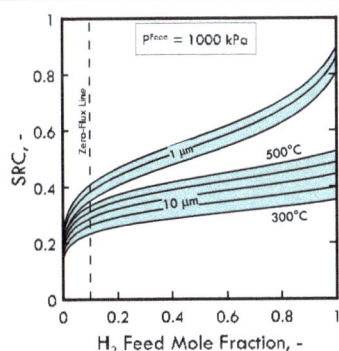

Figure 7: Support Resistance Coefficient (SRC) as a function of the hydrogen mole fraction in the feed for different values of temperature and membrane thickness. P^{Feed}=1000 kPa.

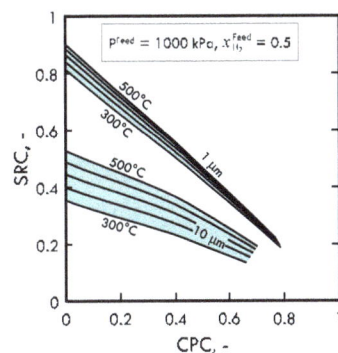

Figure 8: Support Resistance Coefficient (SRC) as a function of the Concentration Polarisation Coefficient (CPC) for different values of temperature and membrane thickness. P^{feed}=1000 kPa, $x_{H_2}{}^{Feed}$=0.5.

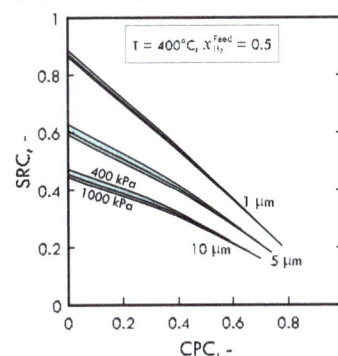

Figure 9: Support Resistance Coefficient (SRC) as a function of the Concentration Polarisation Coefficient (CPC) for different values of membrane thickness and total feed pressure. T=400°C, $x_{H_2}{}^{Feed}$=0.5.

well understood by looking back at the hydrogen profiles depicted in Figure 2, where it is clear that polarisation causes a certain pressure drop before membrane, this in turn inducing a smaller pressure drop in the support, considering that the total pressure drop is determined and fixed by the users.

As a higher value of temperature is considered, SRC becomes higher at a fixed value of CPC. This is because the transport in the Pd-layer is faster whereas, at the same time, the mass transfer on the feed side is weakly influenced by temperature. Therefore, the remaining driving force is redistributed almost completely in the porous support and the sensitivity of SRC to temperature is appreciable.

On the contrary, for Pd-layers as thin as 1 mm, such sensitivity is relatively small. This occurs because, with respect to the other two permeation steps, the transport in the selective layer is fast enough that the hydrogen concentration profile can be considered almost flat, even at 300°C. In these conditions, an increasing polarisation basically causes a driving force re-distribution from the support to the feed film, this determining SRC to vary weakly with temperature than in the case at 10 mm, as the transport in the feed film does not depend on temperature (see Equation 26 reported in Caravella et al., [21]) and that in the support is a non-activated processes whose permeance actually decreases with increasing temperature.

As for the effect of the total feed pressure on SRC at a certain fixed value of hydrogen mole fraction (=0.5), Figure 9 shows that

SRC decreases with increasing total feed pressure, even though such a dependence is found to be weak for all the Pd-layer values considered. This can be understood by considering that the limiting flux related to the external mass transfer in the gas film depends weakly on the total pressure, being approximately proportional to $P^{1/3}$ (see Equation 26 reported in Caravella et al., [21]). Therefore, a higher feed pressure causes the external mass transfer in the feed to be slightly faster, whereas it does not directly affect that in the support, which, thus, becomes relatively slower for increasing feed pressure.

As a general conclusion of the presented analysis, it is withdrawn that, when membrane is used as purifier – i.e., the Pd-based layer is on the high-pressure side and the support on the permeation one – the influence of the porous support strongly depends on the concentration polarisation level.

Therefore, even if the porous support structure were optimised to minimise the mass transfer resistance, the permeation flux through thin selective layer (<10 mm ca.) would be strongly affected by concentration polarisation. Hence, a particular attention should be paid both to a correct evaluation of the overall concentration polarisation level in membrane modules [38] and to the fluid dynamics optimisation in them – like done in several design solutions studied in the literature [36,50-54] – to reduce the effect of the external mass transfer resistance in the feed side, which is found to provide the most serious pressure drop in permeation under mixture conditions.

Conclusions

In this paper, the distribution of the hydrogen permeation driving force across supported Pd-based membranes was systematically evaluated as a function of temperature, total feed pressure and selective layer thickness. For this purpose, first the influence of external mass transfer, internal diffusion and mass transport through an asymmetric five-layered porous support was calculated by using the respective permeation-limiting fluxes. This was required because of the non-linearity of the involved mass transfer resistances.

Then, the Sieverts driving force distribution across membrane was analysed by introducing a novel coefficient, here named as Support Resistance Coefficient (SRC) to quantify both the extent of the driving force lost in the support and the behaviour of its re-distribution over the other permeation steps as function of various operating conditions.

Based on the trend of such a coefficient, which was defined in analogy to the Concentration Polarisation Coefficient (CPC), it was mainly found that the concentration polarisation decreases the influence of the support on permeation. The reason for that was explained by considering that the presence of the external mass transfer resistance represents an additional barrier to permeating flux causing a first hydrogen concentration drop on the feed side that decreases that in the support.

Therefore, it is concluded that the minimisation of the support influence on permeation should be accompanied by the minimisation of the concentration polarisation as well; otherwise the expected flux increase arising from support optimisation could be relatively small in real applications.

Acknowledgements

The "Programma Per Giovani Ricercatori "Rita Levi Montalcini"" granted by the "Ministero dell'Istruzione, dell'Università e della Ricerca, MIUR" is gratefully acknowledged for funding this research.

Notation

m	Number of permeation steps
n	Number of species in mixture
J	Permeating flux, mol s^{-1} m^{-2}
P	Pressure, Pa
T	Temperature, K
x	Mole fraction

Subscripts/Superscripts:

i	Generic species in mixture
j	Generic permeation step
Lim	Limiting (flux)
Mem	Membrane

Acronyms:

CPC	Concentration Polarisation Coefficient
SRC	Support Resistance Coefficient

Greek letters:

a	Influence of a single permeation step, %

References

1. Nam SE, Lee KH (2001) Hydrogen separation by Pd alloy composite membranes: introduction of diffusion barrier. J Memb Sci 192: 177-85.

2. Su C, Jin T, Kuraoka K, Matsumura Y, Yazawa T (2005) Thin Palladium Film Supported on SiO$_2$-Modified Porous Stainless Steel for a High-Hydrogen-Flux Membrane. Ind Eng Chem Res 44: 3053-8.

3. Tong J, Su C, Kuraoka K, Suda H, Matsumura Y (2006) Preparation of thin Pd membrane on CeO$_2$-modified porous metal by a combined method of electroless plating and chemical vapor deposition. J Memb Sci 269: 101-8.

4. Bosko ML, Ojeda F, Lombardo EA, Cornaglia LM (2009) NaA zeolite as an effective diffusion barrier in composite Pd/PSS membranes. J Memb Sci 331: 57-65.

5. Li H, Goldbach A, Li W, Xu H (2007) PdC formation in ultra-thin Pd membranes during separation of H$_2$/CO mixtures. J Memb Sci 299: 130-7.

6. Itoh N, Akiha T, Sato T (2005) Preparation of thin palladium composite membrane tube by a CVD technique and its hydrogen permselectivity. Catal Today 104: 231-7.

7. Zhang X, Xiong G, Yang W (2008) A modified electroless plating technique for thin dense palladium composite membranes with enhanced stability. J Memb Sci 314: 226-37.

8. Peters TA, Kaleta T, Stange M, Bredesen R (2011) Development of thin binary and ternary Pd-based alloy membranes for use in hydrogen production, J Memb Sci 383: 124-134.

9. Peters TA, Kaleta T, Stange M, Bredesen R (2012) Hydrogen transport through a selection of thin Pd-alloy membranes: Membrane stability, H$_2$S inhibition, and flux recovery in hydrogen and simulated WGS mixtures. Catal Today 193: 8-19.

10. Lim H, Oyama ST (2011) Hydrogen selective thin palladium-copper composite membranes on alumina supports. J Memb Sci 378: 179-85.

11. Yun S, Ko JH, Oyama ST (2011) Ultrathin palladium membranes prepared by a novel electric field assisted activation. J Memb Sci 369: 482-9.

12. Dittmar B, Behrens A, Schödel N, Rüttinger M, Franco T, et al. (2013) Methane steam reforming operation and thermal stability of new porous metal supported tubular palladium composite membranes. Int J Hydrogen Energy 38: 8759-71.

13. Kanezashi M, Asaeda M (2006) Hydrogen permeation characteristics and stability of Ni-doped silica membranes in steam at high temperature. J Memb Sci 271: 86-93.

14. Kitiwan M, Atong D (2010) Effects of Porous Alumina Support and Plating Time on Electroless Plating of Palladium Membrane. J Mater Sci Technol 26: 1148-52.

15. Brunetti A, Caravella A, Fernandez E, Pacheco Tanaka DA, Gallucci F, et al.

(2015) Syngas upgrading in a membrane reactor with thin Pd-alloy supported membrane. Int J Hydrogen Energy 40: 10883-93.

16. Fernandez E, Medrano JA, Melende J, Parco M, Viviente JL, et al. (2015) Preparation and characterization of metallic supported thin Pd-Ag membranes for hydrogen separation. Chem Eng J.

17. Gallucci F, Fernandez E, Corengia P, van Sint Annaland M (2013) Recent advances on membranes and membrane reactors for hydrogen production. Chem Eng Sci 92:40-66.

18. Ward T, Dao T (1999) Model of hydrogen permeation behavior in palladium membranes. J Memb Sci 153: 211-231.

19. Catalano J, Baschetti MG, Sarti GC (2009) Influence of the gas phase resistance on hydrogen flux through thin palladium silver membranes. J Memb Sci 339: 57-67.

20. Caravella A, Barbieri G, Drioli E (2008) Modelling and simulation of hydrogen permeation through supported Pd-alloy membranes with a multicomponent approach. Chem Eng Sci 63: 2149-60.

21. Caravella A, Hara S, Drioli E, Barbieri G (2013) Sieverts Law Pressure Exponent for Hydrogen Permeation through Pd-based Membranes: Coupled Influence of Non-Ideal Diffusion and Multicomponent External Mass Transfer. Int J Hydrogen Energy 38: 16229-44.

22. Caravella A, Hara S, Sun Y, Drioli E, Barbieri G (2014) Coupled influence of non-ideal diffusion and multilayer asymmetric porous supports on Sieverts' law pressure exponent for hydrogen permeation in composite Pd-based membranes. Int J Hydrogen Energy 39: 2201-14.

23. Ryi SK, Park JS, Kim SH, Cho SH, Kim DW (2006) The effect of support resistance on the hydrogen permeation behavior in Pd–Cu–Ni ternary alloy membrane deposited on a porous nickel support. J Memb Sci 280: 883-8.

24. Gepert V, Kilgus M, Schiestel T, Brunner H, Eigenberger G, et al. (2006) Ceramic supported capillary Pd membranes for hydrogen separation: potential and present limitations. Fuel Cells 6: 472-81.

25. Huang Y, Dittmeyer R (2007) Preparation of thin palladium membranes on a porous support with rough surface. J Memb Sci 302: 160-70.

26. Huang Y, Shu S, Lu Z, Fan Y (2007) Characterization of the adhesion of thin palladium membranes supported on tubular porous ceramics. Thin Solid Films 515: 5233-40.

27. Ryi SK, Li A, Lim CJ, Grace JR (2011) Novel non-alloy Ru/Pd composite membrane fabricated by electroless plating for hydrogen separation. Int J Hydrogen Energy 36: 9335-40.

28. Sanz R, Calles JA, Alique D, Furoñes L, Ordonez S, et al. (2011) Preparation, testing and modelling of a hydrogen selective Pd/YSZ/SS composite membrane. Int J Hydrogen Energy 36: 15783-93.

29. Lee TH, Park CY, Lee G, Dorris SE, Balachandran UB (2012) Hydrogen transport properties of palladium film prepared by colloidal spray deposition. J Memb Sci 415-416:199-204.

30. Strathmann H (1981) Membrane separation processes. J Memb Sci 9: 121-189.

31. Chern RT, Koros WJ, Yui B, Hopfenberg HB, Stannett VT (1984) Selective permeation of CO_2 and CH_4 through Kapton polyimide: Effects of penetrant competition and gas phase non ideality. J Polym Sci 22: 1061-1084.

32. Rautenbach R, Albrecht R (1989) Membrane Processes. Wiley, New York.

33. Pinho MN, Semiao V, Geraldes V (2002) Integrated modeling of transport processes in fluid/nanofiltration membrane systems. J Memb Sci 206: 189-200.

34. Ahmad AL, Lau KK, Abu Bakar MZ, Shukor SRA (2005) Integrated CFD simulation of concentration polarization in narrow membrane channel. Comput Chem Eng 29: 2087-95.

35. Caravella A, Barbieri G, Drioli E (2009) Concentration polarization analysis in self-supported Pd-based membranes. Sep Purif Technol 66: 613-24.

36. Boeltken T, Belimov M, Pfeifer P, Peters TA, Bredesen R, et al. (2013) Fabrication and testing of a planar microstructured concept module with integrated palladium membranes. Chem Eng Process 67: 136-47.

37. Nekhamkina O, Sheintuch M (2016) Approximate models of concentration-polarization in Pd-membrane separators. Fast numerical analysis. J Memb Sci 500: 136-150.

38. Caravella A, Sun Y (2016) Correct Evaluation of the Effective Concentration

Polarization Influence in Membrane-assisted devices. Case Study: H_2 Production by Water Gas Shift in Pd-Membrane Reactors. Int J Hydrogen Energy, In Press.

39. Caravella A, Scura F, Barbieri G, Drioli E (2010) Inhibition by CO and Polarization in Pd-Based Membranes: A Novel Permeation Reduction Coefficient. J Phys Chem B 114: 12264-76.

40. Catalano J, Giacinti Baschetti M, Sarti GC (2010) Hydrogen permeation in palladium-based membranes in the presence of carbon monoxide. J Memb Sci 362: 221-233.

41. Mejdell AL, Chen D, Peters TA, Bredesen R, Venvik HJ (2010) The effect of heat treatment in air on CO inhibition of a ~ 3 mm Pd-Ag (23 wt.%) membrane. J Memb Sci 350: 371-377.

42. Abir H, Sheintuch M (2014) Modeling H_2 transport through a Pd or Pd/Ag membrane, and its inhibition by co-adsorbates, from first principles. J Memb Sci 466: 58-69.

43. Caravella A, Scura F, Barbieri G, Drioli E (2010) Sieverts Law Empirical Exponent for Pd-Based Membranes: Critical Analysis in Pure H_2 Permeation. J Phys Chem A B 114: 6033-47.

44. Hara S, Ishitsuka M, Suda H, Mukaida M, Haraya K (2009) Pressure-dependent hydrogen permeability extended for metal membranes not obeying the square-root law. Journal of Physical Chemistry B 113: 9795-801.

45. Flanagan TB, Wang D (2010) Exponents for the Pressure Dependence of Hydrogen Permeation through Pd and Pd-Ag Alloy Membranes. J Phys Chem A C 114: 14482-8.

46. Hara S, Caravella A, Ishitsuka M, Suda H, Mukaida M, et al. (2012) Hydrogen diffusion coefficient and mobility in palladium as a function of equilibrium pressure evaluated by permeation measurement. J Memb Sci 421-422: 355-60.

47. Flanagan TB, Wang D (2012) Temperature Dependence of H Permeation through Pd and Pd Alloy Membranes. J Phys Chem A 116: 185-92.

48. Morreale BD, Ciocco MV, Enick RM, Morsi BI, Howard BH, et al. (2003) The permeability of hydrogen in bulk palladium at elevated temperatures and pressures. J Memb Sci 212: 87-97.

49. Caravella A, Hara N, Negishi H, Hara S (2014) Quantitative Contribution of Non-Ideal Permeability under Diffusion-controlled Hydrogen Permeation through Pd-membranes. Int J Hydrogen Energy 39: 4676-82.

50. Cao Z, Wiley DE, Fane AG (2001) CFD simulation of net-type turbulence promoters in a narrow channel. J Memb Sci 185: 157-76.

51. Koutsou CP, Yiantsios SG, Karabelas AJ (2004) Numerical simulation of the flow in a plane-channel containing a periodic array of cylindrical turbulence promoters. J Memb Sci 231: 81-90.

52. Geraldes V, Semiao V, Pinho MN (2003) Hydrodynamics and concentration polarization in RO/NF spiral wound modules with ladder type spacer. Desalination 157: 395-402.

53. Li F, Meindersma W, de Haan AB, Reith T (2002) Optimization of commercial net spacers in spiral wound membrane modules. J Memb Sci 208: 289-302.

54. Boeltken T, Wunsch A, Gietzelt T, Pfeifer P, Dittmeyer R (2014) Ultra-compact microstructured methane steam reformer with integrated Palladium membrane for on-site production of pure hydrogen: Experimental demonstration. Int J Hydrogen Energy 39: 18058-68.

14

Preparation and Characterization of Commercial Polyethyleneterephthalate Membrane for Fuel Cell Applications

Abdel-Hady EE*, Abdel-Hamed MO and Gomaa MM

Physics Department, Faculty of Science, Minia University, Minia, Egypt

Abstract

Commercial Polyethyleneterephthalate (PET) based proton exchange membrane was prepared by UV-radiation grafting of styrene onto PET films. The effect of irradiation time and different concentrations of the monomer on the degree of grafting (D.G.) has been studied. It was found that the DG increases linearly with increasing the irradiation time and monomer concentration, reaching a maximum at a certain level. The effect of chlorosulfonic acid concentrations on ion exchange capacity (IEC) and tensile strength also studied to find the optimum concentration of chlorosulfonic acid for using in the sulfonation process. The range of IEC, 0.2 to 0.775 m mol/g, resulting from treating styrene grafted and sulfonated PET (PET-g-PSSA) membranes with different chlorosulfonic acid levels showed that chlorosulfonic acid is an effective tool to control IEC. Fourier transform infrared (FTIR) spectroscopic analysis confirmed grafting and sulfonation onto PET Films. In addition, thermogravimetric analysis was used to investigate the behavior of the original PET film and PET-g-PSSA membranes. The methanol permeability and the proton conductivity of PET-g-PSSA film with D.G. 166% were found to be 1.2×10^{-8} and 58 m S/cm, respectively, better than those of Nafion 212 membrane measured with the same instruments under the same conditions. Since they have lower cost, higher conductivity and lower methanol permeability, PET-g-PSSA could be better used instead of Nafion in direct methanol fuel cells.

Keywords: PET membrane; UV-radiation grafting; Sulfonation; Proton conductivity

Introduction

Worldwide growing demand for fossil fuels, which are currently the most convenient energy source, is expected to lead to an energy crisis, unless sustainable and alternate fuels become available. Furthermore, their combustion emissions are polluting the environment to the threshold of creating health problems, and the carbon dioxide emissions are implicated in global warming [1]. Direct methanol fuel cell (DMFC) is considered as a highly promising power source. It is based on proton exchange membrane (PEM) fuel cell technology. It possess a number of advantages, such as a liquid fuel, quick refueling, low cost of methanol and the compact cell design, making it suitable for various potential applications, including stationary and portable applications, and it is favored for use as commercial products in automobiles, residential homes, and in portable devices, such as laptops and cell phones. DMFCs are also environmentally friendly. Although carbon dioxide is produced, there is no production of sulfur or nitrogen oxides.

The proton exchange membrane (PEM) is a key device in fuel cells, which acts as proton transferring electrolyte, as well as provides a barrier to the passage of electrons between the electrodes [2]. At the present time, per fluorinated membranes, Nafion are the most commonly used PEM around the world, due to their superior chemical and electrochemical stability, as well as high proton conductivity under fuel cell operating conditions [3]. However, some disadvantages, such as their high preparation cost, decrease in ion conductivity at high temperatures, and their high methanol permeability, severely limit their commercialization in fuel cells. Therefore, much effort has been expended in developing new membranes to circumvent these disadvantages [4-6]. Radiation graft polymerization is one of the promising methods, which enables introduction of active monomer functional group at inner polymer chains in film for the modification of the chemical and physical properties of a wide range of polymer materials. Various kinds of grafting polymerization techniques–including ion-radiation induced, photo-induced (UV and chemical

initiator) and plasma-induced grafting polymerizations–have been developed in the last few decades [7-9]. However, the UV-radiation technique is more available and less expensive than other techniques [10]. In comparison with γ-ray radiation grafting, UV-induced photo-grafting is very simple and safe, and is less damaging to the membranes because significant degradation of the membrane main chains can be avoided [11].

Several studies identify that poly(ethylene-alt-tetrafluoroethylene) (PTFE) [12-14], poly(vinylidene fluoride) (PVDF) [15-17], poly(tetrafluoroethylene-co-perfluoropropyl vinyl ether) (PFA) [18] and cross linked polytetrafluoroethylene (ETFE) [18,19] are promising base polymers. Styrene is the most frequently used monomer due to its high thermal stability and moderate sulfonation process of the aromatic ring [4]. The Polyethyleneterephthalate (PET) based proton exchange membrane for using in fuel cells was successfully prepared by gamma radiation-induced graft copolymerization of styrene monomer onto PET film, and the consequent selective sulfonation of the grafting chain in the film state using chlorosulfonic acid ($CISO_3H$) by Mostak et al. [20].

The present research focuses on developing a non-fluorinated, inexpensive proton exchange membrane. The PET-based PEM was prepared successfully by UV-radiation grafting of styrene onto PET films. Figure 1 shows the chemical structure of the PET monomer,

*Corresponding author: Abdel-Hady EE, Physics Department, Faculty of Science, Minia University, Minia, Egypt, E-mail: esamhady@yahoo.com

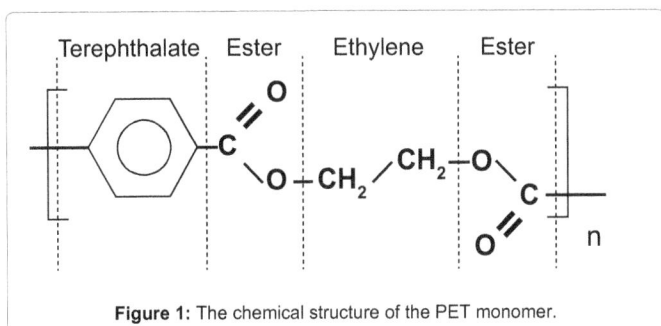

Figure 1: The chemical structure of the PET monomer.

grafted film was then sulfonated by chlorosulfonic acid to form proton exchange membrane. The effects of grafting and sulfonation conditions such as monomer and chlorosulfonic acid concentration, irradiation time were investigated. The characterizations of the membranes, including different chemical and physical parameters such as water uptake, tensile strength, ion exchange capacity, proton conductivity, thermal durability and FTIR spectroscopy were also studied.

Experimental Methods

Materials

Commercial Polyethylene terephthalate (PET) film of 51 μm thickness purchased from Cs Hyde Company, USA, was used as a polymer matrix. Styrene of purity more than 99% (Sigma-Aldrich) was used as the grafting monomer without any further purification, and the photo initiator (benzophenone) was purchased from Oxford laboratory, India. Chloroform was used to remove homopolymer after grafting. 1, 2 Dichloromethane (Lobachemie, India), chlorosulfonic acid (Fluka) and methanol (Fluka) were used as received. Sodium chloride (NaCl) and sodium hydroxide (NaOH) were used for titrimetric analysis and purchased from Adwic. Egypt, Nafion NR 212 (DuPont) was used as a reference material.

Grafting procedure

The PET films of thickness 51 μm were cut into square pieces of known weight, washed with acetone, and then dried in a vacuum oven at 60°C for 1 h. The dried films were placed into a glass ampoule containing monomer solution of known concentration. Photo grafting mixture consists of methanol, which is used as a solvent, benzophenone (BP, photo initiator, with concentration 2%), Styrene (monomer) with different concentrations (25, 30, 35 and 40%) at 50°C. The different parameters affecting the degree of grafting of styrene onto PET were assessed. These parameters include the grafting time and monomer concentration. By changing the grafting time (1-8 hours), PET films with different degree of grafting were obtained. After grafting, polystyrene/PET films were washed several times with a large amount of chloroform to remove any excess unreacted monomer or styrene homopolymer. The process was repeated with fresh chloroform to ensure the complete removal of any residual monomer and homopolymer occluded within the polymer. The grafted films were then dried in a vacuum oven at 80°C, until a constant weight was obtained. The degree of grafting (D.G.%) was calculated using the equation (1)

$$D.G(\%) = \frac{W_{g-W0}}{W_0} \times 100 \tag{1}$$

Where W_g and W_o are the weights of grafted and original PET films, respectively.

Sulfonation

Sulfonation of the PET grafted films was carried out by immersing PET films with different degree of grafting in different concentrations (0.2, 0.5, 0.7 and 1) (v/v)% of chlorosulfonic ($ClSO_3H$) acid in 1-2 dichloromethane. The sulfonation was performed at room temperature for 1 hour. After sulfonation, the membranes were removed from sulfonating solution and immersed in fresh dichloromethane for 3 hours, followed by washing in 1-2 dichloromethane to remove any residual acid, so as to reduce any effect of swelling that may occur if water was used for the initial washing stage, and finally hydrolysis in deionized water at 80°C for 24 hours before measurements. Figure 2 shows the preparation process of the proton-conducting membranes.

Fourier transforms infrared spectra (FTIR)

The functional groups of the grafted membrane were measured using BRUKER ALFA FTIR USA spectrometer with resolution 0.9 cm⁻¹.

Thermal Gravinetric Analysis (TGA)

Thermo-gravimetric analyzer (TGA) instrument model Q50, USA was used to characterize the thermal stability of the membranes with heating rate 10°C min⁻¹ under nitrogen atmosphere.

Tensile strength

The tensile tests were measured at room temperature by UNIVERSAL MATERIALS TESTING MACHINE LLOYD (model LR 5K plus). The measurements were performed at a cross-head speed set at a constant speed of 10 mm/min.

Measurement of ion exchange capacity

Ion Exchange Capacity (IEC) of the sulfonated samples was measured using a typical titration method. The dried membrane in the protonic form was equilibrated with 25 ml of 3 M NaCl solution for 24 hours. A large excess of Na⁺ ions in the solution ensured nearly complete ion exchange. Then, 10 ml of the solution was titrated against 0.05 M NaOH solution using phenolphthalein as indicator. Finally, drops of 0.05 M NaOH solution were added, until the color of solution change from colorless to pink. The IEC was calculated using the following equation:

$$IEC = \frac{0.05 \times n \times V_{NaOH}}{W_{dry}} \tag{2}$$

Where V_{NaOH} (ml) is the volume of the 0.05 M NaOH solution used for titration, n is the factor corresponding to the ratio of the amount of NaCl taken to immerse the polymer to the amount used for titration that is 2.5, and W_{dry} is the dry weight of the polymer electrolyte

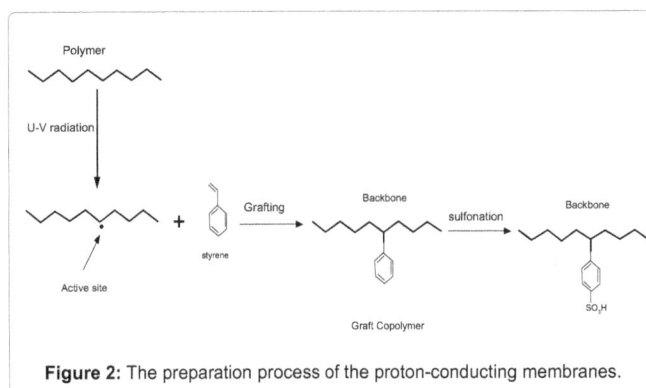

Figure 2: The preparation process of the proton-conducting membranes.

membrane in the protonic form.

Water uptake

The water uptake was determined in the following way: at first, the membrane samples with different degree of grafting after sulfonation were dried under vacuum for 1 hour at 80°C, and then weighted. They were then soaked in deionized water for 2 days, until swelling equilibrium was achieved. The soaked membranes were carefully blotted with a filter paper to remove water on the surface, followed by weight measurement.

$$\text{Water uptake } (\%) = \frac{W_{swollen} - W_{dry}}{W_{dry}} \times 100 \tag{3}$$

Where $W_{swollen}$ is the weight of the swollen membrane and W_{dry} is the weight of the dried membranes.

Proton conductivity

Proton conductivity measurements of the sulfonated grafted membranes were derived from AC impedance spectroscopy measurements over a frequency range of 50 to 5M Hz with an oscillating voltage of 50-500 mV, using a system based on a HIOKI LCR Hi-Tester, Model:3532, Japan. Each membrane sample was cut into sections 2.5 cm×2.5 cm prior to being mounted in the cell. The cell was placed in a temperature controlled container open to air by a pinhole, where the sample was equilibrated at 100% RH at ambient atmospheric pressure and clamped between two electrodes. The proton conductivities of the samples were measured in the longitudinal direction, and were calculated from the impedance data using the following relationship:

$$\sigma = \frac{L}{RS} \tag{4}$$

where σ is the proton conductivity (in Simon/cm), L is the distance between the electrodes used to measure the potential (L=1 cm), S is the membrane cross-sectional surface area (membrane width×membrane thickness) required for protons to penetrate through the membrane (in cm²), and R is derived from the low intersection of the high frequency semicircle on a complex impedance plane with the Re (Z) axis.

Methanol permeability

Methanol permeability measurement was carried out at room temperature using a liquid diffusion cell composed of two compartments containing solution A and B. One compartment A (V_A=50 ml) was filled with 5 mol/L methanol solution, the other compartment (V_B=50 ml) was filled with deionized water only. The tested membrane was immersed in deionized water for hydration before measurements, and then vertically placed between the two compartments by a screw clamp. Both compartments were kept under stirring slightly during the permeation experiments. Amount of methanol diffused from compartment A to B across the membrane was measured over time. The methanol permeability P was calculated by the following equation:

$$P = \frac{KV_BL}{AC_A} \tag{5}$$

Where k is the slope of the straight line plot of methanol concentration in solution B versus permeation time, V_B, L and A are the volume of solution B, the thickness and the effective area of the tested membrane, respectively.

Single cell test

Single cell test was conducted using 2 M methanol aqueous solution in the passive mode (the methanol at the anode and at the cathode,

O_2 supported from the surrounding air without air flow) at room temperature and atmospheric conditions. The cell was connected to a voltammeter and the cell voltage was recorded as a function of time.

Results and Discussion

Effect of monomer concentration

Polyethylene terephthalate (PET) films were placed into a glass ampoule containing monomer solution with different concentrations and irradiated with U-V radiation. The degree of grafting D.G (%) of styrene monomer onto the commercial PET membrane as a function of monomer concentration has been shown in Figure 3. The degree of grafting (D.G.) increased from 89% (at 20 volume % styrene concentration) to 166% (at 30 volume % styrene concentration), because of increasing copolymerization between styrene and PET film. The Degree of grafting then decreased from 166% (at 30 volume % styrene concentration) to 54% (at 40 volume % styrene concentration), because of dominating homopolymer formation that hindered the movement of styrene towered PET films. It was reported that the homopolymerization effect increased the viscosity of the grafting solution, hindering the movement of monomer toward the substance, and consequently, reduced the grafting yield [21,22]. So, the better concentration for preparation grafting of styrene onto PET films by UV radiation technique was found to be 30 volume % styrene concentrations.

Effect of irradiation time

Grafting onto PET film was carried out as a function of irradiation time, keeping the monomer concentration constant at 30% volume styrene concentration. PET films were placed with a styrene solution into glass ampoules. The ampoules were irradiated with grafting solution (30% styrene concentration). Figure 4 shows the UV-photo grafting of styrene into PET films with 30% monomer styrene concentration as a function of irradiation time. As can be seen from the figure, the degree of grafting increases gradually with the increase of radiation time up to 4 hours, then a suddenly increase can be seen

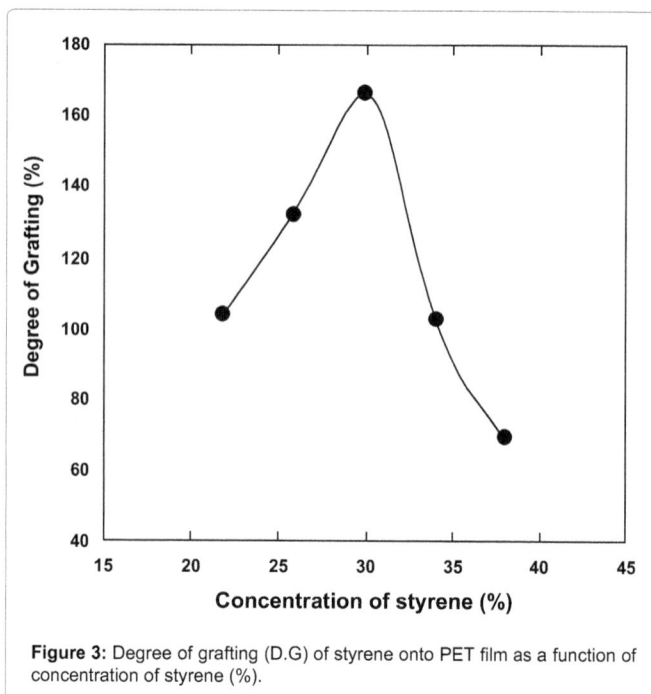

Figure 3: Degree of grafting (D.G) of styrene onto PET film as a function of concentration of styrene (%).

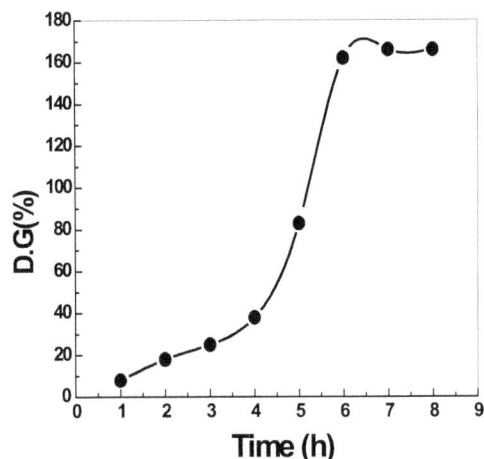

Figure 4: Degree of grafting (D.G. %) of styrene onto the PET film as a function of UV- Irradiation time.

up to 6 hours. After 6 hours radiation time, no appreciable increase in the degree of grafting was observed. This behavior was observed in the simultaneous radiation grafting of styrene into PTFE films, where the grafting increases almost linearly with the increase in the radiation dose, and reasonably high graft levels up to 70% were achieved [23,24]. However, higher irradiation doses are not preferred due to the deterioration of mechanical properties [25]. By increasing radiation time, the number of free radicals formed in the grafting solution also increases; leads to more monomers diffuse to the film resulting in an increase in the degree of grafting. After 6 hours, a saturation of the radical sites can be noticed, i.e. the equilibrium degree of grafting value (166 %) was reached at 6 hours.

Sulfonation and ion exchange capacity

The PET film which has the highest degree of grafting value (166%) was sulfonated with different concentration of chlorosulfonic acid (0.2, 0.5, 0.7, 1) (v/v) % in 1-2 dichloromethane. It was observed that low chlorosulfonic acid ($CISO_3H$) solution concentration give low IEC and increased with increasing concentration of chlorosulfonic acid, (Figure 5), due to the incorporation of the increased number of SO_3H groups into the grafted PET film. At 166% degree of grafting and 1% chlorosulfonic acid concentration, the IEC of the PET-g-PSSA membrane reached a maximum of 0.775 m mol/g, which is comparable to the Nafion 212 membrane (IEC of 0.841 m mol/g), measured with the same instrument under the same conditions. The IEC for PET-g-PSSA (D.G. 166%) shows higher value than the maximum IEC of the PET grafted membrane prepared under UV radiation by Mostak et al. [20]. They found that the IEC has a value of 0.04385 mmol g^{-1} at 12.7% degree of grafting (D.G.). The number of sulfonic acid groups in the membrane increases with the increase in the D.G. At higher styrene concentrations, more benzene rings are in contact with sulfonic acid groups, which results in more sulfonic acid groups in the membrane. However, the efficiency of the sulfonation reaction depends to large extent on whether or not the membrane is grafted through its thickness [26]. If the samples contained a core of ungrafted parts, sulfonation was incomplete at room temperature due to insufficient swelling of the samples and the difficulty of diffusion of the sulfonating agent.

Tensile strength

In the present work, to find the optimum concentration of chlorosulfonic acid $CISO_3H$ for sulfonation of grafted PET films, the

tensile strengths of the sulfonated film which has the highest degree of grafting value (166%) with different concentration of chlorosulfonic acid (0.2, 0.5, 0.7, 1) (v/v) % in 1-2 dichloromethane were measured. Table 1 illustrates the mechanical parameters for Nafion and PET-g-PSSA, with degree of grafting 166% membranes at different concentrations of chlorosulfonic acid. As can be seen from this table, the sulfonation reaction at higher chlorosulfonic acid $CISO_3H$ concentration is accompanied with decreasing tensile strength of the film. This can be explained by the fact that the continuous increases in the content of sulfonic acid in the grafted PET film results in the deterioration of mechanical properties of the membranes, because of highly hydrophilic property of the film [27], which make the films, have low tensile strength [28]. After sulfonation with concentration more than 1% of chlorosulfonic acid, it was observed that the membranes are more fragile. The high IEC and acceptable value of tensile strength at 1% concentration of chlorosulfonic acid make this concentration of chlorosulfonic acid is the optimum concentration for sulfonation of grafted PET films.

Membranes characterization

Fourier transforms infrared spectra (FTIR): Figure 6 shows the FTIR spectra of pure, grafted and sulfonated films with D.G (166%). The original PET film is characterized by a strong band at 730 cm^{-1}, representing the stretching vibration of the C=O-O- (ester group) and the absorption bands at 2870 and 2950 cm^{-1}, representing the symmetric and the asymmetric stretching vibration of CH_2 group of ethylene, respectively, as depicted from spectrum A. The presence of the benzene ring in the PET films is established by the =C—H stretching vibration at 3050 cm^{-1} and the skeletal C=C in-plane stretching vibrations at 1500 and 1600 cm^{-1}, respectively. The Para-substitution of benzene ring is represented by the band at 860 cm^{-1}. The grafting of styrene is confirmed by the mono substitution of the benzene ring of the polystyrene, which is represented by the aromatic out-of-plane C—H deformation band at 730 cm^{-1}. The absorption bands at 2975 cm^{-1} is assigned for the symmetric stretching of CH of the polystyrene and peak intensity is increased at 1450-1600 cm^{-1}, which proves the monomer was grafted. Sulfonation at the grafted PET film was confirmed by showing two peaks at 1146 cm^{-1} (asymmetrical stretch) and 1056 cm^{-1} (symmetrical stretch) for sulfonic acid group. The broad peak at 3410 cm^{-1} assigned

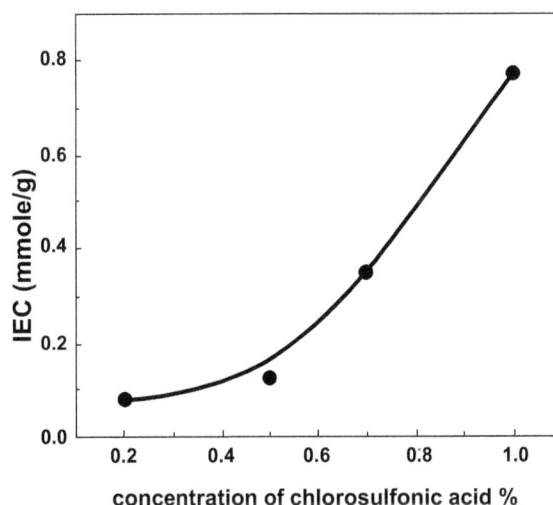

Figure 5: Ion exchange capacity (IEC) as a function of chlorosulfunic acid concentration %.

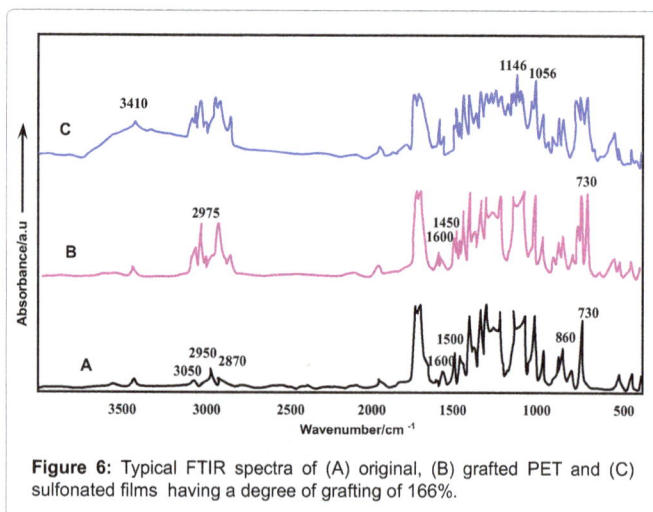

Figure 6: Typical FTIR spectra of (A) original, (B) grafted PET and (C) sulfonated films having a degree of grafting of 166%.

to water molecules involved in hydrogen bonding with the SO_3 groups [29]. In addition, the grafted films were found to be not transparent compared to the original PET film, which was transparent. Moreover, their dimensions were found to be increased compared to that of the original films. All this features together confirm the successful grafting of styrene onto PET films.

Thermal Gravimetric Analysis (TGA): The thermal properties of pure PET, PET-g-PSSA and Nafion membranes were preliminarily evaluated by TGA in N_2 heating at 10°C min^{-1} as shown in Figure 7. For Pure PET, the main weight loss appears at around 450°C, which can be attributed to the degradation of the backbone of the polymers. This temperature decreased in the PET-g-PSSA membranes from 450 to 430°C indicates that the incorporation of styrene makes the PET film vulnerable to thermal degradation due to incorporation of some weak links into the macromolecular chain. There was a gradual weight loss of PET-g-PSSA membrane. The weight loss at 30-300°C is attributed to residual boundary water loss [30], associated with sulfonic acid groups. Decomposition of sulfonic acid groups begins around 300°C and completed up to 400°C; decomposition of the grafted chain occurs above 410°C and followed by PET main chain degradation. Nafion 212 exhibits a similar degradation behavior to that of PET-g-PSSA. The major weight loss around 460°C is due to the degradation of the backbone of Nafion. The data indicate that the PET-g-PSSA membrane is thermally stable within the temperature range for Proton Exchange Membrane Fuel Cell (PEMFC) applications.

Water uptake

The water uptake is an important parameter for fuel-cell membranes, and has a direct connection to the proton conductivity and the dimensional stability. Higher water uptake generates more solvated species, which is needed for high conductivity, but unfortunately, greater water content produces mechanically less stable membrane [31]. Figure 8 shows the water uptake as a function of the degree of grafting of the PET based proton conducting membranes. The higher degree of grafting was associated with a higher water uptake (59%), indicating the presence of hydrophilic sites (sulfonated graft chains) within the hydrophobic PET base films. The same behavior was observed by Abdel-Hady et al. [32]. Using the same instrument and under the same conditions, the water uptake of the Nafion membrane was also measured to be 53%, which is in a good agreement with the obtained value of PET-g-PSSA membrane with degree of grafting

166%.

Proton conductivities

The proton conductivity of the PET-g-PSSA membranes was estimated using impedance diagrams, acquired in the frequency range 50-10^6 Hz. All the samples were in the fully water swollen state. Figure 9 shows the impedance spectra of PET-g-PSSA, with D.G. 166% as an example. It was observed that the resistance decreased with increasing D.G. due to higher content of styrene grafted to the main chain provides more sites for the attack of sulfonic acid group, which is responsible for conduction. Resistance of the membrane obtained from the impedance spectra, and then converted into proton conductivity of the membrane as conduction is the reverse of resistance. Table 2 shows the change of proton conductivity with the degree of grafting at room temperature. The proton conductivity of PET-g-PSSA membrane increases with increasing the degree of grafting due to the increasing in sulfonic acid group. Low degrees of grafting, the polystyrene grafts

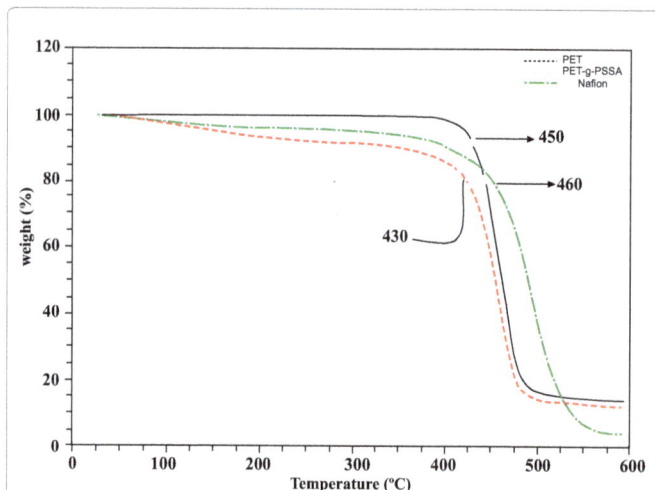

Figure 7: Typical TGA weight loss curves of pure PET, PET-g-PSSA and Nafion membranes.

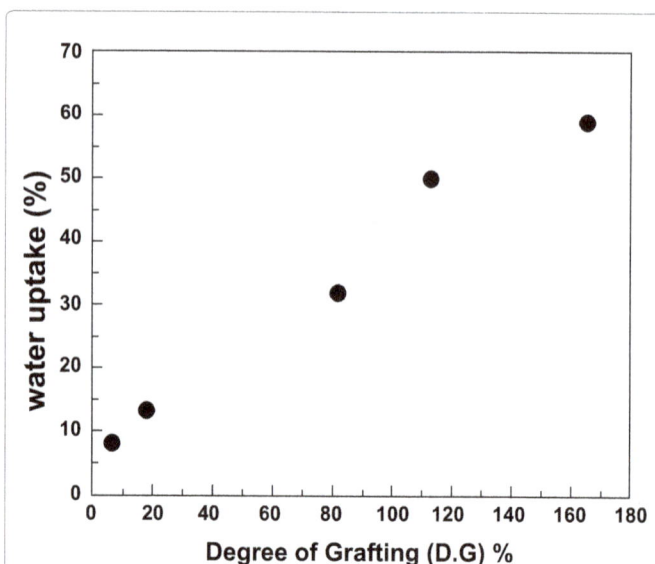

Figure 8: Water uptake as a function of Degree of grafting.

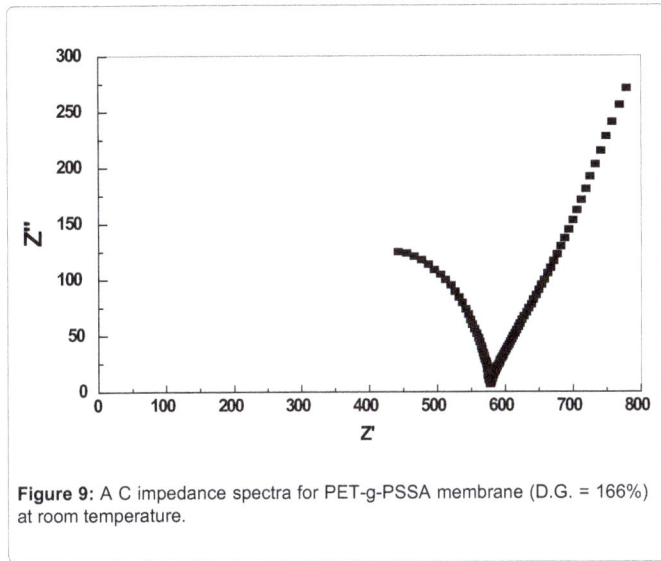

Figure 9: A C impedance spectra for PET-g-PSSA membrane (D.G. = 166%) at room temperature.

are located near the surface of the film, while its middle remains ungrafted, and subsequently, it exerts high local resistance to proton due to inhomogeneous distribution of ion exchange sites over the area of the membrane. Hydration of the membrane (i.e. incorporation of water molecules into the polymer structure) leads to convert of the acid groups into mobile H^+.

The resulting nanophase-separated structure is an interpenetrating network of hydrophobic polymer backbone material providing structural integrity and aqueous domains, allowing proton transport within water-containing channels. The proton conductivity of the material depends on the density of acidic groups. So, it was observed that the films which have high water uptake give high conductivity (Table 2). The maximum value of conductivity for PET-g-PSSA film with D.G. (166%) was observed to be 58 m S/cm. The conductivity of Nafion 212 membrane was also measured to be 52 m S/cm with the same instrument under the same conditions for comparison. The prepared membranes show higher conductivity similar to the Nafion membrane, but they are a lot cheaper compared to Nafion or other equivalent membranes. Also, the observed proton conductivity of the PET-g-PSSA membrane is much better (4 times higher), as compared to that of the PET grafted membrane prepared under UV radiation by Mostak et al. [20].

Effect of temperature on the Proton conductivity

The effect of temperature on conductivity of PET-g-PSSA film with degree of grafting (D.G=166%) is calculated by using impedance diagrams at different temperature ranging from room temperature up to 90°C. Figure 10 shows the impedance spectra of PET-g-PSSA with D.G (166%) different temperatures which converted into proton conductivity. From the impedance diagrams, it was found that the resistance to conducting ions decreases linearly with increasing the temperature up to 90°C. The elevation of temperature favors both the dynamics of proton transport and the structural reorganization of polymeric chains that is attributed to the activation energy of the conducting ions, which increases with temperature resulting in the increased proton conductivity at higher temperatures.

The proton conductivity σ for PET-g-PSSA (D.G=166%) and Nafion membranes at different temperatures are presented in Figure 11. It was observed that the proton conductivity for both membranes

increased linearly as the temperature increases up to 90°C. Also, it was noticed that the PET-g-PSSA with D.G=166% has a proton conductivity ranging from 58 to 97 m Scm⁻¹, which is attractive for fuel cell application. Moreover, the proton conductivity of the Nafion membrane was found to be ranged from 52 up to 147 m Scm⁻¹. The increasing of conductivity with increasing temperature is due to

Materials	Concentration of chlorosulfonic acid %	Young's modulus(MPa)	Tensile strength (MPa)
PET-g-PSSA	0.2	3566 ± 23	60.7 ± 1.2
	0.5	2115 ± 34	41.5 ± 1.1
	0.7	1037 ± 40	8.6 ± 1.3
	1	677 ± 30	7.2 ± 1.2
Nafion NR212	---	368 ± 26	29 ± 1.1

Table 1: The mechanical parameters for Nafion and PET-g-PSSA with degree of grafting 166% prepared from different concentration of chlorosulfonic acid.

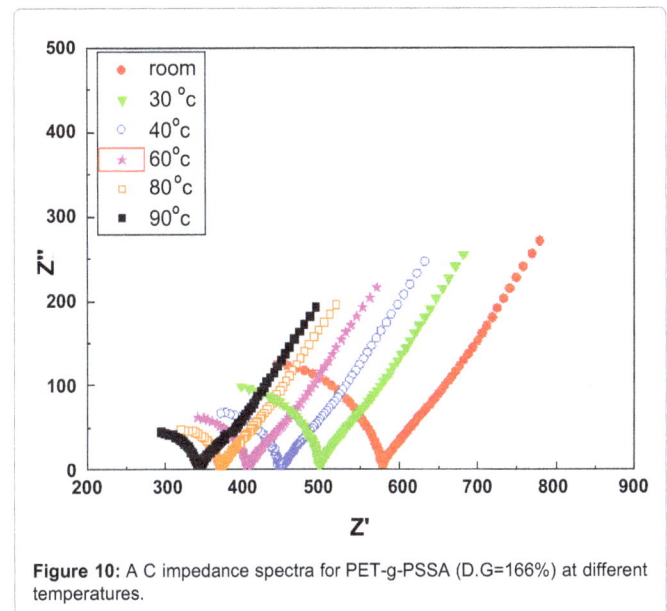

Figure 10: A C impedance spectra for PET-g-PSSA (D.G=166%) at different temperatures.

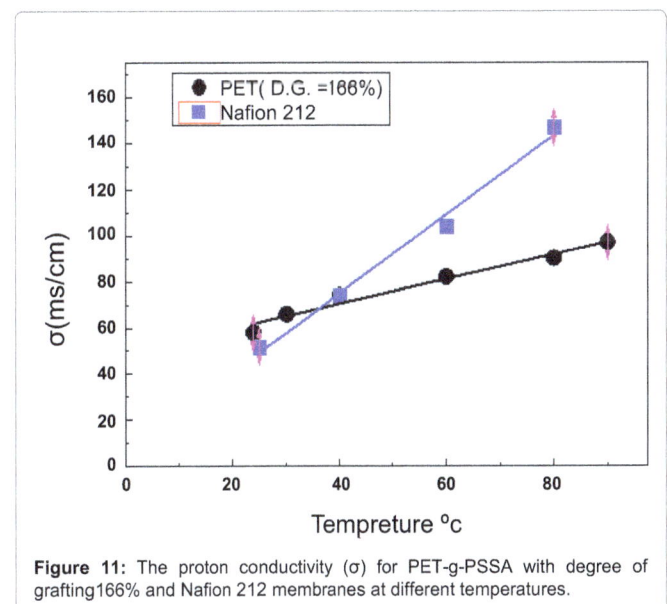

Figure 11: The proton conductivity (σ) for PET-g-PSSA with degree of grafting166% and Nafion 212 membranes at different temperatures.

Materials	D.G %	Water uptake %	Conductivity mS/cm
PET-g-PSSA	8	8	8
	18	13	16.5
	83	32	18
	113	50	46
	166	59	58
Nafion	--	53	52

Table 2: The conductivity and water uptake of PET g-PSSA as a function of D.G at room temperature.

increasing the proton mobility through the membranes. Due to the gradual increase in the proton conductivity for PET-g-PSSA with increase in temperature up to 90°C represents that the membrane can work efficiently up to 90°C.

Activation Energy (E_a)

The activation energy (Ea), the minimum energy required for proton transport across the membrane, was calculated using the Arrhenius equation:

$$\sigma = A \times e^{-Ea/RT} \tag{6}$$

where σ is the proton conductivity (mScm^{-1}), A is a constant proportional to the number of charge carriers, Ea is the activation energy (kJmol^{-1}), R is the universal gas constant (8.314 Jmol^{-1} K^{-1}), and T is the absolute temperature (K). The Arrhenius plot for PET-g-PSSA with degree of grafting 166% and Nafion membranes are shown in Figure 12. The plot of ln (σ) versus the inverse of temperature (1000/T) give a linear function with a slope (- Ea/R). From the plot, the activation energy Ea was calculated for the different membranes and found to be 6.3 kJmol^{-1} and 16.3 kJmol^{-1} for PET-g-PSSA, with degree of grafting 166% and Nafion, respectively.

The proton conduction mechanism in these membranes is known to occur by two routes. The first route is a hopping or jumping mechanism, also known as the Grotthuss model, in which a proton is passed through a channel of water molecules; the protons are transferred from one vehicle to the other by hydrogen bonds. The second route is a vehicle mechanism, where in a proton combines with solvent molecules, producing a complex which diffuses through the membrane. It was reported that for the hopping mechanism, the activation energy for proton conductivity should be around 14-40 kJ mol^{-1} [33,34]. So, in Nafion the hopping mechanism is the predominant, while in PET-g-PSSA, the vehicle mechanism is believed to exist.

Methanol permeability

DMFCs are attractive, especially for portable power applications. Membranes for use in DMFC system must both possess adequate proton conductivity and be an effective barrier for methanol crossover from the anode to the cathode compartment, to prevent loss of methanol and oxidation with oxygen, leading to a mixed potential [31]. One of the Nafion's drawbacks is its high methanol crossover in the DMFC application. This limitation is associated with the microstructure of Nafion. For determination of methanol permeability, the amount of methanol transported through the PET-g-PSSA with D.G=166% and Nafion membranes was estimated repeatedly by spectrometer after drawing calibration curve between the methanol concentration and refractive index. The methanol permeability for PET-g-PSSA with D.G=166% was found to be 1.2×10^{-8} cm^2/s, which is lower than that of Nafion (1.65×10^{-6} cm^2/s).This mean that the PET-g-PSSA with D.G=166% is more reliable for fuel cell applications.

DMFC performance of PET-g-PSSA membrane

The single cell DMFC was performed at room temperature and 1 M methanol solution for 120 hours. As can be seen from Figure 13, the performance curve has two trends; during the first 2 hrs, the output voltage of the single cell increases from 0.45 to 0.57 volt. Such an observation may be attributed to the activation of electrocatalyst and decrease in the internal resistance of the single cell [35]. In the following stage, the voltage of the single cell decreases to 0.51 V. The results showed that the cell voltage is stable at 0.51 V, which is better than obtained by Hasani-Sodrabadi et al. [36]. By considering the steady performance curve of the DMFC single cell, it is expectable that the durability and stability of the PET-g-PSSA membranes to be suitable for methanol fuel cell.

Conclusions

Polymer electrolyte membranes (PEMs) for fuel cells, also termed as proton exchange membranes, have been successfully prepared by UV-radiation grafting of styrene onto PET films, and consequent, selective sulfonation by chlorosulfonic acid. The PEMs displayed excellent thermal, mechanical and electrical properties. The maximum

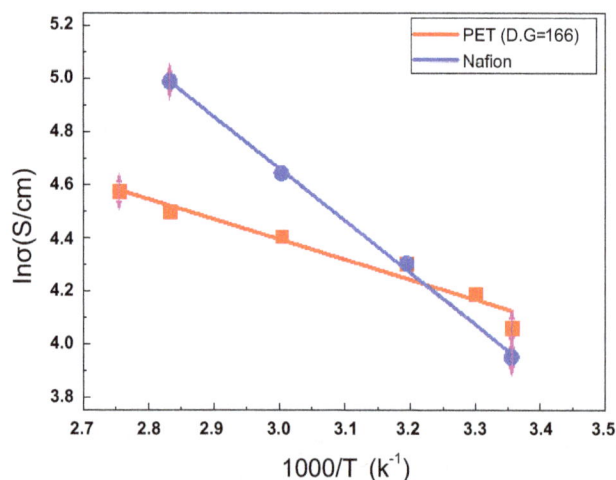

Figure 12: Arrhenius plots of proton conductivity σ for PET–g-PSSA with degree of grafting 166% and Nafion 212 membranes.

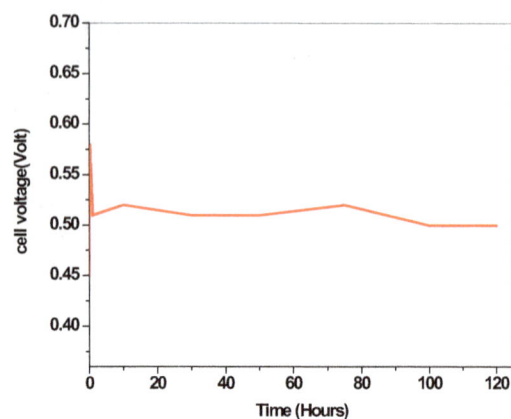

Figure 13: The cell voltage as a function of time.

ion-exchange capacity (IEC) and the proton conductivity of the PEM were measured to be 0.7 mmol g^{-1} and 58 m Scm^{-1}, respectively. The activation energy was calculated as 6 k jmol^{-1} by Arrhenius plot, which reveals that the vehicle mechanism is the predominant conduction mechanism in PET-g-PSSA membranes. Due to their low coast, higher conductivity, suitable water uptake and low methanol permeability, PET-g-PSSA membranes are considered for use in DMFCs as alternatives to Nafion.

Acknowledgements

The authors wish to sincerely thank the STDF of Egypt, (ID220), for financially supporting this project.

References

1. Sopian K, Shamsuddin AH, Nejat VT (1995) Solar hydrogen energy option for Malaysia. Proceedings of the International Conference on Advances in Strategic Technology.

2. Zhou W, Xiao J, Chen Y, Zeng R, Xiao S, et al. (2010) Synthesis and characterization of sulfonated poly (ether sulfone ether ketone ketone) for proton exchange membranes. J Appl Polym Sci 117: 1436

3. Mauritz KA, Moore RB (2004) State of Understanding of Nafion. Chem Rev 104: 4535-4586.

4. Nasef MM, Hegazy EA (2004) Preparation and applications of ion exchange membranes by radiation-induced graft copolymerization of polar monomers onto non-polar films. Prog Polym Sci 29: 499-561.

5. Gubler L, Gursel SA, Scherer GG (2005) Proton exchange membranes prepared by radiation grafting of styrene/divinylbenzene onto poly(ethylene-alt-tetrafluoroethylene) for low temperature fuel cells. Fuel Cells 5: 317.

6. Dworjanyn PA, Garnett JL (1988) Synergistic effects of urea with polyfunctional acrylates for enhancing the photografting of styrene to polypropylene. J Polym Chem Polym Lett Ed 26: 135-138.

7. Hollahan JR (1974) Plasma Chemistry Industrial Application. Wiley, New York, USA.

8. Wilkie CA, Deacon C (1996) Graft copolymerization of acrylic acid on to acrylonitrile-butadiene-styrene terpolymer and thermal analysis of the copolymers. Eur Polym J 32: 451-455.

9. Hasegawa S, Suzuki Y, Maekawa Y (2008) Preparation of poly (ether ether ketone)-based polymer electrolytes for fuel cell membranes using grafting technique. Radiat Phys Chem 77: 617-621.

10. Asano M, Chen J, Maekawa Y, Sakamura T, Kubota H, et al. (2007) Novel UV-induced photografting process for preparing poly(tetrafluoroethylene)-based proton-conducting membranes. J Polym Sci A PolymChem 45: 2624-2637 .

11. Ismail AF, Zubir N, Nasef MM, Dahlan KM, Hassan AR (2005) Physico-chemical study of sulfonated polystyrene pore-filled electrolyte membranes by electrons induced grafting. J Membr Sci 254: 189-196.

12. Li J, Kohei S, Shogo I, Saneto A, Shigetoshi I, et al. (2005) Pre-irradiation induced grafting of styrene into crosslinked and non-cross linked poly tetrafluoroethylene films for polymer electrolyte fuel cell applications II Characterization of the styrene grafted films. Eur Polym J 41: 547-555.

13. Nasef MM, Saidi H, Dessouki AM, El-Nesr EM (2000) Radiation-induced grafting of styrene onto poly(tetrafluoroethylene), PTFE films I Effect of grafting conditions and properties of the grafted films. Polym Int 49: 399-406.

14. Holmberg, Lehtinen T, Näsman J, Ostrovskii D, Paronen M, et al. (1996) Structure and properties of sulfonated poly [(vinylidene fluoride)–g-styrene] norous membranes porous membranes. J Mater Chem 6: 1309-1317.

15. Flint SD, Slade RCT (1997) Investigation of radiation-grafted PVDF-g-polystyrene-sulfonic-acid ion exchange membranes for use in hydrogen oxygen fuel cells. Solid State Ionics 97: 299-307.

16. Elomaa M, Hietala S, Paronen M, Walsby N, Jokela K, et al. (2000) The state of water and the nature of ion clustersincrosslinked proton conducting membranes of styrene grafted and sulfonatedpoly(vinylidene fluoride). J Mater Chem 10: 2678-2684.

17. Kyoung RP, Phil HK, Young CN (2005) Preparation of PFA-g-polystyrene sulfonic acid membranes by the Gamma-radiation grafting of styrene onto PFA film. React Funct Polym 65: 47-56.

18. Brack HP, Bührer HG, Bonorand L, Scherer GG (2000) Grafting of pre-irradiated poly(ethylene-alt-tetrafluoroethylene) films with styrene influence of base polymer film properties and processing parameter. J Mater Chem 10: 1795-1803.

19. Scott K, Taama WM, Argyropoulos P (2000) Performance of the direct methanol fuel cell with radiation-grafted polymer membranes. J Membr Sci 171: 119-130.

20. Mostak Ahmed, Mubarak A Khan, Md Pezwan Miah, Syed Abdul Monim, M Anwar H Khan (2011) Gamma Radiation-induced graft copolymerization of styrene onto polyethyleneterephthalate fims application in fuel cell technology as a proton exchange membrane. J Macromol Sci A Pure Appl Chem 48: 927-936.

21. Nasef MM (2000) Gamma radiation-induced graft copolymerization of styrene onto poly(ethyleneterephthalate) films. J Appl Polym Sci 77: 1003-1012.

22. Kattan M, El-Nesr E (2006) Gamma-Radiation-induced graft copolymerization of acrylic acid onto poly (ethylene terephthalate) films A study by thermal analysis. J Appl Polym Sci 102: 198-203.

23. Nasef MM, Saidi H, Dessouki AM, El-Nesr EM (2000) Radiation-induced Grafting of Styrene onto Poly(tetrafluoroethylene) (PTFE) Films I Effect of Grafting Conditions and Properties of the Grafted Films. J Polym Int 49: 399-406.

24. Nasef MM, Saidi H, Nor HM (2000) Proton exchange membranes prepared by simultaneous radiation grafting of styrene onto poly(tetrafluoroethylene-co-hexafluoropropylene) films i effect of grafting conditions. J Appl Polym Sci 76: 220-227.

25. Rager T (2003) Pre-Irradiation grafting of styrene/divinylbenzene onto poly (tetrafluoroethylene-co-hexafluoropropylene) from non-solvents. Helv Chim Acta 86: 1.

26. Walsby N, Paronen M, Juhanoja, Sundholm (2001) Sulfonation of styrene-grafted poly(vinylidene fluoride) films. J Appl Polym Sci 81: 1572-1580.

27. Zhou W, Xiao J, Chen Y, Zeng R, Xiao S, et al. (2010) Synthesis and characterization of sulfonated poly(ether sulfone ether ketone ketone) for proton exchange membranes. J Appl Polym Sci 117: 1436-1445.

28. Takahashi, Okonogi H, Hagiwara T, Maekawa Y (2008) Preparation of polymer electrolyte membranes consisting of alkyl sulfonic acid for a fuel cell using radiation grafting and subsequent substitution/elimination reactions. J Membr Sci 324: 173-180.

29. Nasef M, Zubir NA, Ismail AF, Dahlan KZM, Saidi H, et al. (2006) Preparation of radio chemically pore-filled polymer electrolyte membranes for direct methanol fuel cells. J Power Sources 156: 200-210.

30. Shoibal Banerjee, Dennis E (2004) Nafion® perfluorinated membranes in fuel cells. Curtin Journal of Fluorine Chemistry 125: 1211-1216.

31. Tauqir A Sherazi, Shujaat Ahmad, Akram Kashmiria M, Michael Guiver (2008) Radiation induced grafting of styrene onto ultra-high molecular weight polyethylene powder and subsequent film fabrication for application as polymer electrolyte membranes Influence of grafting conditions. J Membr Sci 325: 964-972.

32. Abdel-Hady EE, El-Toony MM, Abdel-Hamed MO, Hammam AM (2011) Grafting of styrene onto commercial PTFE membrane and sulfonation for possible use in fuel cell. J Membr Sci Technol 1: 3.

33. Li L, Zhang J, Wang Y (2003) Sulfonatedpoly ether ether ketone membranes for direct methanol fuel cell. J Membr Sci 226: 159-167.

34. Nagarale RK, Gohil GS, Shahi VK (2006) Sulfonated polyether ether ketone / polyaniline composite proton-exchange membrane. J Membr Sci 280: 389-396.

35. Zhai Y, Zhang H, Zhang Y, Xing D (2007) A novel H$_3$PO$_4$/Nafion–PBI composite membrane for enhanced durability of high temperature PEM fuel cells. J Power Sources 169: 259.

36. Hasani-Sadrabadi MM, Dashtimoghadam E, Sarikhani K, Fatemeh S, et al. (2010) Electrochemical investigation of sulfonated poly(ether ether Ketone)/clay nano composite membranes for moderate temperature fuel cell applications. J Power Sources 195: 2450-2456.

15

Sintering of the Immersion-Induced Porous Stainless Steel Hollow Fiber Membranes

Haixia Li, Jian Song and Xiaoyao Tan*

State Key Laboratory of Separation Membranes and Membrane Processes, Department of Chemical Engineering, Tianjin Polytechnic University, Tianjin 300387, China

Abstract

Porous stainless steel (PSS) hollow fiber membranes have been fabricated by an immersion induced phase inversion and sintering technique. This paper is mainly focused on the sintering process. The influences of the sintering atmosphere as well as the sintering temperature and dwelling time on the microstructure, mechanical strength, and the permeation properties of the hollow fiber have been extensively investigated. Experimental results indicate that a H_2-containing sintering atmosphere favors the formation of highly permeable porous PSS membranes with desirable mechanical strength. But the sintering atmosphere with H_2 concentration larger than 25% is not necessary as it yields a marginal improvement on the membrane performances. For a sufficient sintering, the sintering temperature should be controlled between 1050 and 1100°C, which ensures not only a desirable mechanical strength but also good membrane porosity. For cost effective and less time consuming, sintering time can be controlled within two hours. The resultant PSS hollow fibers satisfy the longstanding criteria of porous supports for gas separation membranes in terms of high porosity, high fluid permeance and robustness to withstand stress.

Keywords: Porous stainless steel membrane; Hollow fiber; Phase inversion; Sintering

Introduction

Inorganic membranes possess the advantages of high thermal stability, good chemical resistance and high mechanical strength, and thus have been targeted by a variety of advanced applications at high temperatures or environments involving corrosive chemicals [1,2]. Among them, the porous ceramic membranes have been broadly applied for many years but they generally exhibit some drawbacks such as brittleness and low thermal shock resistance. In contrast, porous metal membranes often made from stainless steel have high mechanical strength, offer good thermal shock resistance and allow for easy integration by welding or brazing in membrane module assembly, and therefore have been extensively used as the porous support to deposit other thin separation films such as Pd, silica and zeolites [3-9].

Most inorganic membranes in applications are usually designed into a planar or tubular configuration. Comparatively, the hollow fiber geometry is more favorable for practical use because it can provide a much higher surface-area-to-volume ratio. In the past decade, a combined phase inversion/sintering technique has been developed to fabricate various ceramic hollow fiber membranes [10-13]. Compared to other methods to produce inorganic hollow fibers, the phase inversion induced hollow fiber membranes also exhibit very significant advantages over the planar or tubular ones such as low permeation resistance due to the resultant asymmetric microstructures and low fabrication costs. Furthermore, the phase inversion/sintering technique is very versatile and can be modified to synthesize porous stainless steel (PSS) hollow fiber membranes [14-17].

So far the limited several studies on the preparation of PSS hollow fiber membranes by the phase inversion/sintering technique are mainly focused on the microstructure formation of the PSS hollow fibers during the phase inversion process. For example, Luiten-Olieman and co-workers reported the variation trend of the PSS hollow fiber microstructures with the spinning conditions and the composition of the spinning suspension [14-16]. We also reported the synthesis of the PSS hollow fiber membranes in spiral configurations by adding special additives in the spinning solution [18]. But the hollow fibers have low

gas permeability due to the low surface porosity despite their apparent porosities are larger than 40%, indicating the majority of pore volume was closed inside the fiber wall. Actually, the membrane properties like microstructure, surface porosity, pore size distribution and mechanical strength that should be critically considered in practical applications are significantly affected by the sintering conditions in the preparation of inorganic membranes [19]. For example, sintering of the PSS hollow fibres under air, CO_2, He and H_2 atmospheres would lead to different microstructures [20]. However, the general evolution of microstructure and performance of the resultant PSS hollow fibers with the variation of sintering conditions has not been much clarified.

The objective of this work was to elucidate the effects of sintering parameters including atmosphere, temperature and dwelling time on the microstructure and properties of the PSS hollow fibers spun by the phase inversion-sintering technique, and to identify the optimal sintering conditions at which high performance PSS hollow fiber membranes can be achieved.

Experimental

Spinning of the PSS hollow fiber precursors

The PSS hollow fiber precursors were spun from a dope mixture via the phase inversion process described in more details elsewhere [10,11]. For preparation of the spinning dope, stainless steel powder (AISI 316L, Fe/Cr18/Ni10/Mo3, mean particle size=3 μm, purchased

***Corresponding authors:** Xiaoyao Tan, State Key Laboratory of Separation Membranes and Membrane Processes, Department of Chemical Engineering, Tianjin Polytechnic University, Tianjin 300387, China
E-mail: tanxiaoyao@tjpu.edu.cn

from Good Fellow Cambridge Limited) and organic additives were added into N-methylpyrrolidone (NMP, AR Grade, >99.8%, Kermel Chem Inc, Tianjin, China) solution followed by stirring for 30 min. Polysulfone (PSf, Udel35000, Solvay USA) was then added in three steps, each separated by 2 h. The mixture was stirred for 36 h untill all the polymer was completely dissolved to form a uniform suspension. The contents of PSf, NMP, 316L steel powder and organic additives in the spinning solution were 6.2, 26.8, 65.4, and 1.6 wt%, respectively. Prior to spinning, the suspension was degassed by applying vacuum for 1 h and left overnight under dry air. The spinning solution was extruded by a syringe pump through a tube-in-orifice spinneret (inner and outer diameter of 1.5/3.0 mm, respectively) into a tap water bath for coagulation. Deionized water as the internal coagulant was pumped through the bore of the spinneret at a flow rate of 8 mL min^{-1}. The PSS hollow fiber precursors were kept in a water bath for 24 h for the thorough removal of NMP solvent.

Sintering of the membranes

Thermal treatment of the hollow fiber membranes was performed in a tubular furnace (GSL-1700X, Hefei Kejin Materials Tech. Co., China). Prior to sintering, the hollow fiber precursor was cut into short pieces of around 20 cm in length and dried in air for 48 h. The hollow fiber precursors were firstly heated with the heating rate of 5°C min^{-1} to 600°C and calcined in static air at this temperature for 60 min to remove the polymer binder and the organic additives. Subsequent sintering was performed in static air or H$_2$-N$_2$ streams with different hydrogen concentrations, while the temperature was further elevated at a heating rate of 3°C min^{-1} to 1000~1200°C, dwelling for 30, 60, 120 or 240 min, respectively. The sintering process ended by cooling down to room temperature at 3°C min^{-1} under the same atmosphere. For each sintering condition, five hollow fiber samples were prepared for characterization and performance test. The average values over the samples were taken to evaluate the sintering effects on the hollow fiber properties.

Characterization

Scanning electron microscope (SEM, JEOL JSM 5600 LV) was used to examine the morphology and the microstructures of the hollow fiber membranes sintered under different conditions. SEM images were taken on both the cross-section and surface of the fibers. X-ray diffraction (XRD) patterns of the PSS hollow fibers were recorded with a Bruker D8 Advance diffractometer using Cu Kα radiation (λ= 0.15404 nm). The hollow fibers were ground into fine powders prior to the XRD measurement. Continuous scan mode was used to collect 2θ data from 10° to 80° with a 0.02° sampling pitch and a scan rate of 2° min^{-1}. The X-ray tube voltage and current were set at 40 kV and 30 mA, respectively.

The pore size distribution of hollow fibers was measured using a mercury porosimeter (Micromeritics, Auto Pore IV 9500) employing mercury pressures between 1.50 and 33000 psi. Prior to measurement, the samples were heat-treated at 200°C for 2 h to remove adsorbed vapor. The weight of each sample for the measurements was around 1.2 g. The evacuation and the equilibrating time were 5 min and 10 s, respectively. In addition, the porosity of the hollow fibers was also calculated from the bulk density estimated by the weight and the corresponding dimensions of the samples.

The mechanical strength of the hollow fibers was measured with a three-point bending instrument (Instron Model 5544) with a crosshead speed of 0.5 mm min^{-1} [21]. Hollow fiber samples were fixed on the

sample holder with a distance of 4 cm. The bending strength, σ_F, was calculated from the following equation:

$$\sigma_F = \frac{8FLD}{\pi(D^4 - d^4)} \tag{1}$$

Where F is the measured force at which fracture takes place; L, D and d are the length (4 cm), the outer diameter and the inner diameter of the hollow fibers, respectively. The values of outer diameter (D) and inner diameter (d) were obtained from the SEM micrographs.

The permeation properties of the PSS hollow fiber membranes were evaluated by the nitrogen and pure water permeation measured in a dead-end mode under constant flux operation. The modules were custom-made by gluing a single fiber into glass tubing with an organic sealant [22]. Three different flux settings were used for each sample. The permeate flow was recorded as a function of the trans-membrane pressure. The pressure drop over the fiber is assumed to be negligible. The gas or water permeance through the hollow fiber was calculated by,

$$P = \frac{N_p}{A_m \Delta p} \tag{2}$$

Where N_p is the molar flow rate of the permeate gas or water, A_m the effective membrane area calculated by $A_m = \frac{\pi(D-d)L}{\ln(D/d)}$, and Δp is the pressure difference across the membrane.

Results and Discussion

In order to obtain pure stainless steel membranes, the organics in the PSS hollow fiber precursors should be removed completely while the stainless steel phase is retained well. The burnout of organics is highly dependent on the composition of organics and the microstructure of precursor, thus much related to the gaseous atmosphere surrounding the precursor. A major concern of the sintering atmosphere is the presence of oxygen as it is required to burn out the organics, but on the other hand it should be used as less as possible to avoid the possible oxidation of iron phase. Thus, in this work, air was used at lower temperature of 600°C to burn out the organics in the hollow fiber precursor, after which inert or reducing gases were introduced to protect the stainless steel hollow fiber during the subsequent high temperature treatment to achieve the desired density and mechanical strength. Figure 1 shows the morphology of the hollow fiber precursor and the fiber sintered at 600°C for 1 h. It can be seen that short finger-like structures have been formed near both the outer and inner walls of the fiber, while a sponge-like structure occurs at the center of the fiber precursor (Figure 1a). This is the special feature of the hollow fiber membranes prepared by phase inversion method using water as both the external and the internal coagulants as the rapid precipitation occurred at both the inner and outer walls close to coagulants resulting in short finger pores but the slow precipitation at the center of the fiber giving the sponge-like structure [10,11]. The stainless steel particles are well dispersed and embedded inside polymeric matrix, as displayed in Figure 1a. After sintering at 600°C for 1 h, the cross-sectional structure of the fiber precursor has been well preserved, as shown in Figure 1(b). The surface SEM image shows the absence of organic binders as they were removed completely during the sintering process (Figure 1b). The stainless steel particles keep their original spherical shape with particle size of around 1-3 microns, indicating sintering has not occurred yet at this temperature.

In order obtain the sintered PSS hollow fiber membranes with

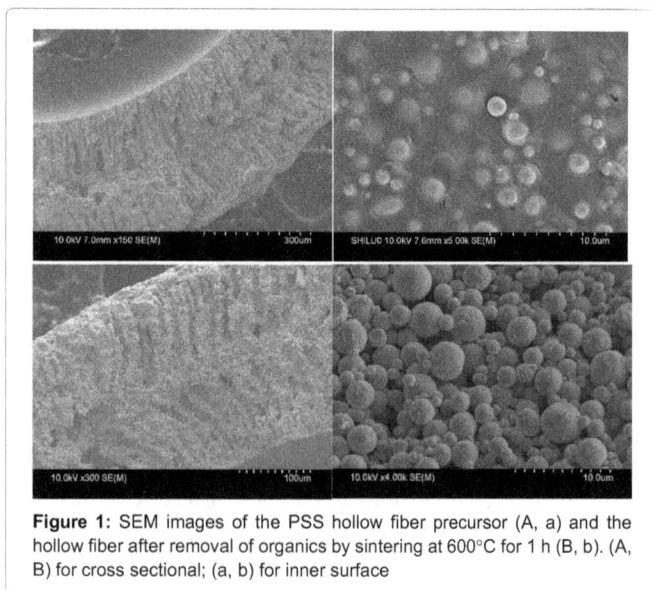

Figure 1: SEM images of the PSS hollow fiber precursor (A, a) and the hollow fiber after removal of organics by sintering at 600°C for 1 h (B, b). (A, B) for cross sectional; (a, b) for inner surface

certain mechanical strength, the sintering temperature was raised to 1050°C at heating rate of 5°C min⁻¹ and dwelling for 2 hours, under different atmospheres with flowing air, nitrogen and hydrogen-containing streams. Figure 2 compares the morphology of the resultant PSS hollow fibers. The general porous structure can be observed in all the sintered samples from the cross sectional areas, but the sandwiched structure is more perfectly preserved when using hydrogen-containing atmosphere although the quantity and the size of the pores have greatly decreased after the sintering process. The significant effect of the sintering atmosphere can be seen more clearly from the inner surface morphologies. For the fiber sintered in air, the inner surface looks quite dense with a few pores in the size of 1-2 microns randomly distributed. The formation of such dense surface structure may be due to the iron oxidation on the surface layer. Compared to metal iron, the formed iron oxides of Fe_2O_3 or Fe_3O_4 would increase the weight by 40%, thus causing the volume swelling and making the surface denser. The formation of iron oxides can be confirmed by the XRD pattern that will be discussed later. For the sample sintered in nitrogen atmosphere, the surface also looks dense but the number of open pores is more than that of the fiber sample sintered in air. The possible reason for this phenomenon is still related to the volume and weight increase caused by the formation of iron nitride formed during the high temperature sintering stage or other impurities like iron oxides or carbonized compounds [23], which also can be evidenced by the XRD results below. However, when H_2-containing atmosphere was applied, the resultant hollow fibers have shown a completely different surface morphology. The stainless steel particles retained their original spherical shape, and connected with each other to form a large quantity of micro pores. Sintering does not occur as deeply as in the air and nitrogen atmospheres, because hydrogen as the most effective reducing atmosphere allows for enhanced oxide removal, which necessitates longer sintering times. Furthermore, the small hydrogen atoms may diffuse into the metal lattice, leading to a puffy structure on the particle surfaces and inhibiting the elimination of final porosity, as shown in the Figure 2 [24]. Nevertheless, the dependence of the porosity with the H_2 partial pressure in the sintering atmosphere cannot be easily perceived just from the SEM image alone. Other characterization techniques are required to better evaluate the effects of the hydrogen

concentration in sintering atmosphere on the properties of the PSS hollow fiber membranes.

Figure 3 shows the XRD patterns of the PSS hollow fiber membranes sintered at 1050°C under different atmospheres, where the XRD pattern of the original stainless steel powders is also presented for comparison. As can be seen, the dominate phase in the original powder is martensite phase (α-Fe) identified at 44.50° and 64.76°, but a minor amount of austenite phase (γ-Fe) is also present identified by the characteristic peaks of 43.54°, 50.70° and 75.8°. After the stainless steel powder is formed into hollow fibers via sintering, the predominant phase will transform from martensite to austenite. Furthermore, the intensity of the characteristic peaks of all the hollow fibers has been reduced significantly. This can be attributed to the re-crystallization effect, i.e., the large grains were broken up into smaller ones during the high-temperature sintering process [25]. Some additional peaks are also identified in the XRD patterns of the hollow fibers, indicating that some impurity phases have been generated after sintering. Types of the impurity phases are highly dependent on the sintering atmosphere. For example, carbon compounds such as $Mo_{12}Fe_{22}C_{10}$ (PDF card code

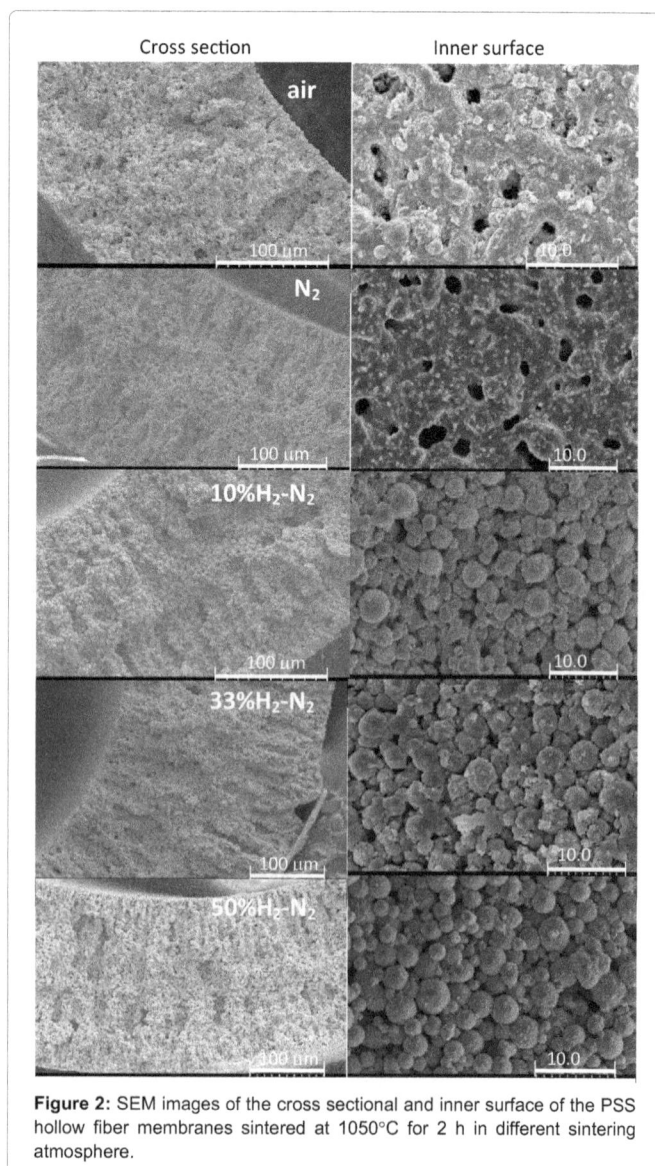

Figure 2: SEM images of the cross sectional and inner surface of the PSS hollow fiber membranes sintered at 1050°C for 2 h in different sintering atmosphere.

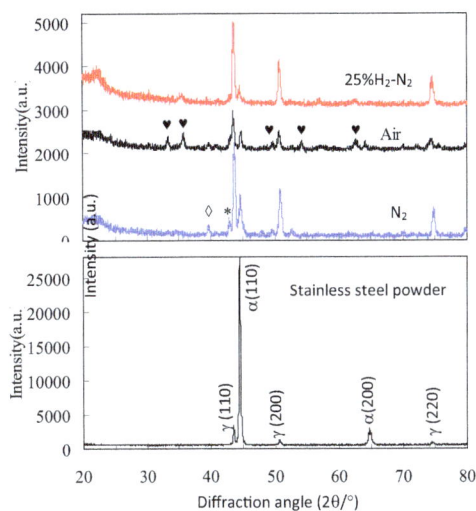

Figure 3: XRD patterns of the original stainless steel powder and the PSS hollow fiber membranes sintered at 1050°C under different atmospheres (impurity phases: * $Mo_{12}Fe_{22}C_{10}$; ◊ Mo_2C; ♥ Fe_2O_3).

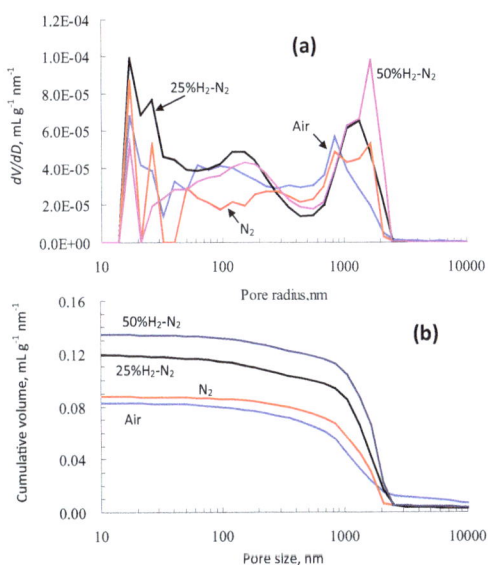

Figure 4: Pore size distribution (a) and cumulative volume (b) of the PSS hollow fiber membranes sintered under different atmospheres.

78-0272) and Mo_2C (PDF card code 77-0720) were detected in the N_2-sintered hollow fibers, whereas the main impurity would be Fe_2O_3 in the hollow fibers sintered in the air atmosphere. Under the H_2-containing reducing atmosphere, the impurities would no longer be observed in the XRD pattern. In the presence of hydrogen, any iron oxides like Fe_2O_3 could be reduced to metal phase, and the formation of iron compounds would be inhibited.

Figure 4 describes the pore size distribution of the hollow fibers sintered at 1050°C under different sintering atmospheres. Despite the atmosphere difference, all the fiber samples exhibit a bimodal pore size distribution (Figure 4a) with the pores ranging from 0.01 to 3 μm. This is a typical feature of the asymmetric membranes fabricated by the phase inversion method [26,27]. Figure 4b shows the cumulative

pore volume distribution, from which it is clear that the majority of porosity is contributed from the pores of 1-3 microns. The pore volume alternation is highly related to the sintering atmosphere. At the same sintering temperature, the sintering atmosphere with larger H_2 partial pressure confers the sample with a larger volume; on the other hand, the sample sintered in oxidant (air) environment possesses the smallest volume. This result is consistent with the observation from the SEM images, and also agrees with the literature report [24]. It suggests that, to be applied as the substrate for supported thin membrane synthesis for fluid separation, the PSS hollow fibers should be sintered in the atmosphere with H_2 since a more porous structure is preferred to minimize the transport resistance through the substrate. Furthermore, although the fibers sintered in the 50% H_2-N_2 atmosphere possess the largest pore volume, the 25% H_2-N_2 atmosphere is more preferably recommended since the resultant fibers have the most micro pores smaller than 0.2 μm while those sintered in the 50% H_2-N_2 atmosphere mainly possess the macro pores between 1~3 μm in diameter. In order to further recognize the role of hydrogen in sintering, the effect of hydrogen concentration in the sintering atmosphere on the porosity, mechanical strength and permeances of both nitrogen and pure water have been investigated and presented below.

Figure 5 plots the porosity and mechanical strength of the prepared PSS hollow fibers against the H_2 concentration in the sintering atmosphere. As displayed, the general trend of porosity change with the H_2 concentration is increasing, which is opposite to the change of mechanical strength. This is a reasonable observation as a higher porosity generally signals a lower mechanical strength. However, as consistently observed, the change of the porosity and mechanical strength with further H_2 concentration increment is very marginal after the H_2 concentration reached up to 25%. In this case, a higher H_2 concentration than 25% in the sintering atmosphere is not very necessary as it will increase the operational cost related to the safety issues during the sintering operation.

Figure 6 displays the N_2 and water permeance of the sintered membranes versus the H_2 content in the sintering atmosphere, where the sintering temperature and the dwelling time were fixed at 1050°C and 2 h, respectively. As expected, both the nitrogen and water permeance increases with increasing the hydrogen concentration in the sintering atmosphere, since the presence of hydrogen induces more porosity especially on the hollow fiber surfaces. For example, when sintered in nitrogen atmosphere, the hollow fiber exhibited a nitrogen permeance of $1.77×10^{-4}$ mol m^{-2} s^{-1} Pa^{-1}, which is of the same order of magnitude as reported in the literature [16]. As the hydrogen content in the sintering

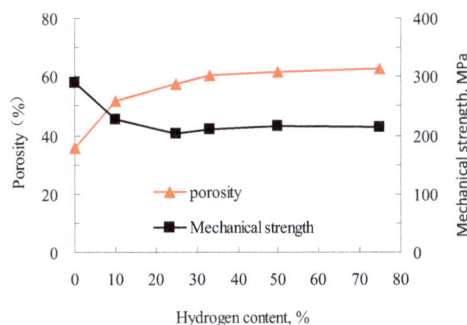

Figure 5: Effect of hydrogen content in the H_2-N_2 sintering atmosphere on the porosity and mechanical strength of the PSS hollow fibers sintered at 1050°C for 2 h.

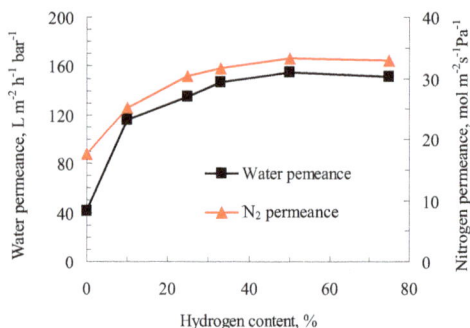

Figure 6: Plots of nitrogen and pure water permeances of the PSS hollow fibers sintered at 1050°C for 2 h against the hydrogen content in the H_2-N_2 sintering atmosphere.

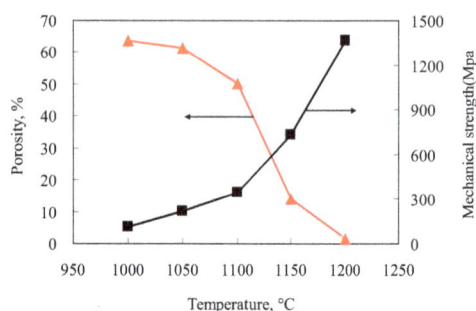

Figure 7: Effect of sintering temperature on the porosity and mechanical strength of the PSS hollow fiber membranes sintered under the 25% H_2-containing atmosphere.

stream was increased to 10% and 25%, the nitrogen permeance would increase by 42.8% and 72.6% to 2.52×10^{-4} and 3.05×10^{-4} mol m^{-2} s^{-1} Pa^{-1}, respectively. These values for the permeance mean that the fibers are suitable as porous support structures for gas separation membranes. Likewise, the water permeance increased significantly from 41.6 L m^{-2} h^{-1} bar^{-1} for fiber sintered in the nitrogen atmosphere to 115.2 (2.79 times) and 135.4 (3.25 times) L m^{-2} h^{-1} bar^{-1} for the fibers sintered in the 10% H_2-N_2 and 25% H_2-N_2 sintering atmospheres, respectively. Again, the permeation results also indicate that the higher hydrogen concentration than 25% in the sintering atmosphere would yield a limited increase in both the nitrogen and water permeance, thus is not necessary for sake of safety and operational cost.

For the H_2-containing atmosphere (25% H_2), other sintering parameters such as the sintering temperature and dwelling time have also been investigated with the results presented in Figures 7-9. Figure 7 illustrates the porosity and mechanical strength of the PSS hollow fibers sintered at different temperatures with a fixed dwelling time of 2 h. In general, there is a trade-off between the porosity or pemeance and the mechanical strength. That is, a larger porosity often mirrors the better performance in fluid permeance but suffers a lower mechanical strength due to the insufficient bonding at lower temperature. An important indication from this figure is that the sintering temperature should not be higher than 1100°C otherwise the porosity would decrease remarkably from 50.3% to 14.1%, although the mechanical strength has also improved significantly. This implies that the permeation properties of the PSS hollow fibers would be deteriorated noticeably

as the sintering temperature increases. As shown in Figure 8, the water permeance decreased significantly from 155.0 L m^{-2} h^{-1} bar^{-1} to 38.3 L m^{-2} h^{-1} bar^{-1}, and then to 2.2 L m^{-2} h^{-1} bar^{-1} as the sintering temperature was increased from 1050°C to 1100°C and 1150°C, respectively. Further increment of the sintering temperature to 1200°C will transfer the hollow fiber into a dense structure, for which no gas permeation could not be detected any more.

Figure 9 shows the porosity and mechanical strength of the PSS hollow fibers sintered at 1050°C under the 25% H_2-containing atmosphere as a function of sintering time. It can be seen that the porosity decreased from 65.1% to 61.7%, while the mechanical strength increased noticeably from 107.9 MPa to 210.9 MPa as the sintering time was extended from 30 min to 90 min, but further extension of the sintering time would hardly change the mechanical strength and the porosity any more. This implies that the fusion and bonding of the stainless particles mainly occurred within the initial 90 min duration. In order to guarantee the complete sintering, two hours of sintering time is sufficient for economical purpose.

Conclusions

Porous stainless steel (PSS) hollow fibers with asymmetric structure have been developed through the immersion-induced phase inversion and sintering technique. The sintering atmosphere has significant influence on the microstructure and properties of the resultant hollow fiber membranes. Experimental results indicate that H_2-containing atmosphere for sintering can lead to the formation of highly permeable

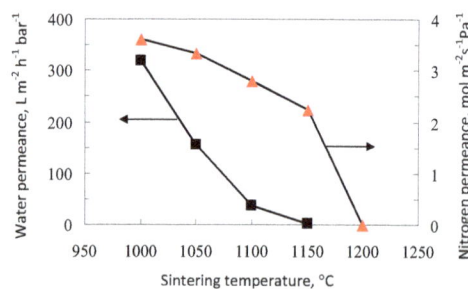

Figure 8: Nitrogen and pure water permeance of the PSS hollow fiber membranes sintered at different temperatures under the 25% H_2-containing atmosphere.

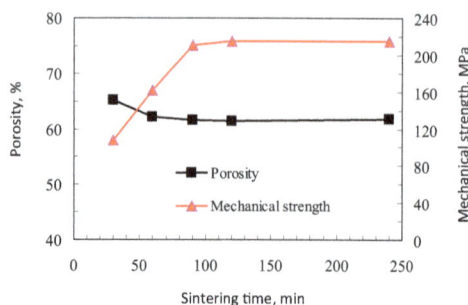

Figure 9: Effect of sintering time on the porosity, mechanical strength of the PSS hollow fiber membranes sintered under the 25% H_2-containing atmosphere.

porous membranes with desirable mechanical strength, but the H_2 concentration in the sintering atmosphere higher than 25% in nitrogen is not necessary as it yields a limited improvement on the membrane performances. For a sufficient sintering, the sintering temperature should be higher than 1050°C so as to yield a desirable mechanical strength, but not higher than 1100°C, otherwise the porosity will be decreased noticeably due to densification. For cost effective and less time consuming, sintering time can be controlled within two hours. The resultant PSS hollow fibers possess the advantages of high porosity, high fluid permeance, robustness to be handled with stress, thereby in addition to be directly applied for microfiltration, they are also suitable to be used as porous supports for next step thin membrane synthesis for gas separations or desalinations.

Acknowledgements

The authors gratefully acknowledge the research funding provided by the National Natural Science Foundation of China (NSFC 21176187), the Research Fund for the Doctoral Program of Higher Education of China (RFDP 20131201110007) and the Program for Changjiang Scholars and Innovative Research Team in University (PCSIRT) of Ministry of Education of China (IRT 13084).

References

1. Burggraaf AJ, Cot L (1996) Fundamentals of Inorganic Membrane Science and Technology. Elsevier Science B.V, The Netherlands.

2. Julbe A, Farrusseng D,Guizard C (2001) Porous ceramic membranes for catalytic reactors - overview and new ideas. Journal of Membrane Science 18: 3-20.

3. Liang W, Hughes R (2005) The catalytic dehydrogenation of isobutane to isobutene in a palladium/silver composite membrane reactor. Catalysis Today, 104: 238-243.

4. She Y, Han J, Ma YH (2001) Palladium membrane reactor for the dehydrogenation of ethylbenzene to styrene. Catalysis Today 67: 43-53.

5. Lin WH, Chang HF (2004) A study of ethanol dehydrogenation reaction in a palladium membrane reactor. Catalysis Today 97: 181-188.

6. Bernal MP, Coronas J, Menendez M, Santamaria J (2002) Coupling of reaction and separation at the microscopic level: esterification processes in a H-ZSM-5 membrane reactor. Chemical Engineering Science 57: 1557-1562.

7. Kong C, Lu J, Yang J, Wang J (2007) Catalytic dehydrogenation of ethylbenzene to styrene in a zeolite silicalite-1 membrane reactor. Journal of Membrane Science 306: 29-35.

8. Haag S, Hanebuth M, Mabande GTP, Avhale A, Schwieger W, et al. (2006) On the use of a catalytic H-ZSM-5 membrane for xylene isomerization. Microporous and Mesoporous Materials 96: 168-176.

9. Brands K, Uhlmann D, Smart S, Bram M, da Costa JCD (2010) Long-term flue gas exposure effects of silica membranes on porous steel substrate. Journal of Membrane Science 359: 110-114.

10. Tan X, Liu S, Li K (2001) Preparation and characterization of inorganic hollow fiber membranes. Journal of Membrane Science 188: 87-95.

11. Tan X, Liu Y, Li K (2005) Preparation of La0.6Sr0.4Co0.2Fe0.8O3-δ hollow fiber membranes for oxygen production by a phase-inversion/sintering technique. Industrial and Engineering Chemistry Research 44: 61-66.

12. Meng B, Tan X, Meng X, Qiao S, Liu S (2009) Porous and dense Ni hollow fibre membranes. Journal of Alloys and Compounds 470: 461-464.

13. Luiten-Olieman MWJ, Raaijmakers MJT, Winnubst L, Bor TC, Wessling M, et al. (2012) Towards a generic method for inorganic porous hollow fibers preparation with shrinkage-controlled small radial dimensions, applied to Al2O3, Ni, SiC, stainless steel, and YSZ. Journal of Membrane Science 407–408: 155-163.

14. Luiten-Olieman MWJ, Winnubst L, Nijmeijer A, Wessling M, Benes NE (2011) Porous stainless steel hollow fiber membranes via dry–wet spinning. Journal of Membrane Science 370: 124-130.

15. Luiten-Olieman MWJ, Raaijmakers MJT, Winnubst L, Wessling M, Nijmeijer A, et al. (2011) Porous stainless steel hollow fibers with shrinkage-controlled small radial dimensions. Scripta Materialia 65: 25-28.

16. Michielsen B, Chen H, Jacobs M, Middelkoop V, Mullens S, et al. (2013) Preparation of porous stainless steel hollow fibers by robotic fiber deposition. Journal of Membrane Science 437: 17-24.

17. Aran HC, Pacheco Benito S, Luiten-Olieman MWJ, Er S, Wessling M, et al. (2011) Carbon nanofibers in catalytic membrane microreactors. Journal of Membrane Science 381: 244-250.

18. Li H, Song J, Tan X, Jin Y, Liu S (2015) Preparation of spiral porous stainless steel hollow fiber membranes by a modified phase inversion–sintering technique. Journal of Membrane Science 489: 292-298.

19. K. Li (2007) Ceramic Membranes for Separation and Reaction. (1stedn), John Wiley and Sons Ltd, West Sussex, England.

20. Rui W, Zhang C, Cai C, Gu X (2015) Effects of sintering atmospheres on properties of stainless steel porous hollow fiber membranes. Journal of Membrane Science 489: 90-97.

21. Yang N, Tan X, Ma Z (2008) A phase inversion/sintering process to fabricate nickel/yttria-stabilized zirconia hollow fibers as the anode support for micro-tubular solid oxide fuel cells. Journal of Power Sources 183: 14-19.

22. Tan X, Liu Y, Li K (2005) Mixed conducting ceramic hollow fiber membranes for air separation. AIChE Journal 51: 1991-2000.

23. Kurgan N (2013) Effects of sintering atmosphere on microstructure and mechanical property of sintered powder metallurgy 316L stainless steel. Materials & Design 52 995-998.

24. Dudek A, Włodarczyk R (2013) Effect of sintering atmosphere on properties of porous stainless steel for biomedical applications. Materials Science and Engineering C 33: 434-439.

25. Umm-i-K, Bashir S, Ali N, Akram M, Mahmood K, et al. (2012) Effect of ambient environment on excimer laser induced micro and nano-structuring of stainless steel. Applied Surface Science 261: 101-109.

26. Kingsbury BFK, Wu Z, Li K (2010) A morphological study of ceramic hollow fibre membranes: A perspective on multifunctional catalytic membrane reactors. Catalysis Today 156: 306-315.

27. Kingsbury BFK, Li K (2009) A morphological study of ceramic hollow fibre membranes. Journal of Membrane Science 328: 134-140.

16

NanoEmulsion for Nanotechnology Size-Controlled Synthesis of Pd (II) Nanoparticles via NanoEmulsion Liquid Membrane

Said N El*, Kassem AT, and Aly HF

Hot Labs, AtomicEnergy Authority, Egypt

Abstract

Palladium nanoparticles with low polydispersity were fabricated via emulsion globules of various sizes with same sauter mean diameter has been proposed for this purpose. In this paper, the important variables affecting sauter mean diameter of the emulsion drops, including injection method of emulsion, stirring speed, oil phase viscosity, composition of inner water phase and solute permeation rate are also systematically investigated of HCl/Co (III) dicarbolide/xylene/HCl forming $(PdCl_4)^{-2}$ in the presence of span 80/85 as the surfactants. The particle size, ranging from 3.2 to 4.2 nm, was optimized be controlled by variation of the surfactant, the agitation time and the concentrations. Nano emulsions were prepared using the spontaneous emulsification mechanism as non-equilibrium systems. XRS, SEM and surface area were used for analysis and the optimized coditions of nanopalladium was 2.5, 1.1 and 0.1 for O/S, glouble diameter and polydispersity respectively. The studies on optimization methods for nano-emulsion gloubles be a required.

Keywords: Emulsion liquid membrane; Globule size; Nano palladium; Spain80/85; Polydispersity

Introduction

Palladium is used in jewelry in dentistry [1,2] and production of surgical [3] instruments. Palladium is also used to make professional transverse flutes [4]. The second biggest application of palladium in electronics is making the multilayer ceramic capacity [5]. Palladiums (and palladium-silver alloys) are used as electrodes in multi-layer ceramic capacitors; Palladium (sometimes alloyed with nickel) is used in connector plating's in consumer electronics. Hydrogen easily diffuses through heated palladium; thus, it provides a means of purifying the gas [6]. Membrane reactors with Pd membranes are therefore used for the production of high purity hydrogen. A large number of carbon-carbon bond forming reactions in organic chemistry are facilitated by catalysis with palladium compounds. The largest use of palladium today is in catalytic converters [7]. In addition palladium, when dispersed on conductive materials, proves to be excellent electro catalysts for oxidation of primary alcohols in alkaline media [8]. These recent applications have made that studies on optimization methods for nano-emulsion preparation of palladium be required. This work is focused on the most recent developments of nano-emulsions as final application products and on the optimization of their preparation. Nano emulsions consist of fine oil-in-water dispersions, having droplets covering the size range of 100–600 nm. In the present work, nano-emulsions were prepared using the spontaneous emulsification mechanism which occurs when an organic phase and an aqueous phase are mixed. The organic phase is a homogeneous solution of oil, lipophilic surfactant and water–miscible solvent, the aqueous phase consists on hydrophilic surfactant and water. An experimental study of nano-emulsion process optimization based on the required size distribution was performed in relation with the type of oil, surfactant and the water–miscible solvent. The results showed that the composition of the initial organic phase was of great importance for the spontaneous emulsification process, and so for the physico-chemical properties of the obtained emulsions. First oil viscosity and HLB surfactants were changed, the most viscous oil gave the smallest droplets size (16 ± 2 nm) HLB required for the resulting oil-in-water emulsion was superior to 8. Second, the effect of water–solvent miscibility on the emulsification process was studied by decreasing xylene proportion in the organic phase. The solvent–xylene proportion leading to a fine nano-emulsion was fixed at 10% of membrane. This phase of emulsion optimization represents an important step in the process of nano size droplet by spontaneous emulsification. In the separation process using emulsion liquid membranes, the polydispersity affects mass transport of metal ions from the external phase to the internal phase because under steady operating conditions, drop size and size distribution are proportional to the interfacial area. The present study aims to assess the optimized conditions nano size of emulsion globules. An advancing reaction front model is considering. The work divided in two parts, first part the optimized conditions of preparation of palladium, and the second part is the suitable conditions for per traction of palladium. The modeling of transport of cations was achieved by advancing stripping model [9]. The carrier mediated transport from high salt content using TBP as membrane was done [10,11].

Nanoemulsion

The formation properties and applications of nano-emulsions are referred as mini emulsions, ultrafine emulsions, and submicron emulsions [12-15]. Nano-emulsion droplet sizes fall typically in the range of 20–200 nm and show narrow size distributions. The most publications on either oil-in-water (Oil/water) or water-in-oil (W/O) nano-emulsions report their formation by dispersion or high energy emulsification methods.

Nano particles and Experimental design

These materials are prepared for permeation metal ions followed by stripping to yield $(PdCl_4)^{-2}$ in stripping phase to yield palladium

*Corresponding author: Said N El, Hot Labs, AtomicEnergy Authority, Egypt
E-mail: neseem.abdel@yahoo.com

metal nanoparticle at the end of the process. The size of such particles depends on the number of metal ions initially loaded into the globules [16]. An orthogonal 24 factorial central composite experimental design with 6 star points (I = 1.68) and 6 replicates at the center point, all in duplicates, resulting in a total of 20 experiments were used to optimize the chosen key variables for the extraction of palladium. The experiments with different agitation speeds are 13000, 14000, 15000, 16000 and 17000 rpm and M/E ratio 0.1, 0.2, 0.3, 0.4 and 0.5(%v/v) and five different carrier concentration 0.01, 0.05, 0.1, 0.15 and 0.2 M of Co (III) dicarbolide of employed simultaneously covering the spectrum of variables for the percentage extraction of palladium in the Central Composite Design. In order to describe the effects of agitation speed (X1), M/E ratio (X2) and carrier concentration (X3) on percentage of palladium extraction, batch experiments were conducted (Table1). Full factorial design matrix of screening experiments and mean droplet diameter measured (Table 2).

Experimental field for a design matrix: variables emulsion properties measured. For production of Pd nanoparticle by emulsion liquid membrane (ELM). The first step is the use of emulsion liquid membrane cell. The

Run	%span80/85	O/S	Addition time (min)	Agitation rate (rpm)	Droplet diameter (nm)
1	0.09 0.1	1.0	1	14000	15.085
2	0.1	1.0	1	14400	16..05
3	0.2	2.0	1	15000	20.0.5
4	0.3	2.0	2	16000	23.45
5	0.4	2.0	2	16500	18.1
6	0.8	2.0	2	17000	25.45
7	1.5	2.5	3	17000	24.65
8	2	2.5	3	16000	26.55
9	2.5	3.0	4	14000	28.3
10	2.8	3.0	5	14000	42.05
11	3	3.5	5	16000	48.15
12	3.2	3.5	5	16000	60.655
13	4	2.5	6	16000	29.38
14	4.5	2.5	6	15000	35.48
15	5	3.5	6	14000	42.618
16	5.5	3.5	6	14000	45.65

Table 1: Full factorial design matrix of screening experiments and mean droplet diameter measured.

Run	S80/S85	O/S	Droplet diameter	Polydispersity
1	4.51	2	12.16	1.5312
2	4.51	3	17.45	2.6576
3	5.12	2	9.7	1.0296
4	5.12	3	16.06	1.0296
5	4.38	2.5	14.8	1.98
6	5.25	2.5	14.16	1.1792
7	4.82	1.79	7.6	0.8096
8	4.82	3.21	14.82	1.2584
9	4.82	2.5	11.1	1.1528
10	48.2	2.50	11.1	0.100
13	4	2.5	16	29.38
14	4.5	2.5	14	35.48
15	5	3.5	14	42.618
16	5.5	3.5	14	45.65

Table 2: Experimental field for a design matrix: variables emulsion properties measured.

second step is using the (F + M). The third step is separation of Feed from membrane. The forth step is de-emulsioncation of membrane (M) by electric charge using electric charge pistole. The fifth step is separation of strip from organic. The sixth step is heating and drying the stripping phase using a heater till 300oC to obtain Palladium nanoparticles and investigated by X-ray and SEM instruments (Figures 3-5). The treatment combinations and responses of two central composite designs that Powder XRD patterns of samples were recorded with a SHIMADZU XD-D1 Diffractometer using Ni-filtered CuKa radiation (k = 1.5406 A°) with the scan rate of 0.1/s. TEM analysis was carried out using a Philips CM12 TEM (transmittance electron microscope), working at a 100 kV accelerating voltage. Samples for TEM analysis were prepared by dispersing palladium, pd nanoparticles in ethanol followed by drop-casting on a copper grid (400 meshes). 9 - (Feed + Membrane), 10 - (Feed + Membrane), 11 - (Feed Membrane) (F + M), 12 - (Feed Membrane) (F + M), 12 - S, 13 – O (oxygen charge pistle), part-B.

Experimental part for kinetic pertraction

Co (III) dicarbolide/xylene, the surfactants of Liquid Emulsion Membrane SPAN 80 and 85 (sorbitol mono-and trioleate) and other chemicals were analytical grade. A turbine type impeller was used for preparation of the liquid membrane with organic; water volume ratio (r$_1$ = 1) and at a mixing rate (14000 – 16000) rpm for (5 – 10) min. Some emulsions remained stable for months. When breakage of emulsions was desired alcohols were used. The extraction of multiple (Water/Oil/water) W/O/W emulsion was performed in a multi stages double-Jackets cell thermo stated temperature 25oC by stirring with magnetic bar (50 x 10 mm) stirrer at (300 - 700) rpm usually with 50 ml of outer (feed) solution. The extraction was followed by sample taking (0.5 - 1.0ml) from solution and measuring by the Atomic absorption spectroscopy type. Atomic absorption/Emission Spectrophotometer/210/VGP, Buck Scientific, USA was used for determination of palladium concentration. The pH values were measured using a pH –meter of the type B-417 HANA Instrument hydrogen ion concentration in the solution. The deviations in the readings were in the range of ± 0.02 at the laboratory temperature 25 ± 2oC the cell used in pertraction of nano palladium metal using emulsion liquid membrane is shown. This design is repre¬sented by a second-order polynomial regression model (Equation 1), to generate contour plots

$$Y = b_0 + b_1 x_1 + b_2 x_2 + b_{11} x_{2\,1} + b_{22} x_{2\,2} + b_{12} x_1 x_2 + \varepsilon . (1)$$

Results and Discussion

Characteristics and properties of nano-emulsions, as non-equilibrium systems independent only on composition but also in the preparation method, although interest in nano-emulsions was developed since about 20 years ago, mainly for nanoparticle preparation, it is in the last years that direct applications of nano-emulsions in consumer products are being developed, mainly in pharmacy. These results were made for studies on optimization methods for nano-emulsion preparation is necessary [12].

Preparation of nano particles from nano emulsion

Palladium chloride (PdCl2) nano-particles with Pd/PdCl2 = 65:35 % were prepared by an emulsion liquid membrane technique under mechanical agitation without the aid of any surface-active agent. SEM images of black particles are shown in (Figure 4 and 5). Aggregates of irregular shaped particles are observed and the size of Pd particles varies from 4 to 33 nm. A nanoparticle also is observed in SEM image as shown in (Figure 1 and 5) before and after preparation of nanoparticles of palladium. Hence, it was difficult to calculate the particle size distribution from SEM images. The (Figure 3) is indexed to be corresponding to (112), (2020), (218), (309), (329) and (418) of Pd metal. Powder XRD pattern of Pd nanoparticles is shown in (Figure 2) the d-spacing corresponding to XRD lines are 1.986, 1.888, 1.299, 1.099 and 1.098 A°. These d-spacing values correspond to (111), (200), (220), (311) and (222) planes [13-16]. Particle size analysis showed a narrow distribution of 25–120 nm range particles with a mean size of 60 nm, thus, confirming their nano-structured nature. Nano-particles of palladium powders were prepared at a significantly low temperature. From the above optimization process different parameters were studied for preparation of nanoparticles and the different parameters were studied as in (Figures 1-3). XRD spectra of palladium nanoparticles (Figure 4 and 5) SEM image of palladium before and after using ELM at 20°C and pH=4. One with O/S, oil/surfactant(80/85) relation the

optimization of three parameters as 3-D was shown in (Figure 6) (X1) is the agitation speed, rpm, M/E ratio %(v/v) which is the volume of membrane/feed, (X2) and carrier concentration (X3) and the plotting of two dimensional contour plots (Figure 7).

Droplet size and polydispersity Emulsion droplet size and polydispersity (intensity based size distributions) were measured by photon correlation spectroscopy (PCS) using Malvern Zetasizer ZS at 25°C. Samples were diluted with water for the measurements The crystallite size was calculated using the Debye–Scherrer formula, $D = 0.9\lambda/\beta$ where D is the particle size, λ the wavelength of the X-ray used, β, θ are the half-width of X-ray diffraction lines and half diffraction angle of 2θ. The crystallite size was found to be between 30 and 8 nm template.

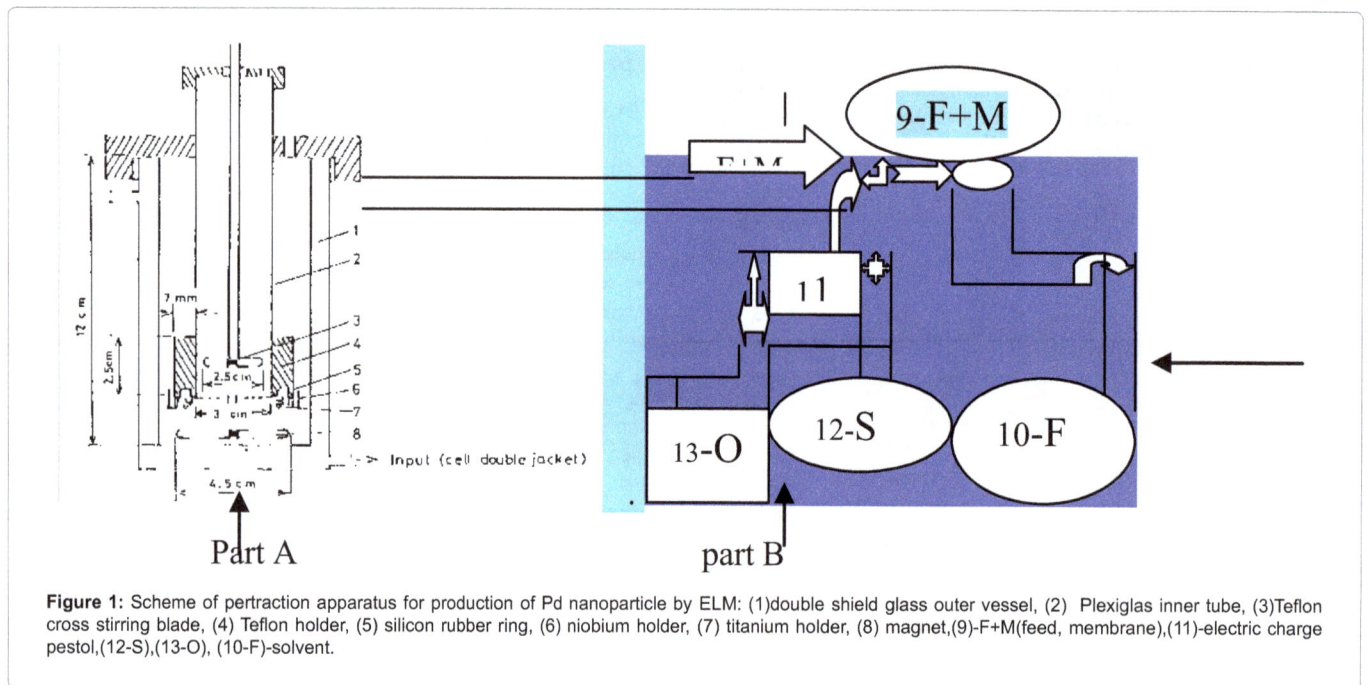

Figure 1: Scheme of pertraction apparatus for production of Pd nanoparticle by ELM: (1)double shield glass outer vessel, (2) Plexiglas inner tube, (3)Teflon cross stirring blade, (4) Teflon holder, (5) silicon rubber ring, (6) niobium holder, (7) titanium holder, (8) magnet,(9)-F+M(feed, membrane),(11)-electric charge pestol,(12-S),(13-O), (10-F)-solvent.

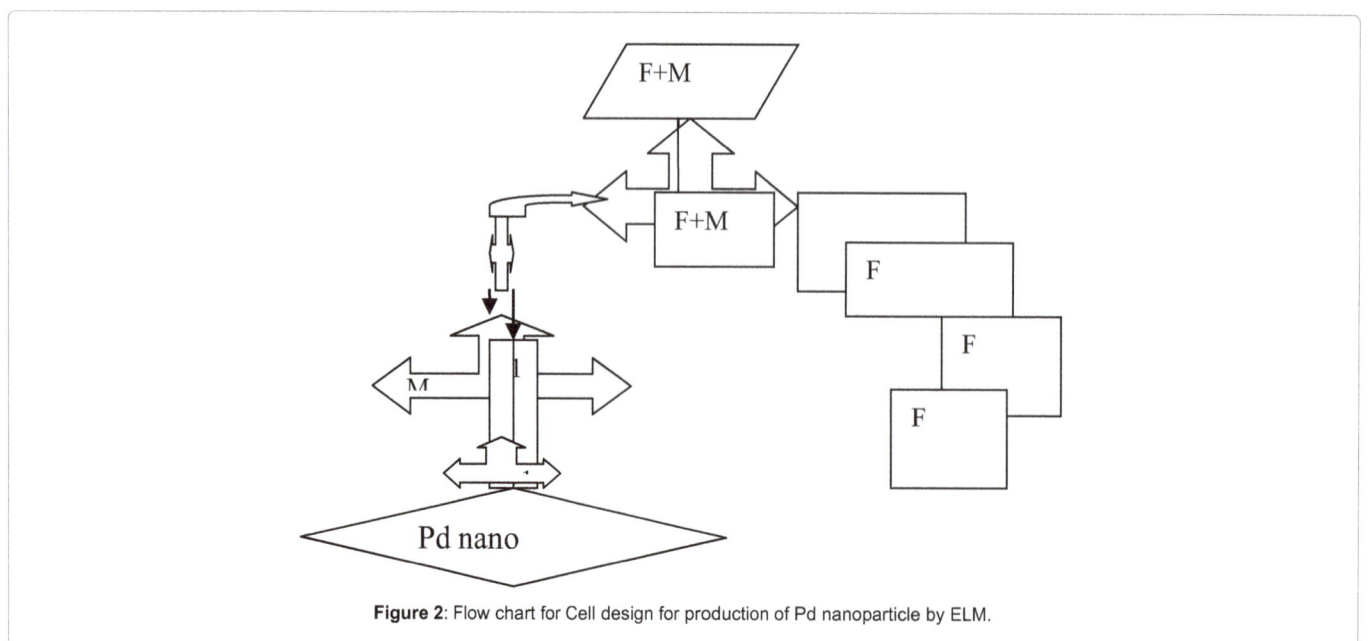

Figure 2: Flow chart for Cell design for production of Pd nanoparticle by ELM.

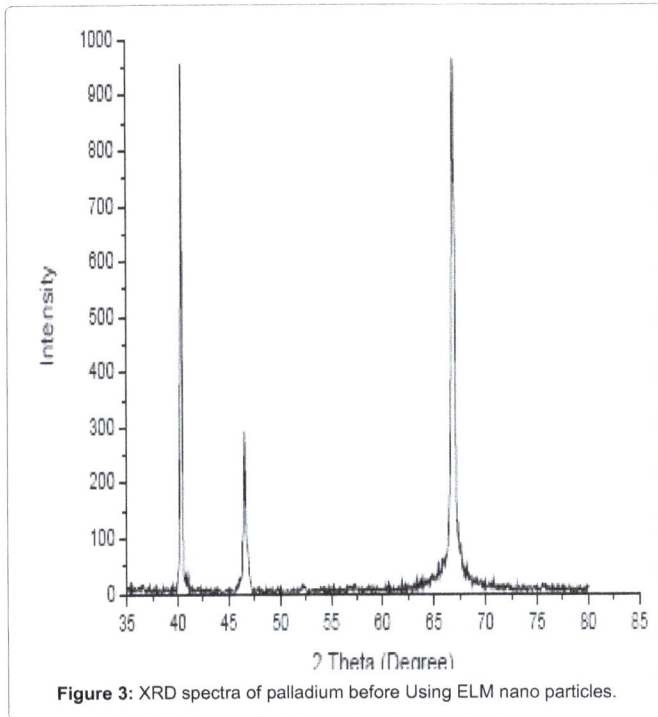

Figure 3: XRD spectra of palladium before Using ELM nano particles.

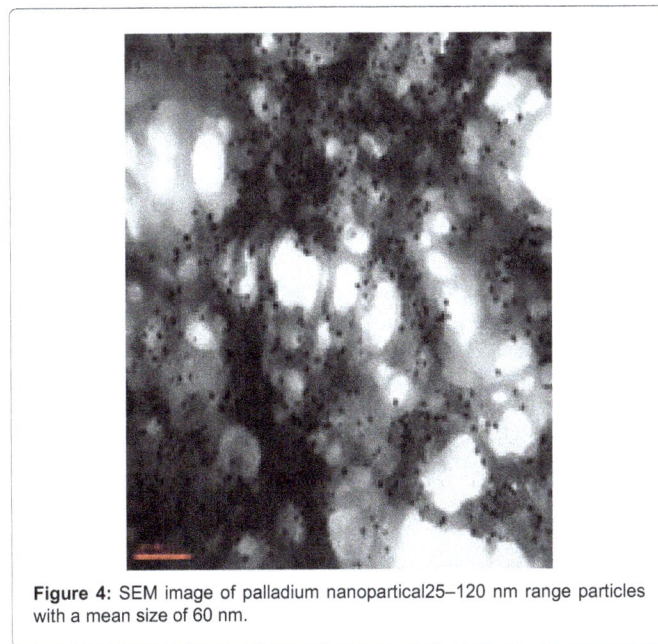

Figure 4: SEM image of palladium nanopartical25–120 nm range particles with a mean size of 60 nm.

Each 3D plot represents the number of combinations of the three variables, showing the effects of agitation speed, M/E ratio and their mutual interaction on extraction of palladium emulsion liquid membrane, ELM using Co (III) dicarbolide as carrier.

Kinetic study

At a constant interface area in double emulsion system and when the concentration of separated element in feed solution is much lower than the concentration of carrier in membrane and their chemical interaction does not change the concentration of carrier at interphase substantially, most of the results on membrane extraction (pertraction) can be approximated by a pseudo-first order kinetic law [11]

$$Log(1 - \frac{R}{R\infty}) = -Kt \tag{1}$$

Or

$$R = R_{\infty}(1-e^{-kt}) \tag{2}$$

Where R is the fraction yield of pertraction .

K is the rate of pertraction min$^{-1.}$

Influence of aqueous phase acidities

Influence of the feed solution acidity from pH (1 to 4). Figure 8 dipicts the effect pH on the extraction of nano particle of Pd (II), it is found that as the pH increases the pertraction of palladium increases and the maximum yield of pertraction reached 0.98 at pH = 4.

Effect of membrane phase

Figure 9 Show the influence of membrane concentration from (0.01M - 0.06M HCl) on the pertraction of Pd^{+2}, as the membrane increases, the extraction yield increases till 0.95 at 0.01M.

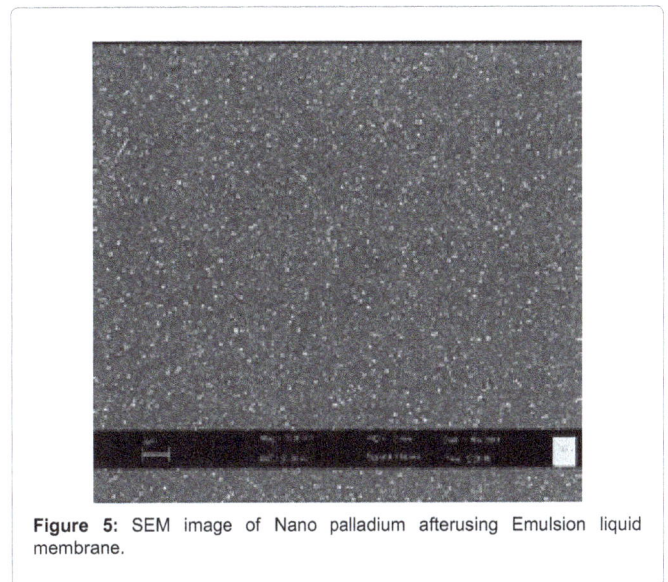

Figure 5: SEM image of Nano palladium afterusing Emulsion liquid membrane.

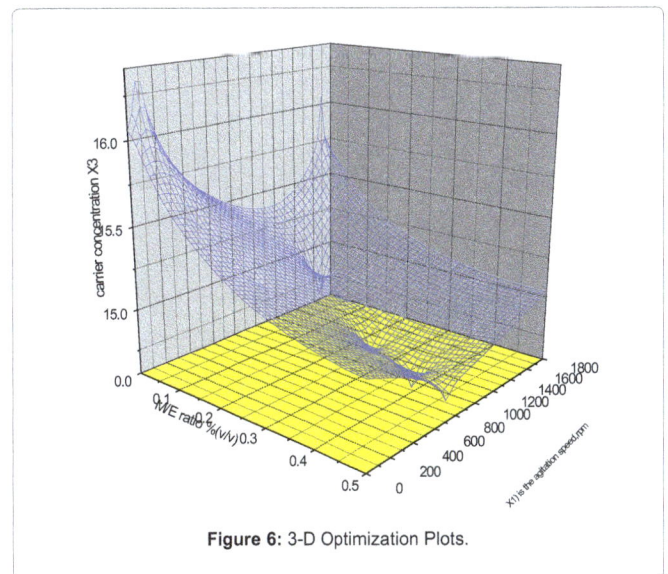

Figure 6: 3-D Optimization Plots.

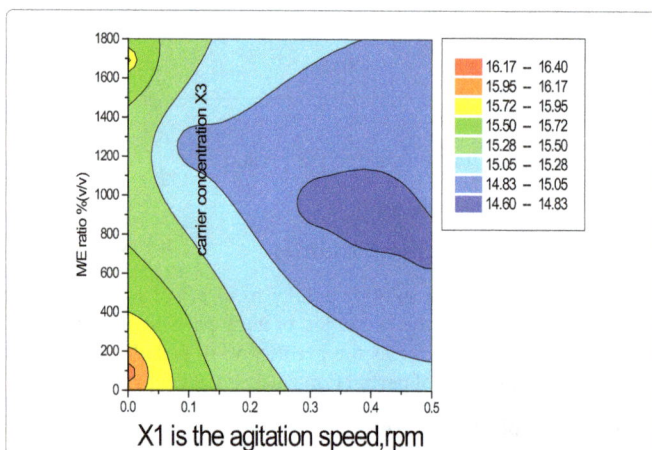

Figure 7: The contour plot.s of (X1), M/E ratio (X2) and X3 and carrier concentration, M.

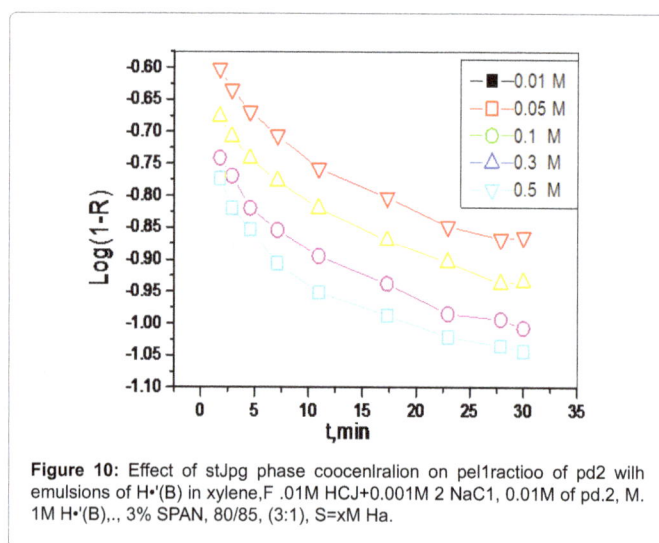

Figure 8: Effect of pH on pertraction of pd^{+2} with emulsion of H$^+$(B)$_2$ in xylene, F=0.001M HCl+0.01 NaCl, 0.01M of pd^{+2}, M=0.01MH^{+1}(B)$_2$, 3% SPAN, 80/85, (3:1), S=0.03M HCl.

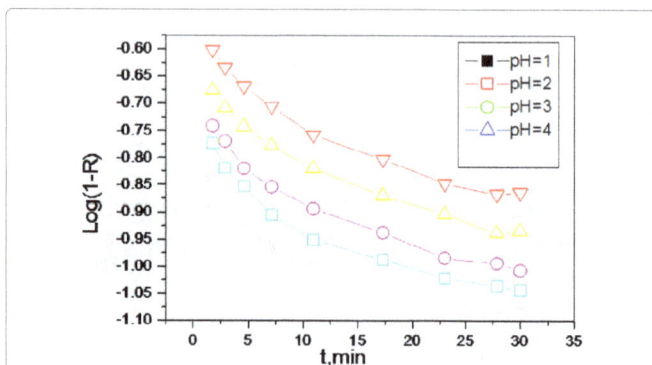

Figure 9: Effect of membrane phase conceniJation on pertraction of pcfwith errulsicms ofW'(B),.m xylene,F=0.01M Ha+().01M Naa,0.01M of pa^2,M=x0.1M W'(B),.,3%SPAN,80/85(3:1) 5={}_03 M Ha.

Effect of stripping phase

Figure 10 Shows the influence of strip concentration from (0.01M -0.5M HCl) on the pertraction of Pd (II). It is interesting to note that the effects of changing strip concentration on the transport of Pd (II) achieve to obtain 99%of pertraction yield were different.

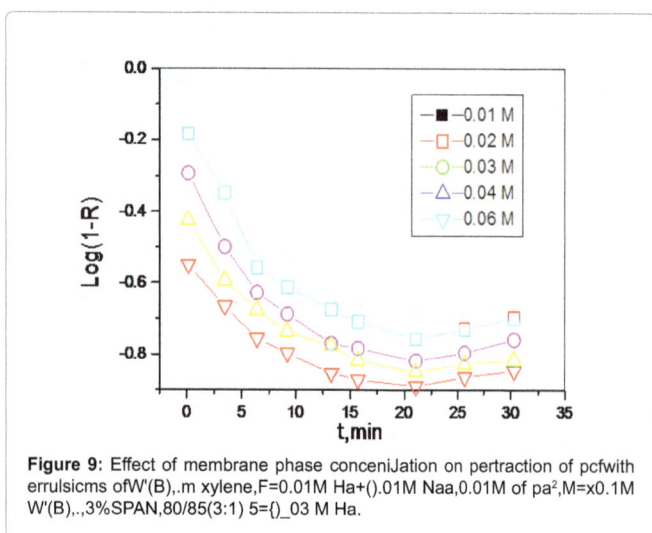

Figure 10: Effect of stJpg phase coocenlralion on pel1ractioo of pd2 wilh emulsions of H•'(B) in xylene,F .01M HCJ+0.001M 2 NaC1, 0.01M of pd.2, M. 1M H•'(B),., 3% SPAN, 80/85, (3:1), S=xM Ha.

Conclusion

The of preparation of palladium by Co(III) dicarbolide from aqueous feeding solution to yield nano particles of palladium in the globule of stripping phase was a chivied and can be studied via optimization at different surfactant concentration, the agitation speed and emulsification time to use nano particles of palladium for pharmaceutical industry. It is found that the emulsion stability was carried out by varying surfactant concentration, agitation speed and emulsification time. The optimum process conditions for the extraction of palladium 1500 rpm, M/E ratio–0.75% (v/v) and carrier concentration 5% (v/v). In nano of ELM, palladium can be optimized through size control of the globule size nano emulsion for nanotechnology. A size-controlled synthesis of Palladium nanoparticles via nano emulsion liquid membrane, ELM, was succeeded. Criterion will depend on particular weight adjudged to either recovery yield, reactants cost and time of performance. There are maxima on neither partial global criterial functions, just local minimum (worst performance) should be avoided to make the process reasonable. A Nano particle size pertraction of palladium was optimized using ELM technique and the effect of different parameters such feeding, membrane and stripping phases were studied. The 3-D optimization and contour plot predict conditions of preparation of nanoparticles of palladium.

References

1. Kristine HE, Gregory MM, Rauch S (2004) Environmental routes for platinum group elements to biological materials a review. J Science of the Total Environment: 21.

2. Hesse, Rayner W (2007) Palladium. Jewelrymaking through history: an encyclopedia Greenwood Publishing Group J: 146

3. S Navaladian (2009) A Rapid Synthesis of Oriented Palladium Nanoparticles by UV Irradiation. J Nanoscale Research Letters: 02

4. Chakraborty M (2003) Effect of drop size distribution on mass transfer analysis of the extraction of nickel (II) by emulsion liquid membrane. Colloids and Surfaces A: Physicochemical and Engineering Aspects: 829.

5. Rajasimman M (2009) Optimization of process parameters for the extraction of chromium (VI) by emulsion liquid membrane using response surface methodology. J of Hazardousn Materials: 830.

6. Grochala W, Edwards PP (2004) Thermal Decomposition of the Non-Interstitial Hydrides for the Storage and Production of Hydrogen. J Chem Rev: 283.

7. Solans C, Izquierdo P, Nolla J, Azemar N, Garcia MJ, et al. (2005) Current Opinion in Colloid & Interface Science 102.

8. JM Gutiérrez, C González, A Maestro, I Solè, CM Pey et.al (2008) Current Opinion in Colloida & Interface Science. 245.

9. El-Said N (2003) Modeling of transport of Cs (137) by emulsion liquid membrane (18C6) in xylene Promoted by ephedrine hydrochloride in stripping phase. J of Membrane Science: 131.

10. Nowier H (2000) Carrier-mediated transport of toxic elements through liquid membranes. J of Membrane Science: 830.

11. Said N (1992) Emulsion liquid membrane extraction of fission products promoted by polyvalent and organic acids. J of Radioanalytical and Nuclear Chemistry Articles: 11.

12. Navaladian S, Viswanathan B, Varadarajan TK, Viswanath RP (2009) A Rapid Synthesis of Oriented Palladium Nanoparticles by UV Irradiation's. Nanoscale Res Lett: 181.

13. Nishi T, Takeichi A, Azuma H, Suzuki N, Hioki T, et al. (2010) Fabrication of Palladium Nanoparticles by Laser Ablation in Liquid. JLMN-J of Laser Micro/Nanoengineering: 3.

14. Pey CM (2006) Optimization of nano-emulsions prepared by low-energy emulsification methods at constant temperature using a factorial design study. J of Colloids and Surfaces A: Physicochemical and Engineering Aspects: 1005.

15. Sathishkumara M, Snehaa K, Kwaka IS, Maoa J, Tripathy SJ et al. (2009) Phyto-crystallization of palladium through reduction process using Cinnamomzeylanicum bark extracts. J of Hazardous Materials: 171.

16. Alexander P, Majewski, Schallon A, Jérôme V, Freitag R, et al. (2012) Dual-Responsive Magnetic Core–Shell Nanoparticles for Nonviral Gene Delivery and Cell Separation. J of Biomacromolecules: 857.

17

Chitosan-Based Anion Exchange Membranes for Direct Ethanol Fuel Cells

Birgit Feketeföldi[1], Bernd Cermenek[2*], Christina Spirk[3], Alexander Schenk[2], Christoph Grimmer[2], Merit Bodner[2], Martin Koller[3], Volker Ribitsch[3] and Viktor Hacker[2*]

[1]Institute for Surface Technologies and Photonics, JOANNEUM RESEARCH Forschungsgesellschaft mbH/Materials, Franz-Pichler-Straße 30, 8160 Weiz, Austria
[2]Institute of Chemical Engineering and Environmental Technology, Fuel Cell Systems Group, Graz University of Technology, NAWI Graz, Inffeldgasse 25C, 8010 Graz, Austria
[3]Institute of Chemistry, University of Graz, Heinrichstraße 28, 8010 Graz, Austria

Abstract

A series of novel cross-linked highly quaternized chitosan and quaternized poly (vinyl alcohol) membranes were successfully synthesized to be applied in alkaline direct ethanol fuel cells. Cross-linking was accomplished using two different cross-linking agents and an additional thermal process to improve both chemical and thermal properties. Equivalent blends of chitosan and poly (vinyl alcohol) membranes with various degrees of cross-linking were prepared by using different amounts of glutaraldehyde and ethylene glycol diglycidyl ether as cross-linkers. To investigate their applicability in direct ethanol fuel cells, the membranes were characterized in terms of their structural properties, chemical, thermal and alkaline stability, ion transport and ionic properties using following methods: Fourier transform infrared spectroscopy, nuclear magnetic resonance spectroscopy, scanning electron microscopy, thermogravimetric analysis, water uptake by mass change, ethanol permeability in the diffusion cell, back titration method (ion exchange capacity) and electrochemical impedance spectroscopy (anion conductivity).

Despite the high degree of quaternization of the applied materials and regardless of the thin film thickness of the blend membranes, the novel cross-linked products displayed outstanding mechanical stability. The lower cross-linked membranes exhibited the best transport and ionic properties with a high anion conductivity of 0.016 S cm^{-1} and a high ion exchange capacity of 1.75 meq g^{-1}, whereas membranes with a higher degree of cross-linking performed superior in terms of reduced ethanol permeability of 3.30·10^{-7} cm^2 s^{-1} at 60°C. The blend membranes - chemically and thermally cross-linked - provide excellent thermal stability with an onset degradation temperature above 280°C and superb alkaline stability in 1.0 M KOH at 60°C for 650 h. Therefore, these composite membranes exhibit high potential for application as alkaline electrolytes in fuel cells.

Keywords: Annealing; Chitosan; Cross-linking; Direct ethanol fuel cells; Ethanol permeability; Poly(vinyl alcohol); Quaternized anion exchange membrane

Introduction

The direct ethanol fuel cell (DEFC) is a promising candidate for supplying portable applications such as laptops and mobile phones with electricity from a liquid fuel, i.e., ethanol [1]. In this process, the chemical energy of ethanol is directly converted into electrical energy. The development and optimization of the fuel cell components such as membranes and catalysts is decisive to enhance their power density.

In general, the use of anion exchange membranes (AEMs) in alkaline fuel cells facilitates: I. The use of non-noble metals such as nickel, cobalt, manganese or iron as electro-catalysts [2,3], II. The reduction of fuel crossover due to the reversed direction of hydroxide ions (electro-osmotic drag) compared to proton conductive membranes [4], III. The suppression of carbonate formation (by substitution of liquid alkaline electrolyte without mobile cations) [3] and, IV. The simplification of water management [3]. Due to alkaline conditions, the kinetics of both ethanol oxidation reaction (EOR) and oxygen reduction reaction (ORR) are enhanced and the costs of the DEFC components (catalyst and membrane) might be reduced [2,5].

An applicable AEM will have feasible transport and ion-exchange capabilities, mechanical and thermochemical stabilities, low degree of alkaline swelling and minor alcohol crossover. As demonstrated previously, a successful procedure utilizes blends of two or more polymers to combine the desirable properties of each component while reducing the shortcomings of the single polymers to obtain solid electrolyte membranes for fuel cells with a high quality [6-10].

Several techniques have been employed to prepare various alkaline anion-exchange membranes for fuel cell application, and distinguished reviews and original articles are available in literature [3,11-13]. Chitosan constitutes a polymer accessible from alkaline or enzymatic deacetylation of chitin of fungal or animal origin. The polymer displays perfect membrane properties and, due to its exchangeable functional groups, the polymer can easily be modified [14]. Poly (vinyl alcohol) (PVA) is an inexpensive semi-crystalline polyhydroxy polymer, and has also excellent film-forming properties and a high density of reactive chemical functions [15].

A common strategy of introducing ionic conductive groups into polymeric matrices is the quaternization with a clear preference for trimethylammonium (-N(CH$_3$)$_3$) groups onto the backbone or side

*Corresponding authors: Bernd Cermenek, Institute of Chemical Engineering and Environmental Technology, Fuel Cell Systems Group, Graz University of Technology, NAWI Graz, Inffeldgasse 25C, 8010 Graz, Austria
E-mail: bernd.cermenek@tugraz.at

Viktor Hacker, Institute of Chemical Engineering and Environmental Technology, Fuel Cell Systems Group, Graz University of Technology, NAWI Graz, Inffeldgasse 25C, 8010 Graz, Austria, E-mail: viktor.hacker@tugraz.at

chains of the polymers [15-21]; this ensures a better chemical stability than other functional groups such as pyridinium, sulfonium and imidazolium [22].

Quaternized polymers have a high degree of hydrophilicity clearly evident in the high degree of swelling, thus resulting in modest membrane strength. To prepare durable membranes for fuel cells in alkaline medium with improved chemical and mechanical stability, suitable cross-linking agents must be chosen to form three dimensional networks of the water soluble polymers [23,24].

Both chitosan and PVA can be cross-linked by various dialdehydes such as glyoxal (GO) or glutaraldehyde (GA) [25,26], and by epoxide compounds such as epichlorohydrin (ECH) or ethylene glycol diglycidyl ether (EGDGE) [27]. Another cross-linking system for chitosan and PVA-derivatives represents 1-ethyl-3-(3-dimethylaminopropyl) carbodiimide (EDC) and N-hydroxysuccinimide (NHS) as catalyst. These compounds have been successfully used for wound dressing and biomedical applications as non-poisonous and biocompatible cross-linking systems [28]. Dialdehydes react with the alcohol groups of PVA and amine groups of chitosan to form stable acetal and imine bonds, respectively. The epoxide compounds form networks with the hydroxide and amine groups of the polymers by opening the epoxide. The carbodiimide compounds, in combination with a catalyst, cross-link primary amines to yield amide bonds. In addition, the cross-linking can be supported by thermal treatment (annealing) [29], which can improve the thermal and chemical properties of the composite membrane.

Preparation of quaternized PVA/chitosan with the aforementioned cross-linking agent GA has been reported to improve the mechanical stability and the electrochemical performance [20,21], whereas the chemical stability of the membranes was not reported although, at present, this constitutes a critical parameter for alkaline DEFCs. By the use of GA, only a limited degree of quaternization (DQ) of chitosan (DQ less than approximately 35%) is allowed in order to obtain a three dimensional network with sufficient chemical properties [30]. To obviate this limitation, at first the synthesis of quaternized PVA and quaternized chitosan with a DQ of 2.6% and 39%, respectively, was performed; afterwards, quaternized chitosan nanoparticles were introduced into the composite membrane to increase the ionic conductivity [31].

To the best of our knowledge, this research is the first attempt to synthesize composite membranes on the basis of highly quaternized chitosan (DQ higher than 80%) and quaternized PVA by cross-linking with two different agents and an additional annealing process. The aim of the present work was to improve the performance of the membrane in highly alkaline media by a novel cross-linking strategy for the membrane with improved chemical and thermal properties. GA and EGDGE were alternately introduced to the membranes, in which the amount of the first crosslinking agents was varied and the second cross-linker was constant. The cross-linked membranes have been characterized by Fourier transform infrared spectroscopy (FT-IR), scanning electron microscopy (SEM) and thermogravimetric analysis (TGA) techniques. Water uptake, anion conductivity, ion-exchange capacity (IEC) and ethanol permeability were measured as a function of the degree of cross-linking under alkaline conditions. Studies on chemical stability of the differently cross-linked membranes were also reported.

Experimental

Materials

Chitosan, poly (vinyl alcohol) (PVA), (2,3-Epoxypropyl) trimethylammonium chloride (EPTMAC), ethylene glycol diglycidyl ether (EGDGE), glutaraldehyde (GA, 25 wt.-% content in distilled water) were supplied in p.a. quality by Sigma-Aldrich Handels GmbH, Austria, and were used without further purification. PVA had a degree of polymerization of 1600 and a hydrolysis degree of 97.5–99.5 mol%. Chitosan had a degree of deacetylation of 92%.

Preparation of HTCC

N-[(2-Hydroxy-3-trimethylammonium)propyl] chitosan chloride (HTCC) was prepared based on a method similar to those already described elsewhere [32]. Briefly, chitosan (3.0 g, 18.6 mmol) was dispersed in deionized water at 80°C. EPTMAC (11.3 g, 74.4 mmol) was added to chitosan suspension. After reaction for 4 h at 80°C, the reaction solution was poured into cold acetone and stirred overnight in a refrigerator. The white precipitated product was collected by filtration and dried at 60°C.

Preparation of quaternized PVA

Quaternized PVA (QPVA) was prepared by a method similar to those described elsewhere [33]. Briefly, PVA (3.0 g, 66.7 mmol) was dissolved in deionized water under stirring at 95°C. Then 2.0 M KOH solution (2.5 g, 4.5 mmol) and EPTMAC (5.0 g, 32.9 mmol) were added. The reaction was allowed to proceed for 4 h at 65°C while stirring. Then the mixture was deposited and washed three times by anhydrous ethanol. The product was dried at 60°C in vacuum to obtain white solid quaternized PVA (QPVA).

Preparation of the cross-linked quaternized Chitosan/QPVA membrane

QPVA was dissolved in deionized water under stirring at 90°C for one hour in order to prepare a 5 wt.-% solution, and the pH-value adjusted to 5.0 with 1.0 N HCl solution. To this solution, a prescribed amount of 10 wt.-% GA was added under stirring (Table 1). After dissolving HTCC in 1 wt.-% acetic acid solution, yielding a 5 wt.-% solution, a prescribed amount of 10 wt.-% EGDGE in ethanol was added (Table 2). The mixture was stirred at 60°C for 1 h. QPVA (5 g) and HTCC (5 g) were mixed and filtered with a 5 μm PTFE membrane filter. The solution was coated onto the surface of a plexiglass petri dish and dried at 60°C overnight. Afterwards, the membrane was dried at 130°C for 2 min.

Two different series of HTCC/QPVA membranes were prepared.

Membrane [a,b,c]	Molar ratio of GA to membrane [mmol mmol⁻¹]	Degree of cross-linking QPVA [%]
HTCC/QPVA-a	0.20/2.2	9.1
HTCC/QPVA-b	0.30/2.2	13.7
HTCC/QPVA-c	0.45/2.2	20.4
HTCC/QPVA-d	0.60/2.2	27.4
HTCC/QPVA-e	0.90/2.2	41.1
HTCC/QPVA-f	1.10/2.2	50.3

[a]All membranes were prepared using a same kind of HTCC and QPVA, the degree of quaternization (DQ) was 90.4% and 9.0%, respectively, determined by elemental analysis.
[b]The molar ratio of EGDGE to composite membrane was 0.3/2.2 mmol mmol⁻¹, having a degree of crosslinking (DC) of 13.1%.
[c]The thickness of membranes was 45 nm, +/- 3 nm dry state.

Table 1: Cross-linking conditions for HTCC/QPVA composite membrane with GA.

Membrane[a,b,c]	Molar ratio of EDGE to membrane [mmol mmol⁻¹]	Degree of cross-linking HTCC [%]
HTCC/QPVA-1	0.20/2.2	8.8
HTCC/QPVA-2	0.30/2.2	13.1
HTCC/QPVA-3	0.43/2.2	19.7
HTCC/QPVA-4	0.60/2.2	26.3
HTCC/QPVA-5	0.90/2.2	39.4
HTCC/QPVA-6	1.10/2.2	49.5

[a]All membranes were prepared using a same kind of HTCC and QPVA, the degree of quaternization (DQ) was 90.4 % and 9.0 %, respectively, determined by elemental analysis.
[b]The molar ratio of GA to composite membrane was 0.3/2.2 mmol mmol⁻¹, having a degree of crosslinking (DC) of 13.7%.
[c]The thickness of membranes was 35 nm, +/- 4 nm dry state.

Table 2: Cross-linking conditions for HTCC/QPVA composite membrane with EGDGE.

In the first series, the membranes were cross-linked with varying amounts of GA, while the amount of the second cross-linking reagent (EGDGE) was maintained constant. These series are referred to as HTCC/QPVA-i (i = a, b, c, d, e and f).

In contrast to the aforementioned membrane series, the HTCC/QPVA-j (1, 2, 3, 4, 5 and 6) membranes were cross-linked with varying amounts of EGDGE, while the concentration of GA remained constant.

Methods

Degree of membrane cross-linking: The degree of membrane cross-linking for QPVA and HTCC with glutaraldehyde (GA) and ethylene glycol diglycidyl ether (EGDGE), respectively, was calculated according to Equation 1 [34].

$$DC(\%) = \frac{10\% \times V_{Agent} / Mw_{Agent}}{5\% \times V_{Q\text{-}Polymer} / Mw_{Q\text{-}Polymer}} \times 100\% \qquad (1)$$

where DC (%) is the molar ratio of GA/EGDGE to the composite membrane in repeat units, V_{Agent} is the volume (mL) of cross-linking agent (GA or EGDGE, 10 wt.-%), $V_{Q\text{-}Polymer}$ is the volume (mL) of quaternized polymer solution (QPVA and HTCC, 5 wt.-%), $M_{w, Agent}$ is the molecular weight of GA (100.12 g mol⁻¹) and EGDGE (174.19 g mol⁻¹), respectively. $M_{w, Q\text{-}Polymer}$ is corresponding to the averaged molecular weight of HTCC/QPVA monomers (228.80 g mol⁻¹).

Water uptake: Water uptake of the membranes was carried out by measuring the change in mass of the membranes before and after hydration. The membranes were immersed in distilled water at room temperature (RT) for 24 h. Then the surplus water was wiped with filter paper and weighted immediately. The percentage water uptake W was calculated using the following relation (Equation 2):

$$W(\%) = \frac{m_{w-}m_w}{m_d} \times 100 \qquad (2)$$

where m_w is the mass [g] of wet membrane and m_d is the mass [g] of dry membrane.

Ion exchange capacity (IEC): The ion exchange capacity (IEC) was measured by back titration. Therefore, membranes (0.2 g) were soaked in 1.0 M KOH for 24 h, washed with deionized water and equilibrated with 40 mL of 0.01 M HCl standard solution for 24 h, followed by potentiometric titration with 0.01 M NaOH standard solution. 40 mL of 0.01 M HCl standard solution was used as blank sample. The IEC value was calculated by the following Equation 3:

$$IEC = \frac{(V_{blank} - V_{membrane}).CHCl}{m_d} \times 1000 \qquad (3)$$

where V_{blank} and $V_{membrane}$ were the consumed volumes [mL] of the NaOH solution for the blank sample and the HTCC/QPVA membrane sample, respectively, C_{HCl} was the concentration [M] of HCl solution, and m_w is the mass [g] of dry membrane sample.

Ethanol permeability: The ethanol permeability was measured by a diffusion cell comprising two compartments [21]. The membrane was clamped between the two compartments. One compartment was loaded with deionized water and the other with 1.0 M ethanol solution. The solutions in each compartment were agitated with magnetic stirrers. And the concentration of permeated ethanol was measured by refractive index (Smartline RI Detector 2300 Knauer).

Anion conductivity: The membrane resistances of five HTCC/QPVA-(2-6) membranes with different degrees of cross-linking of EGDGE (2: 13.1%, 3: 19.7%, 4: 26.3%, 5: 39.4% and 6: 49.5%) and five HTCC/QPVA-(b-f) membranes with different degrees of cross-linking of GA (b: 13.7%, c: 20.4%, d: 27.4%, e: 41.1% and f: 50.3%) were determined by means of electrochemical impedance spectroscopy (EIS) using a Gamry Reference 600 Potenstiostat and a Conductivity Clamp (Bekktech BT110 LLC, Scribner Associates, Southern Pines, NC, USA). For the measurements, a piece of each membrane (b-f and 2-6) in size of 2.5 × 1.0 cm (in dry state) was alkalized in a 1.0 M KOH solution for approximately 24 h. After alkalization, the membrane was rinsed with ultra-pure water (~18 MΩ cm) and then immersed in ultra-pure water for another 24 h. Finally, the full hydrated membrane sample was incorporated into the four electrode configuration of the conductivity clamp and placed into ultra-pure water. The conducted measuring procedure has been reported previously [35-37].

In short, all measurements were carried out at RT in potentiostatic mode and 5 points per decade were recorded. At the beginning of each measurement, the open circuit potential (OCP) was recorded for 50 seconds. Afterwards a sinusoidal alternating current (AC) voltage of 50 mV was applied in a frequency range of 0.1 – 10 000 Hz. The resulting sinusoidal AC current was used to measure the impedance Z of the membrane. The frequency-independent measured resistance R_{tot} of respective membrane was determined from the high frequency range (HFR) intercept in the corresponding Nyquist plots. The real membrane resistance $R_{membrane}$ was calculated using Equation 4 under consideration of the electrolyte resistance $R_{UPW,after}$ (Table 3) [35-37].

$$\frac{1}{R_{tot}} = \frac{1}{R_{membrane}} + \frac{1}{R_{UPW, after}} \qquad (4)$$

For calculating the in-plane anion conductivity (see Equation 5), the thickness T of the respective membrane was measured in wet state by using a micrometer screw (10-fold determination) (Table 3). The parameter d denotes the distance between inner sense electrodes (=0.425 cm) of the conductivity clamp; while W is the width of the membrane in wet state (Table 3) [35-37].

$$\sigma_{membrane} = \frac{d}{R_{membrane}.T.W} \qquad (5)$$

Membrane characterizations: The chemical stability in alkaline medium was evaluated by immersing the membranes in 1.0 M KOH solution at 60°C for different periods of time. The compactness and appearance of leaks of the membranes was visually observed. The chemical stability of the membranes in alkaline medium was also investigated by monitoring the ethanol permeability as a function of immersion in 1.0 M KOH solutions at RT for 24 h and 168 h. The structure of the chemically cross-linked composite membranes was

Membrane	[a]$R_{UPW,before}$ [Ω]	[b]$R_{UPW,after}$ [Ω]	R_{tot} [Ω]	$R_{memb.}$ [Ω]	d [cm]	T [cm]	W [cm]	[c]ρ [Ω·cm]	σ [mS·cm^{-1}]
HTCC/QPVA-b	243796	254656	4752	4842	0.425	0.0065	1.2	89	11.22
HTCC/QPVA-c	219702	229570	6600	6796	0.425	0.0051	1.2	97	10.30
HTCC/QPVA-d	228244	239416	6272	6441	0.425	0.0058	1.2	105	9.48
HTCC/QPVA-e	210739	233521	6392	6572	0.425	0.0070	1.2	130	7.68
HTCC/QPVA-f	199585	230983	7031	7251	0.425	0.0063	1.2	129	7.73
HTCC/QPVA-2	235029	237083	4564	4654	0.425	0.0047	1.2	62	16.09
HTCC/QPVA-3	219180	279603	9787	10143	0.425	0.0032	1.2	90	11.09
HTCC/QPVA-4	257922	253250	7521	7751	0.425	0.0038	1.2	83	12.09
HTCC/QPVA-5	213628	227102	8658	9001	0.425	0.0035	1.2	89	11.21
HTCC/QPVA-6	269105	268272	7791	8024	0.425	0.0046	1.2	104	9.66

[a]Initial resistance of ultra-pure water (UPW)
[b]Resistance of ultra-pure water (UPW) after measurements
[c]Resistivity of membrane

Table 3: Determined anion conductivities of all cross-linked HTCC/QPVA membranes resulting from adjacent measuring parameters.

characterized by Fourier transform infrared spectroscopy (FT-IR). The infrared spectrometer (Bruker Tensor 27) was used to obtain spectra in the range from 4000 to 600 cm^{-1} with a wavenumber resolution of 4 cm^{-1}. Nitrogen microanalysis was performed by elemental analysis (Multiprobe UHV-surface-analysis system, Omicron Nanotechnology). Thermal gravimetric analysis (TGA) was conducted using a TGA/DSC system (NETZSCH STA 449, Germany). Samples were heated from 25 to 600°C at a heating rate of 10°C min^{-1} using a flow of 20 mL min^{-1} nitrogen as inert gas. The morphology of composite membranes was evaluated using a JEOL JSM 5600 scanning electron microscope (SEM) operating at 5 kV. The membranes were fractured in liquid nitrogen and sputtered with gold, then examined at 2000-fold magnifications. ^{1}H-NMR measurements were performed on a nuclear magnetic resonance spectrometer Bruker Avance spectrometer at 300 MHz Ultrashield. QPVA powder was dissolved in 0.1-0.3 vol.% D$_2$O and stirred at 85°C for 1 h. HTCC powder was dissolved in 0.1-0.3 vol.% CD$_3$COOD/D$_2$O.

Results and Discussion

Membrane preparation

For the quaternization of PVA and chitosan, the reagent EPTMAC was used. Figures 1 and 2 shows the ^{1}H-NMR spectra of QPVA and HTCC. The peak at 3.1 ppm was attributed to the methyl protons of the quaternary ammonium salt group. According to ^{1}H-NMR spectra, successful quaternization of chitosan and PVA was confirmed. The nitrogen content of the quaternized polymers was measured by elemental analysis and the calculation of DQ was performed as described elsewhere [19,32]. The degree of quaternization of the resulting quaternized chitosan and QPVA was 90.4% and 9.0%, respectively.

Cross-linking HTCC/QPVA membranes

As mentioned in the experimental section, the novel composite membranes were prepared from chitosan and PVA via quaternization and cross-linking, whereas cross-linking represents the most important step. The hydrophilicity of HTCC and QPVA was increased by introducing quaternary ammonium groups. To reduce water solubility of polymers and to increase chemical, mechanical and thermal stability, the composites were cross-linked in order to form three dimensional polymer networks. The effect of degree of cross-linking on the membranes structure and the chemical/physical properties were investigated. In this work we used GA and EGDGE to cross-link the polymer membranes. The amount of the first crosslinking agents was varied, whereas the amount of the second cross-linker was constant.

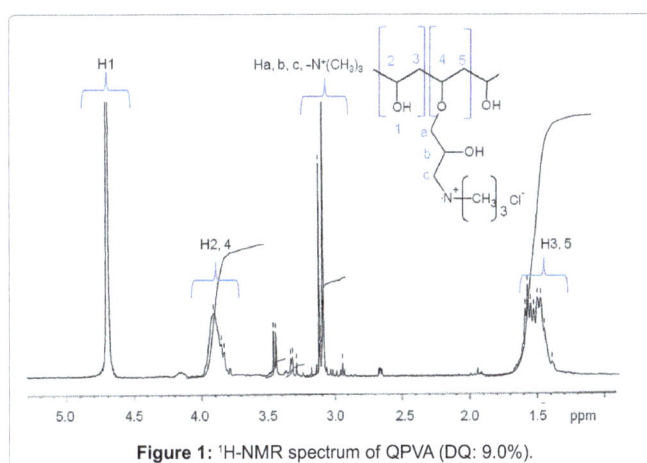

Figure 1: ^{1}H-NMR spectrum of QPVA (DQ: 9.0%).

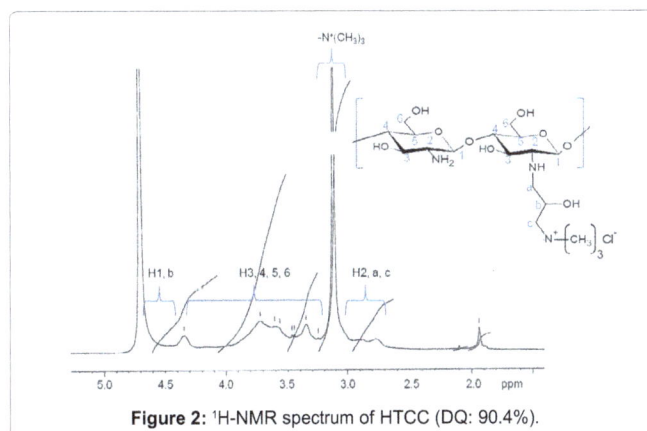

Figure 2: ^{1}H-NMR spectrum of HTCC (DQ: 90.4%).

It is well known that GA is a common cross-linking agent for PVA and chitosan. The network formation takes place between the hydroxide groups of PVA, the amino groups of chitosan and the aldehyde groups of GA. However, a high degree of quaternization (DQ) of the chitosan induced a limitation of free amino groups, which basically caused GA to not work efficiently enough to cross-link the polymer and therefore less mechanically and chemically stable composite membranes were obtained. According to literature [38] higher cross-linked networks can be achieved by reduction of the degree of quaternization of chitosan. Nevertheless a reduction of DQ results in a decrease of ionic exchange capacity and a weaker specific anion conductivity of the membrane.

For the anion conductivity to be as high as possible, a second cross-linking agent - EGDGE - was used in this work. Due to the presence of two reactive epoxy groups, EGDGE facilitates the cross-linking reaction with two amino groups of two different quaternized chitosan molecules and two hydroxyl functions of two PVA molecules during drying process [39]. This covalent linkage allowed the preparation of HTCC/QPVA composite membranes with quaternized chitosan with a DQ higher than 80%. In addition to cross-linking by GA and EGDGE, PVA exhibits hydroxyl groups which are capable of forming hydrogen bonds with the hydroxyl groups of HTCC, thus completing the formation of the polymer network as shown in Figure 3.

The cross-linking conditions for these membranes are summarized in Tables 1 and 2. In the present study, all cross-linked membranes were prepared using HTCCs and QPVAs with a DQ higher than 80% and 5%, respectively. The stability of the membranes toward alkaline media was observed visually from physical appearance (fragmented membranes) of the membranes after immersion in 1.0 M KOH solution for a certain period of time. With a cross-linking density, lower than the selected value, highly swollen membranes with poor mechanical properties were obtained leading to unexaminable membranes. By increasing the amount of cross-linker, the membranes became more stable with an optimum range of 0.3/2.2 mmol mmol^{-1} for GA and 0.43/2.2 mmol mmol^{-1} for EGDGE. The alkaline stability of these membranes sustained on average for 650 h in 1.0 M KOH solution at 60°C. Higher amounts of GA or EGDGE with a molar ratio of 0.6/2.2 mmol mmol^{-1} caused the membranes to exhibit less mechanical characteristics and crumbling behaviour due to the formation of rigid structure with poor hydration. Comparably to EGDGE, the membranes cross-linked by GA exhibited less mechanical strength. The higher cross-linked membranes were merely stable in 1.0 M KOH solution at 60°C for 290 h. It is possible that, by raising the amount of cross-linker, the bond cleavage between HTCC/QPVA and GA/EGDGE increases, thus leading to a binding of water molecules to the polymers. Consequently, the networks became

fragile and instable. All prepared membranes underwent an additional heat treatment at 130°C for two minutes (annealing) after basic drying at 60°C overnight. The membranes showed less elastic properties in dry condition, but a higher alkaline stability in 1.0 M KOH solution at 60°C. A chemical stability for 650 h of all cross-linked membranes was obtained.

Fourier transform infrared spectroscopy (FT-IR)

The FT-IR spectra of the composite membrane HTCC/QPVA and the cross-linked membranes with GA (0.6/2.2 mmol mmol^{-1}), EGDGE (0.6/2.2 mmol mmol^{-1}) and GA/ EGDGE (each with 0.6 mmol) are shown in Figure 4. The two peaks at 840 and 1250 cm^{-1} in the spectra of the composite membranes are characteristics of C-N bonds. The two signals at 1560 and 1650 cm^{-1} for quaternized chitosan are assigned to the N-H bending of primary and secondary amines, respectively [40]. A strong absorption band at 1440 cm^{-1} was recorded, corresponding to the C-H bending of trimethylammonium groups, which shows the introduction of quaternary ammonium salt groups on the backbones of chitosan and PVA, respectively [30]. A broad band around 3300 cm^{-1} was attributed to the hydroxyl groups. It is also noted that a new absorption peak at 2870 cm^{-1} for the cross-linked membranes was observed, corresponding to -CH$_2$ bending with QPVA and HTCC cross-linked by EGDGE and GA.

Water uptake

Membranes used as electrolyte in fuel cells work well within an optimum region for water uptake. The presence of water affects the ionic properties and the chemical and mechanical stability of the membranes. The concurrent swelling is related to the amount of polymer cross-linking agents, the thickness of the dry membrane and the hydrophilic properties due to the quaternary ammonium groups. It is possible that membranes exhibiting too high water uptake restrict the preparation of membrane electrode assembly (MEA) by decreasing contact between the active layer of the electrodes and the membrane. Studies are currently carried out on the assembly of MEA with novel HTCC/QPVA membranes and will be published at a future date.

All membranes were prepared using the same kind and composition of HTCC and QPVA with a DQ of 90.4% and 9.0%, respectively. The amount of ammonium groups of the composite membranes was approximately the same. The thickness of the membranes in dry state varied in a small range between 35 and 45 nm. In this work, the dominant parameter concerning water uptake was the degree of cross-

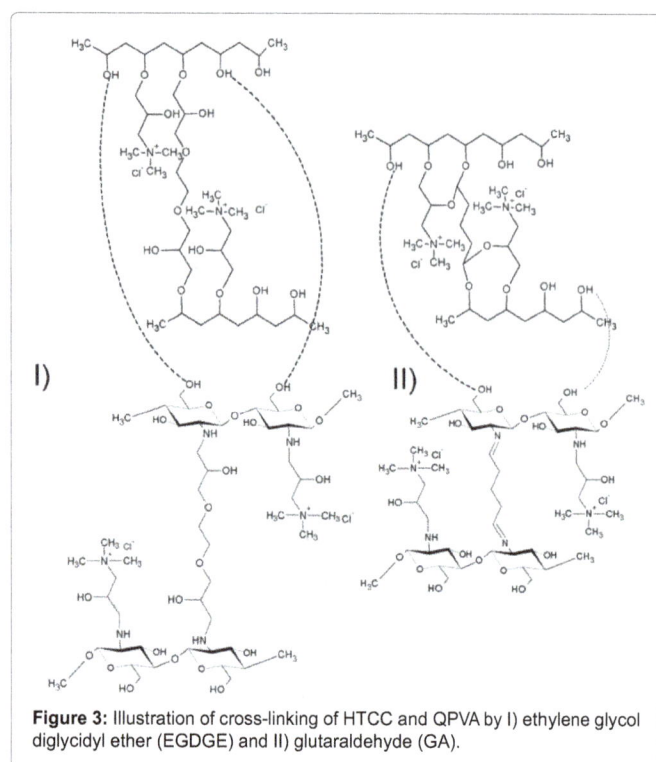

Figure 3: Illustration of cross-linking of HTCC and QPVA by I) ethylene glycol diglycidyl ether (EGDGE) and II) glutaraldehyde (GA).

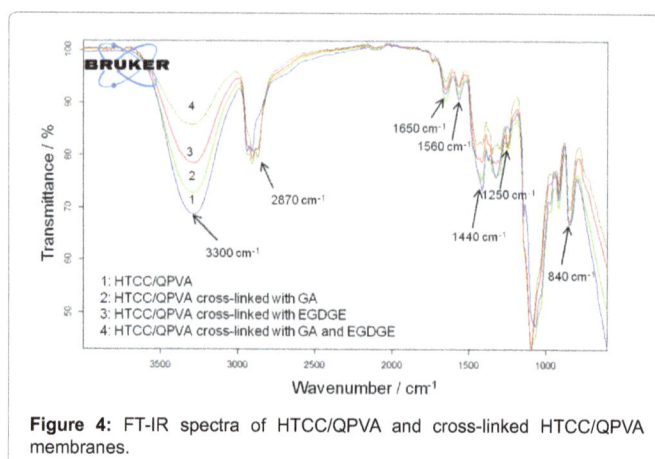

Figure 4: FT-IR spectra of HTCC/QPVA and cross-linked HTCC/QPVA membranes.

linking. The water uptake of the composite membranes cross-linked with GA and EGDGE is given in Figure 5.

A high initial water uptake rate of the composite membranes with a small amount of the first cross-linking agent was obtained (*nota bene*: The second cross-linking agent was constant). It was found that with an increase in DC percent, the water uptake of the membranes gradually decreased from around 198% to 98%. This is a consequence of the formation of cross-linked structures with more GA or EGDGE as cross-linking reagents. According to the results, adding GA in the range between DC of 10% and 50% causes an increase in cross-linking of the composite membrane, evidenced by lower water uptake. The differences in water uptake can be directly attributed to the varying strong interactions in the polymer matrix by the cross-linkers. The swelling properties of the dried (60°C) and the heated (130°C) membranes remained the same over the whole examined range.

Ion exchange capacity (IEC)

The IEC of anion exchange membranes describes the amount of exchangeable ammonium groups in the membrane, whereby a high IEC basically corresponds to high anion conductivity. The IEC of the composite membranes as a function of amount of cross-linking agents (GA and EGDGE) are given in Figure 6. The results indicate that, at a constant DQ, an increase of the amount of cross-linkers leads to a gradually decreased IEC of the HTCC/QPVA membranes. The highest IEC value is reached at a cross-linker amount of 8.8% (EGDGE) and 9.1% (GA), respectively. While the amount of quaternary ammonium groups $(R-N^+(CH_3)_3)$ in composite membranes remained constant, the measurable differences observed among IECs can only be associated with the formation of micro-structural modifications of the membranes. A higher water uptake in membranes generates transferring channels and creates a better clearance for ammonium ions, whereas a high amount of cross-linking agents increases the polymer density and restricts the mobility of the polymer functional groups and the ion transport. It is also noted that between DQ of 10% and 25%, the IEC of the EGDGE-series of membranes is to a lesser extent higher than to the IEC of the GA-series of membranes. The composite membranes exhibit the same behaviour in terms of water uptake and ion exchange capacity. In addition, the alkaline stability of the cross-linked membranes was measured in terms of ion-exchange capacity. For this purpose, the membranes were immersed in 1.0 M KOH solution for 168 h at 25°C. The composite membranes showed 80% – 90% of their initial IEC due to the degradation of ammonium groups by namely Hofmann elimination and nucleophilic substitution [41,42] (Figures 7 and 8).

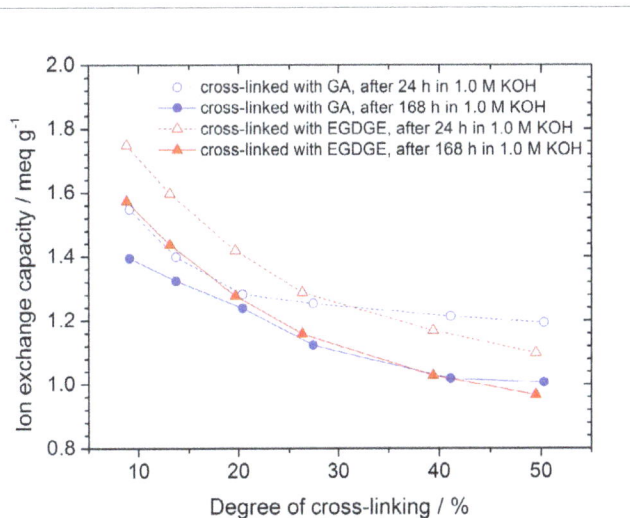

Figure 6: Ion exchange capacity (IEC) of the cross-linked HTCC/QPVA membranes vs. the degree of cross-linking; alkaline treatment for 24 h and 168 h in 1.0 M KOH solution at 25°C.

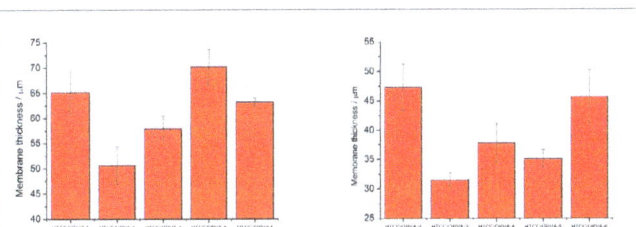

Figure 7: Thickness of the cross-linked HTCC/QPVA membranes (left: b-f with GA; right: 2-6 with EGDGE) in wet state.

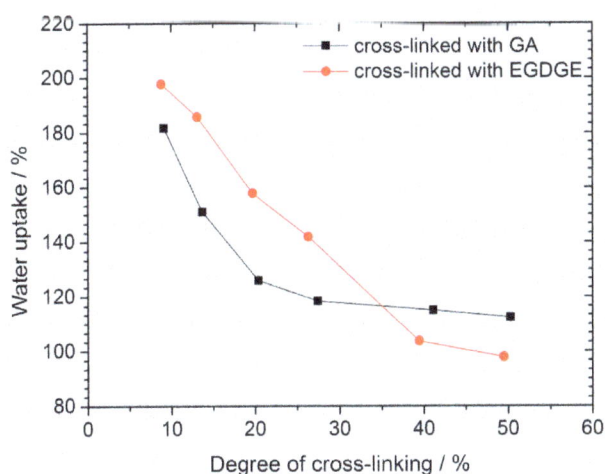

Figure 5: Water uptake of the cross-linked HTCC/QPVA membranes vs. degree of cross-linking.

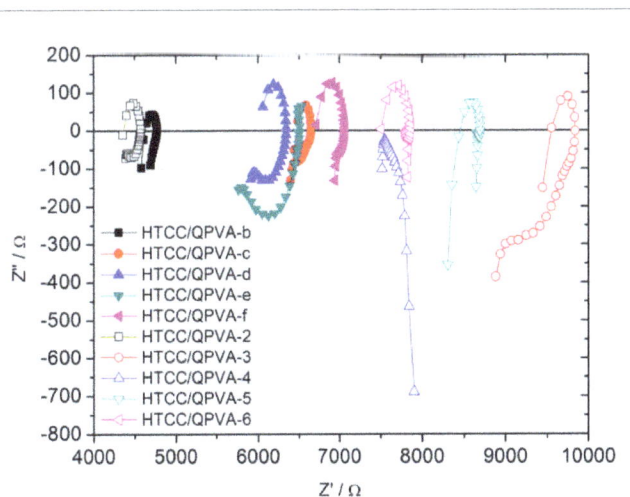

Figure 8: Nyquist plots of the cross-linked HTCC/QPVA membranes (b-f with GA and 2-6 with EGDGE).

Anion conductivity

Figure 9 shows the anion conductivity of the HTCC/QPVA membranes as a function of the degree of cross-linking after being hydrated for 24 h at RT using electrochemical impedance spectroscopy (EIS). During EIS measurements, the synthesized anion exchange membranes typically conduct hydroxide ions (OH^-) but also to a very small extent carbonate/hydrogen carbonate ions (CO_3^{2-}/HCO_3^-) since the used measurement setup was open to air. Air contains approximately 0.04% carbon dioxide (CO_2), which reacts with hydroxide ions to form carbonate [5].

The anion conductivities of all cross-linked membranes range from $7.7 \cdot 10^{-3}$ to $1.61 \cdot 10^{-2}$ S cm^{-1}. It was observed that the anion conductivity decreases with increasing cross-linking density of EGDGE and GA, respectively (Figure 9, right). The highest anion conductivity of the HTCC/QPVA membranes cross-linked with EGDGE and GA was achieved with a cross-linking density of 13.1% and 13.7%, respectively (Figure 9, right). The thickness of the cross-linked HTCC/QPVA membranes with GA (51-71 µm) and EGDGE (32-47 µm) does not influence the resulting anion conductivity (Figure 7 and Table 3).

According to the results, the IEC and the anion conductivity are determined by the water uptake of the membrane. A high IEC, which is attributed an increased number of quaternary ammonium groups and high water content, results in high anion conductivity due to high migration of anions.

When the DQ of the materials is constant, the IEC, the water uptake and the anion conductivity show a similar trend (Figures 5, 6 and 9). With increasing cross-linking density, which is indirectly correlated to water uptake, the structural anion transport channels alter to narrow microscopic membrane pores, creating a higher resistance for the anions and additionally the active amount of the free ammonium ions are reduced.

Ethanol permeability

Figure 10 shows the ethanol permeability of the cross-linked composite membranes by EGDGE as a function of degree of cross-linking (DC) at 20°C, 40°C and 60°C after immersing the membranes in 1.0 M KOH solutions at RT for 24 h and 168 h. The ethanol permeability decreases with increasing DC of the membranes. The cross-linking agent EGDGE formed compact structures and a high resistance for ethanol diffusion, which caused a continuously lower ethanol crossover. The ethanol permeability increases with increasing temperature. The average ethanol permeability of the cross-linked composite membranes was $3.17 \cdot 10^{-8}$ cm^2 s^{-1} at 20°C, $1.99 \cdot 10^{-7}$ cm^2 s^{-1} at 40°C and $5.21 \cdot 10^{-7}$ cm^2 s^{-1} at 60°C. Figure 10 shows also the influence of alkaline treatment for 168 hours at RT on the ethanol permeability of the cross-linked membranes. With increasing duration of alkaline

treatment of the membranes from 24 h to 168 h, the results of the ethanol permeability were not changed at 20°C and were slightly higher at 40°C. These data indicate that the cross-linked membranes have low ethanol permeability and hence a higher tolerance toward ethanol as reported by other studies [21]. However at a temperature of 60°C, an increase of ethanol diffusion of about 20% was obtained for pre-treated membranes, due to induced degradation in alkaline medium [38].

A linear correlation of the results between a degree of cross-linking of 20% and 50% were obtained (Figure 10). Lower cross-linked membranes (DC < 20%) showed less homogeneous structure. Therefore, the diffusion for non-aqueous media (1.0 M EtOH) varied and the results were out of the linear region of the permeability measurements.

Scanning electron microscopy (SEM)

The SEM images for the surface of the annealed membranes cross-linked by GA and EGDGE are shown in Figure 11. The composite membrane without cross-linking shows less compact structure and possible phase separation (Figure 11a). In the network cross-linked by

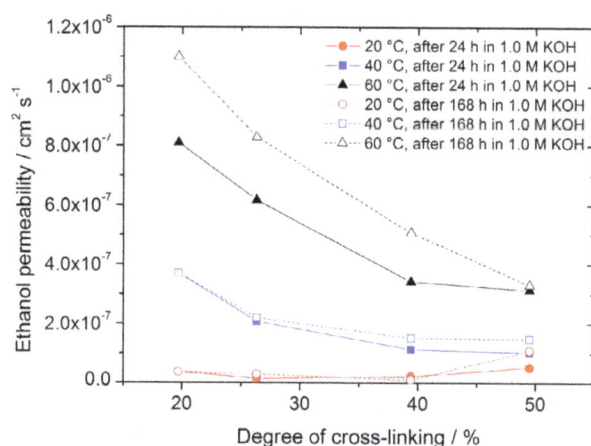

Figure 10: Ethanol permeability of the cross-linked HTCC/QPVA membranes.

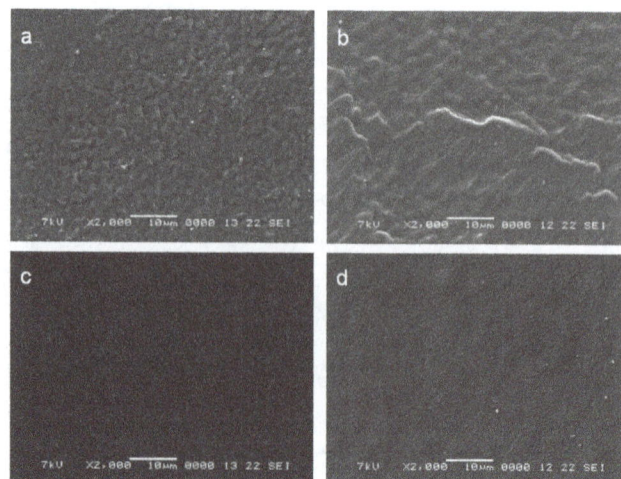

Figure 11: SEM photographs of HTCC/QPVA membranes (a) without cross-linking (b) cross-linked by GA with an amount of 0.6/2.2 mmol mmol^{-1} (c) cross-linked by EGDGE with an amount of 0.6/2.2 mmol mmol^{-1} and (d) cross-linked by GA/EGDGE with an amount of each 0.6/2.2 mmol mmol^{-1}.

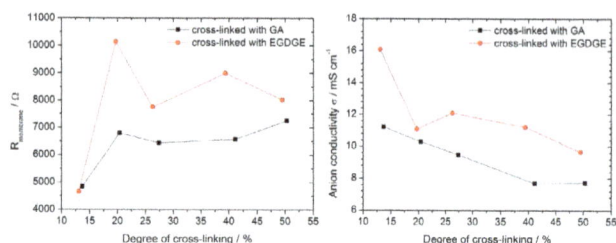

Figure 9: Membrane resistance R$_{membrane}$ (left) and anion conductivity σ (right) of the cross-linked HTCC/QPVA membranes vs. the degree of cross-linking.

GA, a more homogeneous structure was obtained (Figure 11b). The morphology of the HTCC/QPVA membrane cross-linked by EGDGE is shown in Figure 11c. The surface is flat and homogenous and appears denser than the membrane cross-linked by GA. The combination of the cross-linkers EGDGE and GA with an amount of each 0.6/2.2 mmol mmol^{-1} decreases the uniformity of the membrane structure (Figure 11d). A less compact structure was obtained, which may be attributed to an overloading with cross-linker that caused the membrane to become less stable. The results indicate that the polymers are combined well with both GA and EGDGE as cross-linking agents, but the amount of the agents must be selected within an optimal range.

The bright dots on SEM images can be attributed to debris originating from sample particles. As the membrane samples were slightly brittle and can crumble into pieces to a certain extent and, in addition, the samples were not prepared in a cleanroom environment, such particles are frequently observed. Nevertheless, we do not expect that this would affect the experimental results and data interpretation presented in the manuscript.

Thermogravimetric analysis (TGA)

The thermal degradation behavior of the HTCC- and QPVA-powder and HTCC/QPVA composite membranes cross-linked by EGDGE is shown in Figure 12. The decrease at around 100°C is due to the bonded water evaporation of 7.5% of all materials. Pure chitosan is thermally not stable enough and starts to degrade continuously at around 120°C due to the breakdown of side chains and backbone. By addition of QPVA the composite membrane is stable up to 280°C. The quaternized PVA powder shows the same thermal gravimetric properties as the composite HTCC/QPVA membrane cross-linked with EGDGE. This indicates that the thermal stability of the membrane is mainly determined by QPVA.

Two stages of thermal degradation were found for the cross-linked membranes. The onset temperature of degradation (T_d) of the first stage is about 260°C, and the degradation temperature of the onset of the second stage is about 380°C.

Figure 13 shows the TGA curves of the composite membranes cross-linked by EGDGE and GA as well as with a thermal annealing process. In addition to the evaporation of the bonded water, there are two more weight loss regions. The second weight loss occurs within a range of 260-320°C due to the degradation of quaternary ammonium

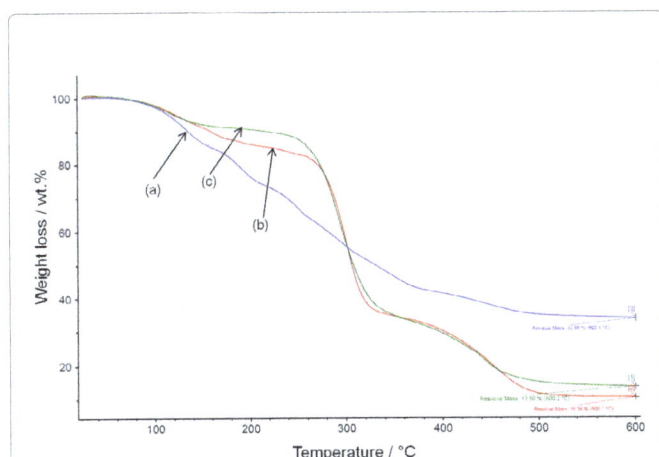

Figure 13: TGA curve of HTCC/QPVA membranes cross-linked by (a) GA, (b) GA annealed, (c) EGDGE and (d) EGDGE annealed.

groups. The weight loss of the membranes cross-linked by EGDGE and GA is about 35% and 45%, respectively. At the third stage at 320-420°C, the degradation is attributed to the break of the C-C backbone of the main polymer chains. The TGA curves for membranes cross-linked by GA and EGDGE were somewhat different in the region between 260-420°C. The decomposition of membranes cross-linked with GA occurs rapidly up to 300°C due to the loss of fewer cross-linked hydroxyl groups of the membrane matrix [43]. At this third stage, the weight loss of the membranes cross-linked with EGDGE was slower. It is possible that the cross-linking reaction with EGDGE has formed more stable membranes with increased thermal stability. The total weight loss of the membranes cross-linked by EGDGE and GA is about 75% and 85%, respectively. The degradation effects of the membranes that were annealed for 2 min at 130°C were less pronounced than of the dried membranes, which can be explained by the formation of stronger three dimensional networks.

Conclusion

Two series of novel quaternized chitosan/PVA membranes cross-linked with different cross-linking agents were successfully prepared. After subjecting the as-prepared membranes to an additional annealing process, the resulting membranes were characterized toward suitability for direct ethanol fuel cells. The composite membranes showed excellent transport and ionic properties for anions and concurrently a low ethanol permeability. The combined cross-linking strategy enhanced the chemical and thermal stability, as well as the long-term stability in alkaline environment. Furthermore, the developed membranes exhibit a high water uptake leading to superior anion conductivity. The cross-linked and annealed HTCC/QPVA membranes were stable in alkaline solution for 650 h at 60°C and showed less chemical degradation affecting ethanol crossover and ion exchange capacity.

This study demonstrates that the developed novel cross-linked membrane systems meet the benchmark of contemporarily applied membrane technology. Future investigations will focus on the screening of additional polymers to be applied as a high quaternized membrane matrix.

Acknowledgement

Financial support by the Austrian Climate and Energy Fund, Austrian Federal Ministry of Transport, Innovation and Technology (BMVIT), The Austrian Research Promotion Agency (FFG) and the IEA research cooperation are gratefully acknowledged.

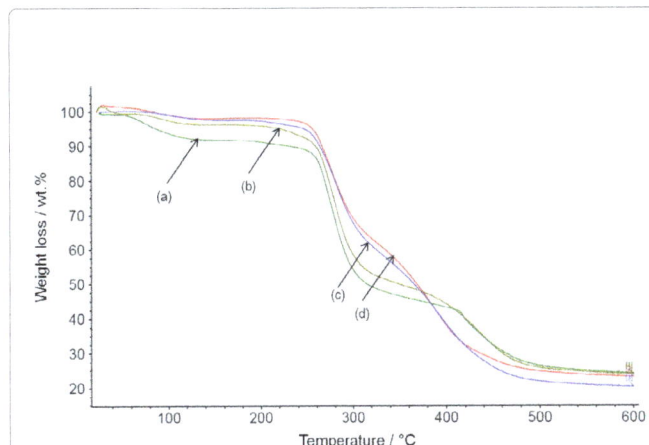

Figure 12: TGA curve of (a) chitosan powder, (b) QPVA-powder and (c) HTCC/QPVA composite membrane.

References

1. Kamarudin MZF, Kamarudin SK, Masdar MS, Daud WRW (2013) Review: Direct ethanol fuel cells. Int J Hydrogen Energy 38: 9438-9453.

2. An L, Zhao T, Li YS (2015) Carbon-neutral sustainable energy technology: Direct ethanol fuel cells. Renewable and Sustainable Energy Reviews 50: 1462-1468.

3. Antolini E, Gonzalez ER (2010) Alkaline direct alcohol fuel cells. J Power Sources 195: 3431-3450.

4. Yu EH, Krewer U, Scott K (2010) Principles and Materials Aspects of Direct Alkaline Alcohol Fuel Cells. Energies 3: 1499-1528.

5. Zhao TS, Li YS, Shen SY (2010) Anion-exchange membrane direct ethanol fuel cells: Status and perspective. Front Energy Power Eng China 4: 443-458.

6. Xiang Y, Yang M, Guo Z, Cui Z (2009) Alternatively chitosan sulfate blending membranes as methanol-blocking polymer electrolyte membrane for direct methanol fuel cell. J Memb Sci 337: 318-323.

7. Buraidah MH, Arof AK (2011) Characterization of chitosan/PVA blended electrolyte doped with NH4I. J Non Cryst Solids 357: 3261-3266.

8. Smitha B, Sridhar S, Khan AA (2005) Synthesis and characterization of poly(vinyl alcohol)-based membranes for direct methanol fuel cell. J Appl Polym Sci 95: 1154-1163.

9. Binsu VV, Nagarale RK, Shahi VK, Ghosh PK (2006) Studies on N-methylene phosphonic chitosan/poly(vinyl alcohol) composite proton-exchange membrane. React Funct Polym 66: 1619-1629.

10. Meenakshi S, Bhat SD, Sahu AK, Sridhar P, Pitchumani S, et al. (2012) Chitosan-polyvinyl alcohol-sulfonated polyethersulfone mixed-matrix membranes as methanol-barrier electrolytes for DMFCs. J Appl Polym Sci 124: E73-E82.

11. Ma J, Sahai Y (2013) Chitosan biopolymer for fuel cell applications. Carbohydrate Polymers 92: 955-975.

12. Couture G, Alaaeddine A, Boschet F, Ameduri B (2011) Polymeric materials as anion-exchange membranes for alkaline fuel cells. Prog Polym Sci 36: 1521-1557.

13. Merle G, Wessling M, Nijmeijer K (2011) Anion exchange membranes for alkaline fuel cells: A review. J Memb Sci 377: 1-35.

14. Gavhane YN, Gurav AS, Yadav AV (2013) Chitosan and Its Applications: A Review of Literature. Int J Res Pharmaceut Biomed Sci 4: 312-331.

15. Moulay S (2015) Review: Poly(vinyl alcohol) Functionalizations and Applications. Polym Plast Technol Eng 54: 1289-1319.

16. Wan Y, Peppley B, Creber KAM, Bui VT, Halliop E (2006) Preliminary evaluation of an alkaline chitosan-based membrane fuel cell. J Power Sources 162: 105-113.

17. Wan Y, Creber KAM, Peppley B, Bui VT (2006) Chitosan-based electrolyte composite membranes II. Mechanical properties and ionic conductivity. J Memb Sci 284: 331-338.

18. Wan Y, Creber KAM, Peppley B, Bui VT (2006) Chitosan-based solid electrolyte composite membranes: I. Preparation and characterization. J Memb Sci 280: 666-674.

19. Xiong Y, Fang J, Zeng QH, Liu LQ (2007) Preparation and characterization of cross-linked quaternized poly(vinyl alcohol) membranes for anion exchange membrane fuel cells. J Memb Sci 311: 319-325.

20. Xiong Y, Liu QL, Zhang QG, Zhu AM (2008) Synthesis and characterization of cross-linked quaternized poly(vinyl alcohol)/chitosan composite anion exchange membranes for fuel cells. J Power Sources 183: 447-453.

21. Yang JM, Chiu HC (2012) Preparation and characterization of polyvinyl alcohol/chitosan blended membrane for alkaline direct methanol fuel cells. J Memb Sci 419-420: 65-71.

22. Vogel C, Meier-Haack J (2014) Preparation of ion-exchange materials and membranes. Desalination 342: 156-174.

23. Berger J, Reist M, Mayer JM, Felt O, Peppas NA, et al. (2004) Structure and interactions in covalently and ionically cross-linked chitosan hydrogels for biomedical applications. Eur J Pharm Biopharm 57: 19-34.

24. Bolto B, Tran T, Hoang M, Xie Z (2009) Cross-linked poly(vinyl alcohol) membranes. Prog Polym Sci 34: 969-981.

25. Gupta KC, Jabrail FH (2006) Glutaraldehyde and glyoxal cross-linked chitosan microspheres for controlled delivery of centchroman. Carbohydrate Research 341: 744-756.

26. Pauliukaite R, Ghica ME, Fatibello-Filho O, Brett CMA (2010) Electrochemical impedance studies of chitosan-modified electrodes for application in electrochemical sensors and biosensors. Electrochimica Acta 55: 6239-6247.

27. Liu R, Xu X, Zhuang X, Cheng B (2014) Solution blowing of Chitosan/PVA hydrogel nanofiber mats. Carbohydrate Polymers 101: 1116-1121.

28. Zhang D, Zhou W, Wei B, Wang X, Tang R, et al. (2015) Carboxyl-modified poly(vinyl alcohol)-crosslinked chitosan hydrogel films for potential wound dressing. Carbohydrate Polymers 125: 189-199.

29. Zhang J, Qiao J, Jiang G, Liu L, Liu Y (2013) Cross-linked poly(vinyl alcohol)/poly (diallyldimethylammonium chloride) as anion-exchange membrane for fuel cell applications. J Power Sources 240: 359-367.

30. Wan Y, Peppley B, Creber KAM, Buia VT, Halliop E (2008) Quaternized-chitosan membranes for possible applications in alkaline fuel cells. J Power Sources 185: 183-187.

31. Liao GM, Yang CC, Hu CC, Pai YL, Lue SJ (2015) Novel quaternized polyvinyl alcohol/quaternized chitosan nano-composite as an effective hydroxide-conducting electrolyte. J Memb Sci 485: 17-29.

32. Wu J, Su ZG, Ma GH (2006) A thermo- and pH-sensitive hydrogel composed of quaternized chitosan/glycerophosphate. Int J Pharmaceutics 315: 1-11.

33. Shin MS, Byun YJ, Choi YW, Kang MS, Park JS (2014) On-site crosslinked quaternized poly(vinyl alcohol) as ionomer binder for solid alkaline fuel cells. Int J Hydrogen Energy 39: 16556-16561.

34. Tian G, Liu L, Meng Q, Cao B (2014) Preparation and characterization of cross-linked quaternised polyvinyl alcohol membrane/activated carbon composite electrode for membrane capacitive deionization. Desalination 354: 107-115.

35. Bodner M, Cermenek B, Rami M, Hacker V (2015) The Effect of Platinum Electrocatalyst on Membrane Degradation in Polymer Electrolyte Fuel Cells. Membranes 5: 888-902.

36. Ranacher C, Resel R, Moni P, Cermenek B, Hacker V, et al. (2015) Layered Nanostructures in Proton Conductive Polymers obtained by initiated Chemical Vapor Deposition. Macromolecules 48: 6177-6185.

37. Dedmond E, Cooper K (2016) Application Note – Effect of Solution Conductivity on In-Plane Membrane Conductivity Measurement.

38. Wang J, He R, Che Q (2011) Anion exchange membranes based on semi-interpenetrating polymer network of quaternized chitosan and polystyrene. J Colloid Interface Sci 361: 219-225.

39. Merle G, Schwan Hosseiny S, Wessling M, Nijmeijer K (2012) New cross-linked PVA based polymer electrolyte membranes for alkaline fuel cells. J Memb Sci 409: 191-199.

40. Peng ZX, Wang L, Du L, Guo SR, Wang XQ, et al. (2010) Adjustment of the antibacterial activity and biocompatibility of hydroxypropyltrimethyl ammonium chloride chitosan by varying the degree of substitution of quaternary ammonium. Carbohydrate Polymers 81: 275-283.

41. Lim SH, Hudson SM (2004) Synthesis and antimicrobial activity of a water-soluble chitosan derivative with a fiber-reactive group. Carbohydrate Research 339: 313-319.

42. Varcoe JR, Poynton SD, Slade RCT (2009) Handbook of Fuel Cells: Advances in Electrocatalysis, Materials, Diagnostics and Durability - Fundamentals, Technology and Applications. Wiley.

43. Ye L, Zhai L, Fang J, Liu J, Li C, et al. (2013) Synthesis and characterization of novel cross-linked quaternized poly(vinyl alcohol) membranes based on morpholine for anion exchange membranes. Solid State Ionics 240: 1-9.

18

Structures and Properties of Oxide Barrier-Film of Anodized Aluminum by Electrochemical Impedance Spectroscopy at the Nano-metere Scale

K. Habib* and K. Al-Muhanna

Materials Science Laboratory, Department ofAdvanced Systems, KISR, P.O.Box 24885 Safat, 1319 Kuwait

Abstract

In this study, the effect of the annealing treatment on electrochemical behavior and the oxide barrier-film thickness of anodized aluminum-magnesium (Al-Mg) alloy were investigated. Electrochemical parameters such as the polarization resistance (R_p), solution resistance (R_{Sol}), alternating current impedance (Z), and the double layer capacitance (C_{dl}) of the anodized Al-Mg alloy were determined in sulfuric acid solutions ranged from -1 % H_2SO_4 by electrochemical impedance spectroscopy (EIS) methods. Then, the oxide film thickness of the anodized Al-Mg alloy was determined from the obtained electrochemical parameters as a function of the sulfuric acid concentration (-1 % H_2SO_4), in the as received sample and annealed sample conditions. The optimum thickness of the oxide film was determined for the as received samples (4.2nm) and for the annealed samples (0.63nm) in sulfuric acid concentrations of 4% and 2% H_2SO_4 respectively. The reason behind the oxide film thickness of the as received samples is greater than the one for the annealed samples, because the former samples are thermodynamically unstable (more chemically active) as compared to the annealed samples. A mathematical model was developed to interpret the mechanism of the oxide film build up on the aluminum substrate. The mathematical model of the oxide film build up on the aluminum substrate was proposed for the next challenge of the present work.

Keywords: Intermetallics; Surface treatments; Anodized Al-Mg alloy; Thin films; Electrochemical impedance spectroscopy (EIS); and Electrical properties.

Introduction

It has been known that the formation of anodic oxide films is a common means of achieving corrosion protection of aluminium and its alloys. For example, the formation of a porous oxide layer, with an underlying barrier (compact) oxide film, is normally achieved in acidic media [1], while barrier (compact) oxide films alone are typically formed in neutral buffer solutions [2]. The porous oxide layers are preferred to the barrier (compact) films because they can form to high oxide thickness and yield excellent abrasion protection. Also, the porous oxide layers provide improved corrosion resistance after chemically sealing of the pores. There is a number of methods for the determination of the oxide film thickness and among those methods are the optical interferometry [1] and the Electrochemical Impedance Spectroscopy (EIS) [2]. The optical interferometry is found very useful of determining the porous oxide-layer thickness of anodized aluminum alloy in a micrometer scale. On the other hand, the Electrochemical Impedance Spectroscopy (EIS) is found very useful technique of determining the oxide barrier-film thickness of anodized aluminum alloy in a nano-meter scale [2]. A detail discussion of both methods are given elsewhere [1,2]. Also, the discussion includes some examples of the determination of the oxide film layer by optical interefrometry and electrochemical impedance spectroscopy (EIS).

In general, the EIS is an alternating current (A.C) method for corrosion studies of metallic, organic, and inter-metallic samples in relatively high resistive conditions. For instance, high resistive solutions, inhibited solutions, organic coated samples, high resistive-oxide forming samples, like in the present case, are considered high resistive conditions. The reason one would use the EIS rather than direct current (D.C) methods for corrosion studies of samples in such high resistive conditions is to avoid erroneous results due to heating of samples in the D.C method in such conditions [3,4]. Normally the results of the EIS consist of two plots. The first plot is the complex plane plot (Nyquist Plots) which is basically the logarithm of the imaginary

impedance (Z_{Image}) versus the logarithm of the real impedance (Z_{real}). The values of the polarization resistance (R_p) and the solution resistance(R_s) can be obtained by the data fitting method of the Randell's semi circle [3,4]. The second plot is the Bode plot which is basically the logarithm of impedance (Z) (Y-coordinate) and the phase (θ) (Y- coordinate) plotted versus the logarithm of the frequency (X-coordinate). Values of the alternating current (A.C) impedance and the double layer capacitance can be obtained from the Bode plot at a low frequency based on the extrapolation of the intersection line at a frequency equal to 0.16 Hz from the x-coordinate in Bode plots, to the y-coordinate in Bode plot.

The objective of this study was to determine the oxide barrier (compact)-film thickness and the electrochemical parameters of anodized aluminium-magnesium (Al-Mg) alloy by A.C techniques. Electrochemical parameters such as the polarization resistance, solution resistance, alternating current (A.C) impedance, and the double layer capacitance of the anodized Al-Mg alloy were determined in sulfuric acid solutions (-1 % H_2SO_4) by electrochemical impedance spectroscopy (EIS) methods. Then, the oxide barrier- film thickness of the anodized Al-Mg alloy was determined from the obtained electrochemical parameters as a function of sulfuric acid concentration (-1 % H_2SO_4) by applying the following relationship [3]:

$$C_{dl} = 1/Z = (e\ e\ A/L) \qquad (1)$$

---

*Corresponding author: K. Habib, Materials Science Laboratory, Department of Advanced Systems, KISR, P.O.Box 24885 Safat, 131 9 Kuwait, E-mail: khaledhabib@usa.net

Where

C_{dl} is the double layer capacitance of the oxide film

Z is the alternating current (A.C) impedance of the oxide film

e is the dielectric constant, 8.4 of the aluminum oxide, Al_2O_3

e is the permittivity of the free space, 8.85×1^{-14} F/cm

A is the exposed surface area of the sample to the aqueous solution

L is the oxide Barrier- film thickness.

Therefore, from Equation No.1, one can determine the oxide barrier- film thickness from the obtained electrochemical parameters, i.e., the double layer capacitance of the oxide film, as the following [3]:

$$L = (ee\, A/C_{dl}) \qquad (2)$$

Equation No.2 will be used in this study to determine the oxide film thickness of the Al-Mg alloy as a function of sulfuric acid concentration (-1 % H_2SO_4).

Experimental Works

In the present investigation, an Aluminum-Magnesium (Al-Mg) alloy was used. The alloy composition comprises of 0.27% Fe, 1.7%Mg, and balanced aluminum, known to be 636 Aluminium alloy. The samples were fabricated in a cylindrical form with a diameter of 2cm, for the electrochemical investigation, according to the standard methods of the American Society for Testing and Materials [4]. Then, some of the samples were solution annealed at 45°C for two hours in an electric furnace, then, the samples were slow cooled in the furnace. Thereafter, all samples (annealed and as received samples) were polished and ground by silicon carbide papers until the finest grade (12) was reached. Then, a set of the annealed and the as received samples were prepared for metallographic examinations, in order to determine the microstructures of the samples. An optical light microscope (Made by Nachet, France) was used for the metallographic examinations.

For determination of the electrochemical parameters of the annealed and the as received samples, all EIS measurements were performed against a saturated calomel electrode (SCE) according to procedures described elsewhere [4]. A standard electrochemical cell was used, the cell made of a 1 cm³ flask, a reference electrode, the saturated Colomel electrode (SCE), a counter electrode, made of platinum wire, and a working electrode, annealed and as received Al-Mg Samples. The exposed surface area of all samples was 3.14 cm². In this study, EIS measurements were conducted using a potentiastat/Galvanostat made by Gamry instruments in order to obtain impedance spectra. The EIS spectra of the annealed and as received samples of Al-Mg alloy were determined in, 2,4,6,8,1 % H_2SO_4. The sulfuric acid solution was diluted by distilled water. Values of the polarization resistance and the solution resistance were obtained by the complex plane plots (Nyquist plots). The complex plane plots (Nyquist Plots) are basically the logarithm of the imaginary impedance (Z_{image}) versus the logarithm of the real impedance (Z_{real}). The values of the polarization resistance (R_p) and the solution resistance (R_s) were obtained by the data fitting method of the Randell's semi circle. Also, values of the alternating current (A.C) impedance and the double layer capacitance were obtained from Bode plots at a low frequency. The alternating current (A.C) impedance and the double layer capacitance were obtained at low frequency based on the extrapolation of the intersection line at a frequency equal to 0.16 Hz from the x-coordinate in Bode plots, to the y-coordinate in Bode plot.

Bode plots are basically the logarithm of impedance (Z) (Y-coordinate) and the phase (θ) (Y- coordinate) plotted versus the logarithm of the frequency (X-coordinate). All the electrochemical parameters of the Al-Mg samples were determined by using Gamry's based software. It is worth noting that there was no delay time at the beginning of each test during the measurements of the electrochemical parameters of the Al-Mg samples in, 2, 4, 6, 8, 1 % H_2SO_4. This step was necessary to avoid any chemical oxidation of the samples. Also, in order to plot the complex plane (Nyquist) and Bode plots, the frequency range was chosen to range between 1Hz to .1Hz. Finally the obtained data of the alternating current (A.C) impedance and the double layer capacitance, of all investigated samples, were used to measure the oxide film thickness of by using equation No. 2, as a function of the sulfuric acid concentration, 2,4,6,8,1% H_2SO_4.

Results and Discussion

Figures 1 and Figure 2 show the microstructures of the Al-Mg alloy in the as received and annealed conditions, respectively. It is obvious from Figure 1 and Figure 2 that the microstructures of the Al-Mg samples in the as received and annealed conditions consists of two distinguished phases. The matrix phase is (Al) phase, white phase. In contrast, the second phase is in a form of second phase particles, β phase (Al_3Mg_2), dark phase. The different between the microstructures of the as received sample (Figure 1) and the annealed sample (Figure 2) are

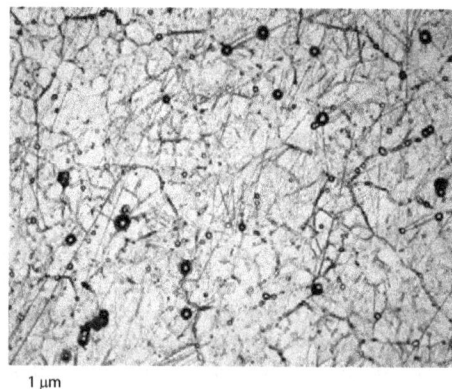

1 μm

Figure 1: shows the microstructure of the Al-Mg alloy in the as received condition. The magnification of the figure is 2 X, of an average grain size of 1 μm of the (Al) Phase.

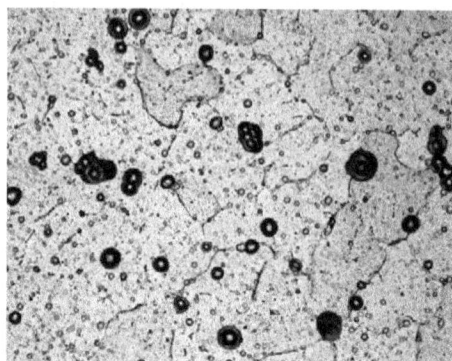

Figure 2: shows the microstructure of the Al-Mg alloy in the annealed condition. The magnification of the figure is 2 X, of an average grain size of 224 μm of the (Al) Phase.

in the grain size of the (Al) phase as well as the size of the second phase particle, the β phase. It is clear that the average grain size of the (Al) Phase is larger in the annealed sample than in the as received condition. The average grain size of the (Al) Phase is found equivalent to 224 μm as compared to 1μm for the as received sample. The average grain size has been determined by the line intersection method according to the standards of the American Society of Metals [5]. In addition, the second phase particle (β phase) is found larger and coarser in the annealed sample than in the as received sample. This occurred because the annealed sample was normalized (slow cooled in the furnace) after the solution annealing treatment for two hours at 45°C in an electric furnace. As a result this led to the precipitation of the second phase particles (β phase), by nucleation and growth, in the (Al) matrix phase. Then second phase particles were enlarged by coalescing with other second phase particles, see (Figure 2). This process eventually led to the coarsening of the second phase particles when the temperature of the furnace reached the room temperature.

The oxide film thickness of the as received samples was obtained by using Equation No.2, and using the obtained values of the double layer capacitance. The obtained data in Table 1 show that as the concentration of the sulfuric acid increased from -4% H_2SO_4, the polarization resistance and the double layer capacitance decreased from 12.22 to 0.6 Kohms and 31.3 to 5.61 μF respectively.

In contrast, as the concentration of the sulfuric acid increased from -4% H_2SO_4, values of the alternating current impedance and the oxide film thickness were observed to increase from 31.95 to 178K ohms and 0.75 to 4.2 nm, respectively. This observation is in agreement with the known electrochemical concept of as the thickness of the oxide layer increases, the resistance (impedance) of the metal increases as well, because the oxide film protects (shields) the base metal from the surrounding environment [3]. Consequently, the values of the double layer capacitance decreased because the double layer capacitance has an inverse proportional relationship with the alternating current impedance, see Equation No1.In the meantime, the polarization resistance was observed to decrease from 12.22 to .6K ohms because of the high tendency of the Al element, in general, to oxidation (anodization) in sulfuric acid solution. On the contrary, the solution resistance was observed to decrease all the way from 1.3K ohms to .65 ohms as the solution concentration increased from to 1% H_2SO_4.This naturally occurred because of increase of the density of the ionic species in the distilled water. In other words, the addition of sulfuric acid from -1 % H_2SO_4 to the distilled water has increased the distilled water conductivity. Furthermore, the obtained data in Table 1, show that as the concentration of the sulfuric acid increased from 4-1 % H_2SO_4 the polarization resistance was observed to increase first from 0.6 to 2.58K ohms, then decrease to 0.45K ohms. On the other hand, the double layer capacitance was observed to increase from 5.61 to 29.61 μF as a function of the increase of the sulfuric acid concentration from 4-1% H_2SO_4. In contrast, as the concentration of the sulfuric acid increased from 4-1 % H_2SO_4, values of the alternating current impedance and the oxide film thickness were observed to decrease from 178 to 33.8K ohms and 4.2 to 0.8 nm, respectively. This observation is in agreement with the known electrochemical concept of as the thickness of the oxide layer decreases, the resistance (impedance) of the metal decreases as well, because the oxide film becomes less protective of the base metal from the surrounding environment. Consequently, the value of the double layer capacitance was observed to increase because the double layer capacitance has an inverse proportional relationship with the alternating current impedance. In addition, Figure 5 shows the Nyquist plots of the annealed samples of Al-Mg alloy in the, 2,4,6,8,1 % H_2SO_4.

From the Nyquist plots the polarization resistance, solution resistance, alternating current (A.C) impedance, and the double layer capacitance of the Al-Mg alloy were determined in the annealed conditions in, 2, 4, 6, 8, 1 % H_2SO_4.

Tabulated values of the polarization resistance, solution resistance, alternating current (A.C) impedance, the double layer capacitance, and the oxide film thickness of the annealed samples of the Al-Mg alloy are given in Table 2.

In the same way, the oxide film thickness of the annealed samples was measured by implementing Equation No.2, and using the obtained values of the double layer capacitance. From the obtained data in Table 2, one can tell that as the concentration of the sulfuric acid increased from -1 % H_2SO_4, values of the polarization resistance and the solution resistance were observed to decrease from 14.71K ohms and 3.36K ohms to a steady state value of .1K ohms and .1K ohms, respectively. In contrast, as the concentration of the sulfuric acid increased from -1 % H_2SO_4, values of the alternating current impedance and the oxide film thickness were observed to increase first from 24.15 to 26.93K ohms and 0.56 to 0.63 nm, respectively. Then, the values of the alternating current impedance and the oxide film thickness were observed to attain a steady state value of around 17K ohms and 0.4 nm, respectively. The electrochemical behavior of the annealed samples was expected to behave as the as received samples as a function of the sulfuric acid concentration.

In contrast, the double layer capacitance was observed to vary several times in a range between 37.1 3 to 62.18 μF as a function of the increase of the sulfuric acid concentration from -1% H_2SO_4.This behavior was unexpected ,because the double layer capacitance should have an inversely proportional relationship with the alternating current impedance. In order to obtain the optimum thickness of the oxide film with respect to the sulfuric acid concentration, plots of the oxide film thickness versus the sulfuric acid concentration were constructed.

Figure 6 shows the relationship between the oxide film thickness and the sulfuric acid concentration, for the as received and annealed samples, of the Al-Mg alloy. The figure show clearly at sulfuric acid concentrations of 4% and 2% H_2SO_4 the optimum thickness of the oxide film was determined for the as received samples (4.2nm) and for the annealed samples (0.63nm), respectively. The reason behind

% H_2SO_4	R_p (Kohms)	R_s (Kohms)	C_{dl} (μF)	Z (Kohms)	L (nm)
	12.22	1. 3	31.3	31.95	.75
2	1.72	. 623	9.36	1 7	2.5
4	.6	.52	5.61	178	4.2
6	1.53	. 57	8. 8	124	2.9
8	2.58	. 53	8.4	119	2.8
1	.45	. 65	29.61	33.8	.8

Table 1: The Electrochemical Parameters and the Oxide Film thickness of Al-Mg alloy in the as Received Condition in Different H_2SO_4 Concentrations.

% H_2SO_4	Rp (Kohms)	Rs (Kohms)	Cdl (μF)	Z (Kohms)	L (nm)
	14.71	3.36	41.4	24.15	.56
2	.2	. 26	37.13	26.93	.63
4	.11	. 12	62.18	16.1	.375
6	.14	. 11	53. 9	18.83	.44
8	.11	. 3	59.13	16.91	.395
1	. 885	. 63	5 .4	19.84	.463

Tabel 2: The Electrochemical Parameters and the Oxide Film thickness of Al-Mg alloy in the annealed Condition in Different H_2SO_4 Concentrations.

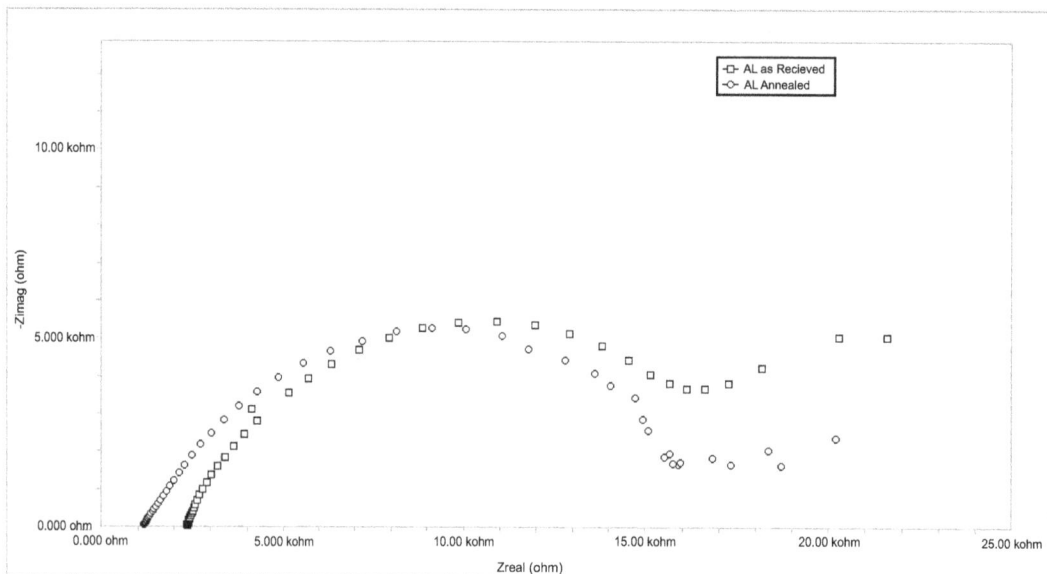

Figure 3: Shows the Nyquist plot of the as received and annealed samples of Al-Mg alloy in % H_2SO_4.

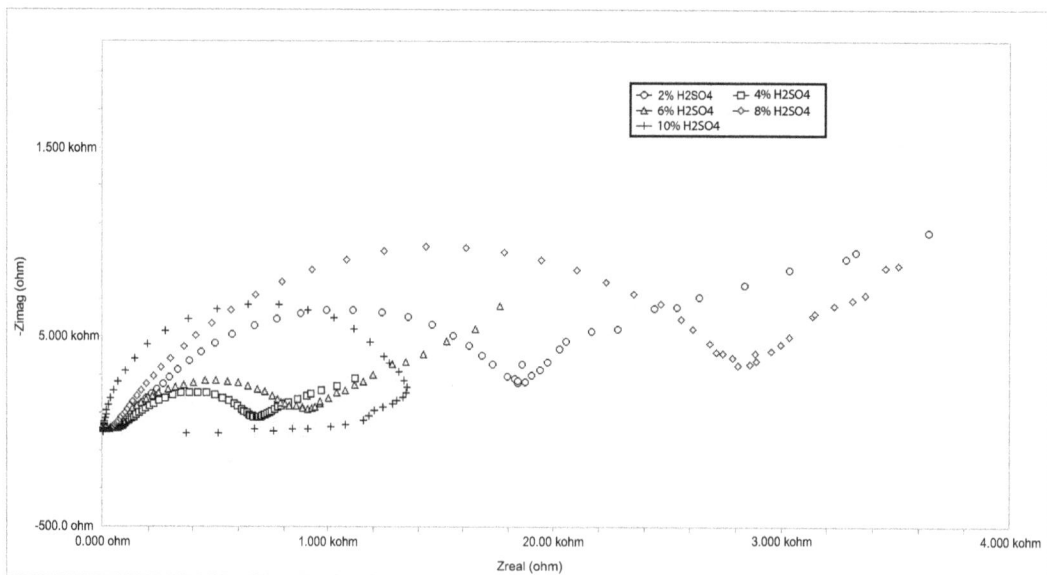

Figure 4: Shows the Nyquist plot of the as received samples of Al-Mg alloy in 2%-1 % H_2SO_4.

the oxide film thickness of the as received samples is greater than the one for the annealed samples, because the former samples are thermodynamically unstable (more chemically active) as compared to the annealed samples. In addition, this observation is in agreement with the fact that the E.I. Spectroscopy is useful technique to determine the electrochemical parameters and the oxide barrier (compact)-film thickness of aluminum alloys [2].

Mathematical model of the oxide film build up on the aluminum substrate

In order to develop a general model of the mechanism of the oxide barrier-film build up on the aluminum substrate during the anodization of the aluminum, one should consider the following:

1-The aluminum substrate always has a thin oxide film, due to the oxidation of aluminum in air.

2-The aluminum half reaction is as the followings:

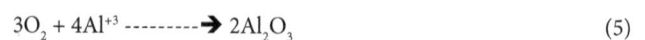

$$2Al^{+3} + 6e ----\rightarrow 2Al \tag{3}$$

$$2Al^{+3} + 2Al^{+3} -------\rightarrow 4Al^{+3} \tag{4}$$

$$3O_2 + 4Al^{+3} ---------\rightarrow 2Al_2O_3 \tag{5}$$

during the anodization of the aluminum in an electrochemical cell made of aluminium electrode/sulfuric acid solution .

3-The oxidation of the aluminum substrate in equation 3 and equation 4 is governed by the applied electrochemical potential (E), described by Nernst's equation 6 as the following:

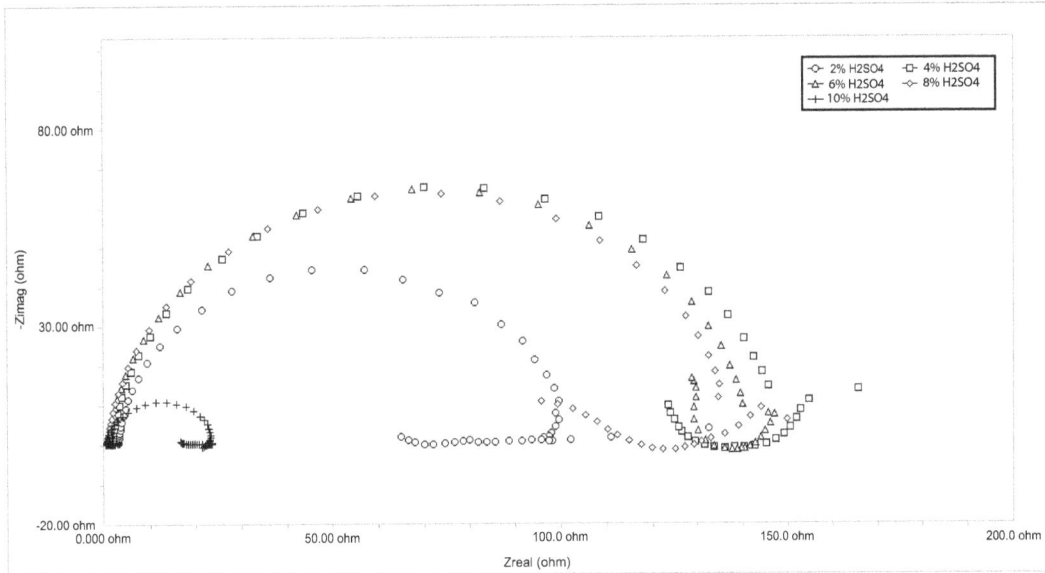

Figure 5: Shows the Nyquist plot of the annealed samples of Al-Mg alloy in 2%-1 % H_2SO_4.

$$E=E^\circ +[RT/nF]\ ln[(Al^{+3})/(Al)] \tag{6}$$

Where

E° is the equilibrium potential.

R is the ideal gas constant.

T is the absolute temperature of the solution.

n is the electronic charge number.

F is Faraday's constant.

(Al^{+3}) is the concentration of Al^{+3}.

(Al) is the concentration of Al.

4-Finally, the mathematical model is not only a function of the electrochemical potential (E) of the oxidation of aluminum substrate, described in equation 3 and equation 4, but also, the mathematical model is a function of a number of independent variables. The mathematical model is a function of the Oxygen transport in the oxide barrier -film above the aluminum substrate, a function of the Oxygen diffusion through the porous oxide- layer above the in the oxide barrier-film, and a function of time. In other words, the mathematical model is a function of Oxygen diffusion flux (N_s), solid diffusion, in the oxide barrier (compacted)- film , a function of the Oxygen flux through the porous Oxide layer (N_p),porous diffusion, and time(t). So, the general form of the model is derived as the following:

$$L=L(E,N_s,N_p,t) \tag{7}$$

L is the oxide barrier-film thickness.

E is the applied electrochemical potential, see in Equation.6.

N_s is the Oxygen flux in oxide barrier-film.

N_p is the Oxygen flux in the porous oxide layer and N_p is equal to the following:

$$N_p = N_v + N_d \tag{8}$$

Where

N_v is the Oxygen viscous flux in the pores.

N_d is the Oxygen diffusion flux in the pores.

So, the supply of Oxygen to the aluminum substrate, as described in equation.5, is actually consists of

$$N_s + N_p = N_s + N_v + N_d = \tag{9}$$

and details of the derivation of equ. 9 are given elsewhere [7&8] as the followings:

First by letting, $N_s + N_p = N_s + N_v + N_d = N_1{}^{7\&8}$,where

$$N_1 = -\frac{1}{RT}\left[K_0 S_1 + \frac{B_0 P}{\mu_1} + Q_1 e^s\right]\frac{dP}{dL} \tag{10}$$

N_1 indicating that the Oxygen is the single component (O_2=1) flowing through the oxide barrier (compacted)- film and through the porous Oxide layer.

Where

R is the ideal gas constant.

T is the temperature of the solution.

K_o is defined as a structural parameter which is equal to:

$$\frac{2}{3}a\ X\left(\frac{\varepsilon^G}{\tau^G}\right) \tag{11}$$

Where "a" is the pore radius and ε^G is the void fraction and τ^G is the tortousity factor; which is equal to the inverse of the void fraction if not known [7]. The superscript G denotes the gas phase.

The unit on K_o is centimeters (cm). S_1 is the velocity of gas molecules defined as:

$$S_1 = \left(\frac{8RT}{\pi M_R}\right)^{\frac{1}{2}} \tag{12}$$

Where M_R is the molecular weight of the Oxygen, (R=O_2), in flux (O_2 = 3.2 x 1^{-2} kg/mol; This first term ($K_o S_1$) therefore represents the contribution of the diffusive porous flux, in which a> τ^G.

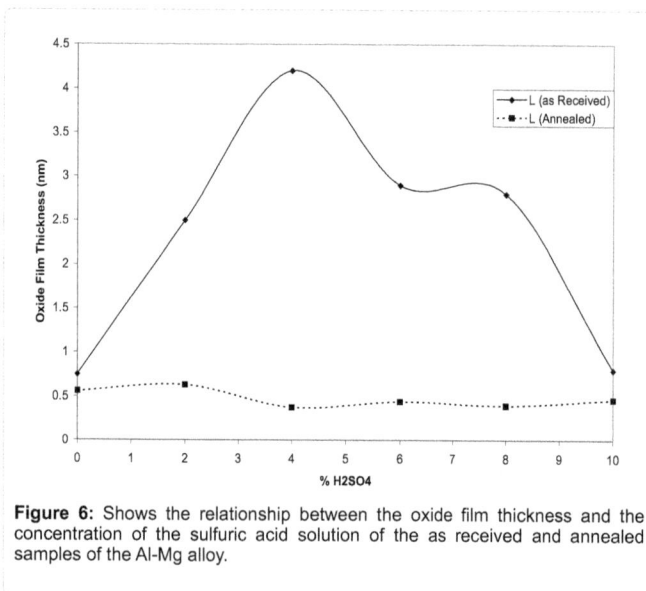

Figure 6: Shows the relationship between the oxide film thickness and the concentration of the sulfuric acid solution of the as received and annealed samples of the Al-Mg alloy.

P is the pressure (units: mmHg ➔ 1333.2 g/cm*s^2) and μ_1 is the Oxygen viscosity

B_0 is the structural parameter defined as:

$$\frac{1}{8}a^2 \times \left(\frac{\varepsilon^G}{\tau^G}\right) \tag{13}$$

This term, $\frac{B_0 P}{\mu_1}$ therefore is the contribution of the viscous flux through the pore, in which a< τ^G.

Additionally $Q_1 e^s$ is the "effective" gas (Oxygen) permeability in the aluminum oxide.

This term thereby represents the solid flux of gas through the oxide barrier-film. The differential $\frac{dP}{dL}$ is the pressure gradient imposed by the system, if that existed during the anodization of the aluminum substrate.

Assuming if the pressure existed during the anodization of the aluminum substrate, then the boundary condition would be of such that at, the oxide barrier-film thickness, L = then P = P (*l*) and when L = *l* then P = P (*l*) where l = a nanometer scale. This then leads to the following simplification:

ΔP = P (*l*) - P (*l*).

Furthermore another simplification can be made:

$$P_{average} = \frac{p(0) - p(l)}{2} \tag{14}$$

Therefore when integrating the N_1 equation leads to the following equation:

$$N_1 = -\frac{1}{RT}\int_0^l \left[K_0 S_1 + \frac{B_0 P}{\mu_1} + Q_1 e^s\right]\frac{dP}{dL} \tag{15}$$

The above mathematical model, equation 7, was given only to describe the mechanism of the processes of the buildup of the oxide barrier-film on the aluminum substrate. The mathematical model will not be used to predict the processes of the buildup of the oxide film on the aluminum substrate by solving the model, using empirical data, because such empirical data are very difficult to obtain on the aluminum/oxide barrier (compacted)-film/porous oxide-layer structure. The solution of equation 7 will be the next challenge of the present work.

Conclusions

The following conclusions are drawn from the present investigation:

The thickness of the oxide film was determined based on the obtained electrochemical parameters by EIS, see Tables 1 and 2 for both conditions. The optimum thickness of the oxide film was detected for the as received samples (4.2nm) and for the annealed samples (0.63nm) in sulfuric acid concentrations of 4% and 2% H_2SO_4, respectively. The reason behind the oxide film thickness of the as received samples is greater than the one for the annealed samples, because the former samples are thermodynamically unstable (more chemically active) as compared to the annealed samples. Consequently, The E.I. Spectroscopy is found useful technique to determine the electrochemical parameters and the oxide barrier (compact)-film thickness of aluminum alloys. In addition, a mathematical model of the oxide film build up on the aluminum substrate was proposed for the next challenge of the present work.

References

1. K.Habib (2001) In Situ Measurement of Oxide Film Growth on Aluminum Samples by Holographic Interferometry. Corrosion Science 43: 449-455.

2. Rk.Ptucek, Rg.Rateick, Vi.Birss (2006) Impedance Characterization of Anodic Barrier Al Oxide Film Beneath Porous Oxide Layer. J Electrochem 153: B304-310.

3. R.Baboian (1987) Electrochemical techniques for Corrosion Engineering NACE Press 58.

4. ASTM, Standard Test Method for measurement of Impedance of Anodic Coating on Aluminum, Annual Book of ASTM standards, B457-67,1994, 179.

5. Metals Handbooks, American Society of Metals, ASM Ninth Edition, metal Park 4:689.

6. A.Bard, L.Faulkner Electrochemical Methods. John Wiley &Sons Inc, New York 5 -51(198).

7. Habib K (1993) General Model of Gas Transport Through Solid Porous Membranes, Metallurgical Transactions A 24: 1527-153.

8. K.Habib, A. Habib (2004) General Model of Hydrogen Transport Through Nanoporous Membranes, Journal of Composites 35:191-195.

19

Modelisation of Membrane Distillation: Mass and Heat Transfer in Air Gap Membrane Distillation

Rochd S*, Zerradi H, Mizani S, Dezairi A and Ouaskit S

Laboratory of Physics of Condensed Matter (URAC10), Ben M'sik Sciences, University Hassan II Casablanca, Morocco

Abstract

Membrane distillation (MD) is receiving recent attention as a technique to efficiently concentrate aqueous solution such as seawater. It has potential benefits of low temperature and pressure operation with high degrees of separation. In this work, the effect of the membrane thickness was studied to produce of the steam flow in the three different mass transfer mechanisms, and in the different possible combinations of its mechanisms in three different temperatures. The results have been carried using a polynomial approximation through MATLAB. A quite important increase in the flow in the model (DGM, Schofield, and KMPT) was observed with a decrease in the stream rating in the model (KMT). However, the Molecular model, DGM model, KMPT model, and Schofield model are not affected by the membrane's thickness. After, we have studied the effect of this parameter (Thickness of the membrane) on the transfer of conduction heat and latent heat.

Keywords: Membrane; Distillation; AGMD; Diffusion; Thickness

List of Symbols

J_v	Local permeate flux at the hot side of membrane in vapor phase [Kg/m².s]
ε	Porosity of the membrane
T	Tortuosity
γ_p	Membrane pore size [m]
M_v	Molar mass [Kg/mol]
R	Universal gas constant [J/mol.K]
T_m	Average temperature of the membrane [°C]
P_m	Average pressure partial of the air [P_a]
δ_m	Thickness of the membrane [m]
P	Pressure [P_a]
$D_{v/a}$	Diffusion coefficient of the vapor in the vapor/air mixture [m²/s]
μ	Dynamic viscosity [Kg/m.s]
P_v	Water vapor pressure [P_a]
C_s	Mole fraction of NaCl
hm	Hot liquid/ membrane interface
M	Membrane
mg	Membrane/ air gap interface
S	Saline
P	Pores
v/a	Vapor/air
C_1	Concentration in the feed part
C_2	Concentration in the air gap part
K	Transfer coefficient [m/s]
D_{eff}	Effective diffusion coefficient [m² s⁻¹]
R_e	Reynold number of the hot solution channel
P_r	Prandlt number
d_h	Half-width of the flow channel [m]
H_m	Membrane length [m]
T_{gf}	Temperature at the interface of hot feed and the membrane [°C]

T_f	Feed bulk temperature [°C]
T_c	Bulk temperature of the coolant [°C]
Q_s	Total heat [KJ/m²h]
Q_c	Flow of heat by conduction [KJ/m²h]
Q_v	Latent heat flux [KJ/m²h]
R_m	Gas constant of membrane [J/mol K]
K_{cm}	Thermal conductivity of the materiel forming the membrane [W/m.K]
K_a	Thermal conductivity of air [W/m.K]
K_m	Thermal conductivity of membrane [W/m.K]
$J_v \Delta h_v$ $J_v \Delta h_v$	Enthalpy of hot solution [J/Kg]
T_h	Hot temperature [°C]

Introduction

In the early of 1980 with the growth of membrane engineering, MD claims to be a cost effective separation process that can utilize low-grade waste and alternative energy sources such as solar and geothermal energy [1]. Membrane distillation is a hybrid process that uses membranes and operates based on evaporation. Unlike most other membrane process, MD does not require a mechanical pressure pump and is not limited by the osmosis pressure [2].

A variety of methods may be employed in MD, such as direct contact membrane distillation (DCMD) in which the membrane is

---

*Corresponding author: Rochd Sanaa, Faculty of Ben M'sik Sciences, Laboratory of Physics of Condensed Matter (URAC10), University Hassan II Casablanca, Morocco, E-mail: rochd.sanaa91@gmail.com

in direct contact with liquid phases in both sides [3-5]. Then, air gap membrane distillation (AGMD) in which an air layer is interposed between the membrane and the condensation surface [6,7]. At that time, a vacuum membrane distillation (VMD) where, a vacuum is applied to increase or establish the vapor pressure difference between the membrane sides and the condensation takes place in an external condenser [8,9]. At that point, sweeping gas membrane distillation (SGMD) in which a stripping gas is used on the cold side to sweep the permeate away, with Condensation in a separate device [10-14].

At this juncture, it is worthwhile to underline that the membrane technique is considered promising since it takes place at temperatures range (30 to 90)°C and can use solar energy [15]. A survey of the state-on-the-art of membrane distillation (MD) and its various and detailed applications was presented by Alklaibi and Lior [16]. Not far, Ding et al. presented a model for predicting the rate of mass transfer in a membrane distillation unit to direct contact (DCMD) [17]. Other researchers, including Meindersma [18], Guijt [19], Payo [14], and Chouikh [20,21] worked on AGMD, but none of them used solar as the energy source. Likewise, Mandiang [22-24] have studied three different types of mass transfer modes through the membrane and possible combinations of this type. Also, Morteza Asghari [23] studied the effect of the thickness of the membrane in transfer mechanism of Knudsen [23].

Our contribution is to redo the results of Mandiang and use them to study the effect of the thickness of the membrane on the production of the steam flow in the three different mass transfer mechanisms, and in different possible combinations of its mechanisms. After, we have studied the effect of this parameter (Thickness of the membrane) on the transfer of conduction heat and latent heat.

Definition of the Method

In air-gap MD, the evaporator channel resembles that in DCMD, whereas the permeate gap lies between the membrane and a cooled walling and is filled with air. The vapor passing through the membrane must additionally overcome this air gap before condensing on the cooler surface. The advantage of this method is the high thermal insulation towards the condenser channel, thus minimizing heat conduction losses. However, the disadvantage is that the air gap represents an additional barrier for mass transport, reducing the surface- related permeate output compared to DCMD. A further advantage towards DCMD is the fact, that volatile substances with a low surface tension, such as alcohol or other solvents can be separated from diluted solutions, due to the fact that there is no contact between the liquid permeate and the membrane with AGMD (Figure 1). A temperatures range of system: T_{hm} = 40°C to 80°C) and (T_{mg} = 10°C to 30°C) [24].

Mechanism of Mass Transfer

The three types of mechanisms of Mass transfer are:

Knudsen diffusion (based on collisions between molecules and the-wall) [24]

This type of distribution is important in systems with high temperature and pressure.

$$J_K = K_K(P_{hm} - P_{mg}) \tag{1}$$

With: $$K_k = \frac{2}{3}\frac{\varepsilon}{\tau}\frac{r_p}{\delta_m}\sqrt{\frac{8M_v}{\pi RT_m}} \tag{2}$$

Molecular diffusion (based on collision between molecules) [24]

This type of distribution is important in systems of intermediate temperature and pressure.

$$J_{M,S} = K_{M,S}(P_{hm} - P_{mg}) \tag{3}$$

With: $$K_{M,S} = \frac{\varepsilon p D_{v/a} M_v}{\tau \delta_m RT_m P_a} \tag{4}$$

Viscous diffusion (based on both types of collisions) [24]

This type of distribution is important in systems of low temperature and pressure.

$$J_p = K_{M,S}\Delta P_v \tag{5}$$

With: $$K_p = \frac{1}{8\gamma_g}\frac{r_p^2\varepsilon}{\tau}\frac{P_m M_v}{\delta_m RT_m} \tag{6}$$

In order to explain the reduction of vapor pressure caused by the dissolved species, Raoul's law [25] may be used.

$$P_{hm} = (1 - C_s)p_v \tag{7}$$

Where C_s is the mole fraction of solute or salinity? The difference in partial pressure of the saturated vapor of both sides of the membrane may be calculated from the law of Antoine [15] using the following equation:

$$p_v = \exp(23.328 - \frac{3841}{T - 45}) \tag{8}$$

The average temperature of the membrane is given by the following equation [22]

$$T_m = \frac{T_{hm} + T_{mg}}{2} \tag{9}$$

Furthermore Qtaishat et al. proposed the expression amount of the steam/air:

PDv/a (Pa.m².s⁻¹) depending on the temperature [26]

$$PD_{v/a} = 1.985.10^{-5} T^{2.072} \tag{10}$$

Modes Combination of Three Flows

DGM model

The model "Dusty Gas" (DGM) shown in Figure 2. In this model, the coefficient of permeability of the membrane due to molecular

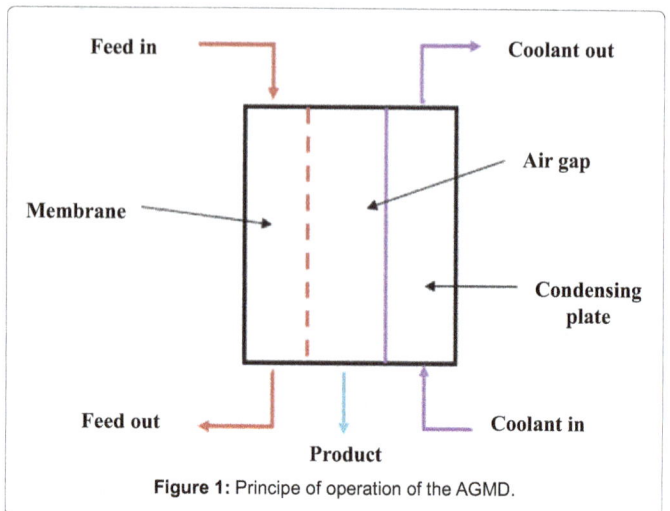

Figure 1: Principe of operation of the AGMD.

diffusion and of Knudsen are combined as resistance in series, where in the potential drops (pressure difference) are additive [22].

$$J_{DGM} = \frac{(J_{M,S} + J_K)J_P}{J_{M,S} + J_K + J_P} \quad (11)$$

Schofield model

In this model, the coefficient of diffusion of Knudsen and Poiseuille are parallel and in series with the Fick's flow (Figure 3) [22].

$$J_{Schofield} = \frac{J_K J_P}{J_K + J_P} + J_{M,S} \quad (12)$$

KMPT model

In this model, the diffusion co-efficient of Molecular Knudsen are in parallel and in series with the poiseuille (Figure 4) [22].

$$J_{KMPT} = \frac{J_K J_{M,S}}{J_K + J_{M,S}} + J_P \quad (13)$$

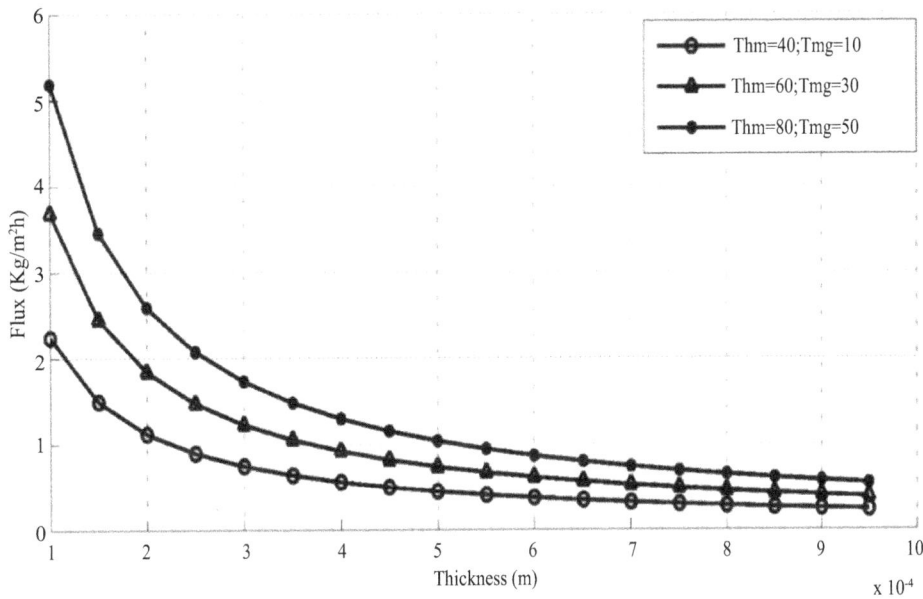

Figure 2: Mass transfer of Knudsen Diffusion according to the membrane thickness in three different temperatures.

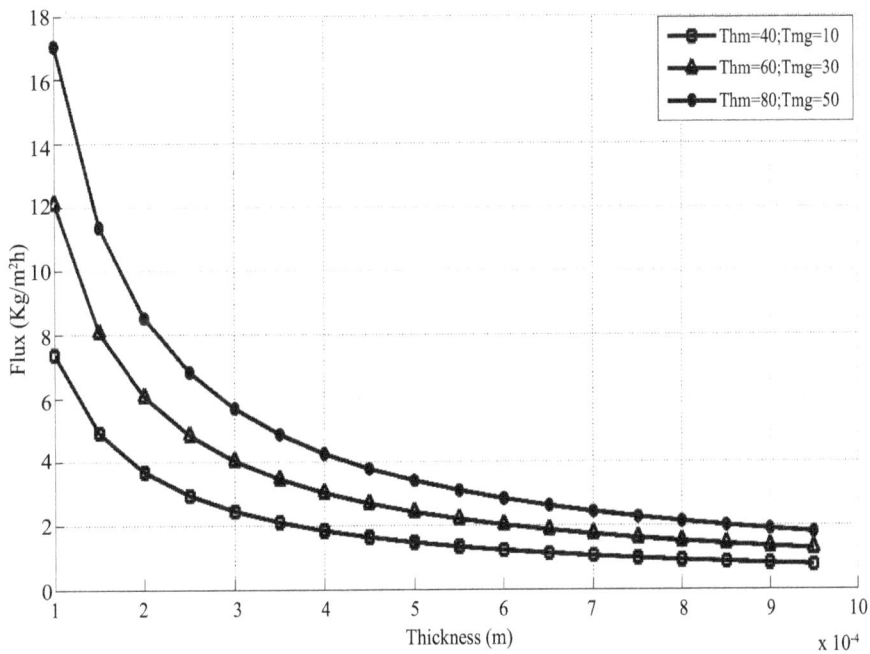

Figure 3: Mass transfer of viscous diffusion according to the membrane thickness in three different temperatures.

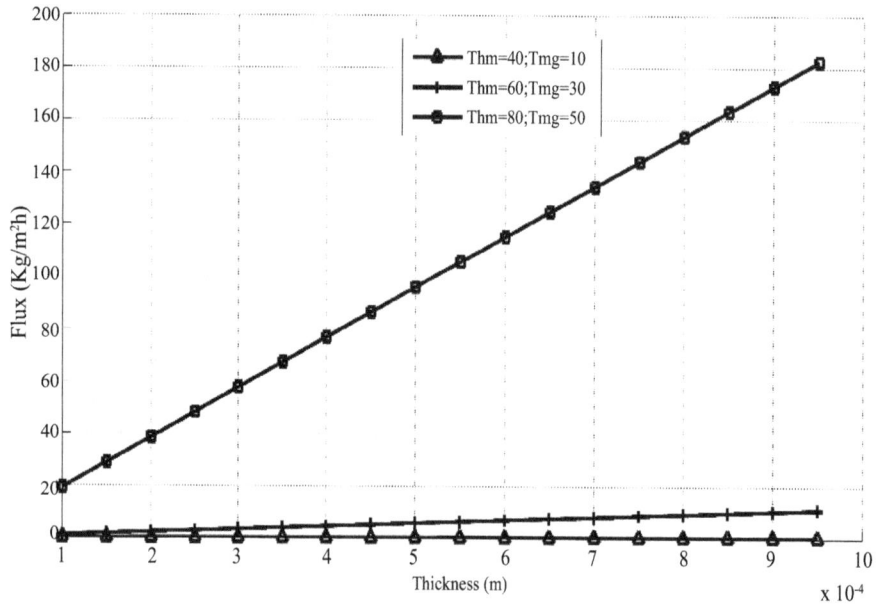

Figure 4: Mass transfer of Molecular Diffusion according to the membrane thickness in three different temperatures.

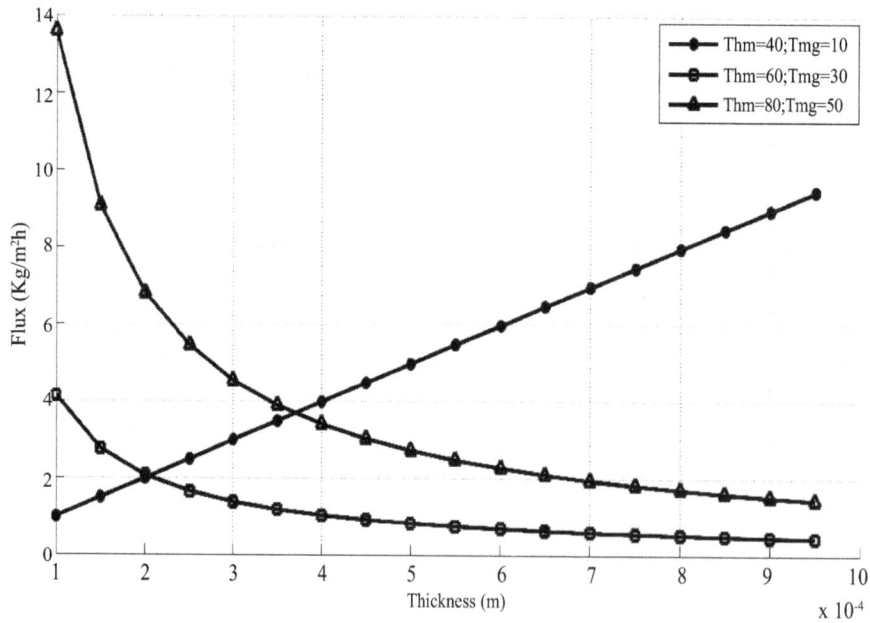

Figure 5: Mass transfer of three different types of diffusion according to the membrane thickness at h =60 °C and =25°C.

KMT model

In this model, the Knudsen and Molecular flows are in parallel (Figure 5) [22].

$$J_{KMT} = \frac{J_K J_{M,S}}{J_K + J_{M,S}} \qquad (14)$$

Transfer of Heat Flow

Two main heat transfer mechanisms occur in the MD system: Latent heat and conduction heat transfer [27]:

$$Q_S = Q_C + Q_v \qquad (15)$$

The flow of heat by conduction (Q_C) is defined by:

$$Q_C = \frac{T_{hm} + T_{mg}}{R_m} = \frac{k_{cm}}{\delta_m}(T_{hm} - T_{mg}) \qquad (16)$$

The thermal conductivity of the membrane k_{cm} is defined by the equation as:

$$k_{cm} = \varepsilon k_a + (1 - \varepsilon)k_m \qquad (17)$$

Where:

k_m: Thermal conductivity of the materiel forming the membrane.

k_a: Thermal conductivity of air.

The latent heat flux (Q_v) of the vapor through the membrane is defined by:

$$Q_v = J_v \Delta h_v \qquad (18)$$

Where $J_v \Delta h_v$: Enthalpy of unit mass of steam (J/Kg).

The temperature of the hot vapor side of the membrane is defined by:

$$T_{hm} = \frac{h_h T_h - J_v \Delta h_v + \dfrac{k_{cm}}{\delta_m} T_{mg}}{h_h + \dfrac{k_{cm}}{\delta m}} \qquad (19)$$

Where the convection coefficient steady correlation is given by the creatz-leveque:

$$h_h = 1.86 (R_e P_r \frac{d_h}{H_m})^{0.33} \qquad (20)$$

The total transfer across the membrane (Q_s) is:

$$\tilde{Q}_S = \tilde{Q}_v + \tilde{Q}_C \qquad (21)$$

Where the x-average conductive heat transfer is:

$$\tilde{Q}_C = \frac{1}{l_m} \int_0^{l_m} Q_c(x) dx \qquad (22)$$

Moreover, the x-average latent heat flux is:

$$\tilde{Q}_v = \frac{1}{l_m} \int_0^{l_m} Q_v(x) dx \qquad (23)$$

Temperature De Polarization

This temperature variation across the membrane can be describe by the temperature polarization factor (TPC) that is defined as [28]:

$$TPC = \frac{T_{hm} - T_{gf}}{T_f - T_c} \qquad (24)$$

Where: T_{hm} is the temperature at the interface of hot feed and the membrane (°C), T_{gf} is the temperature at the interface of hot feed and the membrane (°C), T_f is feed bulk temperature (°C), T_c is bulk temperature of the coolant (°C).

Results and Discussion

Simulation were carried with the following input data: k_m=0.05 W.m^{-1}.K^{-1}; l_m=0.2 m; H_m=0.2 m; d_h=0.002 m; Pr=7.2; Re=192; ka=60 W/m.K. When the temperature is fixed at three different temperatures and the membrane thickness is swept between [1.10^{-4}, 10.10^{-4}] m, following results are observed: Prior to undertaking detailed presentation and discussion of the new results obtained in this study, Figure 2 shows the variation of the vapor of Knudsen diffusion as function of the membrane thickness in three different temperature. In this figure, the decrease of the flux is seen by increasing the thickness such as the flow begins to decrease from the value (1.10^{-4} m) and becomes constant when the value of the thickness of the membrane exceeds (9.10^{-4} m), the maximum value of flux is (4.25 kg/m^{2}h). However, we have seen for the temperature (T_{hm}°C, T_{mg}=50°C a bigger flow of steam production compared to the case temperature (T_{hm} =60°C, T_{mg} =30°C) and (T_{hm} =40°C, T_{mg} =10°C), as in some cases of temperature (T_{hm} =40°C, T_{mg} =10°C), we have seen a very low vapor production.

This observation is in line with Figure 3, which shows the variation of the vapor of viscous diffusion as function of the membrane thickness in three different temperatures. In this figure, the flux decrease is seen by increasing the thickness, such as the flow begins to decrease from the value (1.10^{-4}) and becomes constant when the value of the thickness of the membrane exceeds (9.10^{-4}) m. the maximum value of flux is (13.75 Kg/m^{2}h). However, we have seen for the temperature (T_{hm}=80°C, T_{mg} =50°C) a bigger flow of steam production compared to the case temperature (T_{hm}=60°C, T_{mg}=30°C) and (T_{hm}=40°C, T_{mg}=10°C), as in some cases of temperature (T_{hm}=40°C, T_{mg} =10°C), we have seen a very low vapor production.

Figure 4 shows representative results on the variation of the vapor of molecular diffusion as function of the membrane thickness in three different temperatures. In this figure, an increase of flow depending on the thickness of the membrane of (1.10^{-4}) to (9.10^{-4}) is observed. However, we have seen for the temperature (T_{hm}=80°C, T_{mg}=50°C) a bigger flow of steam production compared to the case temperature (T_{hm}=60°C, T_{mg} =30°C) and (T_{hm}=40°C, T_{mg} =10°C), as in some cases of temperature (T_{hm}=40°C, T_{mg} =10°C), we have seen a very low vapor production.

Now turning our attention to Figure 5, that shows the variation of the vapor of different transfer mechanisms as function of the membrane thickness. An observation for both mechanisms Knudsen and viscous (Figures 2-4), shows a rating decrease of flux depending on the membrane thickness because the thickness of the membrane increases the diffusion path which increases the mass transfer resistance and therefore reduces the flow. Rating decrease of the flow would suggest the weakness of the two mechanisms. For the molecular mechanism (Figures 4 and 5), it is observed, that the flow increases simultaneously with the thickness of the membrane because the molecular flow is strangest among the three types of Mass transfer mechanism in this field of temperature.

Figure 6 also indicates the variation of the vapor of DGM model as function of the membrane thickness in three different temperatures. In this figure, an increase simultaneously of the flow is depending on the thickness of the membrane of (1.10^{-4}) to (9.10^{-4}). However, we have seen for the temperature (T_{hm}=80°C, T_{mg}=50°C) a bigger flow of steam production compared to the case temperature (T_{hm}=60°C, T_{mg}=30°C) and (T_{hm}=40°C, T_{mg}=10°C), as in some cases of temperature (T_{hm}=40°C, T_{mg}=10°C), we have seen a very low vapor production.

Figure 7 describes the variation of the vapor of Schofield model as function of the membrane thickness in three different temperatures. An increase of the flow is depending on the thickness of the membrane of (1.10^{-4}) to (9.10^{-4}). However, we have seen for the temperature (T_{hm}=80°C, T_{mg}=50°C) a bigger flow of steam production compared to the case temperature (T_{hm}=60°C, T_{mg}=30°C) and (T_{hm}=40°C, T_{mg}=10°C), as in some cases of temperature (T_{hm}=40°C, T_{mg}=10°C), we have seen a very low vapor production.

Figure 8 shows the variation of the vapor of KMPT model as function of the membrane Thickness in three different temperatures. In this figure, we show an increase simultaneously of flow depending on the thickness of the membrane of (2.10^{-4}) to (9.10^{-4}). However, we have seen for the temperature (T_{hm}=80°C, T_{mg}=50°C) a bigger flow of steam production compared to the case temperature (T_{hm}=60°C, T_{mg}=30°C) and (T_{hm}=40°C, T_{mg}=10°C), as in some cases of temperature (T_{hm}=40°C, T_{mg}=10°C), we have seen a very low vapor production.

Another time Figure 9 shows the variation of the vapor of KMT model as function of the membrane Thickness in three different

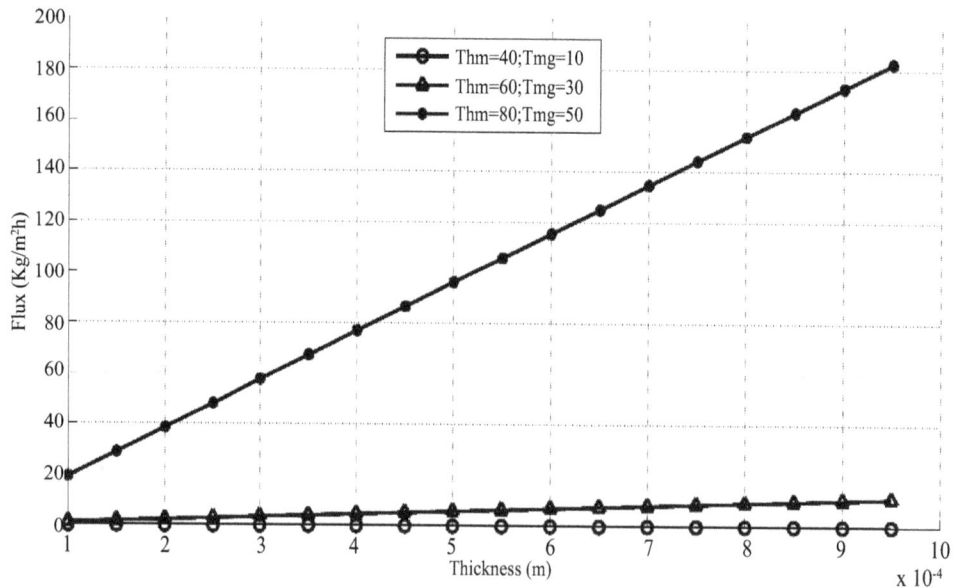

Figure 6: Evolution of Model "DGM" as function of membrane thickness in three different temperatures.

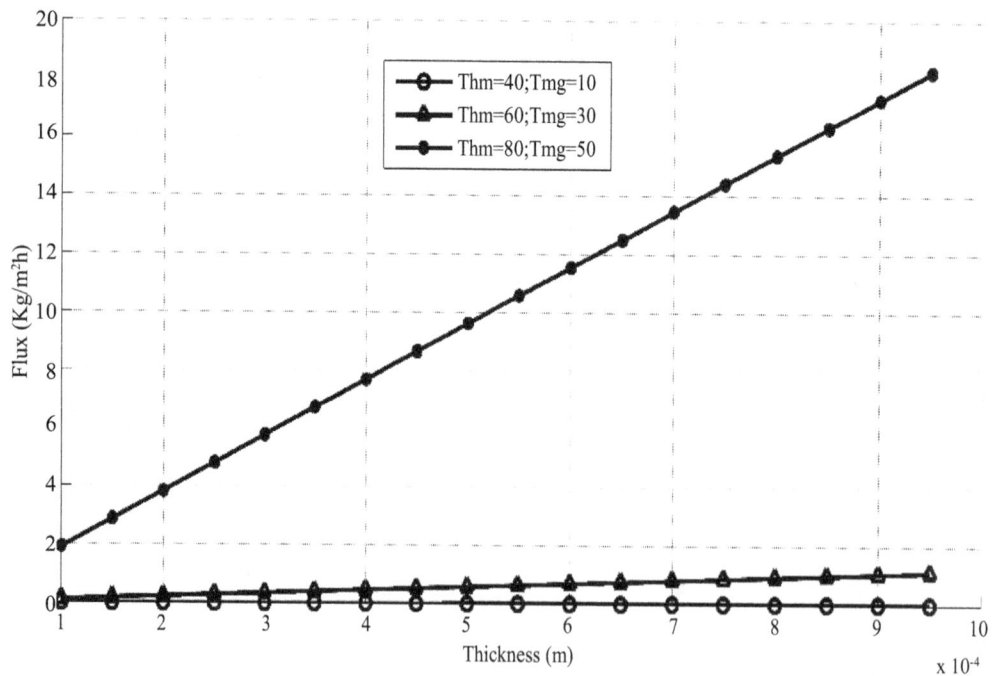

Figure 7: Evolution of Model "Schofiled" as function of membrane Thickness in three different temperature.

temperatures. In this figure, the flux decreases by increasing the thickness such as the flow begins to decrease from the value (1.10^{-4}) and becomes constant when the value of the thickness of the membrane exceeds (9.10^{-4}) m, the maximum value of flux is 4.25 (kg/m^2h), however, we have seen for the temperature $(T_{hm}=80°C, T_{mg}=50°C)$ a bigger flow of steam production compared to the case temperature $(T_{hm}=60°C, T_{mg}=30°C)$ and $(T_{hm}=40°C, T_{mg}=10°C)$, as in some cases of temperature $(T_{hm}=40°C, T_{mg}=10°C)$, we have seen a very low vapor production.

After, the thickness of the membrane is varied in the various possible combinations of flux transfer mechanism (Figures 6-9), an observation shows an increase in the flow in the model (Figures 6-8) and a stream rating decreases in the model (Figure 9). This also indicates that the increased flow as a function of the thickness by the following equation [29]:

$$J = D_{eff} \frac{C_2 - C_1}{\delta_m} = \frac{D_{eff}}{\delta_m}(C_1 - C_2) = K(C_1 - C_2) \quad (25)$$

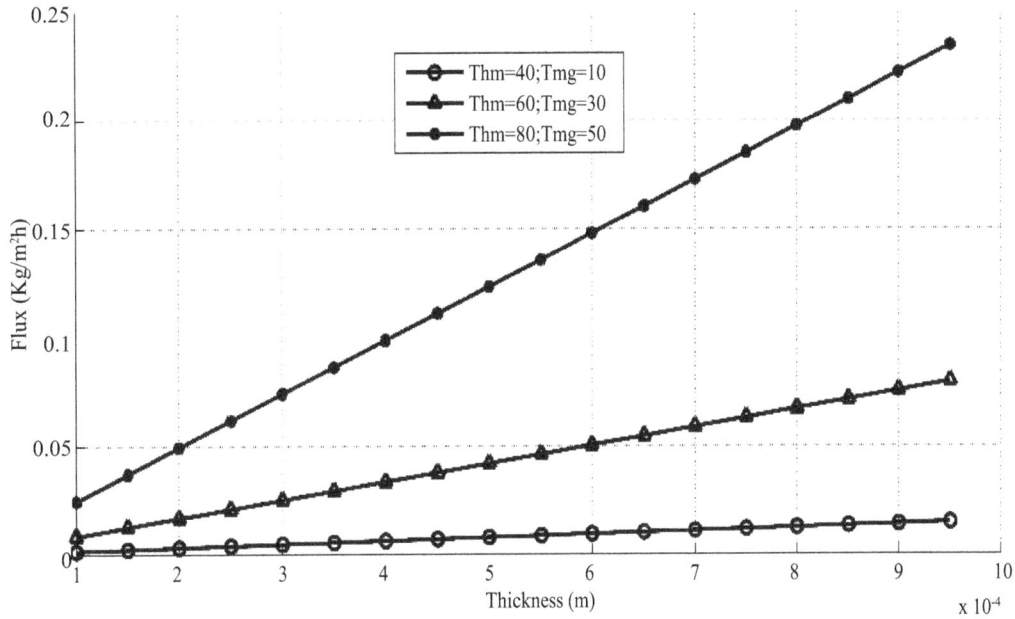

Figure 8: Evolution of model "KMPT" as function of membrane thickness in three different temperature.

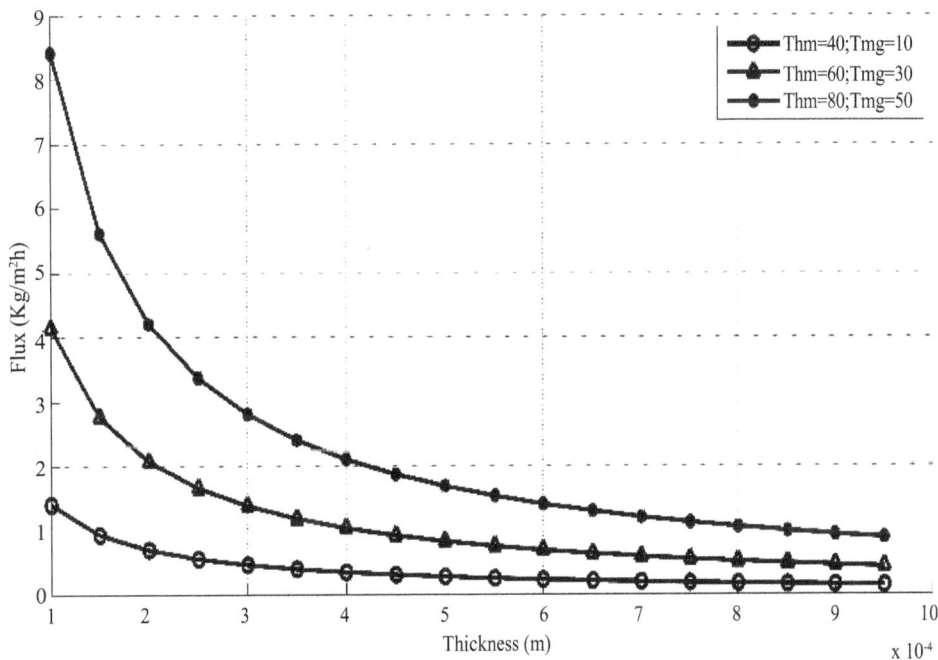

Figure 9: Evolution of model "KMT" as Function of membrane Thickness in three different temperatures.

This equation reflects the diffusion of molecules in a porous and homogeneous system. C_1 and C_2 are the concentrations of both sides of the membrane, K is the transfer coefficient, δ_m is the thickness of the porous membrane.

Figure 10 shows the variation of heat conduction as function of the membrane thickness. In this figure, the heat conduction decreases by increasing the thickness. Furthermore, the flow begins to decrease

from the value (1.10^{-4}) and become constant when the value of the thickness of the membrane exceeds $(8.10^{-4}\,m)$, the maximum value of heat conduction is about $(1,7.10^7\,kg/m^2.s)$. This result is logical because the heat conduction decreases gradually depending on the distance of the heat flow because losing the temperature; it also loses the capacity to conduction.

Figure 11 shows representative results on the variation of latent

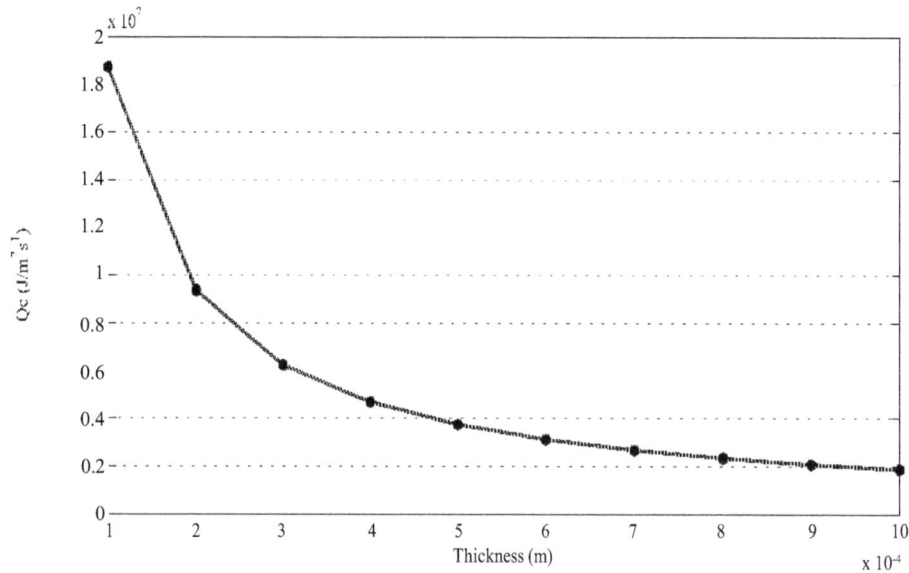

Figure 10: Evolution of the conductive heat as a function of the thickness of the membrane.

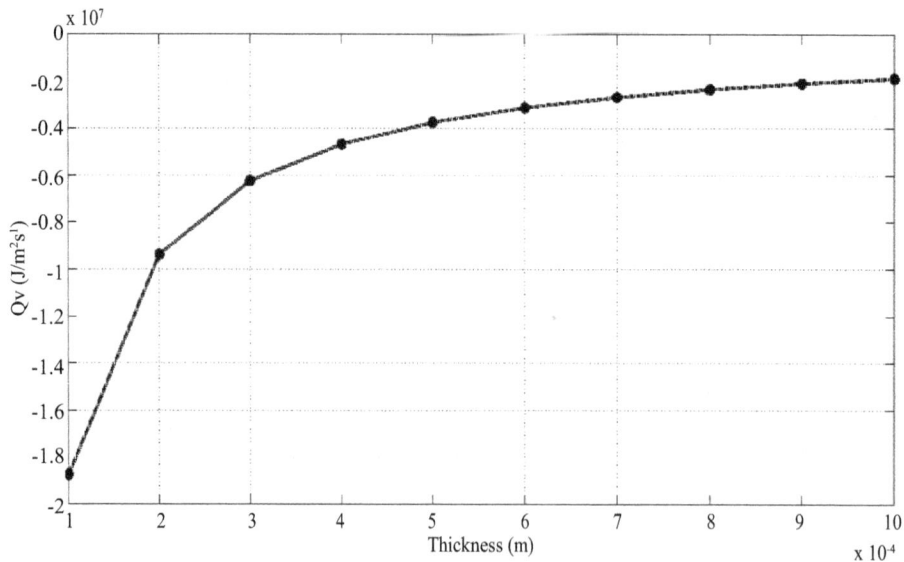

Figure 11: Evolution of the Latent heat as a function of the thickness of the membrane.

heat as function of the membrane thickness. In this figure, a latent heat increases simultaneously with the increase of the thickness of the membrane of $(1.10^{-4}\,m)$ to $(8.10^{-4}\,m)$ and became constant when the value of the thickness of the membrane exceeds $(8.10^{-4}\,m)$. This result is logical because latent heat is released when there is a change of state from, and it has a negative sign because we pass from fusion to condensation.

Figure 12 shows transfer of total heat as function of the membrane thickness. At first, the total heat remains constant up to the thickness $(9.10^{-4}\,m)$ because of compensation of two values of the heat (latent and conduction), then it will increase when the value exceeds $(9.10^{-4}\,m)$.

Conclusion

The broad trends can be summarized as follows, the changes of the steam flow as function of the thickness of the membrane for different flow transfer mechanisms is studied. For both mechanisms Knudsen and viscous, a rating decrease of flux depending on the membrane thickness is observed. For the Molecular mechanism, it is observed, that the flow increases simultaneously with the thickness of the membrane. Later, the thickness of the membrane is varied in various possible combinations of flux transfer mechanism (DGM, Schofield, KMPT, and KMT), an increase of the flow is detected in the model (DGM, Schofield, KMPT) and a stream rating decrease in the model (KMT). We conclude that the flow increases simultaneously with temperature. Finally, before concluding, it is worthwhile to make an observation regarding the Molecular model, DGM model, KMPT model, and Schofield model are not affected by the membrane's thickness. These results were the same conclusion that Mandiang results because the

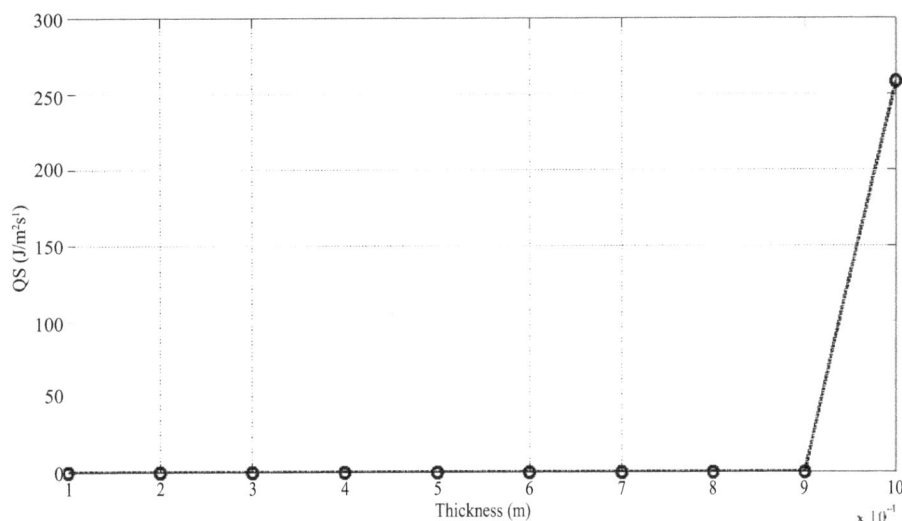

Figure 12: Evolution The total transfer of heat as a function of the thickness of the membrane.

model of Schofield and molecular mechanisms are the most suitable and beneficial for this system. Otherwise, we find that the heat conduction simultaneously decreases depending on the thickness, unlike the latent heat increases by increasing the thickness of the membrane.

References

1. El-Bourawi MS, Ding Z, Ma R, Khayet M (2006) A framework for better understanding membrane distillation separation process. J Membr Sci 285: 4-29.

2. Lawson KW, Lloyd DR (1997) Review Membrane distillation. J Membr Sci 124: 1-25.

3. Lawson KW, Lloyd DR (1996) Membrane distillation. II. Direct contact MD. J Membr Sci 120: 123.

4. Martinez L, Florido-Diaz FJ (2001) Theoretical and experimental studies on desalination using membrane distillation. Desalination 139: 373.

5. Phattaranawik J, Jiraratananon R (2001) Direct contact membrane distillation: effect of mass transfer on heat transfer. J Membr Sci 188: 137.

6. Jonsson AS, Wimmerstedt R, Hanysson AC (1985) Membrane distillation-A theoretical study of evaporation through microporous membranes. Desalination 56: 237.

7. Banat F (1994) Membrane distillation for desalination and removal of volatile organic compounds from water, Ph. D. Dissertation, Mc Gill University, Montreal, Canada.

8. Bandini S, Gostoli C, Sarti GC (1992) Separation efficiency in vacuum membrane distillation. J Membr Sci 73: 217.

9. Sarti GC, Gostoli C, Bendini S (1993) Extraction of organic comportments from aqueous streams by vacuum membrane distillation. J Membr Sci 80: 21.

10. Basini L, D'Angelo G, Gobbi M, Sarti GC, Gostoli C (1987) A desalination process through sweeping gas membrane distillation. Desalination 64: 245.

11. Khayet M, Godino P, Mengual JI (2000) Nature of flow on sweeping gas membrane distillation. J Membr Sci 170: 143.

12. Khayet M, Godino P, Mengual JI. Theory and experiments on sweeping gas membrane distillation. J Membr Sci 165: 261.

13. Rivier CA, Payo GMC, Marison IW, Stocker VU (2002) separation simulations. J Membr Sci 201: 1.

14. Payo GMC, Rivier CA, Marison IW, Stocker VU (2002) Separation of binary mixtures by thermostatic sweeping gas membrane distillation. II. Experimental results with aqueous formicacid solutions. J Membr Sci 198: 197-210.

15. Alkalibi AM (2008) The potential of membrane distillation as a stand-alone desalination process. Desalination 223: 375-385.

16. Alkalibi AM, Lior N (2005) Transport analysis of air gap membrane distillation. J Memb Sci 255: 239-253.

17. Ding Z, Ma R, Fane AG (2002) A new model for mass transfer in direct contact membrane distillation. Desalination 151: 217-227.

18. Meindersma GW, Guijt CM, De Haan AB (2006) Desalination And water recycle by Air Gap Membrane Distillation. Desalination 187: 291-301.

19. Guijt CM, Meindersma GW, Reith T, De Haan AB (2005) Air Gap Membrane Distillation: Separation and purification Technology 43: 233-244.

20. Gil MAI, Payo MCG, Pined CF (1999) Air Gap Membrane Distillation of Surcrose solutions. J Memb Sci 155: 291-307.

21. Chouikh R, Bouguecha S, Shahbi M (2005) Modeling of a Modified Air Gap Distillation Membrane for the Desalination of Sea water 181: 257-265.

22. Mandiang Y, Sene M, Thiam A (2015) Mathematical Modeling And Simulating of coupling parameters Transfers of Steam In A Membrane-Type Solar Still Agmd. J Materiel Sci Eng 4: 1-13.

23. Asghari M (2015) Desalination. 374: 92-100.

24. Mandiang Y, Sene M, Thiam A, Azilinond (2015) Daily Estimate of Pure water in a desalination Unit by Solar Membrane Distillation. J Materiel Sci Eng 4: 170.

25. Ding Z, Ma R, Fane AG (2002) A new model for mass transfer in direct contact membrane distillation. Desalination 151: 217-227.

26. Qtsaishat M, Matsuura M, Kruczek B, Khayet M (2008) Heat and mass transfer analysis in direct contact membrane distillation Desalination 219: 272-292.

27. Sene M, Mandiang Y, Azilinon D (2013) Hybrid Technology Solar Thermal Membrane Desalination of Salt Water for Populations In The Saloum River Delta. Int. Journal of Engineering Research and Application 3: 1091-1102.

28. Fahmi A, Al-Rub A, Banat F (2003) Sensitivity analysis of air gap membrane distillation. SciTechnol 38: 3645-3667.

29. Truskey GA, Yuan F, Katz DF (2004) Transport Phenomena in Bioloigical Systems. NJ Paerson Prentice Hill.

20

Relating Water/Solute Permeability Coefficients to the Performance of Thin-Film Nanofiber Composite Forward Osmosis Membrane

Song Xiaoxiao[1]*, Prince JA[1] and Darren Delai Sun[2]

[1]Environmental & Water Technology, Centre of Innovation, Ngee Ann Polytechnic, Singapore
[2]School of Civil and Environmental Engineering, Nanyang Technological University, Singapore

Abstract

The thin-film nanofiber composite (TNC) forward osmosis (FO) membranes are fabricated and systematically modified by a series of post treatments, such as adding additives into the monomer, NaOH treatment, chlorine treatment and support modification. The post treatments lead to the formation of modified membranes with a wide range of water permeability (A) and solute permeability (B) values. The impact of varied A, B and B/A values on the FO performance are systematically investigated. Furthermore, the value of B/A is related to internal concentration polarization (ICP), external concentration polarization (ECP) and solute leakage, which are firstly proposed in this study. Compared with the Pressure Retarded Osmosis (PRO) orientation (i.e., active layer facing draw solution), the water flux is much lower at FO orientation (i.e., active layer facing feed solution) due to severe loss of effective osmotic pressure, which mainly results from the convective dilution and low mass transfer coefficient in support membrane (i.e., the D/S value). In addition to this, the coupled effect of solute leakage and low D/S value also causes a minor loss of effective osmotic pressure. This is the first study to systematically analyze the B-A relationship of TNC membranes employing different modification techniques and to investigate the impact on the FO performance.

Keywords: Nanofiber composite membrane; Forward osmosis; Internal concentration polarization; External concentration polarization

Introduction

FO process is a low-energy process, which can potentially be applied in water/wastewater treatment and desalination [1]. In a typical FO process, the high-salinity draw solution is separated from the low-salinity feed solution by a semi-permeable membrane. The membrane only allows water to pass through but rejects solutes [2-4]. The selectivity of FO membrane is crucial to maintain the osmotic pressure difference across the membrane. However, the permeability of FO membrane determines the water flux (J_w) the membrane can harvest from preset osmotic pressure difference. Hydration Technologies Inc. (HTI) developed the first commercial flat sheet FO membrane. However the cellulose acetate based membrane has low FO flux because of severe internal concentration polarization (ICP) problem resulted from inefficient solute diffusion in the dense and tortuous membrane structure, additionally, the NaCl rejection is also not sufficient [1]. Later on, polyamide composite membranes have been developed by directly fabricating ultra-thin polyamide selective layers on porous polysulfone (PSU) or polyethersulfone (PES) support membranes. This design allows the individual optimization of active layer and support membrane to achieve high NaCl rejection while keeping resistance reasonably low [5]. These membranes can be fabricated either in hollow-fiber forms [6], flat-sheet form [5,7-9], or resulted from modification of commercial RO membranes using polydopamine [10]. However, the above membranes rely on conventional phase inversion technology to form the support membranes, which results in structure parameter (S value) between 300-3000 μm [11]. Alternatively, thin-film nanofiber composite (TNC) FO membranes with S value as low as 100-150 μm are the ideal candidate membranes to minimize ICP and harvest highest J_w so far [12,13]. While methods to optimize the support membrane have been investigated by the above studies, fewer studies have touched on the optimization of separation abilities of polyamide selective layers [5,11,14]. The fabrication and post treatment methods of polyamide selective layers greatly influence water/solute permeability coefficients and determine the performance of polyamide TNC-FO membranes.

It has generally been understood that the membrane permeability has a trade-off relationship with membrane selectivity [11,15], indicating worse solute rejection at higher membrane permeability. Thus, a pressing need is to determine the relationship between water/solute permeability coefficients, which is a critical constraint for performance of TNC-FO membranes.

The main transportation phenomena, which govern the performance of FO membrane, are external concentration polarization (ECP), ICP and solute leakage [1]. The severity of the transportation phenomena was dependent on membrane intrinsic separation parameters (i.e., A and B value) and structure parameters (S value). Previous study stated that ICP associated with thick and dense support membrane was the main reason to cause flux decline in the HTI membranes [16]. Following that, numerous studies investigated the thin-film polyamide composite membranes in FO orientation (i.e. support membrane facing draw solution) [17]. They also attributed the substantially lower-than-ideal flux, J_W^{FO} (i.e., FO water flux as membrane active layer facing feed solution), to ICP limitation. The severe ICP happens because the inefficient solute diffusion within the conventional phase-inversion support membrane cannot restore the dilution of draw solution by convective J_W^{FO}. However, the flux in PRO orientation, J_W^{FO} (i.e., FO water flux as membrane selective layer facing draw solution), was found to be higher than the J_W^{FO}. This increment

*Corresponding author: Song Xiaoxiao, Environmental & Water Technology, Centre of Innovation, Ngee Ann Polytechnic, Singapore
E-mail: Song_xiaoxiao@np.edu.sg

was attributed to less severe ECP when dilution happens in bulk solution (PRO orientation). Because the cross flow facilitates solute diffusion and reliefs the convective dilution. In addition, the leaked solutes from draw solution into support membrane were believed to accumulate along the direction of convective J_W^{PRO}, resulting in accumulative ICP [16]. However, the analysis results about the phenomena have not been quantified. Neither have the transportation phenomena been related to membrane intrinsic separation parameters (i.e. A and B values). Thus, the elucidation about comparative contribution of above transportation phenomena to limitation of FO performance, especially in the TNC membranes with low S value, is lacking.

The objective of this study is to relate the membrane intrinsic parameters to TNC membranes performance. More specifically, TNC membranes were fabricated with varied hydrophobicity of support membrane [18], IP additives [19], post treatment methods [11] and non-woven substrate [20]. A series of A-B pairs were resulted from above approaches. ICP index (I) and ECP index (E) are defined to quantify the contribution of ICP, ECP and solute leakage to the loss of effective osmotic pressure difference ($\Delta\pi_{eff}$) due to these phenomena. The relationship between the membrane intrinsic separation parameters and the severity of ICP/ECP are then systematically analyzed. This approach helps to guide the general design of cost-effective polyamide FO membrane to maximize the J_W^{FO} while maintaining affordable solute leakage.

Theory

According to definition, ICP either results from building up of feed solute concentration within support membrane in PRO orientation [21] or dilution of draw solute concentration within support membrane in FO orientation [7]. However, the real concentration at the interface of selective layer and support membrane cannot be directly measured [21]. Thus, indexes should be derived to indicate the severity of ICP. Water flux (JW) across the thin film composite FO membrane is given by [21].

$$J_W = A * \Delta\pi_m = A * (\pi_{D,m} - \pi_{F,m}) \tag{1}$$

wherein A is the intrinsic water permeability coefficient of the membrane, $\Delta\pi_m$ is the effective osmotic pressure difference across membrane selective layer. Salt flux (J_s) is given by [22].

$$J_S = B(C_{D,m} - C_{F,m}) \tag{2}$$

wherein B is the salt permeability coefficient, $C_{F,m}$ and $C_{F,m}$ are salt concentration at selective layer surfaces on draw solution side and feed solution side respectively. Van't hoff equation can be applied to relate the NaCl concentration to osmotic pressure [23].

$$\pi = nR_gT * C \tag{3}$$

Taking the ratio of equation (1) and (2) yields [24]:

$$J_W/J_S = A/B * nR_gT \tag{4}$$

Equation (4) is valid to thin film composite membranes, which comply with the solution-diffusion theory [14]. Or, the modeled water flux over reverse salt flux is linearly related to intrinsic water permeability coefficient over salt permeability coefficient.

Considering the mass transfer in FO membrane support membrane (Figure 1), the salt flux is the sum up of two terms: salt diffusion from local salt concentration gradient and dilution by convective J_w [21,23,24]. Thus, the mass transfer equations in FO orientation and PRO orientation can be written as equation (5) and (6), respectively.

$$J_S^{FO} = -D_{eff}\,dC(x)/dx + (-J_W^{FO}C(x)) \tag{5}$$

$$J_S^{PRO} = D_{eff}\,dC(x)/dx + (-J_W^{PRO}C(x)) \tag{6}$$

The boundary conditions for equations (5) are: at $x = 0, C(x) = C_{F,b}$; and at $x = 0, C(x) = C_{F,b}$. Boundary conditions for equation (6) are: at $x = 0, C(x) = C_{F,b}$; and at $x = t_S, C(x) = C_{F,m}$. Solving equations (5) and (6) using above boundary conditions respectively yields:

$$C_{D,m} = C_{D,b}exp(-J_{FO}S/D) - J_S/J_W(1 - exp(-J_{FO}S/D)) \tag{7}$$

$$C_{F,m} = C_{F,b}exp(J_{PRO}S/D) + J_S/J_W(exp(J_{PRO}S/D) - 1) \tag{8}$$

Above equations reveal that actual salt concentrations at the selective layer-support membrane interface at both FO configuration and PRO configuration are determined by two terms: convective flow term and solute leakage term [21].

Here, we define the $\Delta\pi_{eff}$ loss to ICP in support membrane

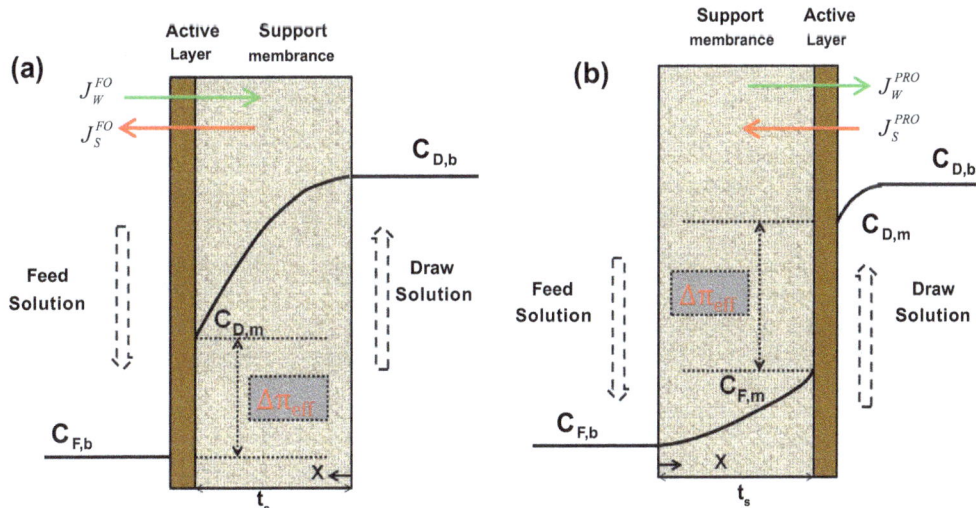

Figure 1: Schematic drawing of solute concentration profile across a TNC membrane at: a) FO orientation and b) PRO orientation.

over apparent driving force as the ICP index (I). The ICP index can be calculated based on its definition and equation (7) and (8):

$$I_{FO} = \frac{C_{D,b} - C_{D,m}}{C_{D,b} - C_{F,b}} = \frac{1}{C_{D,b} - C_{F,b}} * (C_{D,b} + \frac{B}{A * nR_gT}) * (1 - exp(-J_W^{FO}S / D)) \quad (9)$$

$$I_{PRO} = \frac{C_{F,m} - C_{F,b}}{C_{D,b} - C_{F,b}} = \frac{1}{C_{D,b} - C_{F,b}} * (C_{F,b} + \frac{B}{A * nR_gT}) * (exp(J_W^{PRO}S / D) - 1) \quad (10)$$

In the PRO orientation, the negative dilution effect of convective water flux J_W^{PRO}, which happens externally can be similarly defined as the ECP index [21]:

$$E_{PRO} = \frac{C_{D,b} - C_{D,m}}{C_{D,b} - C_{F,b}} = \frac{1}{C_{D,b} - C_{F,b}} * (C_{D,b} + \frac{B}{A * nR_gT}) * (1 - exp(-J_W^{PRO}/k)) \quad (11)$$

Equation 9 to 11 relates the CP indexes, I_{FO}, E_{PRO} and I_{PRO} to intrinsic parameters of membrane (A, B and S values). In FO orientation, the dilution of draw solution locates inside support membrane, thus the I_{FO} is a sum up of two terms: the first term, $C_{D,b} * (1 - exp(-J_W^{FO}S/D)) / (C_{D,b} - C_{F,b})$, represents the coupled effect of diluted draw solution by convective J_W^{FO} and support membrane resistance to diffusion (S/D); while the second term, $B * (1 - exp(-J_W^{FO}S / D)) / (A * nR_gT * (C_{D,b} - C_{F,b}))$, represents the coupled effect of solute leakage with the convective dilution. In PRO orientation, the dilution of draw solution locates in the bulk solution, thus the E_{PRO} is a sum up of two similar terms: the first term $C_{D,b} * (1 - exp(-J_W^{PRO}/k)) / (C_{D,b} - C_{F,b})$, represents the convective dilution in the bulk solution, while the second term, $B * (1 - exp(-J_W^{PRO} / k)) / (A * nR_gT * (C_{D,b} - C_{F,b}))$, represents the coupled effect of solute leakage with the convective dilution by J_W^{PRO}. This is a sum up of two terms: the first term $C_{F,b} * (exp(J_W^{PRO}S / D) - 1)(C_{D,b} - C_{F,b})$, represents the coupled effect of accumulated draw solution by convective J_W^{PRO} and support membrane resistance to diffusion (S/D); while the second term, $B * (exp(J_W^{PRO}S / D) - 1) / (A * nR_gT * (C_{D,b} - C_{F,b}))$, represents the coupled effect of solute leakage and the convective accumulation.

Experiment

Electrospinning of nanofiber support membrane

PAN nanofiber support membranes were electrospun-bound from 8 % w/w PAN (Mw=150,000 g mol[-1]) solution in N,N-Dimethyl Formamide (DMF), using a custom-built device [12]. The optimum concentration of 8% was chosen to avoid spraying droplets (low concentration) and thick fibers (high concentration). The solution was mixed for at least 24 hours at 80°C before use. Nanofibers were formed by drying out from continuously elongated viscous polymer solution jet in electrical field. Typically, a DC voltage of 30 kv was applied between an stainless steel (SS) spinneret orifice (D=0.7 mm) and a SS rotating drum (300 rpm) 15 cm beneath the orifice. A horizontally oscillating actuator (60 cm min[-1]) drove the rotating drum simultaneously. Prior to electrospining, the stainless steel rotating drum (Length × Diameter=20 cm × 10 cm) was wrapped with a Poly (ethylene terephthalate) (PET) non-woven substrate (Ahlstrom Grade 3249) pre-wetted with NMP (Acros organics). The PAN solution was fed to the spinneret with a constant speed of 40 μl min[-1]. Dispersed nanofibers were spun-bounded and assembled into a nascent nanofiber support membrane on the rotating drum. The nascent nanofiber support membrane was sandwiched between two pieces of clear glass (10 mm thick) and heat treated at 120°C, resulting in final nanofiber

support membrane. Chemicals (if not specified) were purchased from Sigma Aldrich.

Modification of nanofiber support membrane

A part of PAN nanofiber support membranes were treated with 1.5 M NaOH (Merck) at 60°C for 1 h. The nitrile group on the side chains of PAN readily hydrolyses to form –COOH [25].

Interfacial polymerization

Interfacial polymerization (IP) was performed on surfaces of the electrospun support membranes. The nanofiber support membranes were immersed into isopropanol for better wetting ability prior to IP. The IP recipe was adpoted from literature with some modifications [19]: Support was taped to a piece of glass with four sides sealed, and then soaked in 3% m-Phenylenediamine (MPD) aqueous solution (with or without triethylamine (TEA, 1.1% v/w) and camphor sulfonic acid (CSA, 2.3% w/w)) for 2 minutes. Air knife (operated at 0.2 Bar, 10 mm away from support membrane surface) was used to drain excess MPD solution from saturated nanofiber support membranes until surfaces became dull. Then, 0.15% w/w trimesoyl chloride (TMC) in hexane was poured onto the MPD saturated support membranes. After 1 minute reaction, the nascent composite membrane was cured at 105°C in oven for 2 minutes. Then the stabilized TNC membranes were substantially washed with DI water and kept at 5°C in DI water until further tests were conducted.

Chlorine modification of polyamide active layer

The NaClO modification is related to NaClO concentration, exposure time, pH value and post treatment. In this work, the modification conditions are chosen based on previous publications [26]. Briefly, sodium hypochlorite (4%-4.99%) aqueous solution was diluted by 40 times (~1000 ppm) and 20 times (~2000 ppm). The pH of the diluted NaClO solutions were adjusted to 7.0 ± 0.2 with 6M HCl. Membranes treated with 1000 ppm NaClO were immersed in 0.1 M NaOH solution for 14 h subsequently. Membranes treated with 2000 ppm NaClO were immersed in 0.1 M NaOH solution for 48 h subsequently. Next the treated membranes were washed with DI water extensively, cured again at 95°C for 2 minutes and kept in 5°C DI water before further experiment.

Designation of membranes

The membranes subject to one or two of the above modification methods are summarized and designated in following table. The "MPD+TMC" refers to IP recipe without additives during fabrication. The "substrate removal" refers to removal of PET non-woven substrate before any test of performance. The procedures of other three categories can be found in above experimental sub-sections (Table 1).

Membrane Designation	MPD +TMC	NaOH Treatment	Amine Additive	NaClO Wash (1000ppm)	NaClO Wash (2000ppm)	Substrate Removal
TNC-1	√					
TNC-2		√				
TNC-3			√			
TNC-4			√	√		
TNC-5				√		
TNC-6					√	
TNC-7						√
TNC-8	√					√

Table 1: Designation of membranes.

Determination of water permeability coefficient

Water permeability coefficient (A) is determined by a bench scale cross-flow RO unit [12]. Spacers (sterlitech, Sepa CF feed spacer, 17 mil) were inserted. The cross flow velocity (CFV) was 15 cm/s. The loaded membrane was subject to pure water compaction at 10 Bar for at least 3 h before conduction of experiment. Then the pressure was lowered down to a series of values (ΔP) and the corresponding pure water flux J_W^{RO} was measured. A is the slope by linear fitting ΔP as a function of J_W^{RO} (ΔP=A* J_W^{RO}) [7].

Determination of membrane performance in FO orientation

0.5-2 M NaCl and DI water (CF,b0$\approx$0) was employed as draw solution and feed solution respectively. The nanofiber membrane was loaded in a custom-built FO module (effective membrane area=23.76 cm^2), with support membrane facing draw solution (FO orientation) and active layer facing feed solution. The schematic diagram of the FO module was shown in Figure A1 (Appendix). Spacers (Sepa CF feed spacer, 17 mil thickness) were set in the 2 mm deep troughs on both of feed and draw sides. CFV on both feed and draw side was maintained at 16.7 cm s^{-1}. While equilibrium was reached, the conductivity and weight change of feed and draw solutions were continuously monitored and logged into computer every 2 minutes. The J_W^{FO} and $C_{F,b}$ can be derived from the monitored data [12].

Determination of solute permeability coefficient (B) and membrane structure parameter (S)

The salt permeability (B) and structure parameter (S) was calculated based on FO water flux, J_W^{FO}, and FO solute flux, J_S^{FO}. As such, according to mass balance equation:

$$C_{F,b}(V_0 - J_W^{FO} * A_m * t) = C_{F,b0} * V_0 + J_S^{FO} * A_m * t \quad (12)$$

Wherein $C_{F,b}$ is the bulk feed solute concentration at time t. V_0 is the original feed solution volume. A_m is the membrane area and $C_{F,b0}$ is the initial bulk feed solute concentration. J_S^{FO} is the FO solute flux, which can be rewritten as [11]:

$$J_S^{FO} = B * C_{d,b} * \exp(- J_W^{FO} S/D) \quad (13)$$

Wherein $C_{D,b}$ is the bulk draw solute (NaCl) concentration, D is the solute diffusion coefficient [27]. S is the membrane structure parameter, which can be calculated from Equation 14 [12]:

$$S = \left(D/J_W^{FO}\right) * ln\left((B + A * \pi_{D,b})/(B + J_W^{FO})\right) \quad (14)$$

Wherein $\pi_{D,b}$ is the osmotic pressure of bulk draw solute (NaCl). The above equation 1, 2 and 3 have three unknown variables, thus J_S^{FO}, B and S can be solved.

Determination of membrane performance in PRO orientation

Keeping all other parameters the same with the test in FO orientation, the PRO orientation refers to the reverse membrane orientation—polyamide active layer facing draw solution (PRO orientation) and porous nanofiber support membrane facing feed solution. Accordingly, the water flux in PRO orientation, J_S^{PRO} and solute flux in PRO orientation, J_S^{PRO} can be derived from the monitored data.

Membrane characterization

Micro-images were obtained employing a Zeiss Evo 50 Scanning Electron Microscopy (SEM). High resolution SEM images were obtained using a Joel JSM-7600F thermal field emission SEM (FESEM). Exposed cross-sections were obtained by fracturing the samples immediately after flash-freezing by liquid nitrogen. An EMITECH SC7620 sputter coater was utilized to coat all samples with gold for 45-60 seconds.

Results and Discussion

Membrane structure and morphology

Figure 2 shows the comparison between FESEM images showing top surfaces of phase-inversion support membrane (left) and nanofiber support membrane (right), respectively. Compared with nanofiber support membrane, phase inversion support membrane showed smaller and tortuous pores on the surface. The narrow and tortuous pores in the nano-range results in overall low surface porosity [28]. This pore-structure of support membrane is not favorable for: i) the diffusion of draw solutes, and ii) the contact between draw solution and selective layer, thus is not the best candidate for FO applications [12]. By contrast, the nanofiber membranes have a uniformly distributed porosity of near 80% along the entire support membrane thickness, favoring the contact between draw solution and selective layer [12,13]. Furthermore, the inter-connected and less-tortuous pore structure is favorable for convective solute diffusion and thus alleviates ICP within the support membrane.

Figure 3 shows SEM images of surfaces of TNC-1, TNC-3, TNC-5 and TNC-6 membranes (a-d) and typical cross section image of TNC membranes (e-f). After IP was performed, surfaces of the nanofiber support membrane have been bonded with a dense polyamide layer. The morphology of the polyamide layer is typically "ridge and valley" characteristics of rough IP surface (Figure 2a). By visual observation from the SEM image, numerous "leaf-like" structures appear at the surface. With the addition of amine (TEA), the leaf-like structures can hardly be seen and the surface shows smoother and "nodular-like" morphology (Figure 2b). Similar observation has been found by Ghosh et al. [19]. When the TNC membrane is treated with 1000 ppm and 2000 ppm NaClO, the surface becomes rougher as more "leaf-like" micro-structures, can be observed in the SEM image with the same magnification (Figure 2c and 2d). The change of surface morphology has been linked with possible change of membrane performance in previous publications [29]. The cross-section view of the membrane presents a thin and dense polyamide active layer been successfully synthesized and closely bonded with the nanofiber support membrane (Figure 2e and 2f).

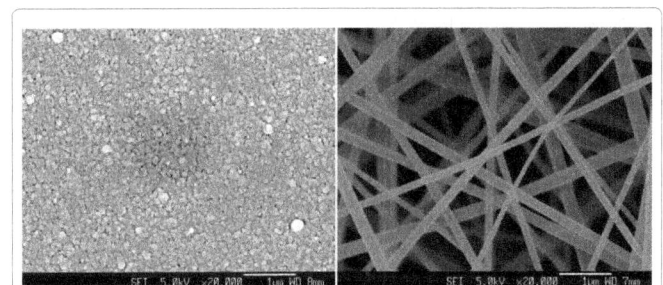

Figure 2: Top surface FESEM images of phase-inversion resulted PAN support membrane (left) and electrospun PAN nanofiber support membrane (right).

intrinsic parameters of membranes before and after hydrolysis are plotted in Figure 5. As support membrane b treated by NaOH (TNC-2), the A value of membrane increased from 1.23 ± 0.08 L m^{-2} h^{-1} bar^{-1} to 1.39 ± 0.12 L m^{-2} h^{-1} bar^{-1} while the B and S values were maintained at nearly the same level. Note that the alkali treatment does not alter the structure of the PAN nanofiber membrane, thus the change of FO performance upon alkali treatment should be attributed to the enhancement of hydrophilicity. After removal of PET non-woven substrate (TNC-8), the A value further increases to 1.43 ± 0.10 L m^{-2} h^{-1} bar^{-1}.

Role of IP additive in the selective layer: Additives into interfacial polymerization recipe substantially influence the polyamide active layer morphology and membrane intrinsic separation parameters [19,29]. The intrinsic parameters of membranes with (TNC-3) or without (TNC-1) additives are compared in Figure 6. Also, the NaClO post-treated membranes with (TNC-4) without (TNC-5) additives are

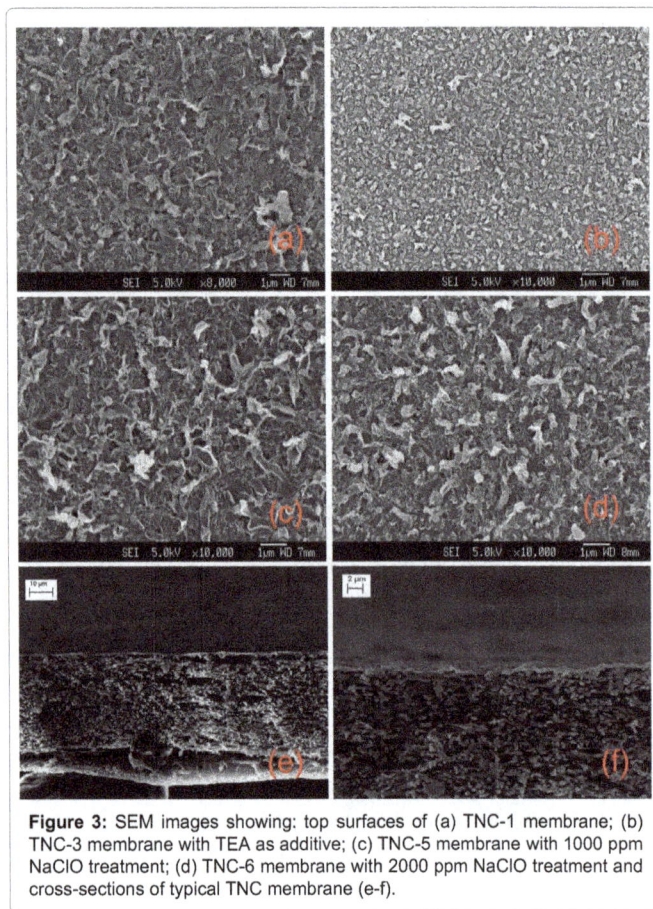

Figure 3: SEM images showing: top surfaces of (a) TNC-1 membrane; (b) TNC-3 membrane with TEA as additive; (c) TNC-5 membrane with 1000 ppm NaClO treatment; (d) TNC-6 membrane with 2000 ppm NaClO treatment and cross-sections of typical TNC membrane (e-f).

Factors that influences permeability coefficients (A and B value) in TNC-FO membranes

Role of hydrophilicity of nanofiber support membrane: Hydrophobicity of membrane support membrane inhibits the membrane wetting ability and exacerbates ICP [18]. In our previous work, the unique pore structure of nanofiber support membranes have been proven to favor the water to pass through the membrane and alleviates the ICP [12,30]. However, the impact of hydrophobicity of nanofiber support membrane on the FO performance of TNC membranes has not been studied [31-33].

In this work, PAN, a polymer with alterable hydrophobicity and widely been used to fabricate commercial ultrafiltration (UF) membranes, was selected to fabricate the nanofiber support membrane [31,34]. Figure 4 shows the comparison of FT-IR adsorption spectrums before and after alkali treatment. Peaks at 1452 cm^{-1} and 2243 cm^{-1} are the characterization absorption peaks of PAN, attributing to stretching of -C≡N and bending of -CH2 respectively. Strong new adsorption peaks at 1569 cm^{-1} and 1405 cm^{-1} (overlapped with 1452 cm^{-1}) indicates the formation of carboxylic groups during hydrolysis of PAN [34,35]. Concurrently, contact angle has been reduced from 41° to 11° upon the alkali treatment (Figure 4, inserted pictures). The colour of the support membrane turned from white to golden yellow through the treatment. Similar observations of hydrophilicity increase upon alkali modification have been found in PAN phase-inversion membranes [34,35]. It has been revealed that the alkali treatment is an efficient solution to enhance the hydrophilicity of PAN because of efficient conversion of -C≡N groups to –COOH groups [36]. The membrane

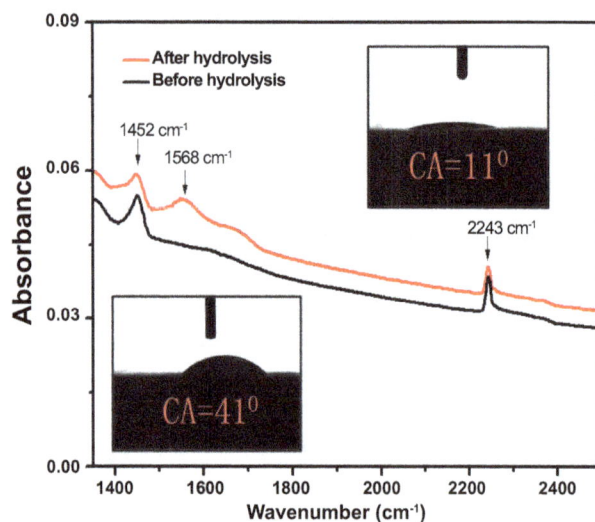

Figure 4: FT-IR adsorption spectrums before (black line) and after (red line) alkali treatment. Inserted images show the according contact angle.

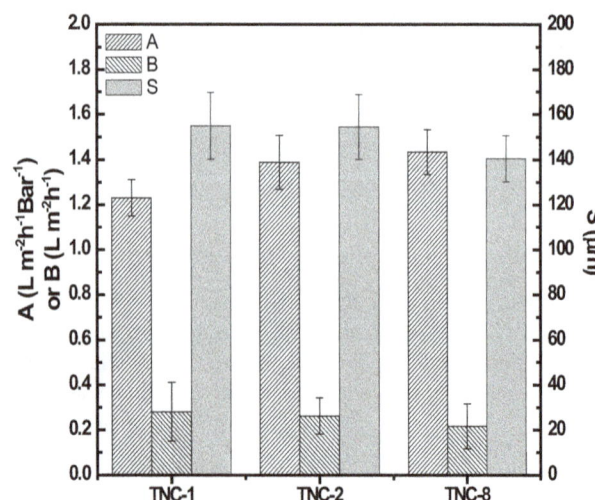

Figure 5: The membrane intrinsic parameters of TNC membranes before (TNC-1), after hydrolysis (TNC-2) and further removal of PET nonwoven substrate (TNC-8).

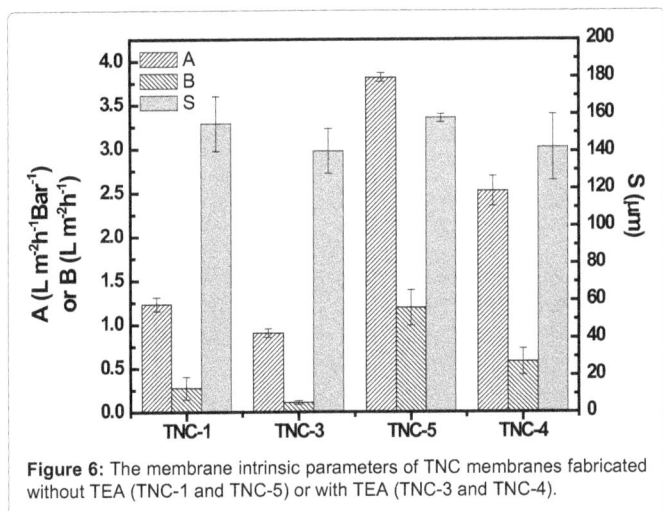

Figure 6: The membrane intrinsic parameters of TNC membranes fabricated without TEA (TNC-1 and TNC-5) or with TEA (TNC-3 and TNC-4).

compared in the same figure. In both cases, membranes fabricated with TEA-CSA as additives have lower A, B and B/A values compared with that without the additive. When membranes were not treated with NaClO, the addition of the TEA-CSA decreased B/A from the value of 0.28 to 0.11. When membranes were treated with 1000 ppm NaClO, the TEA-CSA decreased B/A from 0.31 to 0.23. Lower B/A values indicate better ability for membrane to separate water and solutes [14]. The TEA is generally believed to facilitate IP reactions by releasing free amine groups and neutralizing HCl produced [19,37]. The function of CSA is likely to prevent the microstructures in the polyamide layer from collapsing during the high temperature curing [19]. It has been shown in Figure 2b that the addition of TEA in IP recipe results in a smoother "nodular-like" morphology, this feature may relate to its lower water and solute permeability, as shown by a previous study of polyamide membranes [29].

Role of chlorine modification of membrane selective layer: A series of water permeabilities can be obtained by treatment applying different concentration of NaClO [21]. It is found recently that a combination of N-chlorination and hydrolysis processes can result in the increase of membrane permeability [38,39]. The chlorine treatment increases water permeability at the expense of some salt passage. The "trade-off" relationship between water permeability and solute permeability has been studied by several studies [11,15,40]. Study of the relationship between water/solute permeability gives insightful knowledge to optimize membrane for various FO applications. In Figure 7, the membrane treated with mild conditions of chlorine, designated as "TNC-5", has tripled A value (3.82 L m^{-2} h^{-1} bar^{-1}) compared to membranes with no NaClO treatment (TNC-1, 1.23 L m^{-2} h^{-1} bar^{-1}). The trade-off is the quadrupled B value of TNC-5 (1.19 L m^{-2} h^{-1}) compared to TNC-1 (0.28 L m^{-2} h^{-1}). The harsher chorine treatment yields even higher A and B value of 5.31 L m^{-2} h^{-1} bar^{-1} and 3.86 L m^{-2} h^{-1} respectively. This result is consistent with previous publications that NaClO is effective to enhance water permeability of polyamide layer [21,38]. Note that as A value increases, B value increases much faster. The trade-off relationship is discussed in the following section.

Role of PET non-woven substrate: The thick and dense non-woven PET substrates for pressure-driven membranes, such as reverse osmosis membranes, were considered to cause substantial water flux loss in osmotically-driven processes [18,20]. Alternatively, thinner and more porous non-woven PET substrate has been used for higher performance [7]. In this study, the role of PET substrate on FO

membrane is also investigated (Figure 8). After carefully removal of non-woven substrate, the A value was increased from 1.23 L m^{-2} h^{-1} bar^{-1} (TNC-1) to 1.41 L m^{-2} h^{-1} bar^{-1} (TNC-7).

Trade-off relationship of membrane intrinsic water permeability and solute permeability

In previous studies, it has been proposed that B is linearly related with A^3 for polyamide membranes [11,15]. Membranes with higher A value facilitates the clean water production at a higher rate and thus is land-saving; however, subject to the trade-off relationship, the concurrent increases of B value result in more severe leakage of draw solutes. Thus, the efficiency of the FO system is determined by the design of FO membrane to achieve high A value while maintaining affordable B value. A and B values of all TNC membranes are plotted in Figure 9. It was found that all the A-B pairs can be fitted with the function of $B=KA^3$ ($k=0.025$ bar^{3}m^{4}h^{2}L^{-2}). Note that the above "trade-off" relationship can be re-arranged to be:$B/A=KA^2$. This implies that any attempt to increase the water permeability will suffer from concurrently increase B/A value and sacrifice the selectivity. The B/A value have been employed to indicate the selectivity of membrane [14]. Thus in this paper, B/A is employed as the main variable to reflect the

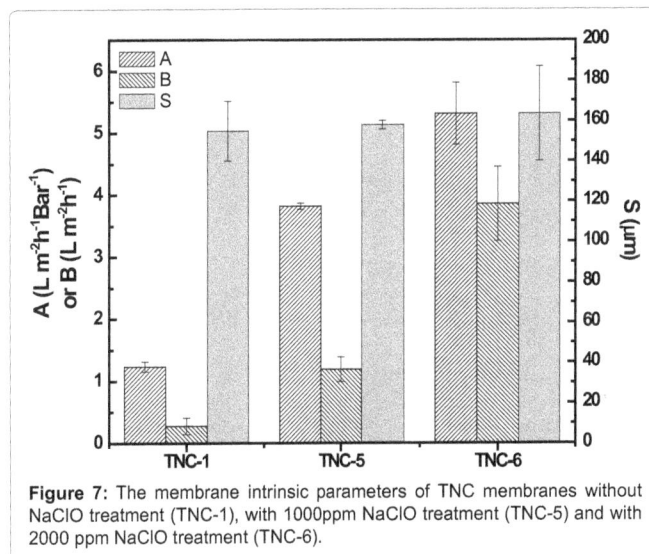

Figure 7: The membrane intrinsic parameters of TNC membranes without NaClO treatment (TNC-1), with 1000ppm NaClO treatment (TNC-5) and with 2000 ppm NaClO treatment (TNC-6).

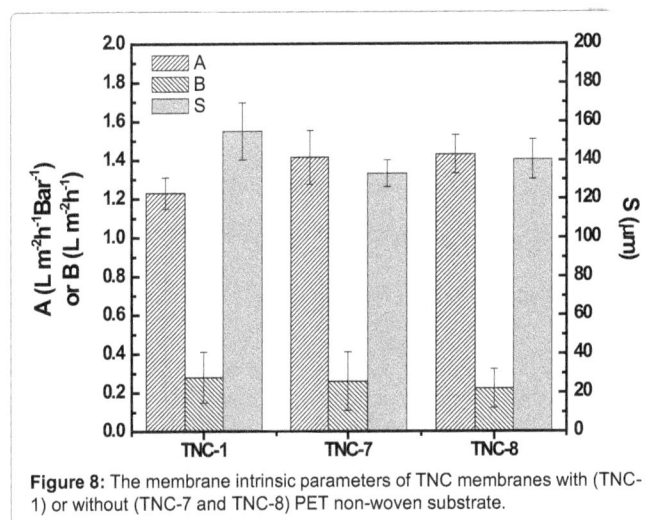

Figure 8: The membrane intrinsic parameters of TNC membranes with (TNC-1) or without (TNC-7 and TNC-8) PET non-woven substrate.

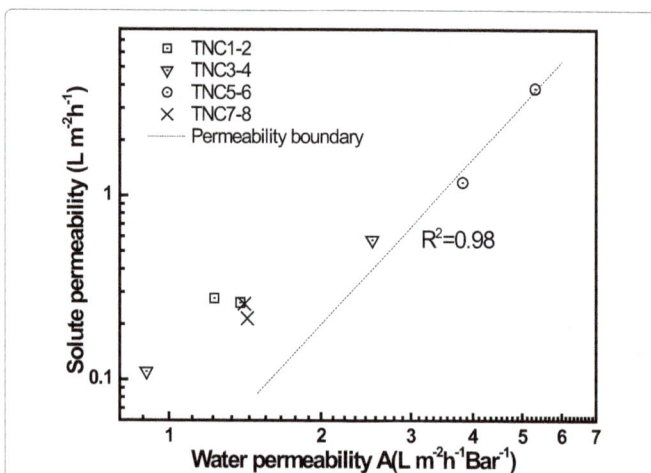

Figure 9: Trade-off relationship revealed by all TNC membranes fabricated in this study. All the A-B pairs can be fitted in the empirical relationship of $B = k * A^3$ ($k = 0.025\ Bar^3 m^4 h^2 L^{-2}$).

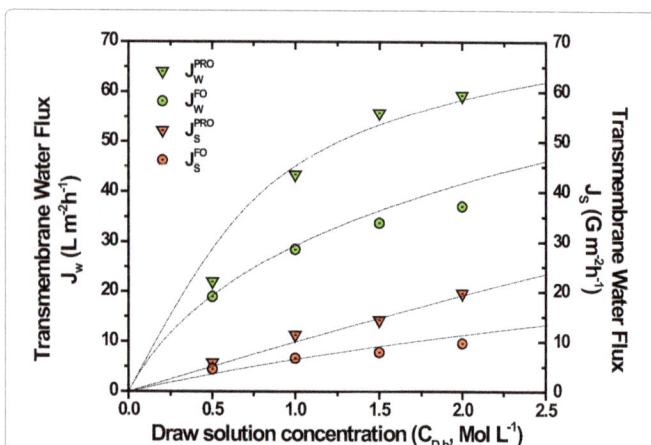

Figure 10: Performance of TNC-2 membrane in FO and PRO orientation (vertical axis) as a function of $C_{D,b}$. $C_{D,b}$ (NaCl) was varied from 0.5 M to 2 M while the feed solution concentration was maintained at 1 mM.

membrane selectivity and is related to the FO performance of TNC membranes.

Relating TNC membrane intrinsic parameters with FO performance

Typical behavior of water/solute flux as CD, b increases: The typical FO water flux of the TNC-FO membrane (TNC-2) is plotted in Figure 10. The green triangles represent the experimental water flux (J_W^{PRO}) in the PRO orientation. The concentrations of NaCl draw solution range from 0.5 M NaCl to 2 M NaCl. The resulted J_W^{PRO} ranged from 22.0 L m^{-2}h^{-1} to as high as 59.3 L m^{-2}h^{-1}. While switched to FO orientation (green circles, J_W^{FO}), the resulted J_W^{FO} was from 19.0 L m^{-2}h^{-1} to 37.1 L m^{-2}h^{-1} over the same range of concentration. The lower J_W^{FO} compared with J_W^{PRO} was believed to be attributed by more severe ICP at FO orientation [1]. It was also believed that the lower-than-expect water flux of FO membrane was mainly contributed by ICP rather than by ECP [41]. With the increase of $C_{D,b}$, both of J_W^{PRO} and J_W^{FO} increases initially at low $C_{D,b}$, but then leveled off

at high $C_{D,b}$. This phenomenon has been described as "self-limiting" water flux behavior, and was believed to be caused by ICP induced by inefficiency of solute diffusion in support membrane [16]. The solute fluxes increase with similar trends of water fluxes: at PRO orientation, J_S^{PRO} increased from 5.79 g m^{-2}h^{-1} to 19.69 g m^{-2}h^{-1} over the solute range tested; while at FO orientation, J_S^{FO} increased from 4.49 g m^{-2} h^{-1} to 9.71 g m^{-2}h^{-1}.

Relating water/solute flux to water/solute permeability coefficients: In Figure 11, the value $nR_gT^*J_S/J_W$ (vertical axis) is plotted against the B/A value (horizontal axis). At low B/A values, all the data points lies near the line where $nR_gT^*J_S/J_W=B/A$. This suggests that flux behavior fulfills the S-D model suggested by Equation 4 very well [14,34]. This behavior is consistent with the validation of using the B/A as the indicator of the separation ability of the membrane, as the J_S/J_W stands for the unit of solute sacrificed to produce per unit of clean water. However at high B/A values, the values of $nR_gT^*J_S/J_W$ tends to deviate from fitted line. The bigger value of $nR_gT^*J_S/J_W$ than B/A, especially at high membrane permeabilities, may reflect that J_S/J_W is no longer linearly related with B/A when solute rejection is low. The non-linear increase of $nR_gT^*J_S/J_W$ is likely to be cause by 1) the non-ideal solution-diffusion transportation of the solutes and 2) the coupled effects of the reverse solute permeation and the internal concentration polarization.

Relating membrane performance to water/solute permeability coefficients: For all the membranes, the summarized membrane performance (J_W^{PRO} , J_S^{PRO} , J_W^{FO} and J_S^{FO} on vertical axis) is plotted as a function of B/A (horizontal axis) in Figure 12. When $C_{D,b}$=0.5 M NaCl (upper left), the J_W^{PRO} and J_W^{FO} generally increase as B/A increases. This can be interpreted as higher A value of membrane allows more water to pass through the membrane. When $C_{D,b}$=2 M NaCl (upper right), the J_W^{PRO} and J_W^{FO} increases first and then decreases while B/A value is excessively high. For example, J_W^{PRO} and J_W^{FO} for TNC-5 (B/A=0.36) is 91.2 L m^{-2}h^{-1} and 60.9 L m^{-2}h^{-1} respectively. However when B/A was further increased to 0.73 (i.e. TNC-6), the value of J_W^{PRO} and J_W^{FO} decreased to 86.9 L m^{-2} h^{-1} and 57.9 L m^{-2}h^{-1}, respectively. This phenomenon of J_W, due to a synergistic effect of solute leakage and convective J_W induced concentration polarization, will be analyzed and explained later. The

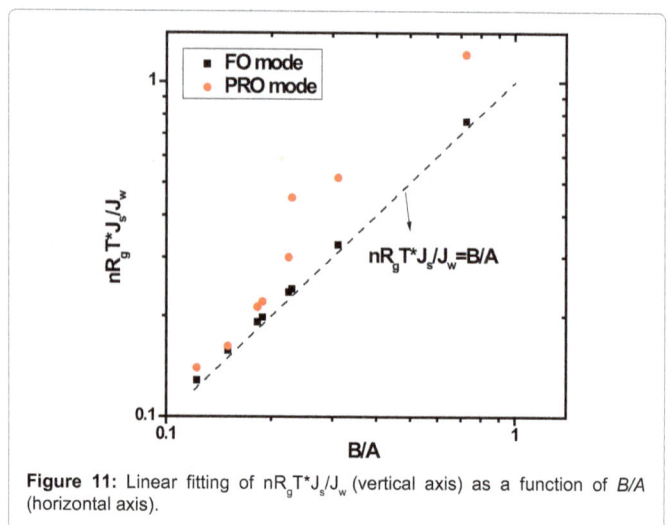

Figure 11: Linear fitting of $nR_gT^*J_S/J_W$ (vertical axis) as a function of B/A (horizontal axis).

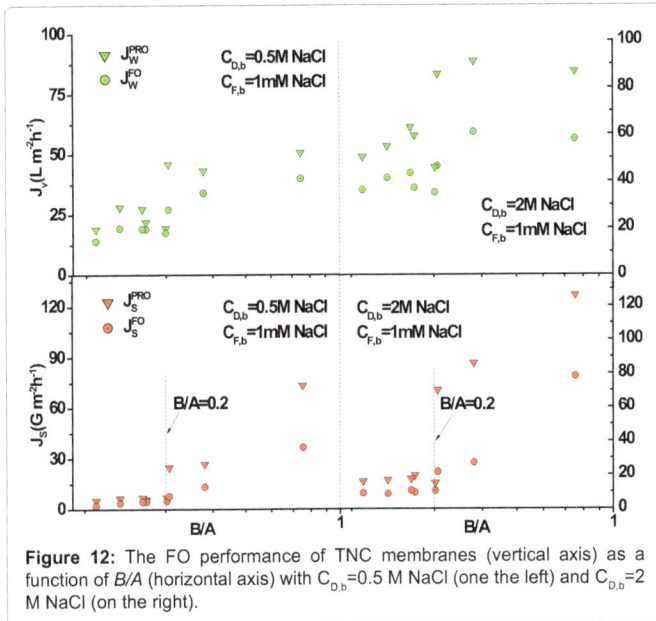

Figure 12: The FO performance of TNC membranes (vertical axis) as a function of B/A (horizontal axis) with $C_{D,b}$=0.5 M NaCl (one the left) and $C_{D,b}$=2 M NaCl (on the right).

solute fluxes, J_S^{PRO} and J_S^{FO}, constantly increased with bigger B/A value. It is worth to note that beyond B/A=0.2, only a little increase of of B/A results in substantial increase of both J_S^{PRO} and J_S^{FO}. To conclude, although the more permeable (higher B/A value) enjoys higher J_W, this enhancement generally reaches a plateau when B/A is larger than 0.2. Concurrently, J_S was observed to increased substantially beyond B/A=0.2 and resulted in limiting, or even deterioration effect for J_W. Thus from economical angle, FO membranes should generally have B/A<0.2 to minimize the uneconomical loss of draw solutes due to reverse diffusion (Supplementary Figure 1).

Relating membrane intrinsic properties with ICP and ECP indexes

Using experimentally determined J_S, J_W, A, B, S values, the ICP and ECP indexes of FO orientation and PRO orientation can be derived by Equation 9, 10 and 11. Study of the CP indexes reveals the severity of undermined solute concentration by the according ICP and ECP phenomena. In the PRO orientation, both of ICP and ECP indexes are modeled [21,42]. However, in FO orientation, only the ICP is modeled while the ECP is considered paltry [16]. The value of ICP index is defined as the portion of $\Delta\pi_{eff}$ loss, which is either due to the dilutive ICP (I_{FO}, FO orientation) or concentrative ICP (I_{PRO}, PRO orientation), divided by the osmotic pressure difference between the bulk draw and feed solutions. The value of ECP index is defined as the portion of $\Delta\pi_{eff}$ loss due to external dilution, divided by the bulk osmotic pressure difference. The I_{FO}, I_{PRO} or E_{PRO} is plotted as a function of B/A in Figure 13. At both draw solution concentrations, I_{FO} generally increased with bigger B/A. This is because: a) bigger B/A values indicate more solute leakage, leading to exacerbated dilutive ICP; b) according to the "trade-off" relationship between A and B values, higher water permeability always accompanies enhanced B/A value. As a result, higher convective water flux J_W exacerbated the dilutive ICP, as can be interpreted in Equation 11 that the ICP factor (i.e. $1 - exp(-J_W^{FO} S / D)$) is increased [16,43]. Similar explanations can be made to the trends of E_{PRO} and I_{PRO}.

The E_{PRO} (green solid circles) and I_{FO} (red solid circles) exhibited

similar trends and magnitude. The formulas in Equation 11 and Equation 13 to calculate the two indexes are identical, as the two phenomena both result from dilution by the convective J_W. However, the physical meaning of the ECP and ICP factor is different: the k in the ECP factor, $(1 - exp(-J_W^{PRO}/k))$, stands for mass transfer coefficient in bulk solution; while the D/S in the ICP factor, $(1 - exp(-J_W^{FO}/(D/S)))$, stands for mass transfer coefficient in support membrane. The resistance of the support membrane always results in lower mass transfer coefficient than in the bulk solution, thus the value of J_W^{FO} is constantly lower than the value of J_W^{PRO} [16]. Since the D value was kept as constant (i.e. solute type and temperature kept the same) [27], the deviation between J_W^{FO} and J_W^{PRO} is determined by the S value of the FO membrane. When support membrane is thick and dense (i.e. big S value), the ICP factor is much bigger than ECP factor, resulting in more severe dilution in the support membrane. The mass transfer coefficient in support membrane, D/S, is inefficient to restore the solute concentration in the support membrane, resulting in much lower J_W^{FO} compared to J_W^{PRO} [16,18]. This scenario happens in conventional composite polyamide membranes fabricated by phase-inversion methods with S value ranging from 300-3000 μm [11]. However, when support membrane is porous and thin (i.e. small S value), the ICP factor approaches ECP factor, resulting in less discrepancy between J_W^{FO} and J_W^{PRO}. This explains why J_W^{FO} and J_W^{PRO} for TNC membranes ($S\sim150$μm) shows less disparity than the equivalent for conventional polyamide composite membranes [5-7,44]. As a result, the I_{FO} and E_{PRO} have similar magnitude, as they are reflections of J_W^{FO} and J_W^{PRO} respectively.

Compared with I_{FO} and E_{PRO}, the value of I_{PRO} is substantially smaller, especially at low B/A value (Figure 13). For example, when B/A=0.19 (TNC-2 membrane), the I_{PRO} was merely 0.007, while I_{FO} and E_{PRO} was 0.42 and 0.35, respectively. Even at the highest B/A value investigated in this paper (0.73, TNC-6 membrane), the I_{PRO} was merely 0.16, while I_{FO} and E_{PRO} was 0.70 and 0.52, respectively. It was observed that the significance of I_{PRO} literally increases with higher B/A value. Because $C_{F,b}$ was kept constantly at 1 mM, the I_{PRO} increase should be contributed by more severe leakage of solutes into support membrane according to Equation 12. However, the substantially smaller value of I_{PRO} compared with E_{PRO} reveals that the external convective dilution causes more severe $\Delta\pi_{eff}$ loss compared with internal concentration

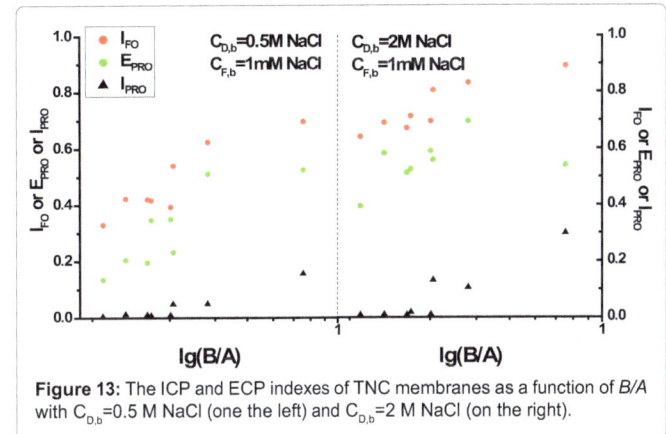

Figure 13: The ICP and ECP indexes of TNC membranes as a function of B/A with $C_{D,b}$=0.5 M NaCl (one the left) and $C_{D,b}$=2 M NaCl (on the right).

polarization. This implies that the high J_W has become the main restriction of membrane performance.

To conclude this section, when dilute feed solution was used, $\Delta\pi_{eff}$ loss to coupling of convective dilution with mass transfer coefficient in support membrane (i.e., D/S value) is the most severe, resulting in lower J_W^{FO}. $\Delta\pi_{eff}$ loss to coupling of convective dilution with mass transfer coefficient in bulk solution (i.e., k value) is less severe, resulting in higher J_{PRO}. $\Delta\pi_{eff}$ loss to coupling of leakage of solutes with mass transfer coefficient in support membrane (i.e. D/S value) is minor.

Conclusion

A series of TNC membranes have been fabricated/modified by means of NaOH treatment, IP additives, chlorine post treatment, or removal of non-woven PET substrates. These membranes have varied water/solute permeability. It was found that A-B pairs of all the membranes can be fitted into the empirical relationship of $B = k * A^3$ ($k = 0.025$ $Bar^3 m^4 h^2 L^{-2}$). That means any increase of water permeability of polyamide composite membrane accompanies concurrent increase of solute permeability. This "trade-off" relationship between water/solute permeability is closely related to the FO membrane performance. When B/A is low, both of J_W and J_S is low; when B/A is enhanced, J_W will be increased substantially at the beginning but eventually level off and even be decreased. This results from a) the increase of J_W also exacerbate the convective dilution ("self-limiting" phenomenon); and b) bigger B/A value deteriorates the membrane selectivity, thus more severe leakage of solutes exacerbate the ICP within support membrane. The comparison between ICP index in FO orientation, I_{FO}, ECP index in PRO orientation, E_{PRO}, and ECP index in FO orientation, I_{PRO}, reveals that convective dilution by J_W itself is the main restriction of further J_W increase. The coupling of convective dilution with diffusion resistance in support membrane (S value) resulted in further decreased J_W^{FO} compared to J_W^{PRO}. However, the decrease has been minimized because of low S value of TNC membranes (~150 μm). The leakage of solutes into support membrane causes accumulative ICP, however its value is minor compared with convective dilution.

Acknowledgements

This work was supported by the Prime Minister's Office of Singapore via an initiative called The Enterprise Challenge under award number P00579/1273, Singapore Environment & Water Industry (EWI) Development Council under award number MEWR 621/06/166. The authors also acknowledge the financial support from Public Utilities Board of Singapore under the R&D grant (IDD90301/001/75).

References

1. Cath TY, Childress AE, Elimelech M (2006) Forward osmosis: Principles, applications, and recent developments. J Mem Sci 281: 70-87.

2. Pan SF, Dong Y, Zheng YM, Zhong LB, Yuan ZH (2017) Self-sustained hydrophilic nanofiber thin film composite forward osmosis membranes: Preparation, characterization and application for simulated antibiotic wastewater treatment. J Mem Sci 523: 205-215.

3. Yang Q, Lei J, Sun DD, Chen D (2016) Forward Osmosis Membranes for Water Reclamation. Sep Purifi Reviews 45: 93-107.

4. Huang L, Arena JT, McCutcheon JR (2016) Surface modified PVDF nanofiber supported thin film composite membranes for forward osmosis. J Mem Sci 499: 352-360.

5. Wei J, Liu X, Qiu C, Wang R, Tang CY (2011) Influence of monomer concentrations on the performance of polyamide-based thin film composite forward osmosis membranes. J Mem Sci 381: 110-117.

6. Wang R, Shi L, Tang CY, Chou S, Qiu C, et al. (2010) Characterization of novel forward osmosis hollow fiber membranes. J Mem Sci 355: 158-167.

7. Yip NY, Tiraferri A, Phillip WA, Schiffman JD, Elimelech M (2010) High Performance Thin-Film Composite Forward Osmosis Membrane. Environ Sci Technol 44: 3812-3818.

8. Widjojo N, Chung TS, Weber M, Maletzko C, Warzelhan V (2011) The role of sulphonated polymer and macrovoid-free structure in the support layer for thin-film composite (TFC) forward osmosis (FO) membranes. J Mem Sci 383: 214-223.

9. Wang KY, Chung TS, Amy G (2012) Developing thin-film-composite forward osmosis membranes on the PES/SPSf substrate through interfacial polymerization. AIChE J 58: 770-781.

10. Arena JT, McCloskey B, Freeman BD, McCutcheon JR (2011) Surface modification of thin film composite membrane support layers with polydopamine: Enabling use of reverse osmosis membranes in pressure retarded osmosis. J Mem Sci 375: 55-62.

11. Yip NY, Elimelech M (2011) Performance Limiting Effects in Power Generation from Salinity Gradients by Pressure Retarded Osmosis. Environ Sci Technol 45: 10273-10282.

12. Song X, Liu Z, Sun DD (2011) Nano Gives the Answer: Breaking the Bottleneck of Internal Concentration Polarization with a Nanofiber Composite Forward Osmosis Membrane for a High Water Production Rate. Adv Mater 23: 3256-3260.

13. Bui NN, Lind ML, Hoek EMV, McCutcheon JR (2011) Electrospun nanofiber supported thin film composite membranes for engineered osmosis. J Mem Sci 386: 10-19.

14. Tang CY, She Q, Lay WCL, Wang R, Field R, et al. (2011) Modeling double-skinned FO membranes. Desalination 283: 178-186.

15. Geise GM, Park HB, Sagle AC, Freeman BD, McGrath JE (2011) Water permeability and water/salt selectivity tradeoff in polymers for desalination. J Mem Sci 369: 130-138.

16. McCutcheon JR, Elimelech M (2006) Influence of concentrative and dilutive internal concentration polarization on flux behavior in forward osmosis. J Mem Sci 284: 237-247.

17. Tiraferri A, Yip NY, Phillip WA, Schiffman JD, Elimelech M (2011) Relating performance of thin-film composite forward osmosis membranes to support layer formation and structure. J Mem Sci 367: 340-352.

18. McCutcheon JR, Elimelech M (2008) Influence of membrane support layer hydrophobicity on water flux in osmotically driven membrane processes. J Mem Sci 318: 458-466.

19. Ghosh AK, Jeong BH, Huang X, Hoek EMV (2008) Impacts of reaction and curing conditions on polyamide composite reverse osmosis membrane properties. J Mem Sci 311: 34-45.

20. Loeb S, Titelman L, Korngold E, Freiman J (1997) Effect of porous support fabric on osmosis through a Loeb-Sourirajan type asymmetric membrane. J Mem Sci 129: 243-249.

21. Yip NY, Tiraferri A, Phillip WA, Schiffman JD, Hoover LA, et al. (2011) Thin-Film Composite Pressure Retarded Osmosis Membranes for Sustainable Power Generation from Salinity Gradients. Environ Sci Technol 45: 4360-4369.

22. Lee KL, Baker RW, Lonsdale HK (1981) Membranes for power generation by pressure-retarded osmosis. J Mem Sci 8: 141-171.

23. Phillip WA, Yong JS, Elimelech M (2010) Reverse Draw Solute Permeation in Forward Osmosis: Modeling and Experiments. Environ Sci Technol 44: 5170-5176.

24. Tang CY, She Q, Lay WCL, Wang R, Fane AG (2010) Coupled effects of internal concentration polarization and fouling on flux behavior of forward osmosis membranes during humic acid filtration. J Mem Sci 354: 123-133.

25. Muthumareeswaran MR, Agarwal GP (2014) Feed concentration and pH effect on arsenate and phosphate rejection via polyacrylonitrile ultrafiltration membrane. J Mem Sci 468: 11-19.

26. Song X, Liu Z, Sun DD (2013) Energy recovery from concentrated seawater brine by thin-film nanofiber composite pressure retarded osmosis membranes with high power density. Energy Environ Sci 6: 1199-1210.

27. Achilli A, Cath TY, Childress AE (2010) Selection of inorganic-based draw solutions for forward osmosis applications. J Mem Sci 364: 233-241.

28. Strathmann H, Kock K (1977) The formation mechanism of phase inversion membranes. Desalination 21: 241-255.

29. Kim SH, Kwak SY, Suzuki T (2005) Positron Annihilation Spectroscopic Evidence to Demonstrate the Flux-Enhancement Mechanism in Morphology-Controlled Thin-Film-Composite (TFC) Membrane. Environ Sci Technol 39: 1764-1770.

30. Gibson PW, Schreuder-Gibson HL, Rivin D (1999) Electrospun fiber mats: Transport properties. AIChE J 45: 190-195.

31. Feng C, Khulbe KC, Matsuura T, Gopal R, Kaur S, et al. (2008) Production of drinking water from saline water by air-gap membrane distillation using polyvinylidene fluoride nanofiber membrane. J Mem Sci 311: 1-6.

32. Tasuku O, Ding B, Sone Y, Shiratori S (2007) Super-hydrophobic surfaces of layer-by-layer structured film-coated electrospun nanofibrous membranes. Nanotechnology 18: 165607.

33. Thavasi V, Singh G, Ramakrishna S (2008) Electrospun nanofibers in energy and environmental applications. Energy Environ Sci 1: 205-221.

34. Qiu C, Qi S, Tang CY (2011) Synthesis of high flux forward osmosis membranes by chemically crosslinked layer-by-layer polyelectrolytes. J Mem Sci 381: 74-80.

35. Zhang G, Yan H, Ji S, Liu Z (2007) Self-assembly of polyelectrolyte multilayer pervaporation membranes by a dynamic layer-by-layer technique on a hydrolyzed polyacrylonitrile ultrafiltration membrane. J Mem Sci 292: 1-8.

36. Gupta ML, Gupta B, Oppermann W, Hardtmann G (2004) Surface modification of polyacrylonitrile staple fibers via alkaline hydrolysis for superabsorbent applications. J App Poly Sci 91: 3127-3133.

37. Yung L, Ma H, Wang X, Yoon K, Wang R, et al (2010) Fabrication of thin-film nanofibrous composite membranes by interfacial polymerization using ionic liquids as additives. J Mem Sci 365: 52-58.

38. Do VT, Tang CY, Reinhard M, Leckie JO (2012) Degradation of Polyamide Nanofiltration and Reverse Osmosis Membranes by Hypochlorite. Environ Sci Technol 46: 852-859.

39. Kwon YN, Leckie JO (2006) Hypochlorite degradation of crosslinked polyamide membranes: II. Changes in hydrogen bonding behavior and performance. J Mem Sci 282: 456-464.

40. Lau WJ, Ismail AF, Misdan N, Kassim MA (2012) A recent progress in thin film composite membrane: A review. Desalination 287: 190-199.

41. McCutcheon JR, McGinnis RL, Elimelech M (2006) Desalination by ammonia–carbon dioxide forward osmosis: Influence of draw and feed solution concentrations on process performance. J Mem Sci 278: 114-123.

42. Su J, Chung TS (2011) Sublayer structure and reflection coefficient and their effects on concentration polarization and membrane performance in FO processes. J Mem Sci 376: 214-224.

43. Li W, Gao Y, Tang CY (2011) Network modeling for studying the effect of support structure on internal concentration polarization during forward osmosis: Model development and theoretical analysis with FEM. J Mem Sci 379: 307-321.

44. Zhang S, Wang KY, Chung TS, Chen H, Jean YC, et al. (2010) Well-constructed cellulose acetate membranes for forward osmosis: Minimized internal concentration polarization with an ultra-thin selective layer. J Mem Sci 360: 522-535.

21

Selective Permeation of CO_2 through Amine Bearing Facilitated Transport Membranes

Panchali Bharali, Somiron Borthakur and Swapnali Hazarika*

Chemical Engineering Group, Engineering Science & Technology Division, CSIR-North East Institute of Science and Technology, Jorhat, Assam, India

Abstract

The membranes containing facilitated transport groups for Carbon dioxide had been prepared by immobilizing PAMAM (Polyamidoamine) (Generations 0,1,2,3,4) dendrimer into the polymeric membranes. The dendrimer incorporated membranes were prepared by the phase inversion method. The permeation abilities of the membranes for pure CO_2 and binary mixture of CO_2/N_2 were calculated. The effects of feed gas pressure on the permeability of the membranes were studied. The results of the permeation experiments showed that PAMAM dendrimer (Generation 4) composite membrane possessed a better CO_2 permeability and selectivity over N_2 than the any other membranes composite with other generations of dendrimer (Generations 0,1,2,3).

Keywords: Facilitated transport; Membranes; PAMAM dendrimer; Phase Inversion

Introduction

The increase in the concentration of various hazardous gases in the atmosphere results in numerous environmental problems like global warming, greenhouse effect etc. A large number of research studies have been carried out on the process of capturing and storage of CO_2 from gaseous mixture [1,2]. Membranes and membrane processes are not a recent invention. The preparation of synthetic membrane and their utilization on a large industrial scale however are a more recent development which has rapidly gained a substantial importance due to the large number of practical application. Now a days; membranes are used to produce potable water from sea to clean industrial effluent and recover valuable constituent, purify or fractionate macromolecular mixture in food and drug industries and to separate gas and vapours. They are also key component in energy conversion system and in artificial organs and drug delivery devices. Membrane technology is one of the most interesting technologies for its applications in various fields including Biological applications [3-6]. Lin et al. [5] in his research study reported the substrate selectivity of Lysophospholipid Transporter (LplT) involved in Membrane Phospholipids remodelling in *Escherichia coli*. Tong et al. [6] carried out their research study on the structural insight into substrate selection and catalysis of Lipid Phosphate Phosphates in the cell membrane. Besides it's applications in different fields membrane separation technology can be used for the gas separation purpose as it is one of the cost effective process and easy to operate. Because of high permeability and selectivity, the facilitated transport membranes are very useful for gas separation applications. There are various kinds of membranes containing facilitated transport groups, such as ion-exchange membrane, fixed carrier membrane and liquid membranes etc. [1]. Because of the higher permeability and selectivity fixed carrier membranes are more suitable for gas separation. A few researchers have been reported about the fixed carrier membranes for CO_2 separation [2,7]. Works on gas separation applications using membranes composite with PAMAM dendrimer are limited. However, PAMAM dendrimer immobilized liquid membranes are reported by distinguished researchers [8,9]. In this research paper, we have been emphasized on gas separation behaviour of PAMAM dendrimer (G-0,G-1,G-2,G-3,G-4) composite solid membranes due to its stability and regeneration.The dendrimer composite membranes selectively permeated CO_2 because of the reaction between $-NH_2$ and CO_2 producing a carbamate ion and a protonated base [8].

Thus for separation of CO_2 molecules the reactions occurred as follows:

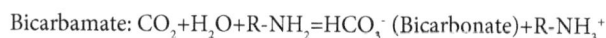

Bicarbamate: $CO_2+H_2O+R-NH_2=HCO_3^-$ (Bicarbonate)$+R-NH_3^+$

Carbamate: $CO_2+2(R-NH_2)=R-NHCOO^-$ (Carbamate)$+R-NH_3^+$

Sarma et al. [9] reported the performance of Poly (amidoamine) (PAMAM G-0) dendrimer and ionic liquid composite membranes which provide a challenging way for CO_2 separation performances. Belmabkhout et al. [10] reported that amine bearing pore expanded MCM-41 (Mobil Composition of Matter, No 41) Silica can be used for gas separation applications for CO_2/CH_4. Jin Huang et al. [11] reported the concept of facilitated transport mechanism for capture of CO_2 from a gaseous mixture CO_2/H_2. Duan et al. [12] reported the CO_2 separation performance of a poly(amidoamine) (PAMAM) dendrimer composite membrane containing hyaluronic acid (HA) in a chitosan (CTS) gutter layer. The novelty of our research work lies in the performance of membrane containing Amine compound (PAMAM dendrimer, G-0,G-1,G-2, G-3, G-4) for separation of CO_2 from CO_2/ N_2 gaseous mixture, describing facilitated transport mechanism for separation of CO_2 and a structure activity relationship study of dendrimers with gas permeability.

Materials and Methods

Materials

Polysulfone (average molecular weight 22000) was obtained from Aldrich Chemicals was used as the membrane material. Polyethylene Glycol (PEG) (average molecular weight 400) used as additive was obtained from Central Drug House (P) Ltd. Bombay. N-Methyl-

***Corresponding author:** Swapnali Hazarika, Chemical Engineering Group, Engineering Science & Technology Division, CSIR-North East Institute of Science and Technology, Jorhat, Assam, India, E-mail: shrrljt@yahoo.com

Pyrollidine (NMP) obtained from Rankem, India. Commercial Dendrimer (0,1,2,3,4 generations) were purchased from Sigma Aldrich Chemical Ltd was used as membrane material. Water was obtained from Millipore water system (Type 1).Carbon di oxide (99.99%) and Nitrogen (99.99%) was supplied by M.S Gas Centre, Jorhat, India.

Methods

Preparation and characterization of polymeric membrane: The flat sheet membranes were prepared by dissolving Polysulfone (PSF) in NMP at room temperature ($28°$-$30°C$).To this solution, a definite amount of PEG-400 was added and the solution was kept on stirring to prepare the casting solution. When the solutions become clear a definite amount of Dendrimer (0,1,2,3,4 generations) was added individually for the preparation of different membranes and kept it on stirring until it became a clear solution. Then the solutions were casted on glass films and the glass films were placed in water (coagulation) bath for about 24 hours. Then the flat sheet membranes were dried at room temperature ($28°C$). Then the finally prepared membranes were analyzed experimentally by IR analysis (PERKIN Elmer system 2000), X-ray Diffraction analysis (JDX-11P-3A, JEOL, Japan), TGA-DTA analysis (PERKIN Elmer PC series, DSC 7). The morphology of the membranes were studied by Scanning Electron Microscope (LEO 1427 VP,UK) analysis etc. For each type of membrane we tested three numbers of membrane samples to get the maximum accuracy. The schematic diagram of membrane fabrication system and gas separation experiments shown in the Figures 1 and 2.

Gas permeation experiment: The gas permeation experiments were carried out in a membrane cell. The permeation cell was consists of Cylinders, membrane cell, rotameter. The membrane was fitted into the Plate and Frame module. Area of the membrane in the cell was 30 cm². The Permeance i.e the pressure normalized flux, through the membranes can be obtained by equation (1).

$$(P/I)=Q/ (A.\Delta P) \tag{1}$$

Where Q is the measured volumetric flow rate (at standard pressure and temperature), P is permeability, l is membrane skin layer thickness, A is effective membrane area and ΔP is the Pressure difference across the membrane. The common unit of Permeance is GPU (1 GPU=10^{-6} cm³ (STP)/cm² s cm Hg). The ideal gas separation factor or Selectivity is given by equation (2)

$$Selectivity (\alpha)=P_i/P_j \tag{2}$$

Where P_i is permeation of i th component (CO_2) and P_j is permeation of j th (N_2) component.

Results and Discussions

Characterization of membrane

IR analysis: The FTIR analysis of the membranes were performed in order to confirm the immobilization of the PAMAM dendrimer (G-0-G-4) molecule into the membrane as well as the facilitated transport mechanism occurs in the membranes during the permeation of CO_2 through the membranes. The prepared membranes were analysed experimentally by IR analysis (PERKIN Elmer system 2000). In FTIR analysis; 12 numbers of scans were used obtain each spectrum. The FTIR spectrum of the Polysulfone membrane (without PAMAM dendrimer) is shown in Figure 3. The FTIR spectra obtained for polysulfone, dendrimer (Generations 0, 1, 2, 3, 4) composite membranes before and after permeation process are shown in Figure 4. For each membrane (Figures 3 and 4) Polysulfone gives two broad peaks at 3400

cm^{-1} and 2330 cm^{-1} attributed to O-H group and C-H (-CH$_3$) stretching respectively. A broad peak at 3380 was observed in case of dendrimer for N-H stretching. Other IR peaks for dendrimer were found to be 2919 cm^{-1}, 1647, 1194 for C-H stretching(-CH$_2$-), C=O and C-N respectively which were not present in the membrane prepared without adding PAMAM dendrimer (Figure 3). As there was no –NH$_2$ linkages in the membrane prepared without adding PAMAM dopant it did not facilitate the transportation of CO$_2$ through it. It was observed that for the membrane the IR peaks were obtained at 3244, 1278, 2800, 1770 cm^{-1} due to N-H, C-N, C-H, C=O str. for presence of dendrimer in the membrane. After the permeation process, all membranes showed an additional peak at 3500 cm^{-1} which was due to the formation of amide linkage of the dendrimer molecule with CO$_2$ as a result of facilitated transport mechanism of dendrimer for CO$_2$ separation.

XRD analysis: X-Ray Diffraction (XRD) study of the various membranes was done for the structural analysis of all the membranes. Figure 5 shows the XRD patterns of the membranes prepared with various generations of dendrimer molecule (Generation- 0, 1, 2, 3, 4). All the membranes showed a broad peak in the range of 1.30-3.69. From

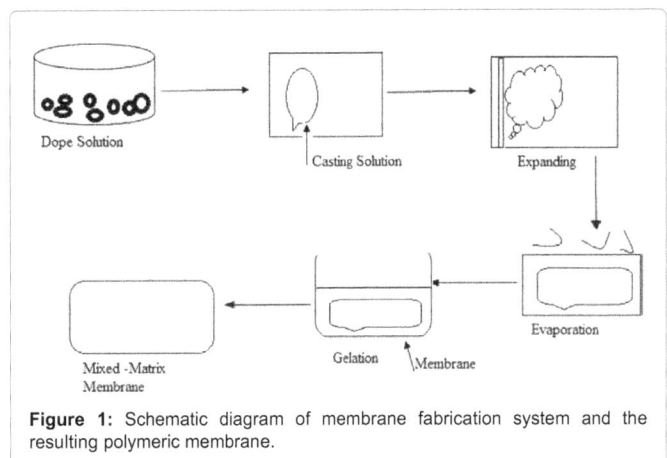

Figure 1: Schematic diagram of membrane fabrication system and the resulting polymeric membrane.

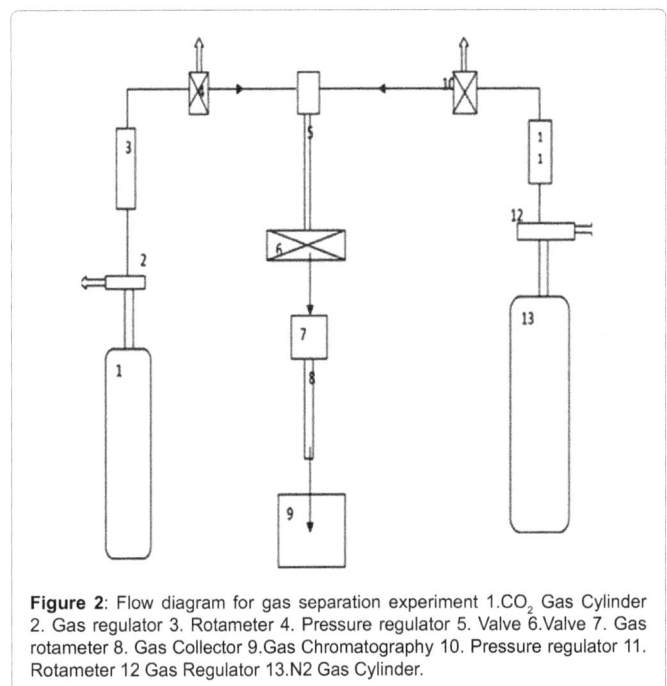

Figure 2: Flow diagram for gas separation experiment 1.CO$_2$ Gas Cylinder 2. Gas regulator 3. Rotameter 4. Pressure regulator 5. Valve 6.Valve 7. Gas rotameter 8. Gas Collector 9.Gas Chromatography 10. Pressure regulator 11. Rotameter 12 Gas Regulator 13.N2 Gas Cylinder.

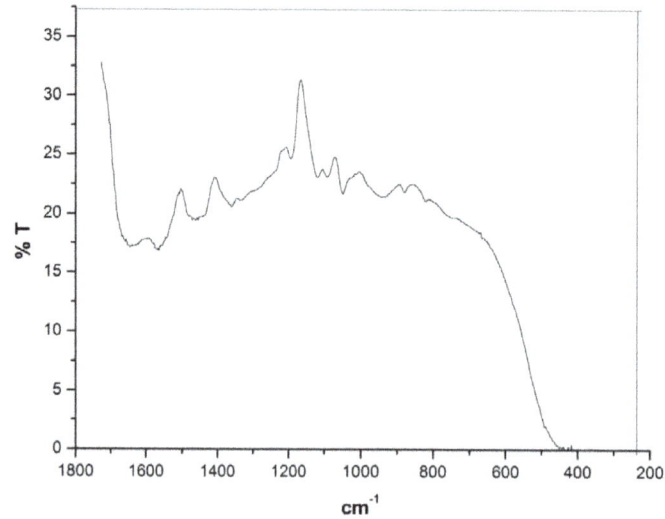

Figure 3: I.R Analysis of Polysulfone membrane (Before permeation).

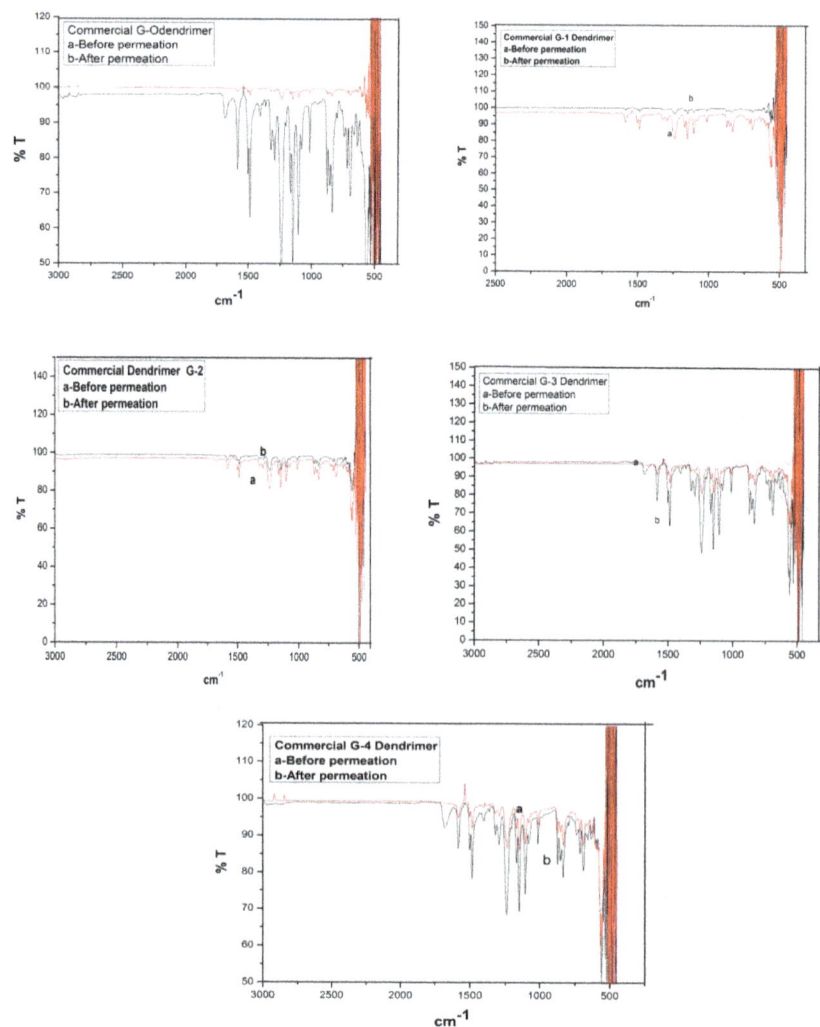

Figure 4: IR analysis of the membranes before / after permeation of CO_2 (Each spectrum was obtained after the completion of 12 numbers of scans).

the patterns of the membranes it was seen that the peak position does not depend on the casting solutions. The broad peaks of the membranes were observed due to complete homogeneity and compatibility among the components of the membrane. However, compared to other membranes, the pattern of G-4 dendrimer composite membrane showed a broadband peak with a decrease in the peak intensity. The broad peak of the composite membrane indicated the amorphous state of the membranes. The value of Full Width Half Maximum value (FWHM) given in Table 1 indicates the value of relative degree of the phase. The FWHM value was maximum in the membranes prepared by using G-4 dendrimer composite membrane and minimum for the G-0 dendrimer composite membrane.

Thermal properties: Thermogravimetric analysis (TGA) and Differential Thermal Analysis (DTA) of the membranes prepared with Dendrimer (G-0,1,2,3,4) are shown in the Figure 6a and 6b. TGA-DTA studies of the membranes were done under the atmosphere of N_2 at a heating rate of 100°C min⁻¹. The thermal decomposition curves of the membranes are shown in Figure 6a and the derivative curves of the membranes are shown in Figure 6b. From the TGA curves it was observed that all the membranes are stable up to a temperature of 480-500°C. Thus; it was observed from the TGA curves that all the membranes prepared by various generations of dendrimer possessed almost similar extent of thermal stabilities. From the DTA curves of the membranes it was observed that all the membranes possessed an exotherm which was obtained around 530-560°C.

Morphological study of membranes: The morphological studies of the membranes were carried out by Scanning Electron Microscopic (SEM) analysis. From the SEM images it was observed that the membranes prepared with various generations of Dendrimer undergo morphological changes before and after permeation process. From Figure 7, it was seen that the morphology of the membranes differ before and after permeation process. This can be attributed to the formation of some gates on the surface of the membranes due to the reaction between CO_2 and dendrimer molecule during the permeation process. The membranes prepared before the permeation process had the dense uniform morphological structure, whereas the membranes after the permeation process possess some pinholes which were observed as a consequence of the facilitated transport mechanism of the membranes.

Effect of feed gas pressure on permeability: For the transportation of gases through the fixed- carrier solid membranes, the flux can be expressed in terms of Fick's law with an effective diffusion coefficient $D_{eff}(p)$ that is dependent on pressure:

$$J=-D_{eff}(p)dp/dx \qquad (3)$$

The pressure dependence of the diffusion coefficient for gas transport through the membranes can be evaluated from the steady – state permeability. Koros, Chan and Paul in 1977, expressed the equation as

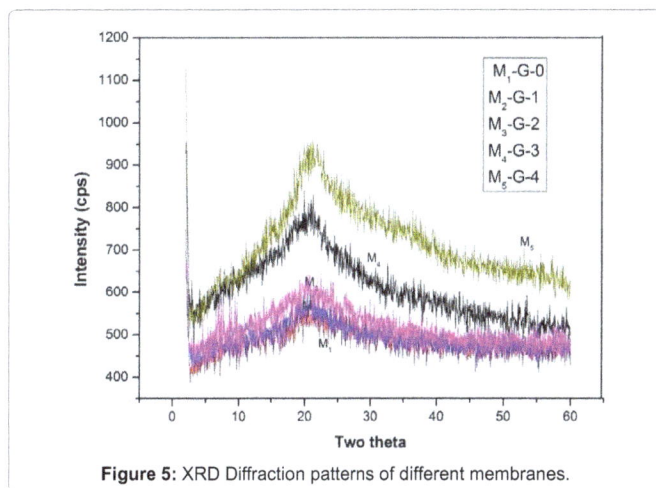

Figure 5: XRD Diffraction patterns of different membranes.

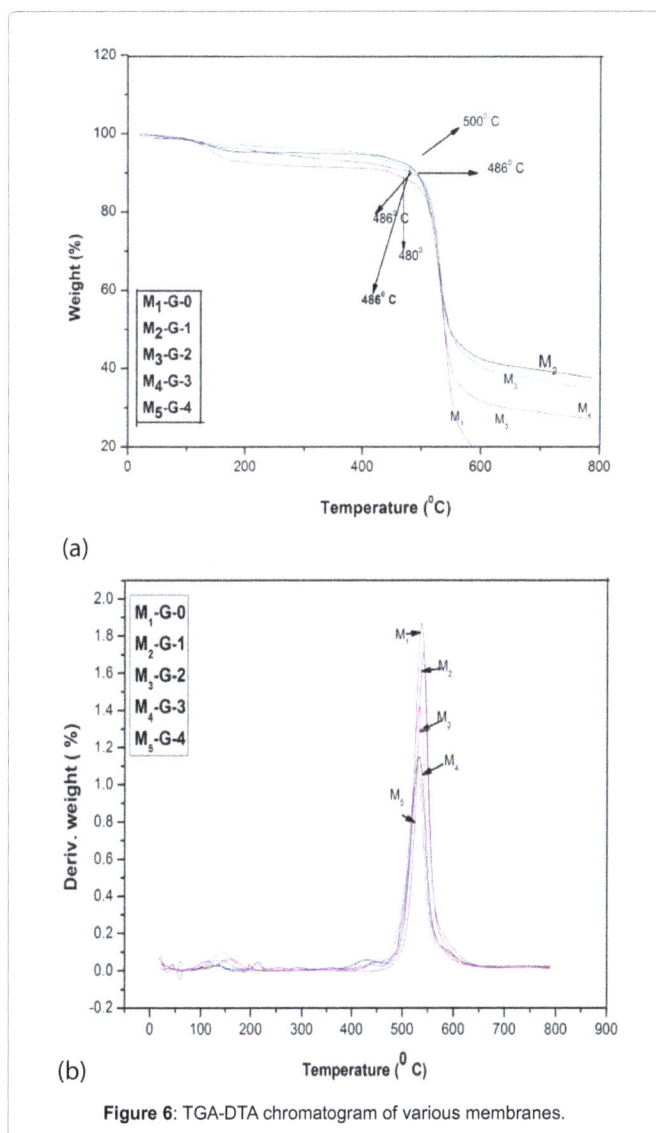

(a)

(b)

Figure 6: TGA-DTA chromatogram of various membranes.

$$P_{Po}=-\int_{p0}^{p1} Deff(p)\,dp \qquad (4)$$

Where $P_1=0$ is assumed for the case where the penetrant pressure is

Dendrimer-Membranes	2Θ (⁰)	d (spacing)	Wavelength (A⁰)	FWHM (cm)
G-4	28	5.15	1.79	9.2
G-3	26	4.30	1.79	8.5
G-2	24	3.87	1.79	7.3
G-1	20	3.24	1.79	4.6
G-0	18	2.76	1.79	2.51

Table 1: XRD Analysis of membrane.

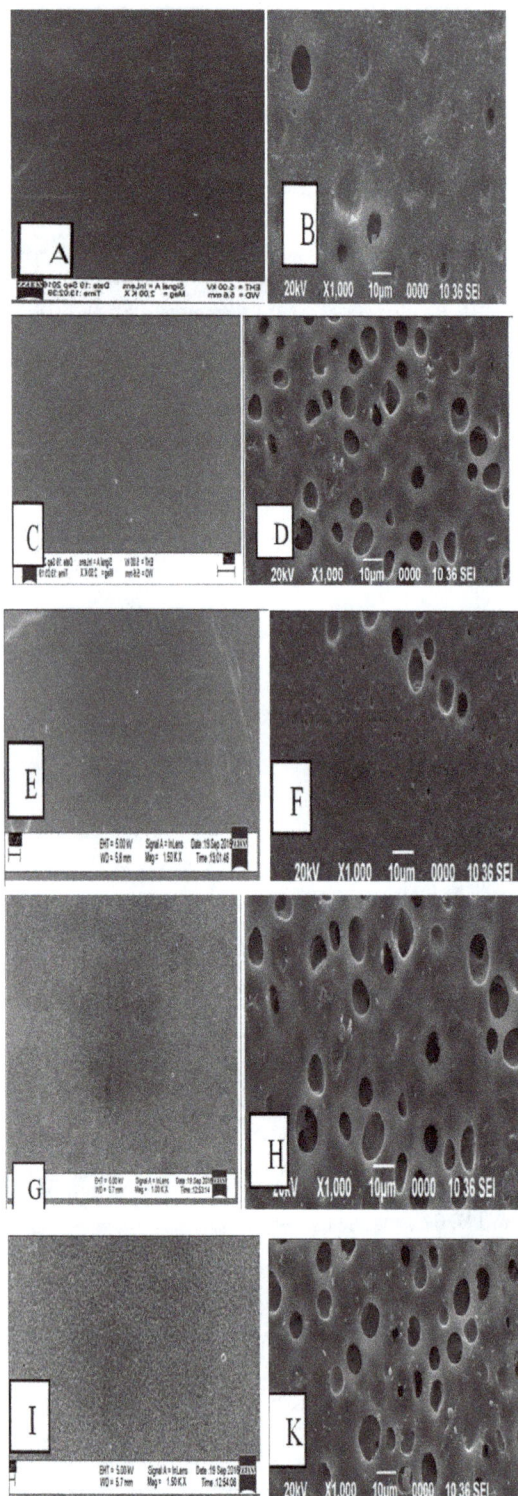

Figure 7: SEM Photographs of membranes for before (left panel) and after (right panel) permeation process:A. PAMAM Dendrimer (G-0)(Before permeation) B. PAMAM Dendrimer (G-0)(After permeation) C. PAMAM Dendrimer (G-1)(Before permeation) D. PAMAM Dendrimer (G-1)(After permeation)E. PAMAM Dendrimer (G-2)(Before permeation) F. PAMAM Dendrimer (G-2)(After permeation) G. PAMAM Dendrimer (G-3)(Before permeation) H. PAMAM Dendrimer (G-3)(After permeation) I. PAMAM Dendrimer (G-4)(Before permeation) K. PAMAM Dendrimer (G-4)(After permeation).

zero at the downstream face. Then by applying the Leibenitz rule, the equation becomes

$$D_{eff}(P_0) = P_0[dp/dc]_{c0} + P[dp/dc]_{c0} \qquad (5)$$

From the permeability vs pressure variation plots it is observed that the permeability of the membranes for CO_2 gas decreased with the increase in the feed gas pressure. The permeability vs feed gas pressure plot obtained here followed the gas diffusion model for the permeability of CO_2 gas through the membranes.

The permeabilities of the membranes for CO_2 decreased with the increase in feed gas pressure. The membranes showed a continuous decrease in CO_2 permeabilities up to a pressure of 50 cm Hg. Figures 8 and 9 showed the effects of feed gas pressure on the CO_2 permeabilities of the membranes from pure CO_2 and a mixture of CO_2 and N_2. The decrease in the Permeance of the membranes was observed as a consequence of the facilitated transport mechanism of dendrimer molecules incorporated into the membranes. The facilitated transport mechanism is schematically represented in Figure 10. According to

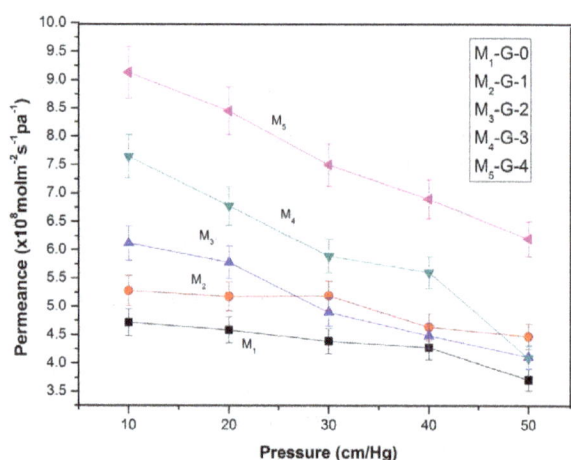

Figure 8: Effect of feed gas pressure on permeability of dendrimer membrane for pure CO_2 .The error bar corresponds to mean of the percent (%) errors of the datas of five independent experiments.

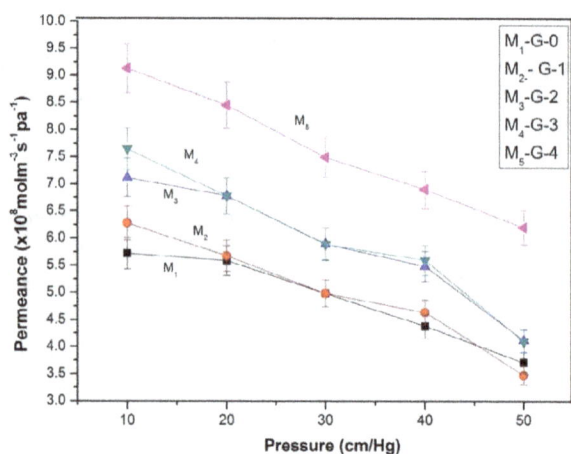

Figure 9: Effect of feed gas pressure on permeability of dendrimer membrane for mixture of CO_2/N_2 as feed gas. The error bar corresponds to mean of the percent (%) error of the datas of five independent experiments.

the facilitated transport mechanism CO_2 was transformed into small and mobile ion HCO_3^- ion. Thus it enhances the transportation of CO_2 through the membrane. Similar variations were observed for Polysulfone and Cellulose Acetate composite membranes and PSF/PDMS composite membrane for separation of CO_2 as reported by Mansoori et al. in previous literature [11,13].

Effect of feed gas pressure on selectivity: The selectivity of membranes for CO_2 decreased with the increase in feed gas pressure. The effects of cross-linking of different generations dendrimer on membrane selectivities for CO_2 with the feed gas pressure are shown in the Figure 11. It may happen due to increase in cross-linking of higher generations dendrimer molecules which form more numbers of gates for CO_2 permeation. The selectivity order of the Silica and Zeolite composite membrane also follow the similar variation as reported by Shehu et al. [14] . The effect of pressure on selectivity of CO_2 over CH_4 also shows the similar behaviour as reported by Zhang et al. [1]. The selectivity of CO_2/CH_4 is comparable with that of Cellulose Acetate membranes reported in literature [11].The effect of pressure on selectivity of CO_2 over N_2 also shows the similar behaviour as reported

by Gil et al. (Supplimentory Figure 1) [15]. In Table 2, literature datas of CO_2 permeation by different membranes are given.

Structure – Activity relationship of dendrimers with permeability: The permeabilities of dendrimer membranes are expected to correlate with their molecular structure. Accordingly the estimated values of permeabilities were plotted against the number of NH_2 group present in the dendrimers of different generations and are shown in Figure 12 which indicates increase in permeabilities with amine groups present in the dendrimer. More number of amine groups presents in the dendrimer increases the number of molecular gates for CO_2 permeation via facilitated transportation of CO_2 gas through the membrane. Thus the membrane prepared by G-4 dendrimer showed maximum permeability for CO_2 gas. In order to analyse the effect of dendrimer polysulfone molar ratio in the composite membrane the permeability values were calculated for the membranes with different dendrimer polysulfone molar ratio and are shown in Table 3. From the table it is seen that the values of permeabilities increases with increase in dendrimer polysulfone molar ratio. However, increase in concentration of dendrimer in the casting solution decreases the membrane forming capacity due to high saturation of dendrimer in the solution. Hence, for better performance of the membrane for CO_2 permeation the dendrimer polysulfone molar ratio is observed as 0.5 to 10.

Conclusion

Membranes for CO_2 separation had been prepared and were characterized by IR, XRD, TGA-DTA, SEM Analysis. The permeability of the membranes for CO_2 had been studied using the membranes. The facilitated transport mechanism of dendrimer molecule (G-0-G-4) incorporated into the membranes which facilitated the transportation of CO_2 through the membranes had been discussed in this paper. The effect of feed gas pressure on the permeabilities and selectivities of the membranes had been also observed. It was observed that permeability and selectivity of the membrane increased with increase in the amine groups present in the dendrimer molecule due to the formation of CO_2 molecular gates by facilitated transport mechanism.

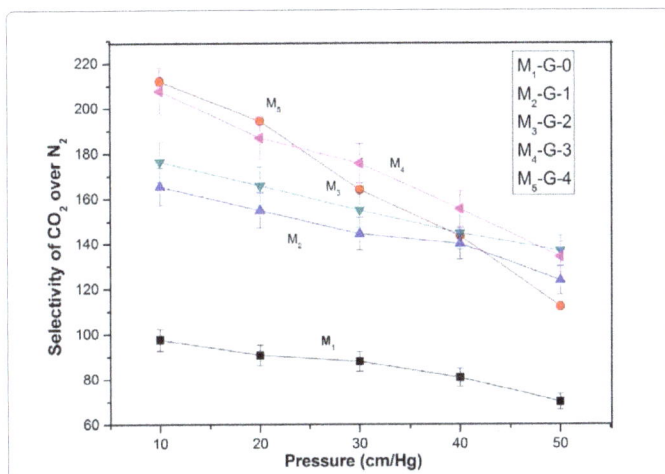

Figure 10: Effect of feed gas pressure on selectivity of dendrimer membrane for CO_2/N_2 mixture .The error bars corresponds to mean of the percent (%) error of experimental datas for five independent experiments.

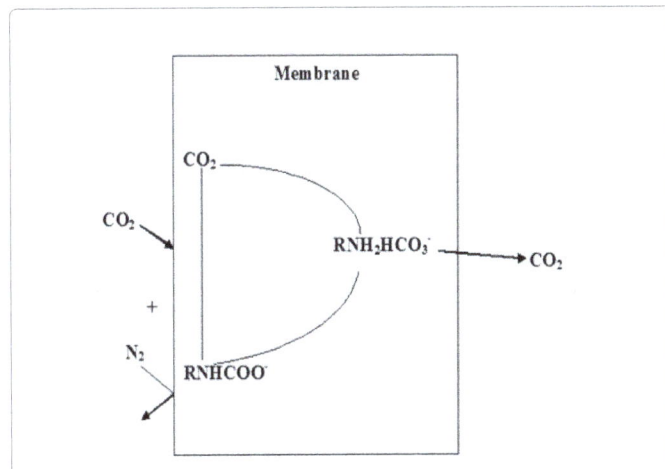

Figure 11: Facilitated transport mechanism of PAMAM dendrimer molecule.

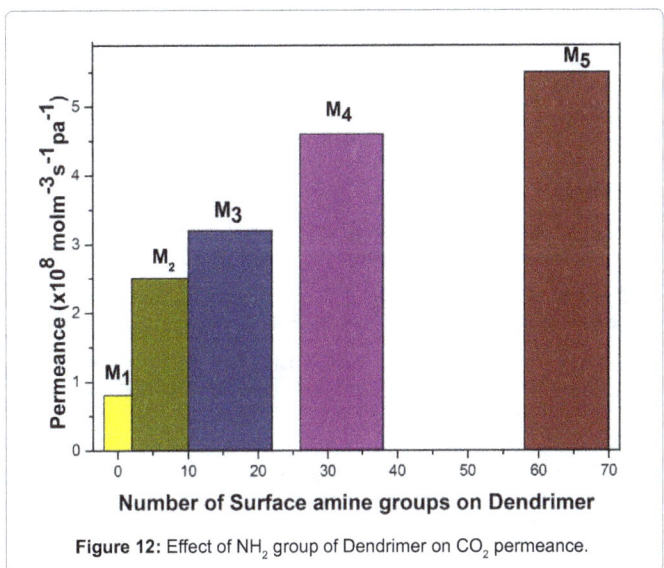

Figure 12: Effect of NH_2 group of Dendrimer on CO_2 permeance.

Sl. No.	Types of Membranes	CO$_2$ permeance P$_{CO2}$ (x10^{8}molm^{-2}s^{-1}pa^{-1})	Separation factor (α)	References
		30	200	
1	DDR- Type membranes	4.6	400	16
	Zeolite- T membranes			17
2	SAPO-34 membranes	20	36	
3	Polyvinylalcohol membrane	6.196	493	18
	PAMAM (G-0)			8
4	Polysulfone –dendrimer membrane	4.1	700	
5				9
	G-0			
	G-1			
6	G-2	1.5	98	
	G-3	5.4	165	
	G-4	5.6	178	Current work
		6.5	210	
		8	215	

Table 2: CO$_2$ separation performances of reported membranes.

Dendrimer	Dendrimer Polysulfone molar ratio	CO$_2$ Permeabilities (x10^{8}molm^{-3}s^{-1}pa^{-1})
G0t	0.1: 10	0.67
	0.2: 10	0.73
	0.3:10	0.8
	0.4:10	1.23
	0.5:10	1.5
G1	0.1:10	1.76
	0.2:10	1.82
	0.3:10	2.91
	0.4:10	3.56
	0.5:10	5.4
G2	0.1:10	3.12
	0.2:10	4.21
	0.3:10	4.67
	0.4:10	4.98
	0.5:10	5.6
G3	0.1:10	4.54
	0.2:10	4.89
	0.3:10	5.12
	0.4:10	6.12
	0.5:10	6.5
G4	0.1:10	2.56
	0.2:10	3.23
	0.3:10	4.13
	0.4:10	7.4
	0.5:10	8

Table 3: Effect of Dendrimer Polysulfone molar ratio on CO$_2$ Permeation.

References

1. Zhang Y, Wang Z and Wang SC (2002) Selective permeation of CO$_2$ through new facilitated transport membranes. Desalination 145: 385-388.

2. Zhang Y, Wang Z, and Wang S (2002) A study on facilitated transport membranes for removal of CO$_2$ from CH$_4$. Fuel Chem Preprint 47: 73- 74.

3. Fuerst O, Lin Y, Granell M, Leblanc G, Padros E, et al. (2015) The Melibiose Transporter of Escherichia coli Critical Contribution of LYS-377 to the structural organization of the interacting substrate binding sites. Journal of Biol Chem 290: 16261-16271.

4. Lin Y, Fuerst O, Granell M, Leblanc G, Fonfría VL, et al. (2013) The substitution of Arg149 with Cysfixes the melibiose transporter in an inward-open conformation. Biochimica et biophysica acta 1828: 1690-1699.

5. Lin Y, Bogdanov M, Tong S, Guan Z, Zheng L (2016) Substrate Selectivity of Lysophospholipid Transporter LplT involved in Membrane Phospholipid Remodeling in Escherichia coli. Journal of Biol Chem 291: 2136-2149.

6. Tong S, Lin Y, Lu S, Wang M, Bogdanov M, et al. (2016) Structural Insight into Substrate Selection and Catalysis of Lipid Phosphate Phosphatase PgpB in the Cell Membrane. Journal of Biol Chem 291: 18342-18352.

7. Tong Z, Winston Ho WS (2016) Facilitated transport membranes for CO$_2$ separation Capture. Sep Sci Technol 6: 1-12.

8. Huang J, Zou J, Winston Ho WS (2007) Carbon Dioxide Capture using a CO$_2$ Selective Facilitated Transport Membrane. Ind Eng Chem Res 47: 1261-1267.

9. Kovvali S, Chen H, Sirkar KK (2000) Dendrimer Membranes: A CO$_2$-Selective Molecular Gate. J Am Chem Soc 122: 7594-7595.

10. Belmabkhout Y, Serna-Guerrero R, Sayari A (2010) Adsorption of CO$_2$-Containing Gas Mixtures over Amine-Bearing Pore-Expanded MCM-41 Silica: Application for Gas Purification. Ind Eng Chem Res 49: 359-365.

11. Donohue MD, Minhas BS, Lee SY (1989) Permeation behaviour of CO$_2$/ CH$_4$ mixtures in Cellulose Acetate membranes. J Membrane Science 42:197-214.

12. Duan S, Chowdhury Firoz A, Kai T, Kazama S, Fujioka Y (2008) PAMAM dendrimer composite membrane for CO$_2$ separation : addition of hyaluronic

acid in gutter layer and application of novel hydroxyl PAMAM dendrimer. Desalination 234: 278-285.

13. Mansoori SAA, Pakizeh, A. Jomekian (2011) CO_2 Selectivity of a new PDMS/PSf Membrane Prepared at Different Conditions. J Membra Sci Technol 5: 500-513.

14. Shehu H, Okon E, Orakwe I, Gobina E (2016) Study of the Selectivity of Methane over Carbon Dioxide Using Composite Inorganic Membranes for Natural Gas Processing. J Adv Chem Eng 6: 1-5.

15. Francisco Gil J (2011) Separation of Carbon Dioxide from Nitrogen Using Poly vinyl alcohol)-Amine Blend Membranes. Int J Mol Sci 46: 1245-1250.

16. Himeno S, Tomita T, Suzuki K, Nakayama K, Yajima K, et al. (2007) Synthesis and permeation properties of a DDR-Type Zeolite Membrane for separation of CO_2/CH_4 Gaseous Mixtures. Ind Eng Chem Res 46: 6989-6997.

17. Cui Y, Kita H, Okamoto K (2004) Preparation and gas separation performance of zeolite T membrane. J Mater Chem 14: 924-932.

18. Poshusta JC, Tuan VA, Pape EA, Noble RD, Falconer JL (2000) Separation of gas mixtures using SAPO-34 membranes. AIChE J 46: 779-789.

22

Experimental Evaluation of Corrugated Feed Channel of Direct Contact Membrane Distillation

Mabrouk A[1]*, Elhenawy Y[2] and Moustafa G[2]

[1]*Qatar Environment and Energy Research Institute, Hamad Bin Khalifa University, Qatar Foundation, Qatar*
[2]*Port Said University, Egypt*

Abstract

Membrane distillation is a hybrid process in which the separation process is based on both thermal potential and membrane characteristics. However, some technical challenges such as thermal boundary layer builds up (temperature polarization) resulted in low mass flux. In this study, the direct contact membrane distillation is equipped with corrugated feed channel to create fluid mix for mass flux improvement. A lab scale flat sheet membrane distillation is assembled with corrugated feed channel to suppress the thermal boundary layer in the vicinity of the membrane wall. The flat sheet PTFE- membrane of a pore size of 0.45 μm and porosity 65% is considered in the present study. The effect of feed channel gap height has been investigated at different values of flow rate, feed temperature, and feed salinity.

The experiments showed that the corrugated feed channel has a dominant effect of improving of the mass flux and the thermal efficiency rather than gap height and operating conditions. The experimental results showed the water flux and thermal efficiency of the corrugated feed channel module is 44% and 33% higher than the original module respectively.

Keywords: Membrane distillation; Heat transfer; Corrugated surface; Gap; Desalination

Nomenclature

A: Membrane area, m^2

Cp: Specific heat capacity of fluid ($J\,kg^{-1}K^{-1}$)

LHv: Latent heat of vaporization of water ($J\,kg^{-1}$)

Nm: Water flux ($kg\,m^{-2}s^{-1}$)

EE: Energy efficiency (%)

T: Temperature (°C)

m: Mass flow rate (kg/sec.)

Subscripts

f: Feed

fi, fo: Entrance, outlet of feed

m: Membrane, or membrane surface

p: Permeate

pi, po: Entrance, outlet of permeate

pm: On permeate-side membrane surface

Introduction

Membrane distillation (MD) is a thermally based separation process in which hydrophobic membrane assist phase change to let only vapor pass through membrane wall. This process may be used not only for desalination of seawater and brackish water but also concentration valuable substances. Membrane distillation differs than RO in the way of separation mechanism and the dominant driving force. On the other hand, earlier work [1] of describing the separation process considered membrane distillation is an infinite stage flash evaporation since each pore acts as a separate stage [1]. The MD has a unique characteristic rather than the mature technology Reverse Osmosis (RO) since it works at a low pressure feed. The MD also has advantage rather than the conventional distillation methods such as Multi-Stage Flash (MSF) and Multi Effect distillation (MED) such that it utilizes low-grade heat and has no corrosion problems.

In addition, the MD can implement different non-volatile solutes under a wide range of concentration with a separation factor of 99.9%. The separation factors depend on the solution, membrane and operating conditions [2-4]. However, using MD modules (flat sheet, spiral wound, hollow fibers) in industrial large scale application is a particular challenge due to the lack of uniform flow conditions, high pressure drop, and low packing density within module and lack of energy recovery.

The Temperature/concentration polarization is also considered a technical challenge which affecting the fluxes of the MD process [2,5-8]. From the process level prospective, increasing feed temperature would increase the mass flux however scale formation and deposit on the membrane surface would restrict the MD performance. Creating turbulence by pumping high feed flow rate in the feed channel showed an adequate increasing in the mass flux, however, specific pumping energy will become a particular concern.

Using lab scale direct contact MD (10 cm × 25 cm) cell, experimental evaluation has been conducted to investigate the effect of the feed flow hydrodynamics on the temperature polarization in PVDF (0.2 μm as mean pore size, 192 μm as thickness, 48% overall porosity, and feed

*Corresponding author: Abdel Nasser Mabrouk, Scientist, Desalination and Water Reuse, Qatar Environment and Energy Research Institute, Qatar E-mail: aaboukhlewa@qf.org.qa

channel height of 1.3 cm) [6]. As the Reynolds number increases higher than 2000, the temperature polarization coefficient increases higher than the value of 0.84, which enhances the process performance [6].

Experimental study of applying a vacuum pressure on the feed and permeate channel of DCMD has been investigated [9]. This approach increased mass transport of water due to higher permeability and due to a total pressure gradient within the membrane [9]. The vacuum pressure is controlled in each MD channel by throttling valve at the channel entrance. However, some doubt about throttling process irreversibility and its effect on decreasing the temperature due to throttling.

Membrane distillation crystallization process has been tested [10]. This study showed that the MD can operate at a high concentration at feed temperature of 50 and 60°C with fluxes up to 20 LMH. Operating in batch concentration, the flux gradually declined due to the vapor pressure suppression and the concentration polarization existence. The experiments showed that the membrane wall temperature is about 5 -10°C lower than the bulk temperature due to the existence of the temperature polarization which hinder the salt saturation and scale deposition formation [10].

A novel composite membrane distillation was prepared by blending the hydrophilic polysulfone with hydrophobic surface [11]. The hydrophobic surface is indirect contact to the hot feed solution however, the hydrophilic surface is in direct to permeate channel. The hydrophilic polymer concentration in blend is optimized to get a permeate flux of 43% higher than that the commercial PTFE membrane with 99.9% NaCl salt rejection [11].

The effect of the thermal and the concentration boundary layers in aqueous lithium bromide solution is experimentally investigated using commercial PTFE membrane (0.2 μm as mean pore size, 80 μm as thickness, 75% overall porosity) [7]. The results showed that the concentration boundary layer can be neglected only when concentration less than 5 wt% (50,000 ppm). However, the thickness of the thermal boundary layer was found larger than the concentration boundary layer which has prominent effect in the whole concentration range.

The integrated a small unit of direct contact membrane distillation equipped with a solar absorber is designed and investigated [12]. The absorber inserted between glass and the membrane to assist and to compensate the energy loss in the feed channel due to evaporation within membrane. The device performance is compromised under wide range of the hot stream temperature (30 -50°C) while absorb solar energy. The maximum trans-membrane flux is 4.1 kg/(m² h) with high purity which is 16.6% higher than that of the traditional membrane distillation operating under the same operating conditions without solar assist. However, the additional capital cost and complexity of solar absorber need to be evaluated.

For the same non-woven support layer, the PTFE membrane is superior to PVDF membrane and the scrim support layer is superior to the non-woven support layer based membrane [13]. This work highlighted that the support layer structure has significant effect rather than pore size on the membrane performance. The results shows for the thinner support and active layer, the mass flux is obtained as 46 LMH at 80°C which is comparable with that of the RO. However, long term operation in order to check fouling of scale deposit and wettability needs to be evaluated.

The process performance of different DCMD membranes with and without backing structure is experimentally and theoretically investigated [14]. The backing material reduces both mass flux and thermal efficiency significantly due to the impact of the effective diffusion path length and the complex thermal resistance. The thicker membrane reduces the flux with no influence on the thermal efficiency while the backing orientation towards the condenser side is advantageous [14].

An experimental evaluation to measure the effect of a spacer in the feed channels of the membrane is carried out [15]. The results showed that the fluxes can be enhanced by improving the mass transfer coefficient in terms of higher porosity-tortuosity factor. However in case of a poor convection heat transfer in liquid the spacers create turbulence and improve the mass transfer [15]. The effect of spacers (hydrodynamic angles range of 70 -90° and void of 60 -70%) on the mass flux for DCMD is experimentally investigated [8]. The channel filled spacer enhances the mass flux by 31 -41% as a result of the created turbulence and the eddy flow in the vicinity of the membrane wall. This turbulence flow enhances the temperature polarization coefficient which reached to the unity [8].

Corrugated PVDF membrane was fabricated for improving flux stabilization in membrane distillation [16]. The corrugated composite flat sheet membrane showed much higher flux stability. The corrugated membrane operated with a minimal flux reduction and significantly less foulant accumulation and less salt deposition which attributed to the improved flow dynamics at the membrane surface and increased membrane surface induced bu surface corrugation.

The main objective of this study is to evaluate a new corrugated feed channel equipped with direct contact membrane distillation module. The aim of implementing a corrugated feed channel is to improve the flow and the hydrodynamic conditions in the vicinity of the membrane surface. A lab scale setup is assembled in order to conduct and to test the new module under different feed channel gap, different flow rate, and different feed temperature. For the sake of fair comparison, the new module is tested under the same operating condition of a traditional configuration.

Experimental Set up Description

A lab scale flat sheet membrane of 30 cm×30 cm dimensions with effective membrane area of 28 cm×28 cm is considered in the present experiments. The laminated PTFE membrane has a pore size, porosity and thickness of 0.45 μm, 0.65 and 127 μm, respectively. As shown in Figures 1 and 2, the hot streams are represented by red color while the permeate streams are represented by blue color. The hot water pump (Figure 2(2)) is circulating the hot fluid between electrical heater (Figure 2(1)) and the membrane module (Figure 2(6)). The hot fluid is directed to the membrane module through top port opening. The hot feed temperature decreases in the feed channel due to the loss of associated energy to the vapor permeate through the membrane. The electrical heater (Figure 2(1)) is compensating the temperature loss and increase the hot stream temperature to its initial temperature. The feed temperature varies between 42 - 68°C. The cold water pump (Figure 2(8)) is circulating the cold water between the chiller (Figure 2(7)) and the membrane module (Figure 2(6)). The temperature of the chilled water get increase through the permeate channel due to condensate of the vapor permeate. The chiller role is to restore the chilled stream to its initial permeate temperature. The chiller is used to control the permeate temperature in the range 20 – 30°C. Flow rates of the feed and permeate are measured using rotameter. The temperature and the pressure are measured using thermocouple of type T and pressure transducers, respectively. The mass flux is determined by the weight of

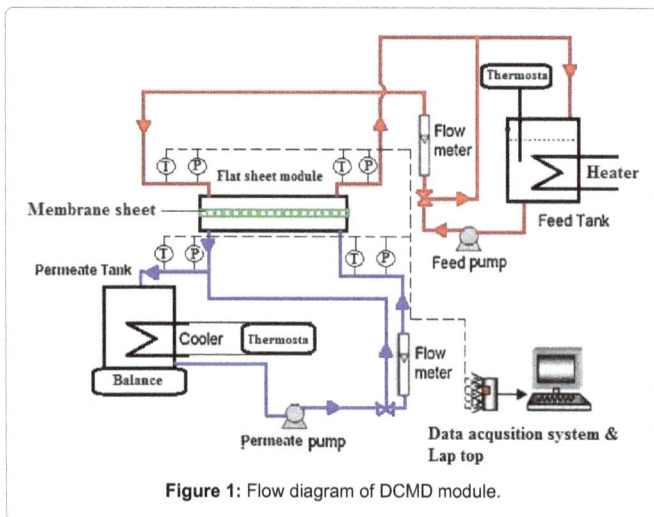

Figure 1: Flow diagram of DCMD module.

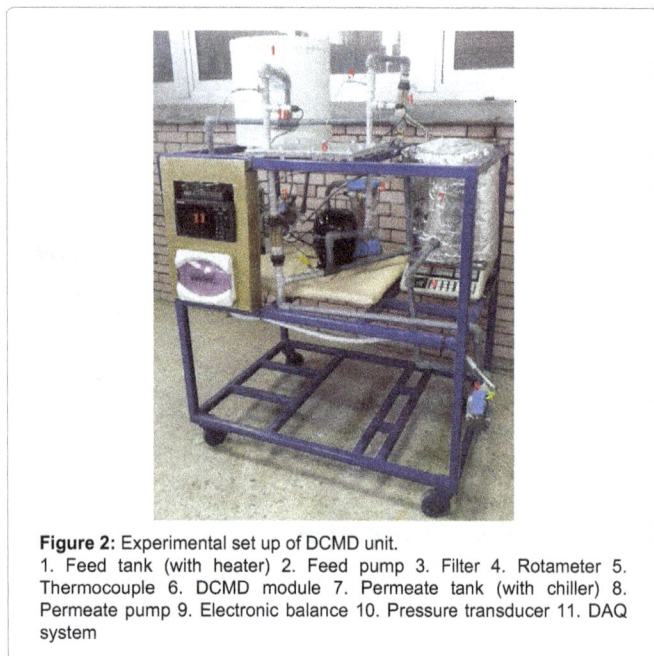

Figure 2: Experimental set up of DCMD unit.
1. Feed tank (with heater) 2. Feed pump 3. Filter 4. Rotameter 5. Thermocouple 6. DCMD module 7. Permeate tank (with chiller) 8. Permeate pump 9. Electronic balance 10. Pressure transducer 11. DAQ system

the distillate using weighing balance. Brine and permeate concentration are measured by TDS meter model. The data is recorded using DATAQ Instruments (Graphtec Model: GL220_820APS, Graphtec). In order to investigate the effect of feed salinity, NaCl with 45 g/l is used as a feed solution. The hot feed flow rate varies between 30 l/min (Reynolds no. = 15.7) to 180 l/h (Reynolds no. = 94) which indicates that the flow regime in the feed channel is laminar. While the permeate flow rate varies between 30 - 120 l/h which indicates that the flow regime in the permeate channel is also laminar.

Figure 3a shows the 3 D drawing of the main parts of the corrugated feed channel DCMD module. The bottom plate, rubber, membrane sheet, spacer and top plates are assembled together using 12 pieces of Allen screw bolt and nuts. The overall size of each plate is 350 × 350 × 6 mm. Figure 3b shows the upper corrugated plate where slots are created. The width of each slot is 10 mm and depth is 3 mm. Figure 3c shows the lower plate which is kept flat without corrugation. Each plate has holes which are machined to assemble both plate with membrane sheet in between using bolts and nuts. The spacer is equipped in

between the top plate and the membrane sheet with a rubber gasket frame. Similarly, the rubber gasket thickness creates the permeate channel height between the bottom plate and the membrane sheet. The module was a rectangular thin channel with a hydraulic diameter of 5.93 mm that define the height of the flow channel of 3 mm. This means that the corrugated feed channel would increase the Reynolds number by approximately two times of the traditional module. The rubber gasket of 1.5 mm is also used in the experiment to create gap of 1.5 mm height. The spacer foil thickness is 0.8 mm. Using different rubber gasket thickness enables to create different gap heights. Similarly, the rubber frame inserted and fixed between the lower plate and the other face of the membrane sheet to create the permeate channel. For the sake of comparison under the same operating conditions, the corrugated module would be compared with traditional module. In order to show the effect of corrugation surface design and quantify its significant comparable to the traditional module under different feed flow rates, and different feed temperatures, the experimental test starts by using low feed salinity (450 ppm) to avoid the effect of fouling effect. Then after, the experimental work is conducted under high feed salinity of 45000 ppm.

The efficiency of the MD system is calculated by using equation 1. The bulk temperature of the feed is used for calculating the latent heat of evaporation. The thermal efficiency of MD unit is given by [3]:

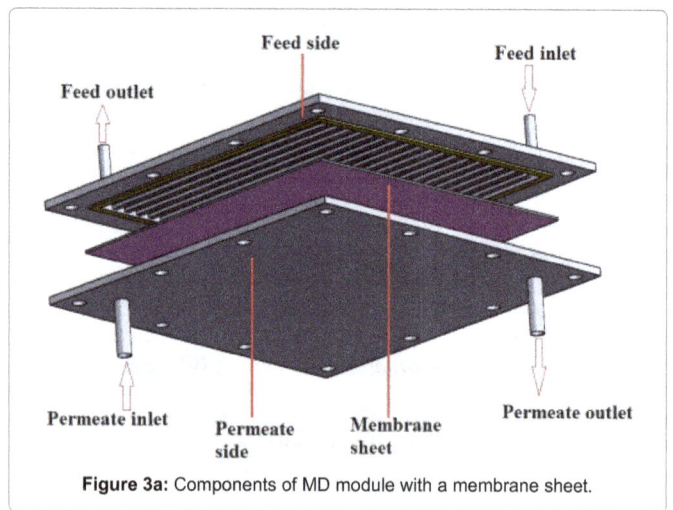

Figure 3a: Components of MD module with a membrane sheet.

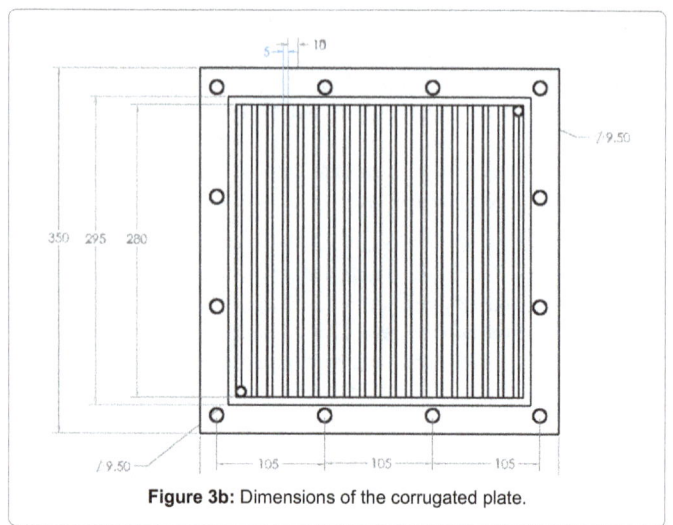

Figure 3b: Dimensions of the corrugated plate.

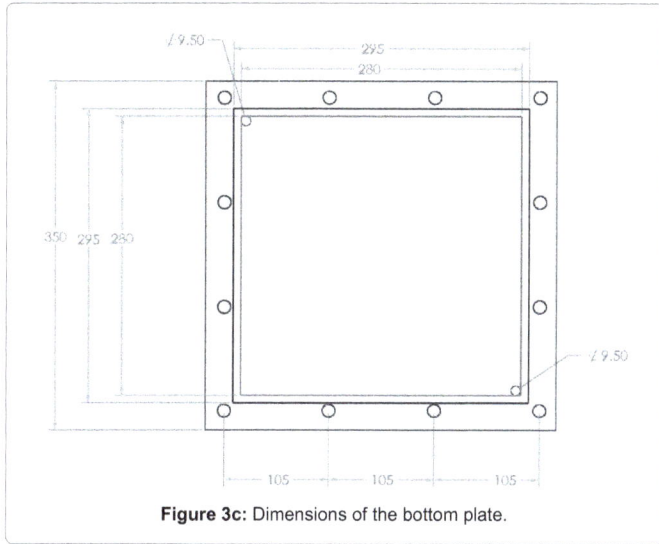

Figure 3c: Dimensions of the bottom plate.

$$E_E = \frac{N_m \, A \, LH_v}{m_p \, CP_p \left(t_{po} - t_{pi}\right)} \qquad (1)$$

Where, A is the effective membrane area, m_p is the mass flow rate of the permeate, N_m is the water flux, LH_v is the latent heat of evaporation of water at the average temperature and CP_p is the heat capacity of water at the average temperature.

LH_v can be calculated as follow [6]:

$$LH_v(T) = 1850.7 \, 2.8273 \, T \, 1.6 \, 10^{-3} T^2 \qquad (2)$$

The mass flux is calculated as:

$$N_m = \frac{\Delta W}{A \times t} \qquad (3)$$

Where, ΔW the difference in weight of permeate tank (kg), t is the time (h)

Results and Discussion

Effect of the feed corrugated channel on the mass flux

The feed flow rate and feed temperature are an important operating parameter due to its effect on the water flux, thermal efficiency and the temperature polarization. The effect of both feed flow rate and feed temperature on the mass flux is illustrated in Figures 4 and 5.

Figure 4 shows the mass flux variation at different feed temperature variation. The feed flow rate is fixed at 0.5 l/min and the permeate is fixed at 1 l/m. The mass flux increases by decreasing the gap height. The experimental measurements show that at gap = 3 mm, the average water flux of the corrugated module of DCMD is 32% higher than that of the original module, while the gap decreased to 1.5 mm, the average water flux is 37% higher than that of the original module.

Figure 5 shows the effect of feed temperature and gap variation on the water flux while the feed flow rates fixed at 3 L/min. The experimental results show that at gap = 3, the mass flux of the corrugated module is 34% higher than that of the original module. However, at gap of 1.5 mm, the corrugated module enhanced the average water flux by 42% when it is compared with original module.

Figures 6 and 7 show the water fluxes for various permeate temperatures. The Feed solution was kept constant at 3 l/min and feed

salinity at TDS= 0.45 g/l. The permeate temperature varies from 20 to 31°C however, for feed solution temperature side is kept constant at 68°C.

Figure 6 shows that the water flux increases when permeate temperature is decreased. The corrugated module enhanced in the average water flux of DCMD by 35% at 3 mm gap. This enhancement occurs at permeate flow rate of 2 L/min.

Figure 7 shows the significant effect of the corrugated feed channel on the mass flux. The measured data shows that the corrugated module leads to an average water flux of DCMD by 44% at gap = 1.5 mm higher than that of the original module. This occurs at the permeate flow rate is 2 L/min. The permeate temperature varies from 20 to 31°C while the feed temperature is kept constant at 68°C and the feed flow rate is controlled at 3 L/min.

Table 1 shows the summary of the experimental results of the membrane distillation mass flux as a results of varying the gap height, the feed flow rate and the permeate flow rate. The percentage

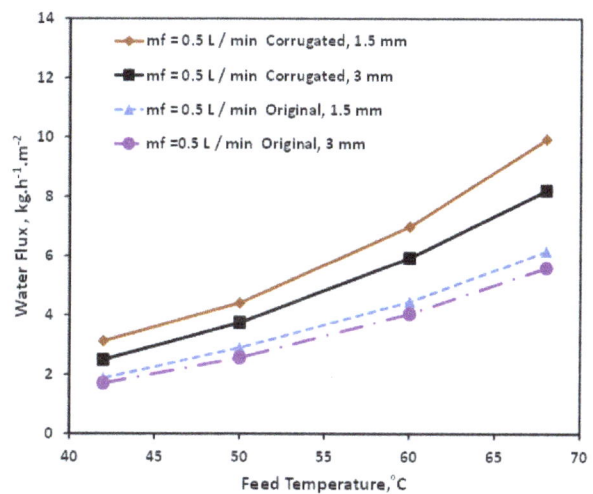

Figure 4: Water flux variation and gaps (mf = 0.5L /min, mp =1 l /min, tp = 20 °C, TDS = 0.45 g/L).

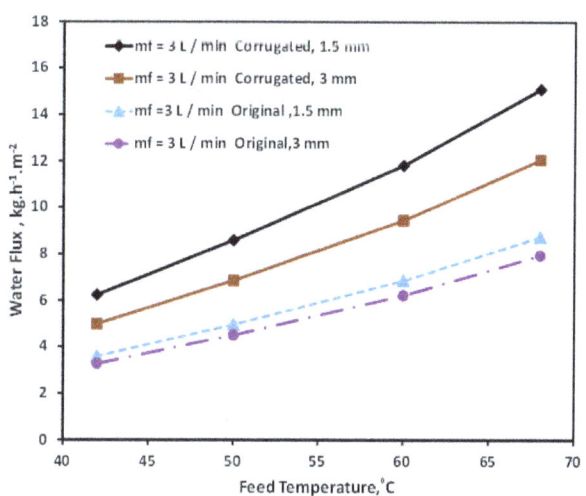

Figure 5: Water flux variation (mf =3L /min ,mp =1 L /min ,tp=20 °C, TDS =0.45 g/L).

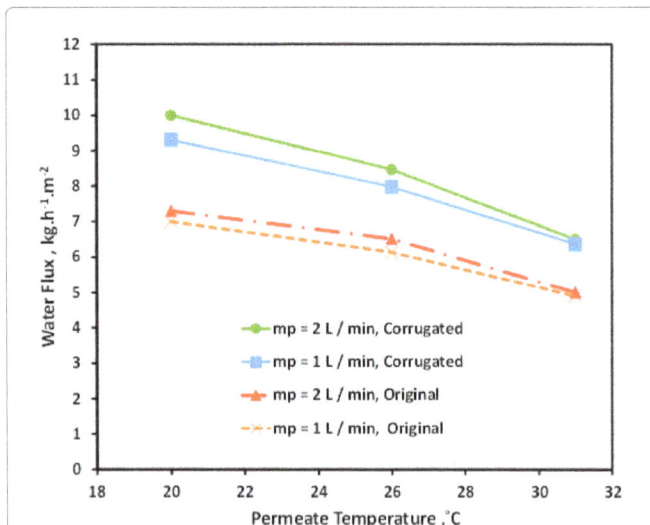

Figure 6: Effect of different permeate flow rate and temperature at gap = 3 mm on water flux (mf = 3L /min, tf = 68°C, TDS=0.45 g/L).

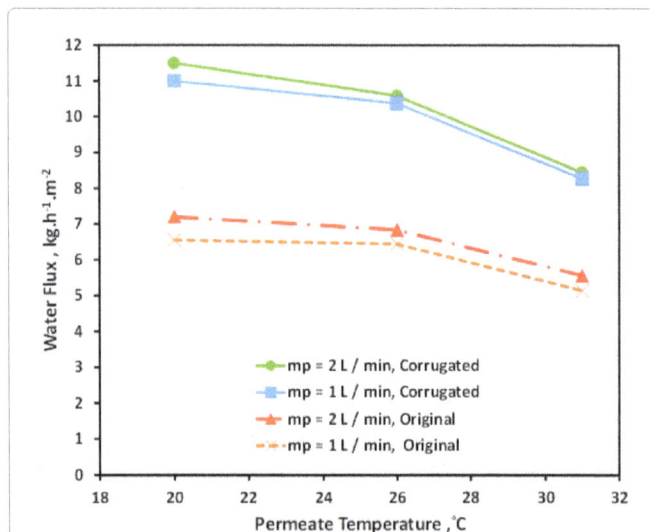

Figure 7: Effect of different permeate flow rate and temperature at gap = 1.5 mm on water flux (mf = 3 l/min, tf = 68°C, TDS = 0.45 g/L).

Feed flow rate, L/min	Gap height, 3 mm	Gap height 1.5 mm	Permeate flow rate , L/min
0.5	32 %	37 %	1
3	34 %	42 %	1
3	35 %	44 %	2

Table 1: Summarized results of % enhancement of mass flux.

enhancement in mass flux is calculated at average temperature. The first line in Table 1 shows that at a gap height of 3 mm, the contribution of the corrugated feed channel is 32% enhancement in mass flux and when the gap height is reduced to 1.5 the effect reached to 37%. This indicates the corrugated feed channel effect is dominant rather than decreasing the feed channel height. At a gap height of 1.5 mm, when the feed flow rate increases from 0.5 up to 3 l/min (6 times), the percentage enhancement of mass flux increases from 37% to 42%. This indicated that the contribution of the feed flow rate increase (up 6 times) has only 5% while the corrugated feed channel contribution is calculated

as 37%. Table 1 shows also, that at a gap height of 1.5 mm, the increase of permeate flow from 1 up to 2 l/min, the mass flux enhancement varies from 42% to reach 44%. This results show that the contribution of doubling the permeate flow rate is only represented by 2% while the contribution of the corrugated channel is 42% which reflects the dominated effect of the corrugated feed channel design.

Effect of the feed corrugated channel on the thermal energy efficiency

Figure 8 shows the effect of the feed temperatures on the thermal energy efficiency of the original and corrugated modules. The feed flow rate is controlled at 0.5 L/min, while the permeate flow rate is fixed at 1 L/min. The hot feed concentration is measured as 0.45 g/L is used. The feed temperature varies from 42 to 68°C for feed side while the permeate inlet temperature is kept constant at 20°C. Different gap height of 1.5 mm and 3 mm are carried out.

As shown in Figure 8, the energy efficiency increases as the feed temperature increases. The energy efficiency of the corrugated module is higher than the original module. At a gap height of 3 mm, the experimental results showed the energy efficiency of the corrugated efficiency is 23% higher than the original module. This is because of created eddies along the corrugated surface which minimize the thermal boundary layer. Decreasing the gap height to 1.5 mm, the energy efficiency of the corrugated module is 26% higher than the original module.

Figure 9 shows comparison between the energy efficiency of the original and that of the corrugated modules at constant feed flow rate of 3L/min and permeate flow rate of 1 l/min. At the gape height of 3 mm, the average energy efficiency of the corrugated module is 28% higher than that of the original module. However, at the gap height of 1.5 mm, the energy efficiency is 30% higher than that of the original module.

Conducting experiment at various permeate flow rates, and the permeate temperature varies from 20 to 31°C while, the temperature of the feed is fixed constant at 68°C and feed flow rate of 3 L/min, show that increasing the permeate temperature decreases the energy efficiency as shown in Figures 10 and 11. This is due to decreasing the driving force of vapor pressure difference.

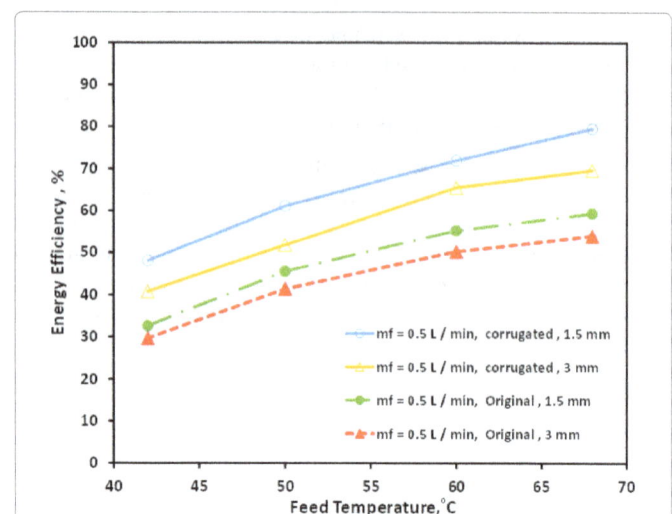

Figure 8: Energy efficiency with different feed temperature and different gaps (Feed flow rate = 0.5 l/m, permeate flow rate = 1 l /min, permeate temperature = 20°0C, Feed TDS = 0.45 g/L).

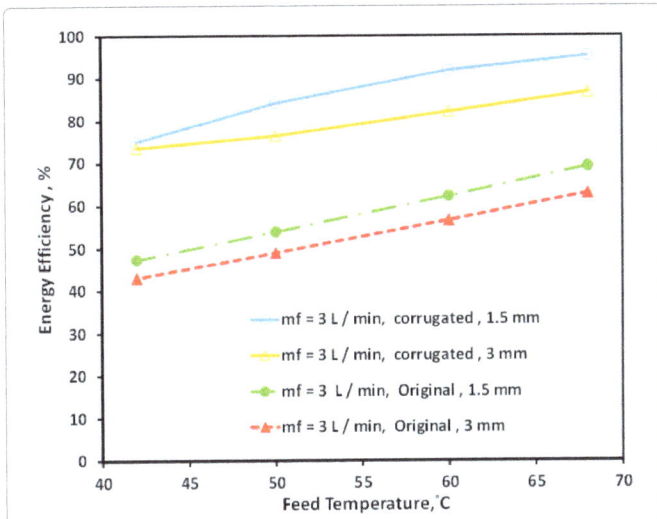

Figure 9: Energy efficiency with different feed temperature and different gaps (Feed flow rate =3 l/m, permeate flow rate = 1 l/min, permeate temperature = 20°C, Feed TDS =0.45 g/L).

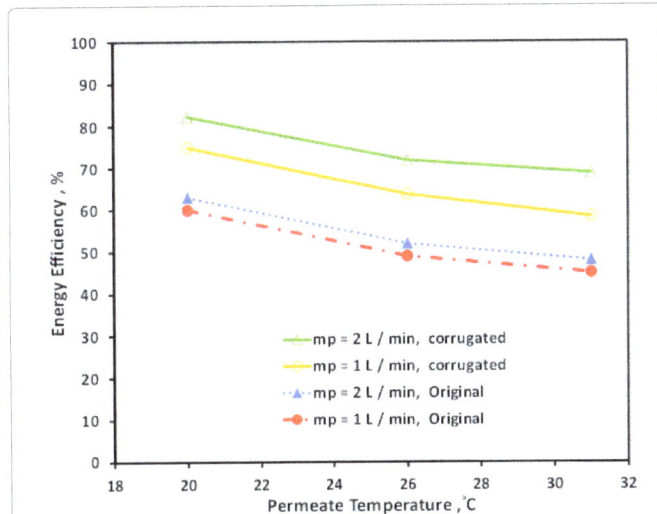

Figure 10: Effect of energy efficiency with different permeate flow rate and temperature at gap = 3 mm (mf = 3L /min, tp = 20°C, TDS = 0.45 g/L).

Figure 10 shows that the corrugated module enables to enhance the energy efficiency of DCMD by 31% higher than that of the original module. This is due to enhancing in the mass transfer as a result of created turbulence flow in the feed channel.

Figure 11 shows the permeate temperature varies from 20 to 31°C while the feed flow rate and temperature are fixed at 3 L/min and 68°C respectively. The energy efficiency increases as the permeate flow increases due to enhancement of the mass transfer. The energy efficiency enhanced as the gap height decrease; this is due to the increase of the flow velocity which decreases the thermal boundary layer. At a channel height of 1.5 mm and permeate flow at 2 l/min, the energy efficiency of the corrugated module is 33% higher than that of the original module.

Table 2 summarized the experimental results of thermal efficiency as a result of varying the feed channel (gap) height, the feed flow rate and the permeate flow rate. The percentage enhancement in thermal efficiency is calculated at average temperature. At a gap height of 3 mm,

the contribution of the corrugated feed channel is 23% enhancement in mass flux. When the gap height is reduced to 1.5 mm the combined effect reached to 26%. This indicates that the corrugated feed channel effect is dominant rather than decreasing the feed channel height. At a gap height of 1.5 mm, Table 2 also shows that the increase of the feed flow rate from 0.5 up to 3 l/min (6 times), the percentage enhancement of thermal efficiency increases from 26% to 30%. This indicated that increases the feed flow rate up 6 times has only 4% while the 26% represents the corrugated feed channel contribution. At the gap height of 1.5 mm, the increase of permeate flow from 1 up to 2 l/min, the thermal efficiency enhancement varies from 30% to reach 33% which reflects the dominated effect of the corrugated feed channel. The different between the middle case (second row in Tables 1 and 2), the permeate flow rate is 1, L/min while the last case (third row in Tables 1 and 2), the permeate flow rate is 2, L/min in order to investigate the effect of the permeate flow rate on the water flux and thermal efficiency while feed flow rate is fixed at 3 L/min. the results showed the contribution of increasing the permeate flow rate is considerably small as shown in Tables 1 and 2.

Effect of feed concentration (TDS) on the water flux

The feed concentration plays a significant role in the energy efficiency of the system. The feed concentration directly affects the viscosity and activity of fluid. It directly affects the flow and vapor pressure of the solution and increases temperature polarization [3,7]. General rule of thumb, increasing concentration, leads to a decrease in the water flux due to decrease in the vapor pressure difference. This can also be true for the energy efficiency as a result of decreasing the flux; consequently, the energy efficiency will decrease.

Figure 12 shows that as the feed concentration increases, the mass flux decreases as well. Different feed temperatures are also implemented along with different concentration are used to conclude the effect of

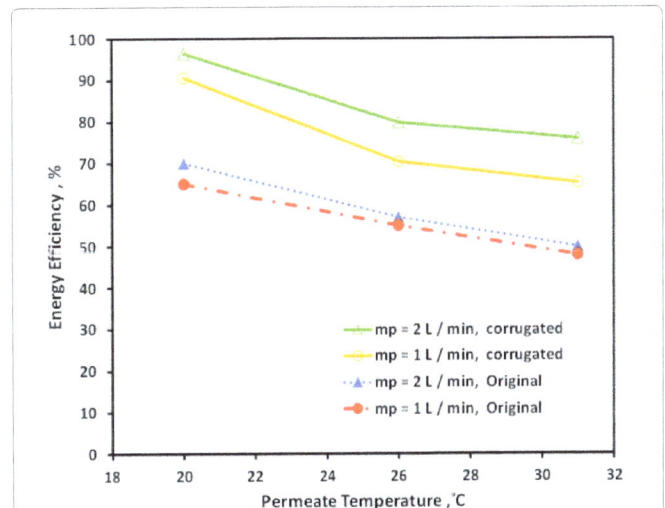

Figure 11: Effect of energy efficiency with different permeate flow rate and temperature at gap = 1.5 mm (mf = 3L /min, tp = 20°C, TDS = 0.45 g/L).

Feed flow rate, L/min	Gap height, 3 mm	Gap height, 1.5 mm	Permeate , L/min
0.5	23 %	26 %	1
3	28 %	30 %	1
3	31 %	33 %	2

Table 2: Summarized results of % enhancement of thermal efficiency.

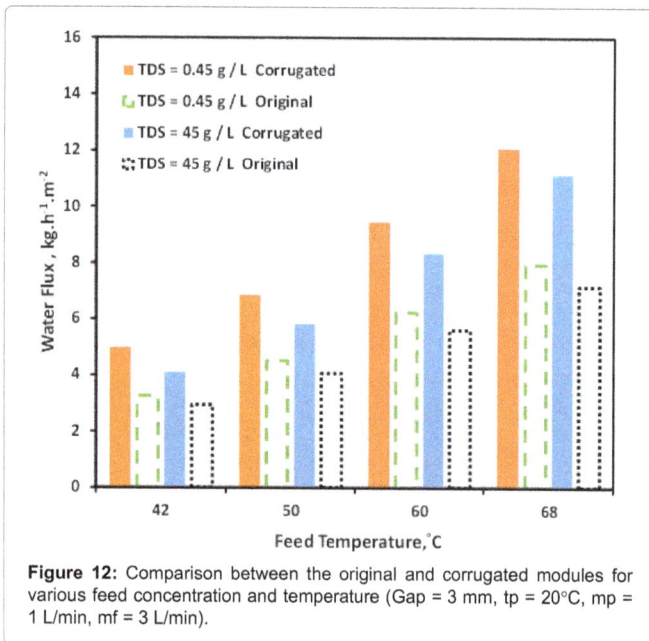

Figure 12: Comparison between the original and corrugated modules for various feed concentration and temperature (Gap = 3 mm, tp = 20°C, mp = 1 L/min, mf = 3 L/min).

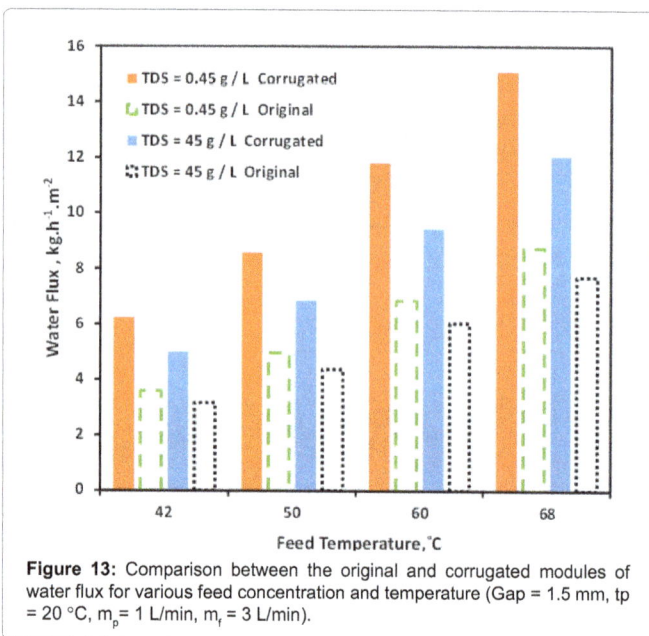

Figure 13: Comparison between the original and corrugated modules of water flux for various feed concentration and temperature (Gap = 1.5 mm, tp = 20 °C, m_p = 1 L/min, m_r = 3 L/min).

channel (gap) height, feed flow rate, temperature and feed salinity are experimentally investigated. The experiments showed that the corrugated feed channel has dominant effect of improving of the mass flux and the thermal efficiency rather than gap height and operating conditions. The experimental results showed that at the feed flow rate of 3 l/min and 68 °C while the permeate flow rate of 2 l/min and 20 °C, the contribution of the corrugated feed channel module at 1.5 mm gaps is summarized as follow:

- The average water flux is 44% higher than the original module.

- The average thermal energy efficiency is 33% higher than the original module.

References

1. Jönsson AS, Roland W, Harrysson AC (1985) Membrane distillation-A theoretical study of evaporation through microporous membranes. Desalination 56: 237-249.

2. El-Bourawi MS, Ding Z, Ma R, Khayet M (2006) A framework for better understanding membrane distillation separation process. J Memb Sci 285: 4-29

3. Khayet, Mohamed, Matsuura T (2011) Membrane distillation: principles and applications. Elsevier

4. Alklaibi AM, Lior N (2006) Heat and mass transfer resistance analysis of membrane distillation. J Memb Sci 282: 362-369.

5. Termpiyakul P, Jiraratananon R, Srisurichan S (2005) Heat and mass transfer characteristics of a direct contact membrane distillation process for desalination. Desalination 177: 133-141.

6. Tzahi YC, Adams VD, Childress AE (2004) Experimental study of desalination using direct contact membrane distillation: a new approach to flux enhancement. J Memb Sci 228: 5-16.

7. Chen, Tsung-Ching, Ho CD (2010) Immediate assisted solar direct contact membrane distillation in saline water desalination. J Memb Sci 358: 122-130.

8. Kharraz JA, Bilad MR, Arafat HA (2015) Flux stabilization in membrane distillation desalination of seawater and brine using corrugated PVDF membranes. J Memb Sci 495: 404-424.

9. Tun CM, Fane AG, Matheickal JT, Sheikholeslami R (2005) Membrane distillation crystallization of concentrated salts—flux and crystal formation. J Memb Sci 257: 144-155.

10. Qtaishat M, Khayet M, Matsuura T (2009) Novel porous composite hydrophobic/hydrophilic polysulfone membranes for desalination by direct contact membrane distillation. J Memb Sci 341: 139-148.

11. Sudoh M, Takuwa K, Iizuka H, Nagamatsuya K (1997) Effects of thermal and concentration boundary layers on vapor permeation in membrane distillation of aqueous lithium bromide solution. J Memb Sci 131: 1-7.

12. Tsung-Ching C, Ho CD (2010) Immediate assisted solar direct contact membrane distillation in saline water desalination. J Memb Sci 358: 122-130.

13. Zhang J, Dow N, Duke M, Ostarcevic E, Gray S (2010) Identification of material and physical features of membrane distillation membranes for high performance desalination. J Memb Sci 349: 295-303.

14. Martinez L, Rodríguez-Maroto (2007) Effects of membrane and module design improvements on flux in direct contact membrane distillation. Desalination 205: 97-103.

15. Winter D, Koschikowski J, Düver D, Hertel P, Beuscher U (2013) Evaluation of MD process performance: effect of backing structures and membrane properties under different operating conditions. Desalination 323: 120-133.

16. Phattaranawik J, Jiraratananon R, Fane AG, Halim C (2001) Mass flux enhancement using spacer filled channels in direct contact membrane distillation. J Memb Sci 187: 193-201.

concentration on water flux. The permeate temperature is fixed at 20°C, while the feed flow rate is 3 l/min. It is seen from Figure 12, that the average water flux of the corrugated module of DCMD is 34% higher than that of the origin module (at feed concentration of 45 g/L).

Figure 13 shows the effect of feed concentration on water flux at the gap 1.5 mm. The experimental comparison shows that at the feed concentration 45 g/l and gap of 1.5 mm, the average water flux of the feed channel corrugated DCMD module is 44% higher than that of the original module.

Conclusions

New corrugated feed channel is suggested and equipment with flat sheet direct contact membrane distillation in order to enhance the mass flux and thermal energy efficiency. The effect of the feed

23

Insight in to the Initial Stages of Silica Scaling Employing a Scanning Electron and Atomic Force Microscopy Approach

Bogdan C Donose*, Greg Birkett and Steven Pratt

The University of Queensland, School of Chemical Engineering, St Lucia, 4072, QLD, Australia

Abstract

The performance of reverse osmosis (RO) desalination can be limited by membrane scaling. Of particular concern is silica scale, which once deposited on the membrane is extremely difficult to remove. In this work, the deposition of silica-rich nanoparticles was considered. A novel *in situ* sample preparation method was developed for a microscopy investigation into the deposition and adhesion of the silica-rich nanoparticles. The method involves placing a clean silica wafer in agitated brine to collect particles to simulate initial stages of scaling. The 'scaled' surfaces were characterized by scanning electron microscopy (SEM) and atomic force microscopy (AFM). Model brines, with varying nanoparticle, cation, and organic composition and concentration were tested, as well as reject brine from a full-scale operational water treatment facility.

Microscopy revealed that silica-rich nanoparticles were deposited from all waters, with smaller nanoparticles more readily attaching to the wafer compared to larger ones. The presence of organics increased nanoparticle adhesion whereas divalent cations (Ca^{2+} and Mg^{2+}) decreased nanoparticle adhesion. These results have implications for the evaluation, selection and operation of RO pre-treatment processes and chemical dosing strategies, particularly the requirement for weak acid cation ion exchange (WAC-IX) and anti-scalant chemicals, respectively.

Keywords: Reverse osmosis; Brine; Silica; Scaling; Colloids; Electron and atomic force microscopy

Introduction

Reliable, safe and cost effective operation of reverse osmosis (RO) membrane desalination technology is vital to support development activities, urban and potable water supplies and agricultural activities in many parts of the world. In Australia, for example, the development of natural gas from coal seams offers tremendous economic opportunity, with development generating brackish 'produced water' which operators can manage using reverse osmosis (RO) membrane desalination to treat water to a quality suitable for a range of beneficial reuse such as agriculture. A waste brine stream is generated during the treatment process which requires storage and further treatment, management and subsequent disposal. A key operational objective can be to maximize the recovery ratio of the plant so to minimize the volume of brine requiring storage prior to, or during, treatment and disposal. The performance of the RO plants can be limited by silica derived membrane scale, which can be difficult to remove. Scaling is a common operational issue with RO membranes as documented in extensive academic and industry literature [1-6]. Groundwaters, including produced waters, are a complex matrix containing silica, hardness ions (e.g. calcium and magnesium), organics and other metal ions, with concentrations varying widely across regions. Our research group has recently shown that produced water has significant concentrations of silica-rich nanoparticles [7,8] and nanoparticles have been shown in the literature to have an adverse impact on the performance of membrane desalination [9-15]. A challenge in understanding the scaling mechanisms is the relatively slow kinetic of the process. Characterisation normally requires long-term experiments to make macro-scale observations. For this reason a new methodology was developed to quickly assess interactions between nanoparticles and membrane surfaces.

The interplay between silica-rich nanoparticles and the other components of produced water is not well understood. For example, experiments [10] have shown that the membrane flux decline during combined scaling and biofouling depends on solution chemistry and colloidal particle size. The combined fouling behaviour cannot be extrapolated from fouling by colloidal materials or dissolved organic matter alone. Synergistic effects were also found in fouling tests combining extracellular polymeric substances extracted from *Pseudomonas aeruginosa* PAO-1 and calcium ions [16]. These synergistic effects make selection and operation of infrastructure to mitigate scaling very challenging. For example, current literature-supported practice is to have weak acid cation exchange (WAC-IX) before the RO unit to reduce hardness in the RO feed stream. Some operations in the natural gas industry in Queensland (Australia) report that WAC IX operation has increased the frequency of RO membrane scaling events, with irreversible silica scale formation observed on RO membranes. Other operators have used IX without incident to the membranes but it is not clear if this operation is optimal.

In this work a novel *in situ* sample preparation method was developed for use with atomic force and electron microscopy to investigate the deposition and adhesion of silica-rich nanoparticles. The method simulates the initial stages of silica scale formation, whereby nanoparticles bind to silica-rich nanoparticles that are caught on membrane surfaces. The application of the method is demonstrated accompanied by a preliminary assessment of water matrix effects on particle adhesion. Synthetic produced water (with varying nanoparticle,

*Corresponding author: Bogdan C Donose, The University of Queensland, School of Chemical Engineering, StLucia, 4072, QLD, Australia
E-mail: b.donose@uq.edu.au

cation and organic content and composition) and brine from an operational reverse osmosis desalination facility are tested.

Materials and Methods

The experimental program involves depositing nanoparticles that are present in water samples on to silicon wafers (the silica trap), and then microscopy examination of the silicon wafers to assess the extent of deposition.

Silica trap

Silicon wafers (coated with a naturally grown silica layer of 2-4 nm) were diced (5×5 mm), washed in 10% H_2SO_4, rinsed in ethanol and deionised (DI) water and then cleaned (30 minutes) using a UV/Ozone unit (BioForce).

The wafers were immersed in the water samples in individual 50 ml plastic beakers for 24 hours under constant agitation (250 rpm) on an orbital shaker. After this step, the wafers were removed from the plastic beakers and immersed in DI water to rinse. Excess water was then removed from their surface using the corner of an extra low

lint Kimwipes and wafers were placed in a vacuum oven for 8 hours at 70°C. Samples were stored in a high vacuum cabinet prior to mounting and SEM/AFM analysis. All steps undertaken for sample preparation and analysis are schematically presented in Figure 1.

Water samples

Table 1 summarises the types of water that were analyzed. Brine samples from a produced water RO desalination facility were typically collected on-site for the characterization of chemical composition. Water is separated from the natural gas at the well site and sent to the water management facility. The produced water then undergoes microfiltration followed by RO desalination.

Physical and chemical characterization of silica and other potential scalants were conducted using various spectroscopic and analytical techniques. Historical water quality data obtained from the RO desalination facility was used to generate an overview for a period of seven years. The RO process operates at around 80% recovery, generating a brine stream which is about 5 times as concentrated when compared to the RO feed. The water quality data of RO feed and brine

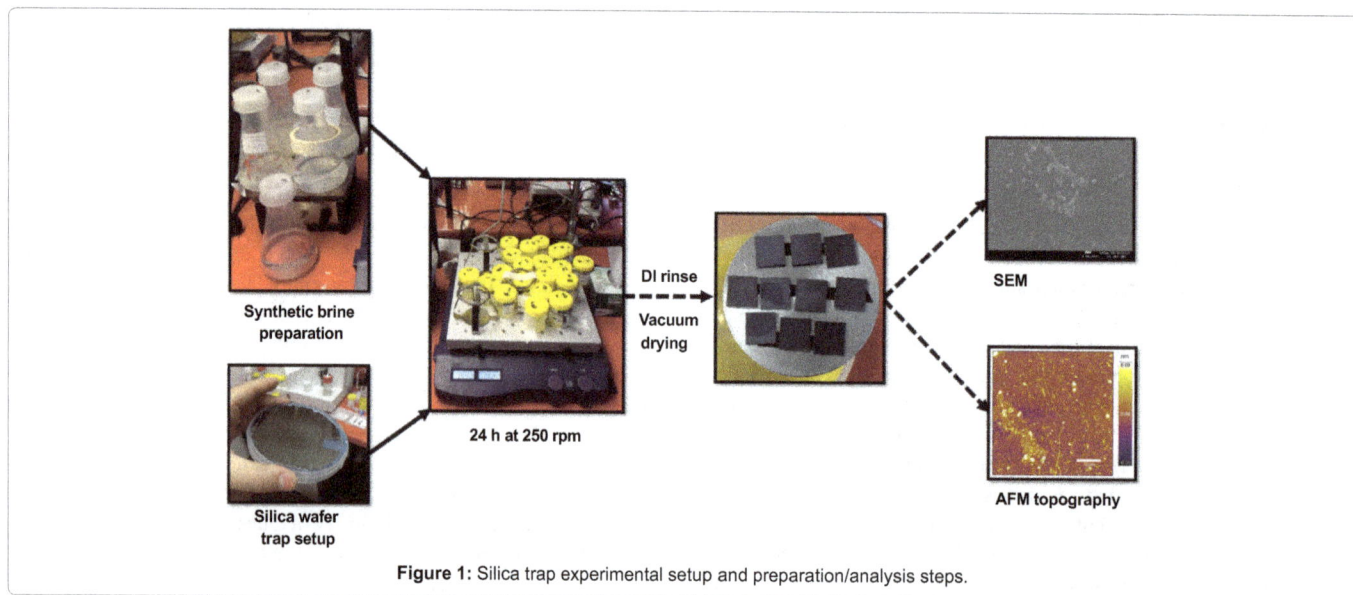

Figure 1: Silica trap experimental setup and preparation/analysis steps.

Experiment	Type of water	Characteristics
1	Model RO feed	pH, cations, organic surrogate, nanoparticles
2	Model brine	pH, cations, organic surrogate, nanoparticles
3	Real brine	Water matrix

Table 1: Waters tested employing silica trap method and their characteristics.

Measurement	Unit	RO feed		Brine		Conc. factor factor
		Average	St. dev.	Average	St. dev.	
Alkalinity (carb.)	mg/L	197.7	61.0	543.8	155.7	2.7
Calcium	mg/L	9.0	4.1	45.4	10.0	5.0
Magnesium	mg/L	2.5	1.2	13.1	2.5	5.2
pH	-	9.0	0.1	8.7	0.1	-
Potassium	mg/L	16.8	9.7	83.2	25.4	4.9
Silicon	mg/L	20.6	10.1	48.5	16.1	2.3
Sodium	mg/L	2557.8	877.4	12644.0	2062.4	4.9
TDS	mg/L	6565.5	677.5	27519.6	4281.8	4.1
TOC	mg/L	9.5	10.6	40.0	26.4	4.2

Table 2: Overview of RO feed and brine matrix profile.

is shown in Table 2. This shows that the produced water is rich in sodium and bicarbonate, with potassium, calcium, and magnesium also present. The silica concentration in the produced water feed is relatively low, but the recovery process increases the silica concentration to near saturation. Such a concentration increase can lead to a dramatic change in the silica particulate fraction: before RO approx. 7% and after RO approx. 35% [7].

Synthetic produced waters, RO feed and brine, were prepared using DI water produced by a Millipore Academic unit (resistivity lower than 18 MΩ·cm), silica nanoparticle standards (Sigma, Ludox 12 nm in diameter and Nissan chemical, Snowtex ST-20L, 60 nm in diameter). $CaCO_3$, K_2CO_3, $MgSO_4$, NaCl were of analytical grade and were used without prior purification. 1.0 M HCl and 1.0 M NaOH were used for pH adjustment.

Using the historical RO desalination facility data as a guide, two types of synthetic waters were used: brine and produced water feed. Concentration factors of 5 were imposed for all ions. Considering silicon is present as dissolved and particulate silica, a concentration factor of 25 for silicon was chosen to cover the observed ranges of particulate and dissolved Si, i.e. to account for a fivefold increase in silica particulate fraction seconded by a silicon concentration factor of five. In this way the chance to observe attached nanoparticles by microscopy techniques was increased.

Table 3 shows the program for the model RO feed and model brine experiments. The table shows the types of samples used in this study as well as the pH values of the synthetic waters in which silica wafer traps were immersed. The conditions tested were: presence or absence of silica nanoparticles or dissolved silica (leftmost columns in Table 4) and different combinations of hardness ions, organics surrogate in deionised water (top row). As silica surrogates, Ludox (diameter=12 nm), Snowtex (diameter=60 nm) and dissolved silica in HF were employed.

The real brine samples were treated in the same manner for microscopy analysis.

Scanning electron microscopy (SEM) and atomic force microscopy (AFM)

SEM was performed by employing a JEOL 7001 Field Emission Gun at 10 mm working distance, spot size 8 or 9 and 5 kV-10 kV accelerating voltage.

AFM micrographs were recorded employing an MFP-3D built on an Eclipse Ti-U Nikon inverted microscope, placed on a Herzan anti-vibration table which was housed in a TMC enclosure and a Cypher (Asylum Research/Oxford Instruments) using Multi 75 DLC (Budget Sensors, Bulgaria) probes of nominal contact radius less than 15 nm and nominal spring constant of 2 N/m. All samples were imaged at 0.7-1 Hz scan rate in AC mode in air over an area of 2×1 µm.

For both microscopy methods samples were imaged as received, post vacuum deposition (as described in above in 2.1 Silica trap), in order to prevent interference potentially introduced by a conductive layer. In this way the errors of scanning are limited only to accelerating voltage and spot size (for SEM) and probe contact radius (for AFM).

Results and Discussion

Experiment 1: Model RO feed

SEM and AFM micrographs for synthetic RO feed produced water are presented in Figures 2 and 3. Both microscopy techniques confirm that nanoparticles were trapped on the wafer. All samples exhibit adhered structures. Several features are captured by SEM at a larger scale, allowing for a general sample overview, while AFM reveals more three-dimensional details. These features include alginate networks (AFM NO-1), salt crystals (SEM NO-3) and nanocrystals (AFM LU-4), as well as nanoparticles (SEM SN-1, AFM SN-4).

As described in the experimental section, after exposure to the water sample under constant agitation, wafers are rinsed with DI water, the excess water is removed and then subjected to thermally assisted

Ion	Model RO feed (mg/L)	Model brine (mg/L)
Na⁺	2200	11400
Mg²⁺	2	10
Ca²⁺	8	40
CO₃²⁻	72	36
K⁺	39	183
SO₄²⁻	9	48
Si⁴⁺	4	100
Alginate	20	100

Table 3: Model brine and RO feed water matrices (with all ions present).

Model feed	Na mg/L	Ca mg/L	Mg mg/L	Alg mg/L	Silica nanoparticles and pH			
					NO	ST	LU	SN
1	2200	8	2	20	8.2	8.8	8.7	8.7
2	2200	8	2	-	8.6	8.9	8.9	8.5
3	2200	-	-	20	8.9	8.9	8.9	8.9
4	2200	-	-	-	8.0	8.9	9.0	8.2
5	-	-	-	-	8.7	8.7	8.8	8.9
Model brine								
1	11400	40	10	100	8.5	8.6	8.6	8.9
2	11400	40	10	-	8.5	8.6	8.5	8.7
3	11400	-	-	100	8.8	8.8	8.5	8.8
4	11400	-	-	-	8.8	8.8	8.5	8.9
5	-	-	-	-	8.7	8.7	8.9	8.7
				TOC (mg/L)	Si (mg/L)		pH	
Real feed	2557.8	9.0	2.5	9.5	20.6		9	
Real brine	12644.0	45.4	13.1	40.0	48.5		8.7	

Table 4: Sample matrix and pH for synthetic scaling and real waters. Sample codes: NO=no silica; ST=dissolved silica; LU=Ludox silica; SN=Snowtex. The numbered columns and labels show the presence/absence of major ions in solution.

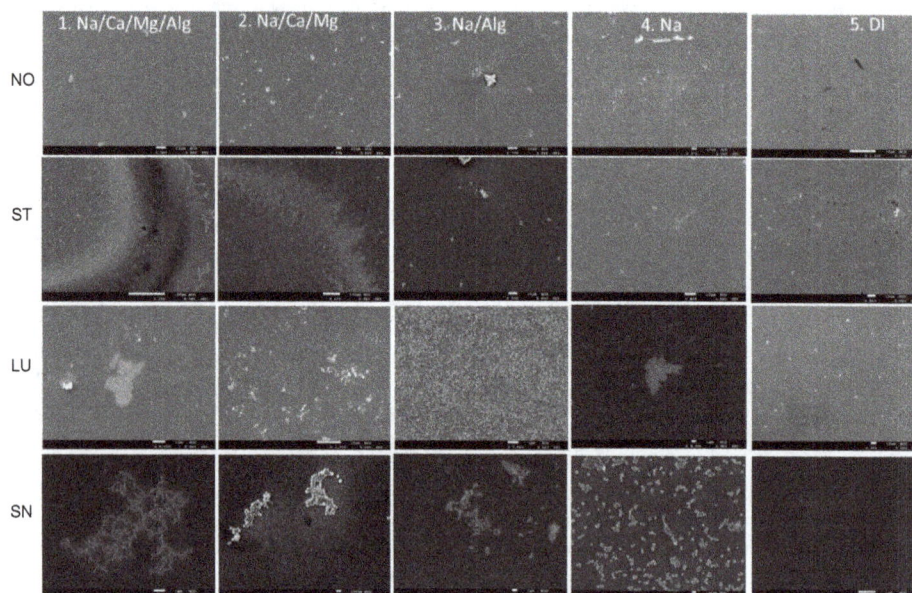

Figure 2: Representative SEM micrographs of the wafer surfaces after exposure to synthetic RO feed waters and selected features marked by red arrows (Scale bars: NO-1 to NO-5, 10 µm; ST-1 and 2, 100 µm; ST-3 to 5, 10 µm; LU-1, 3 and 4, 1 µm; LU-2 and 5, 10 µm; SN-1, 3 and SN5, 1 µm; SN-2 and 4, 100 nm.

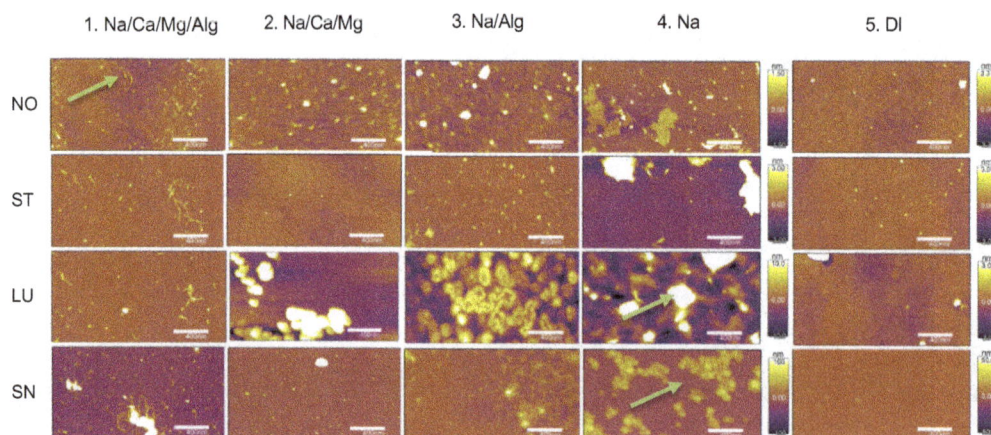

Figure 3: Representative AFM micrographs of the wafer surfaces after exposure to synthetic RO feed waters and selected features marked by green arrows (Scale bars: 400 nm): NO-1: dendritic structures; LU-4: large salt crystals; SN 4: isolated NPs clusters.

vacuum drying. This approach poses significant challenges in choosing a representative area for analysis, especially when samples exhibit salt drying patterns leading to imaging artefacts such as in SEM ST-1 and SEM ST-2 (Figure 2).

In the presence of cations and alginate the dendritic alginate structures (from AFM) and aggregates with colloids (SEM and AFM) are apparent (Figures 2 and 3: column 1. Na/Ca/Mg/Alg). When alginate is removed (column 2. Na/Ca/Mg) the dendritic structures disappear and patches of aggregates can be observed. Removing the divalent cations (column 3. Na/Alg) results in a compact multilayered mat of Ludox nanoparticles being deposited on the trap surface. It is also shown that SN nanoparticles exhibit a different behaviour, suggesting that smaller NPs are more easily attached to the surface. In the presence of just sodium (column 4. Na) different attachment behaviour between SN and LU NPs was observed, with smaller particles attaching more readily. The benchmark (column 5. DI water) confirms little or no attachment as result of electrostatic repulsion.

Experiment 2: Model brine

SEM and AFM micrographs for synthetic brine are presented in Figures 4 and 5. Similar to the case of model RO feed, both microscopy techniques confirm that the trapping approach was successful. For the model brine the adhered structures were more abundant. This was expected because concentration of ions in the brine was increased by a factor of 5 compared to the RO feed, except for silicon which was increased by a factor of 25 to reach 100 mg/L. Again, the observed features included alginate networks (AFM NO-1), salt crystals (SEM NO-3) and nanocrystals (SEM NO-4), as well as nanoparticle dense mats (SEM SN-4, AFM SN-4).

Differences in the adhesion of nanoparticles are most pronounced when comparing both SEM and AFM for the LU4 and SN4 samples. The larger Snowtex particles form layers with voids, while Ludox mats are compact and possibly multi layered. Even at high silicon

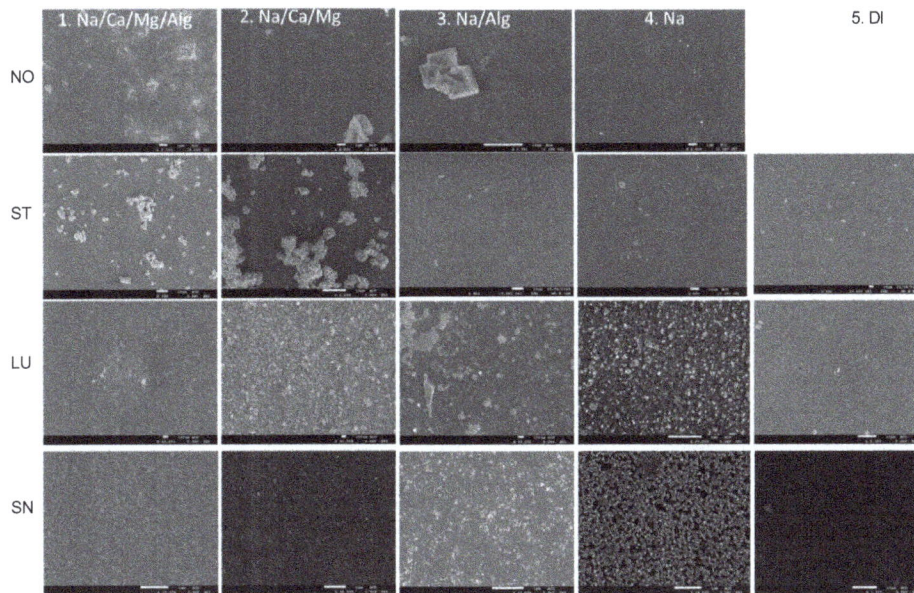

Figure 4: Representative SEM micrographs of the wafer surfaces after exposure to synthetic brine waters and selected features marked by red arrows (Scale bars: NO-1, 2 and 4, 1 µm; NO-3, 10 µm; ST-1 to 5, 10 µm; LU-1 to 3, 100 nm; LU-4, 1 µm; LU-5, 10 µm; SN-1 to 4, 1 µm; SN-5, 10 µm).

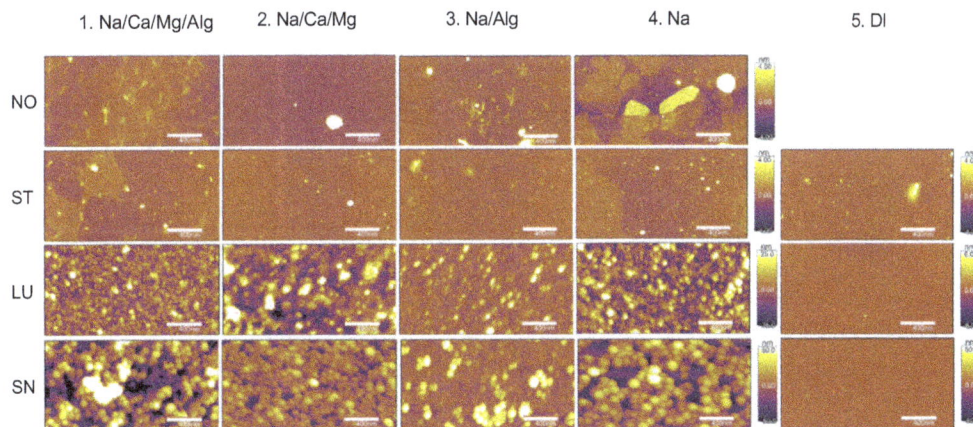

Figure 5: Representative AFM micrographs of the wafer surfaces after exposure to synthetic brine waters and selected features marked by green arrows: NO-1 dendritic structures, SN-4 voids in the Snowtox layer.

concentrations (100 mg/L) benchmark tests with DI water show again a total lack of adhesion due to electrostatic repulsion.

A summary of distinguishing features observed on all samples is presented in Table 5.

Experiment 3: Real brine

Microfiltered brine from a RO desalination facility was also considered. Data is presented in Figure 6, along with, for comparison, traps exposed to synthetic produced waters containing all ionic components and grouped as: no silica, standard dissolved silica, Ludox nanoparticles and Snowtex nanoparticles.

The figure shows that traps exposed to microfiltered brine exhibit a combination of all surface elements presented on the traps from the model waters: dendritic structures (squares), colloidal aggregates (triangles) and salt crystals (circles). The topographical differences between RO feed and brine in samples containing all the ingredients

(mono and divalent cations and organic surrogate/alginate) are highlighted.

There are three types of effects emerging from the existing data:

1. Colloids size is a determining factor in deposition and aggregation as visualised in synthetic brine, with Ludox, the smallest colloids, being easily attached and forming multilayered colloidal mats in comparison with Snowtex which attached as loose monolayers.

2. The presence of alginate (as surrogate for organic matter) in the matrix contributes to an enhancement of colloids/particulates attachment and aggregation.

3. When cations are removed from the RO feed an enhancement of silica deposition on the trap wafer is observed in model feed water.

Discussion

As previously postulated [17], the adhesion between solid particles

Model feed	1 Na/Ca/Mg/Alg	2 Na/Ca/Mg	3 Na/Alg	4 Na	5. pH adjusted DI water
NO (without silica)	Alg. network	Salt nanocrystals	Alg. networks and colloids	-	Clear surface
ST (dissolved silica)	Alg. islands	Clear surface	Alg. network and nanocrystals	-	Few colloids
LU (Ludox nanoparticles)	Alg. network	Large crystals	Colloids mat	Colloids mat	Clear surface
SN (Snowtex nanoparticles)	Colloids and alg. networks	Isolated colloids	Isolated alg. embedded colloids	Dispersed colloids	Clear surface
Model brine	1	2	3	4	5
NO (without silica)	Alg. network	-	Alg. network &colloids	Salt crystals	Clear surface
ST (dissolved silica)	Silica and alginate islands	Aggregates	Alginate network and colloids	Crystals and colloids	Colloids and aggregates
LU (Ludox nanoparticles)	Colloids dense mat	Colloids mat and aggregates	Alginate embedded colloids and islands	Colloids mat &aggregates	Clear surface
SN (Snowtex nanoparticles)	Colloids loose multilayer	Colloids loose layer and aggregates	Alginate embedded colloids	Colloids loose layer	Clear surface

Table 5: Sample matrix and distinguishing features on traps exposed to synthetic RO feed and brine.

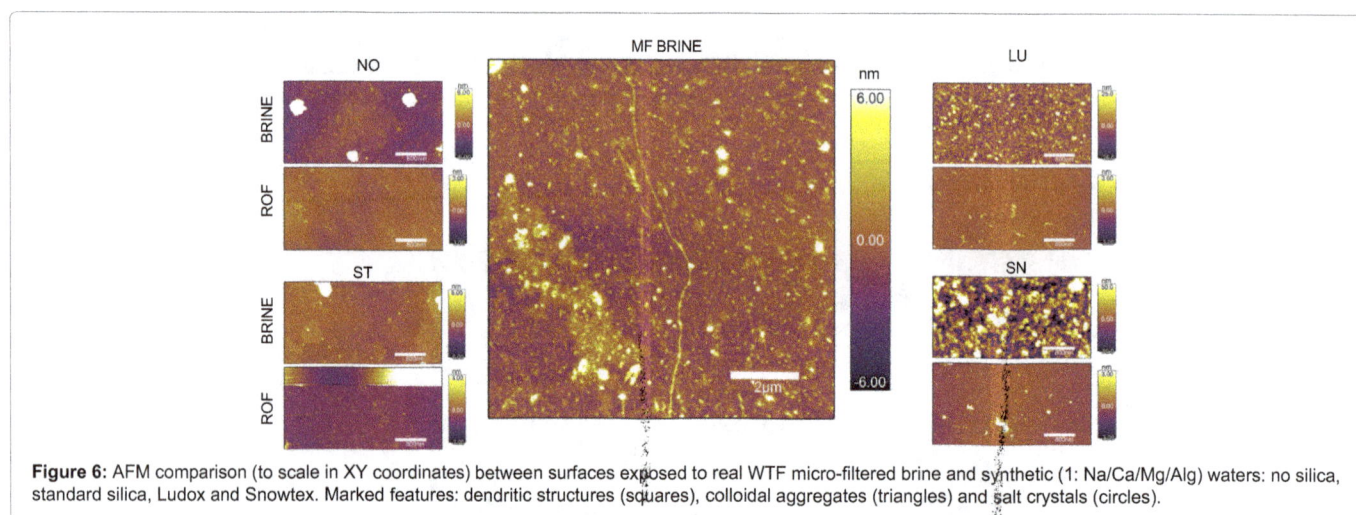

Figure 6: AFM comparison (to scale in XY coordinates) between surfaces exposed to real WTF micro-filtered brine and synthetic (1: Na/Ca/Mg/Alg) waters: no silica, standard silica, Ludox and Snowtex. Marked features: dendritic structures (squares), colloidal aggregates (triangles) and salt crystals (circles).

and solid surfaces is determined by van der Waals, electrostatic and deformational forces [18]. For particles in the nanoscale domain the long range electrostatic force becomes less important and van der Waals, structural and hydration forces dominate interaction and challenge to the limit existing analytical tools such as surface force apparatus and AFM. Over the recent decades of nanoscale force measurements, silica and mica surfaces were tested typically in electrolyte solutions of controlled pH and salt concentration to learn how van der Waals, adhesion, solvation, structural and friction forces behave [19-27]. To our knowledge, there is no comprehensive model that describes interactions in complex fluids such as brines resulting from water treatment. In an experimental system such as the one designed and analysed here, as a proxy to the brine or pre-brine medium, the roughness of the substrate was chosen on purpose to be at its minimum in order to reduce its contribution. The propensity of smaller nanoparticles to attach to the smooth silica could be attributed to the increasing role of van der Waals interactions and to the reduced hydrodynamic effect acting synergistically toward immobilising very small colloids.

The role of alginate in the enhanced colloid/particulates attachment has been previously tested [28] in high ionic strength environments to show that the polyelectrolyte nature of alginate reduces steric repulsion and favours adhesion. Similarly, its ability to enhance adhesion has

been harnessed to boost the impact and efficiency of drug delivery systems [29]. In certain experimental conditions [30,31] alginate was shown to be able to restore flux in fouled membranes, based on its ability to interact with silica.

Without having a definitive answer with regard to the effect of the absence of cations on the silica deposition, namely an increasing amount of deposited colloids, this behaviour could be linked to the ability of certain cations to reduce friction between silica surfaces [21,22]. Also, the role of pH, needs to be further assessed, especially in the vicinity of pH 9 [25] where silica has an accelerated dissolution rate and where hydration and hydrogen bonding and depolymerisation play such an important role.

Conclusions

Silica wafer samples prepared by immersion for 24 hours (under constant agitation) in different types of water are suitable to trap nanoparticles, nanocrystals and organic dendritic structures and their aggregates, which later on, can be subjected to microscopy. Considering silica-silica interactions in different media, the new method allows taking snapshots of the early stages of scaling, replacing a rough RO polyamide substrate [31] by a flat wafer (roughness less than 1 nm as measured by AFM), more suitable for SEM and AFM. The combination of two microscopies and samples' preparation enables obtaining information which otherwise would be inaccessible.

Acknowledgements

This research is supported and funded by the UQ Centre for Coal Seam Gas, which includes industry members APLNG, Arrow Energy, QGC and Santos Ltd. Deirdre Walsh, Paul Wybrew and Josh Davies are acknowledged for reviewing the manuscript. This work was performed in part at the Queensland node of the Australian National Fabrication Facility, a company established under the National Collaborative Research Infrastructure Strategy to provide nano and micro-fabrication facilities for Australia's researchers. The authors acknowledge the facilities and the scientific and technical assistance of the Australian Microscopy & Microanalysis Research Facility at the Centre for Microscopy and Microanalysis (The University of Queensland).

References

1. Antony A, Low JH, Gray S, Leslie G (2011) Scale formation and control in high pressure membrane water treatment systems: A review. J Mem Sci 383: 1-16.

2. Duong HC, Gray S, Duke M, Nghiem L (2015) Scaling control during membrane distillation of coal seam gas reverse osmosis brine. J Mem Sci 493: 673-682.

3. Sheikholeslami R, Tan S (1999) Effects of water duality on silica fouling of desalination plants. Desalination 126: 267-280.

4. Butt FH, Rahman F, Baduruthamal U (1995) Pilot plant evaluation of advanced vs conventional scale inhibitors for RO desalination. Desalination 103: 189-198.

5. Butt FH, Rahman F, Baduruthamal U (1995) Identification of Scale Deposits through Membrane Autopsy. Desalination 101: 219-230.

6. Butt FH, Rahman F, Baduruthamal U (1997) Characterization of foulants by autopsy of RO desalination membranes. Desalination 114: 51-64.

7. Zaman M, Birkett G, Pratt C, Stuart B, Pratt S (2015) Downstream processing of reverse osmosis brine: Characterisation of potential scaling compounds. Water Res 80: 227-234.

8. Zaman M, Masuduz, Greg B, Stuart B, Pratt S (2013) Silica Removal from Coal Seam Gas Brine Using Activated Alumina, in CHEMECA: Australasian Conference on Chemica Engineering pp: 700-703.

9. Fang Y, Duranceau S (2013) Study of the Effect of Nanoparticles and Surface Morphology on Reverse Osmosis and Nanofiltration Membrane Productivity. Membranes 3: 196-225.

10. Li QL, Elimelech M (2006) Synergistic effects in combined fouling of a loose nanofiltration membrane by colloidal materials and natural organic matter. J Mem Sci 278: 72-82.

11. Park JY, Lim S, Park K (2013) A new approach for determination of fouling potential by colloidal nanoparticles during reverse osmosis (RO) membrane filtration of seawater. J Nanopart Res 15: 1548.

12. Zhu X, Elimelech M (1997) Colloidal Fouling of Reverse Osmosis Membranes: Measurements and Fouling Mechanisms. Environ Sci Technol 31: 3654-3662.

13. Belfort G, Davis RH, Zydney AL (1994) The Behavior of Suspensions and Macromolecular Solutions in Cross-Flow Microfiltration. J Mem Sci 96: 1-58.

14. Faibish RS, Elimelech M, Cohen Y (1998) Effect of interparticle electrostatic double layer interactions on permeate flux decline in crossflow membrane filtration of colloidal suspensions: An experimental investigation. J Colloid Interface Sci 204: 77-86.

15. Fane AG, Fell CJD (1987) A Review of Fouling and Fouling Control in Ultrafiltration. Desalination 62: 117-136.

16. Herzberg M, Kang S, Elimelech M (2009) Role of Extracellular Polymeric Substances (EPS) in Biofouling of Reverse Osmosis Membranes. Environ Sci Technol 43: 4393-4398.

17. Chow TS (2003) Size-dependent adhesion of nanoparticles on rough substrates. J Phys Condens Matter 15: L83-L87.

18. Israelachvili JN (2011) Adhesion and Wetting Phenomena, in Intermolecular and Surface Forces. Third Edition, Academic Press: San Diego pp: 415-467.

19. Vakarelski IU, Ishimura K, Higashitani K (2000) Adhesion between Silica Particle and Mica Surfaces in Water and Electrolyte Solutions. J Colloid Interface Sci 227: 111-118.

20. Donose BC, Taran E, Vakarelski IU, Shinto H, Higashitani K (2006) Effects of cleaning procedures of silica wafers on their friction characteristics. J Colloid Interface Sci 299: 233-237.

21. Donose BC, Vakarelski IU, Higashitani K (2005) Silica surfaces lubrication by hydrated cations adsorption from electrolyte solutions. Langmuir 21: 1834-1839.

22. Donose BC, Vakarelski IU, Taran E, Shinto H, Higashitani K (2006) Specific effects of divalent cation nitrates on the nanotribology of silica surfaces. Ind Eng Chem Res 45: 7035-7041.

23. Ducker WA, Senden TJ, Pashley RM (1991) Direct Measurement of Colloidal Forces Using an Atomic Force Microscope. Nature 353: 239-241.

24. Hampton MA, Donose BC, Taran E, Nguyen AV (2009) Effect of nanobubbles on friction forces between hydrophobic surfaces in water. J Colloid Interface Sci 329: 202-207.

25. Taran E, Donose BC, Vakarelski IU, Higashitani K (2006) pH dependence of friction forces between silica surfaces in solutions. J Colloid Interface Sci 297: 199-203.

26. Biggs S, Cain R, Page NW (2000) Lateral Force Microscopy Study of the Friction between Silica Surfaces. J Colloid Interface Sci 232: 133-140.

27. Raviv U, Klein J (2002) Fluidity of Bound Hydration Layers. Science 297: 1540-1543.

28. De Kerchove AJ, Elimelech M (2007) Impact of Alginate Conditioning Film on Deposition Kinetics of Motile and Nonmotile Pseudomonas aeruginosa Strains. Appl Environ Microbiol 73: 5227-5234.

29. Hu L, Sun C, Song A, Zheng X, Chang D, et al. (2014) Alginate encapsulated mesoporous silica nanospheres as a sustained drug delivery system for the poorly water-soluble drug indomethacin. Asian J Pharm Sci 9: 183-190.

30. Higgin R, Howe KJ, Mayer TM (2010) Synergistic behavior between silica and alginate: Novel approach for removing silica scale from RO membranes. Desalination 250: 76-81.

31. Donose BC, Subhash S, Marc P, Yvan P, Jurg K, et al. (2013) Effect of pH on the ageing of reverse osmosis membranes upon exposure to hypochlorite. Desalination 309: 97-105.

24

Heavy Metal Separation from Industrial Effluent and Synthetic Mixtures using Newly Synthesized Composite Material

Mir Shabeer Ahmad[1], Ayaz Mahmood Dar[1], Shafia Mir[1], Shahnawaz Ahmad Mir[2] and Noor Mohd Bhat[2]*

[1]*Department of Chemistry, Government Degree College, Kulgam, J&K, India*
[2]*Department of Chemistry, Bundelkhand University, Jhansi, UP, India*

Abstract

A stable composite cation exchange adsorbent for the treatment of heavy metal ions has been synthesized by sol-gel method and characterized by FTIR, XRD, SEM, TGA and TEM analysis. Ion-exchange capacity, pH titration, elution behavior and distribution studies have been also carried out to determine the primary ion-exchange characteristics of the composite. The material shows exchange capacity of 1.49 meq g^{-1} (for Na$^+$). The composite material exhibits improved ion-exchange capacity, chemical and thermal stability. It can be used up to 300°C with 87.5% retention of initial ion-exchange capacity. The pH titration data reveals bifunctional behaviour of composite. On the basis of distribution coefficient (Kd), the material has been selective for Cd(II), Ba(II), Hg(II) and Pb(II) ions. A number of important and analytically difficult quantitative separations of metal ions have been achieved using columns packed with this exchanger. The composite cation exchanger has been applied for the treatment of sewage water and synthetic mixture successfully.

Keywords: Nano-composite; Ion exchange; Selectivity; Separation

Introduction

Any undesirable change in the water when happens, we call it a water pollution. There are so many causes of this water pollution. Presence of toxic metal ions like lead, mercury, cadmium, etc. is one of them, which has become a major environmental problem that is gradually affecting our ecosystem adversely and getting accumulated in living organisms, due to this human beings are suffering from large number of diseases. The toxic heavy metals are released from anthropogenic activities such as metallurgical, galvanizing, metal finishing, electroplating, mining, power regeneration, electronic devices manufacturing and tannery industries [1]. These metals, when present in the water more than the permissible limit, become unsafe to the health. So, it is indispensable to remove these heavy metals from the water before they are supplied for potable and others useful purposes.

The number of procedures have been introduced and developed for the removal of these heavy metals e.g. solvent extraction, adsorption, preconcentration, reverse osmosis and ion exchanger [2]. Ion exchangers have several advantages over other methods because it is relatively clean, energy efficient and show selectivity for certain ions even in the solution of low concentration of the target ion. Furthermore, it has better exchange capacity, high removal efficiency, fast kinetics [3,4] and can also be utilized in metal recovery process, which are of economical importance [5]. A number of organic and inorganic ion-exchangers have been developed but there are certain limitations existing with them. The main drawbacks associated with organic ion-exchangers are of poor mechanical strength and low thermal stability under high radiation conditions while inorganic ion-exchangers are not suitable for column applications. Hence, in order to overcome all the above limitations, composite ion-exchangers have been synthesized by incorporating organic polymer with the matrix of inorganic precipitate. The composite materials show better exchange capacity, granulometric properties, reproducibility, and chemical stability along with thermal stability and also possess better selectivity for heavy metals compared to pure inorganic and organic materials. These composite ion-exchange materials can also be used as a catalyst [6,7] ion-exchanger [8-10], ion selective electrode [11] and also finds a large number of applications in pollution control and water treatment

[12]. In continuation of the previous work [13], we herein describe synthesis, characterization and analytical applications of polyaniline Zr(IV) molybdophosphate composite

Experimental

Materials and methods

The reagents used for the synthesis of composite were Zirconium oxychloride CDH (India), sodium molybdate CDH (India), orthophosphoric acid, N-methylaniline and ammonium persulfate from Merck (India). All other reagents and chemicals were of analytical grade. A digital pH meter of Elico (EL-10, India) was used in pH measurements. Fourier Transform-IR Spectrophotometer (Perkin Elmer (1730, USA) was used to record FTIR spectra using KBr disc method. Thermal Analysis (TGA/DTA) was carried out by (DTG-60 H; C305743 00134 Schimadzu, Japan) at a rate of 10°C min^{-1} in nitrogen atmosphere. An X' Pert PRO analytical Diffractometer (PW-3040/60 Netherlands, Holland with Cu-Kα radiation λ=1.5418A°) was used for X-ray diffraction (XRD) measurement. The morphology of composite material was characterized by scanning electron microscopy, SEM (SEM; LEO, 435 VF). Transmission Electron Microscopy (TEM) analysis was carried out by Jeol H-7500 Microscope. Atomic Force Microscopy (AFM, Veeco; Digital Instruments; Innova) was used with a typical resonance frequency of ca 300 Hz. The UV-Visible spectrophotometric experiments were carried out using a Shimadzu UV-1601 spectrophotometer. To determine the ion-exchange capacity, 1.0 g dry cation exchanger in H$^+$ form was packed in a column (1.0 cm

*Corresponding author: Noor Mohd Bhat, Department of Chemistry, Bundelkhand University, Jhansi, UP, India, E-mail: noormohd830@gmail.com internal diameter) fitted with glass wool at the bottom. The nitrates of metal ions were used as eluent to elute the H$^+$ ions completely from the cation exchanger column. The effluent was titrated against a standard solution of 0.1 M NaOH.

Preparation of reagents

Solutions of Zirconium oxychloride (0.25 M), orthophosphoric acid (0.25 M), sodium molybdate (0.25 M) solutions were all prepared in demineralised water (DMW) while a 10% solution (v/v) of aniline and 0.1 M potassium persulphate were prepared in a 1 M HCl solution.

Preparation of polyaniline Zr(IV) molybdophosphate

Poly N-methylaniline gel was synthesized using the same method [14]. The inorganic precipitate of Zr(IV) molybdophosphate was prepared by mixing 0.25 M solutions of orthophosphoric acid, sodium molybdate, and Zr(IV) oxychloride steadily with continuous stirring at 25°C for 1 h whereby a yellow gel type slurry was obtained. The pH of the solution was maintained by adding a dilute solution of HCl or HNO$_3$. The resulting yellow precipitate formed was kept overnight in the mother liquor for digestion. The poly N-methylaniline Zr(IV) molybdophosphate composite material was prepared by mixing of inorganic precipitate and polyaniline gel (in 1:1 volume ratio) with continuous stirring for 1 h at 25°C. The resultant d ark green gel obtained was kept for 24 h at room temperature for digestion. The supernatant liquid was decanted, and the gel was filtered under suction. The excess acid was removed by washing with DMW, and the material was dried in an oven at 50°C. The dried material was grounded into small granules, sieved and converted into H+ form by treating with 1.0 M nitric acid solution for 24 h with occasional intermittent shaking and replacing the supernatant liquid with fresh acid. The excess acid was removed after several washings with DMW and finally dried in an oven at 50°C. By applying the above-mentioned chemical route, a number of samples of poly N-methylaniline Zr(IV) molybdophosphate composite

were synthesized under different conditions of mixing volume ratios of reactants. On the basis of good yield, highest ion exchange capacity (for K+ ions), sample A-9 was selected for detail studies. The proposed structure of poly N-methylaniline Zr(IV) molybdophosphate is shown in Figure 1.

Ion exchange capacity

To determine ion exchange capacity, one gram of the exchanger (H$^+$ form) was taken into a glass column (0.5 cm, internal diameter) plugged with glass wool at the bottom. The length of bed was approximately 1.5 cm in height. Alkali and alkaline earth metal nitrates (0.1 M) were used to elute H$^+$ ions from the cation-exchanger. The flow rate of column is maintained at 1.0 mL min^{-1}. The collected effluent was titrated against standard solution of 0.1 M NaOH.

pH titration

In order to determine the nature of the ionogenic group, pH titrations were performed in various alkali and alkaline earth metal chlorides and their corresponding hydroxides using Topp and Pepper method [15]. In this method 0.5 g of the exchanger (H$^+$ form) is taken in each of several 250 mL conical flasks which were followed by the addition of equimolar solution of 0.1 M solutions of alkali or alkaline earth metal chlorides and their corresponding hydroxides in different volume ratios. The final volume is adjusted to 50 mL to maintain the ionic strength.

Effect of eluent concentration

To find out the optimum concentration of the eluent for complete elution of H$^+$ ions, a fixed volume (250 mL) of NaNO$_3$ solution of different concentrations were passed through the columns, containing 1.0 g of the exchanger (H$^+$ form) with a flow rate of 0.5 mL min^{-1}. The effluents were titrated with a standard solution of 0.1 M NaOH to find the H$^+$ ions eluted out.

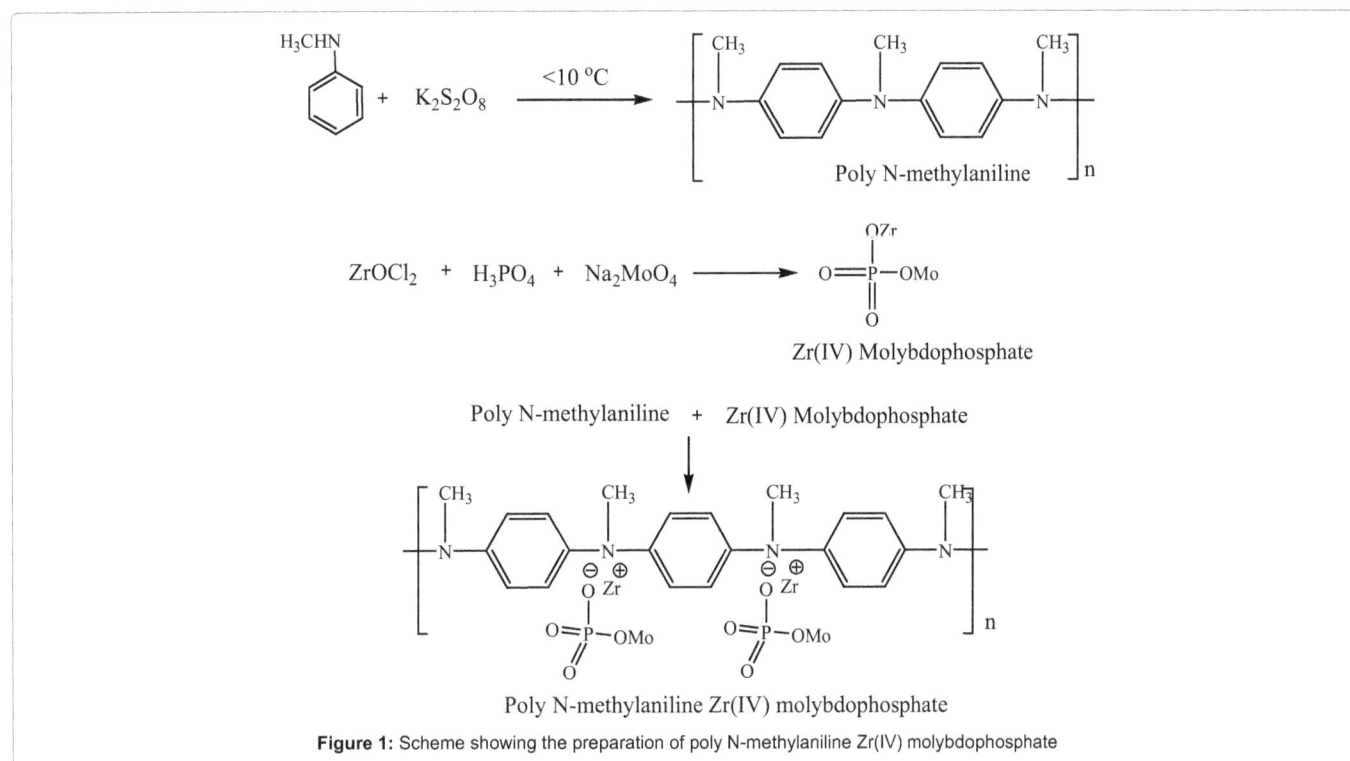

Figure 1: Scheme showing the preparation of poly N-methylaniline Zr(IV) molybdophosphate

Elution behavior

A column containing 0.5 g of exchanger (H+ form) was eluted with 1.0 M sodium nitrate solution. The effluent was collected in 10.0 mL fractions. Each fraction of 10.0 mL was titrated against a standard solution of sodium hydroxide.

Selectivity (sorption) studies

The distribution coefficient (Kd) of metal ions was determined by batch method in different solvents of analytical interest. Distribution coefficient is actually used to access the overall ability of the material to remove the ions of interest. The various portions of (300 mg each) of poly N-methylaniline Zr(IV) molybdophosphate (H+ form) were taken in Erlenmeyer flasks and titrated with 30 mL of different metal nitrate solution in the required medium and subsequently shaken for 6 h in temperature controlled shaker at 25°C to attain the equilibrium. The concentration of metal ion before and after the equilibrium was determined by titration against a standard solution of 0.01 M di-sodium salt of EDTA. The distribution coefficients (Kd) were calculated using the equation:

$$\frac{\text{milli equivalent of metal ions / gm of ion}-\text{exchanger}}{\text{milli equivalent of metal ions / mL of solution}\left(mLg^{-1}\right)}$$

$$K_d = I - F/F \times V/M \text{ mL g}^{-1}$$

where I is volume of EDTA used for metal ion solution without treatment with exchanger. F is the volume of EDTA consumed by metal ion left in solution phase after treatment. The sorption of metal ions involves the ion exchange of the H+ ions in exchanger phase with that of metal ions in solution phase

For example:

$$2R-H^+ + M2^+ \rightleftharpoons R2-M + 2H^+$$

Exchangephase Solutionphase   Exchangephase Solutionphase

where R=Poly N-methylaniline Zr(IV) molybdophosphate

Quantitative separations of metal ions in synthetic binary mixtures

Quantitative separations of some selective metal ions were achieved on columns of poly N-methylaniline Zr(IV) molybdophosphate. 1 g of exchanger (H+ form) was packed in a glass column (0.5 cm, internal diameter) with a glass wool support at the bottom. The column was washed thoroughly with demineralised water and the mixture of two metal ions (each with initial concentration 0.1M) was loaded on it and allowed to pass through the column at a flow rate of 1 mL min-1 till the solution level was just above the surface of the composite material. The process was repeated twice or thrice to ensure the complete absorption of metal ions on the bed. The separation of metal ions is achieved by collecting the effluent in 10 mL fractions and titrating against the standard solution of 0.01M di-sodium salt of EDTA.

Selective separation of metal ion from a synthetic mixture

Selective separation of Cd^{2+} and Hg^{2+} from the synthetic mixtures containing (Fe^{3+}, Cu^{2+}, Al^{3+}, Sr^{2+}, Ca^{2+}, Zn^{2+}, Ba^{2+}) and (Ca^{2+}, Sr^{2+}, Fe^{3+}, Cu^{2+}, Cd^{2+}, Hg^{2+}) was achieved on poly N-methylaniline Zr(IV) molybdophosphate columns. The amount of the Cd^{2+} and Hg^{2+} ions in the synthetic mixture was varied keeping amount of the other metal ions constant.

Results and Discussion

The number of samples of poly N-methylaniline Zr(IV) molybdophosphate were prepared by sol-gel method for the development of composite ion exchanger (Table 1). The ion-exchange capacity of the synthesized material depends upon the pH and mixing ratio of the reagents. It is clear from Table 1 that with increasing pH (0.6-2.0) of the reaction mixture, ion-exchange capacity of composite material decreases (because at higher pH, the hydrolysis of composite material occurs). In order to screen the working ability of the composite exchanger, ion-exchange capacity for some monovalent and divalent cations were determined (Table 2) which follows the sequence $Li^+ < Na^+ < K^+$ and $Mg^{2+} < Ca^{2+} < Sr^{2+} < Ba^{2+}$, respectively. The ion exchange capacity for alkali and alkaline earth metal ions increases with decrease in hydrated ionic radii [16].

Hence, it may be suggested that the ions with smaller hydrated radii enter the pores of the exchanger more easily, resulting in higher sorption [17]. The interesting feature shown by the material is that ion-exchange capacity for alkaline earth metal ions was found to be greater than alkali metal ions.

It was observed from Figure 2 that for complete elution of H+ ions (from 1.0 g of exchanger), the optimum concentration of the eluent was found to be 1.0 M. It was also observed (Figure 3) that the rate of exchange (H+ ions) is quite fast in the beginning and only 120 mL of $NaNO_3$ solution (1.0 M) is sufficient for complete elution of H+ ions from the column containing 1.0 g exchanger. From this experiment it is found that the efficiency of the column is quite satisfactory.

The pH titration curves for alkali and alkaline earth metals show two inflection points which indicate bifunctional strong cation exchange behavior (Figures 4 and 5). Initially alkali and alkaline earth metal chlorides are added in the absence of base. There is a sharp decrease in pH due to the release of hydrogen ions.

S. no	A mol L-1	B mol L-1	C % v/v	Mixing ratio v/v/v	Temp.	pH	Appearance of bead	Exchange capacity	Yield (g)
A-1	0.1	0.1	–	01:01	25 ± 2 °C	0.6	Yellow	0.5	1.82
A-2	0.1	0.1	–	01:02	25 ± 2°C	1	Yellow	0.85	2.35
A-3	0.1	0.1	–	02:01	25 ± 2°C	1.4	Yellow	0.72	1.88
A-4	0.1	0.1	–	01:01	25 ± 2°C	1.7	Yellow	0.71	2.08
A-5	0.1	0.1	20	01:01:01	25 ± 2°C	0.6	Green	0.00	0.00
A-6	0.1	0.1	20	01:01:01	25 ± 2°C	1	Green	1.63	2.71
A-7	0.1	0.1	20	01:01:01	25 ± 2°C	1.4	Green	1.37	2.93
A-8	0.1	0.1	20	02:01:01	25 ± 2°C	1.7	Green	1.31	3.42
A-9	0.1	0.1	20	01:02:01	25 ± 2°C	2	Green	1.51	3.64

A. Zr(IV) oxychloride, B. Sodium molybdate, C. Stock solution of 20% polyN-methylaniline

Table 1: Conditions for the synthesis of poly N-methylaniline Zr(IV) molybdophosphate cation exchanger.

Exchanging ions	Ionic radii (A°)	Hydrated Ionic radii (A°)	Ion exchange capacity (meq g-1)
Li+	0.68	3.4	1.12
Na+	0.97	2.76	1.49
K+	1.33	2.32	1.35
Mg2+	0.78	7	1.06
Ca2+	1.43	5.9	0.91
Sr2+	1.27	6.3	0.98
Ba2+	1.43	5.9	1.1

Table 2: Ion-exchange capacity of various exchanging ions on poly N-methylaniline Zr(IV) molybdophosphate cation exchanger.

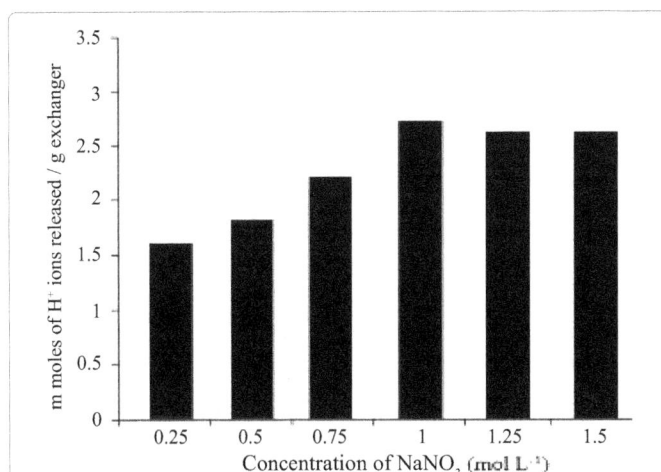

Figure 2: Effect of eluent concentration on ion-exchange capacity of poly N-methylaniline Zr(IV) molybdophosphate.

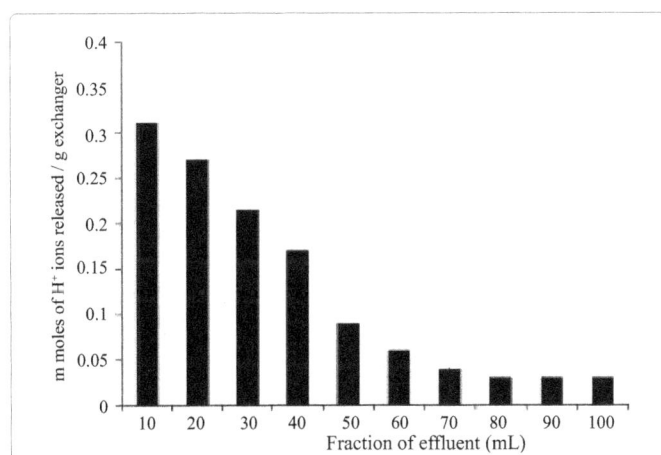

Figure 3: The elution behaviour of poly N-methylaniline Zr(IV) molybdophosphate nano-composite.

The effect of heating on the composite material at different temperatures for 1 h (Table 3) indicates that the ion-exchange capacity and physical appearance of the exchanger changed as temperature increased. The material appears to be thermally stable up to 300°C as it retains significant ion-exchange capacity. From this observation it is clear that the composite cation -exchanger possessed potential thermal stability and ion-exchange capacity.

TGA and DTG curves (Figure 6) of the nano-composite show a weight loss of mass (about 11%) continuous up to 150°C, which is due

to the removal of water molecules [18]. The weight loss in the region of 200-300°C accounts for the loss of interstitial water molecules due to the condensation of -OH groups. Further, weight loss from 320°C to 500°C may be due to complete decomposition of the organic part of the exchanger. A gradual decrease in weight loss beyond 500°C may be due to the formation of metal oxide. The main purpose of the TGA experiment is to study the thermal degradation and stability of composite material.

The Figure 7 depicts the FTIR spectrum of poly N-methylaniline Zr(IV) molybdo phosphate composite which indicates the characteristic absorption peaks shown by the composite material. The three composites show broad absorption peaks at ν 3450-3467 cm-1 for (OH) and ν 32450-3390 cm-1 for (NH). The strong absorption at ν

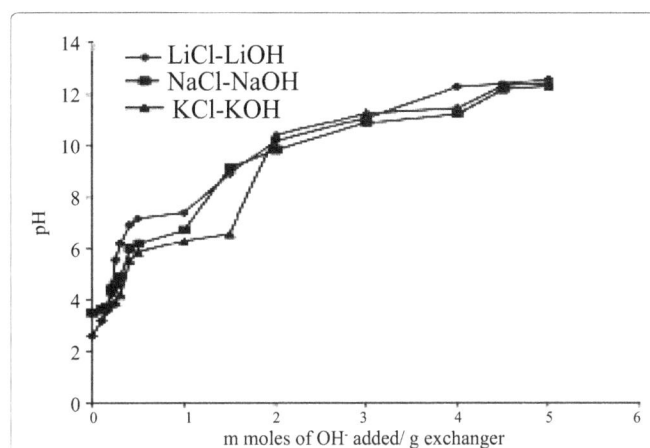

Figure 4: pH-titration curves of poly N-methylaniline Zr(IV) molybdophosphate with various alkali metal hydroxides.

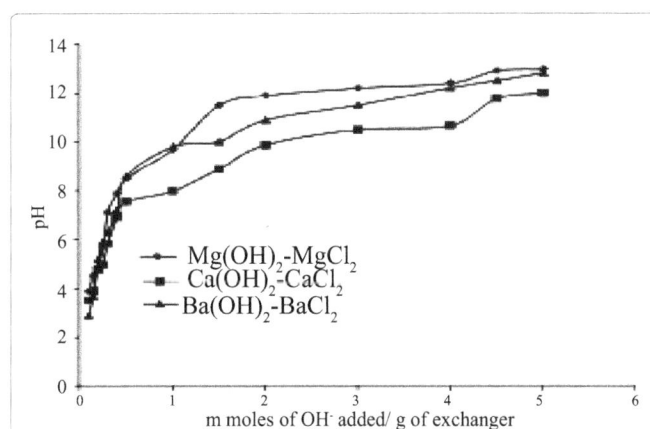

Figure 5: pH-titration curves of poly N-methylaniline Zr(IV) molybdophosphate with various alkaline earth metal hydroxides.

Temperature (°C)	Color	IEC (meq g-1) for Na+	% retention of IEC
50	Green	1.49	100
100	Green	1.38	94.6
200	Dark Green	1.33	90.3
300	Black	1.28	87.5
400	Black	1.14	78.7
500	Muddy color	0.91	69.4

Table 3: Effect of temperature on the ion-exchange capacity of poly N-methylaniline Zr(IV) molybdophosphate cation exchanger on heating time for 1 h.

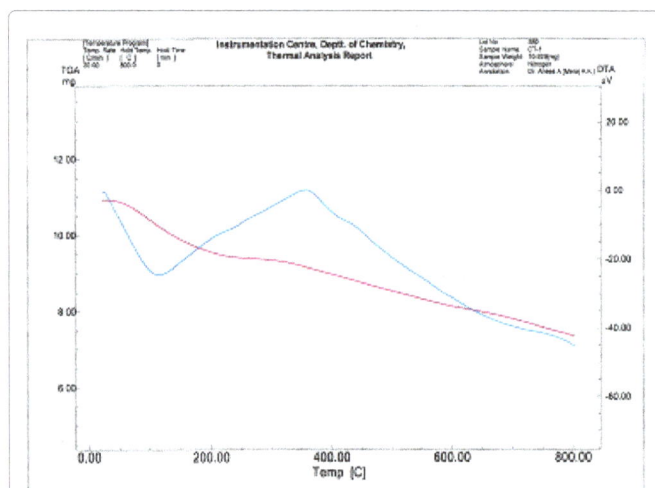

Figure 6: TGA curve of poly N-methylaniline Zr(IV) molybdophosphate cation exchanger.

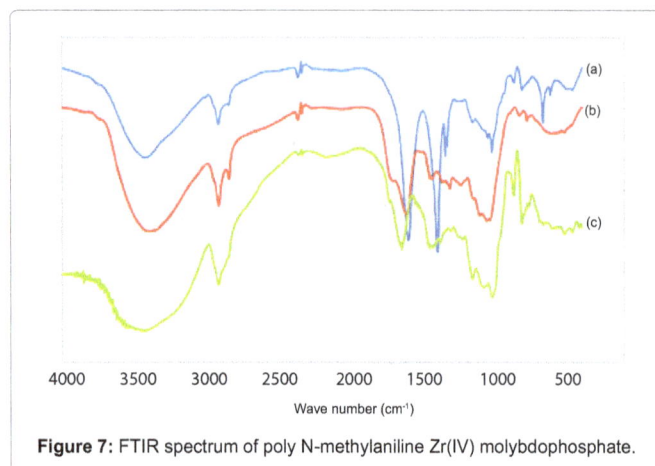

Figure 7: FTIR spectrum of poly N-methylaniline Zr(IV) molybdophosphate.

1590-1641 cm^{-1} (C=C), 3050-3100 (Aromatic), 1220-1245 (P=O) and the weak absorption peaks at ν 530 cm^{-1} and 460 cm^{-1} corresponds to (Zr-N) and (Zr-O).

A typical powder X-ray diffraction (XRD) pattern of nano-composite at room temperature is shown in Figure 8. The analysis shows that the nano-composite is formed in single phase with tetragonal crystal symmetry. All the peaks are very well matched with the crystal structures and no indication of a secondary phase is found. The lattice parameters calculated from the XRD pattern are given in Table 4, which are quite close to the values reported in the literature [19-21]. The maximum deviation that occurred between the observed and calculated values of interplanar spacing (d) remains below 0.0011 Å. In order to calculate the particle size (D) of the nano composite, Scherer's formula was used which is as follows:

$$D = \frac{0.9\lambda}{B\cos\theta} \tag{1}$$

Where λ is the X-ray wavelength (1.5418 Å), B is the full width at the half-maximum (FWHM) of the most intense peak and θ is the diffraction angle.

The particle size of the nano composite was found to be ~19.2 nm, estimated from the line width of the (101) XRD peak. Well-crystallized

diffraction peaks were observed, from which the calculated d-values are in good agreement with those given in the standard data (JCPDS, 36-1451) for nano composite (Table 4). This suggests that nano-composite has crystallized in a tetragonal symmetry. Well-crystallized diffraction peaks were also observed for the nano-composite from which cell parameters, lattice type and d-values were calculated (Table 4) which were also in good agreement with those given in the standard data (JCPDS, 36-1451). The reason for uniform values of particle size of nano-composite is the preparation mode.

The surface morphology of composite A1-A4 was investigated by using SEM micrographs. The SEM images for the composite A1-A4 is with a width of 14 mm and magnification of 1000X (Figure 9).

The signals that derive from SEM depict information about the external morphology (texture), crystalline structure, homogeneity, thickness and orientation of materials making up the nano-composites. The SEM study of nano composites A1-A4 confirm the semi-crystalline

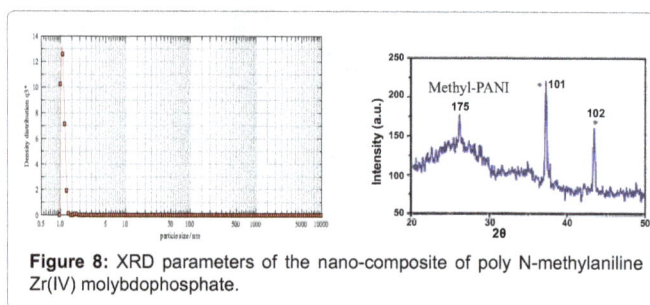

Figure 8: XRD parameters of the nano-composite of poly N-methylaniline Zr(IV) molybdophosphate.

Crystal system Method	Tetragonal Powder
Cell parameters	a=6.8841 b=6.8841 c=13.421 α=90.000 β=90.000 γ=90.000
2θ min-max	19.573-73.298
Particle size	19.2 nm
Lattice Type	p

Table 4: The XRD parameters of the nano-composite prepared in this study.

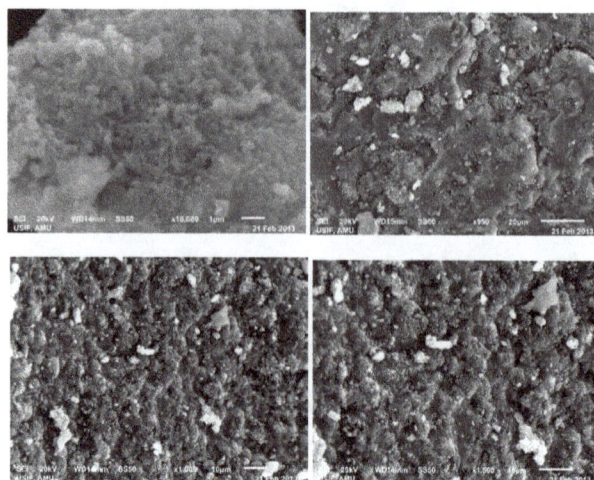

Figure 9: SEM of poly N-methylaniline Zr(IV) molybdophosphate at 1000X and 14 nm width.

nature of the material and it was found that the morphology of the exchanger gets changed after it binds with an inorganic part [Zr(IV) molybdophosphate] with the matrix of poly N-aniline as shown in Figure 9. As evident from TEM (Figure 10) the particle size in the range of 19.2 nm and with the average particle size of 19.6 nm, 19.2 nm and 19.4 nm.

In order to explore the metal-ion separation potential of the composite cation exchange material, distribution coefficient values (Kd) for different metal ions were checked in different solvent systems (Table 5). It has been observed that on decreasing the concentration of different solvents, the uptake of metal ions was sharply increased (Table 5). The data obtained in Table 5 indicate that metal ions such as Cd(II), Ba(II), Pb(II) and Hg(II) ions showed high uptake (high values of distribution coefficient).

On the basis of higher Kd values, the material was found to be selective for Cd(II), Ba(II), Hg(II), and Pb(II) ions. Some analytically important separations (e.g. $Zn^{2+}-Pb^{2+}$, $Ca^{2+}-Hg^{2+}$, $Fe^{3+}-Hg^{2+}$, $Zn^{2+}-Ba^{2+}$, $Ni^{2+}-Ba^{2+}$, $Fe^{3+}-Cd^{2+}$ and $Ca^{2+}-Pb^{2+}$) of metal ions were carried on the columns of poly N-methylaniline Zr(IV) molybdophosphate composite exchanger (Table 6). The possible reason for 100% recovery of Fe^{3+} may be due to the weak attachment or adsorption of metal ions on the adsorbent.

The Tables 7 and 8 show results of selective separation of Cd^{2+} and Hg^{2+} from the synthetic mixtures of other metals. It is clear that the separations are quite sharp and recovery is quantitative and reproducible. The practical utility of the composite material was demonstrated by separating Cd(II) and Hg(II) from synthetic mixtures.

Figure 10: TEM image of poly N-methylaniline Zr(IV) molybdophosphate nano composite.

Metal ion	DMSO 0.4M	DMSO 0.2M	DMSO 0.1M	HNO₃ 0.4M	HNO₃ 0.2M	HNO₃ 0.1M	AcOH 0.4M	AcOH 0.2M	AcOH 0.1M
Ba^{2+}	1107	1233	1390	1703	1922	2093	1911	2073	2153
Bi^{3+}	481	531	605	108	221	387	161	302	418
Mg^{2+}	133	177	222	231	246	381	200	300	389
Zn^{2+}	187	300	371	172	208	243	210	236	254
Ca^{2+}	123	139	153	105	120	138	133	182	237
Cd^{2+}	383	471	604	533	718	961	1042	1193	1311
Cu^{2+}	214	259	374	160	216	351	102	190	232
Ni^{2+}	177	257	382	197	281	369	129	234	391
Hg^{2+}	881	1072	1390	833	1062	1241	718	919	1127
Al^{3+}	133	312	419	176	251	439	202	251	372
Pb^{2+}	243	301	381	431	652	791	1243	1351	1507
Fe^{3+}	98	152	285	87	173	209	100	202	314
Th^{4+}	325	409	502	204	344	471	217	298	346
Sr^{2+}	100	121	141	102	110	122	97	112	128

Table 5: Distribution coefficients (mL g⁻¹) K_d of metal ions on poly N-methylaniline Zr(IV) molybdophosphate cation exchanger in different solvent systems.

S. No.	Metal ion separation (mg)	Amount loaded (mg)	Amount found (mg)	% recovery	Volume of eluent used	Eluent used (mL)
6	Fe^{3+}	5.86	5.34	91	70	2 M HCl
	Cd^{2+}	19.31	18.27	94	60	4 M HCl
4	Zn^{2+}	6.34	6.15	97	60	4 M HNO₃
	Ba^{2+}	15.23	12.03	81	90	2 M HCl
2	Ca^{2+}	4.233	3.92	92	60	0.1 M HCl
	Hg^{2+}	20.51	18.76	91	90	0.4 M AcOH
1	Zn^{2+}	5.821	5.21	89	50	0.1 M AcOH
	Pb^{2+}	20.11	19.33	96	70	0.2 M HCl
5	Ni^{2+}	5.94	5.86	98	60	0.1 M HNO₃
	Ba^{2+}	14.36	13.71	95	80	2 M HCl
7	Ca^{2+}	4.83	4.47	92	70	0.1 M AcOH
	Pb^{2+}	19.28	18.65	97	80	0.4 M HCl
3	Fe^{3+}	5.53	5.79	100	70	2 M HCl
	Hg^{2+}	20.31	19.25	95	70	4 M HCl

Table 6: Quantitative separation of metal ions from a binary mixture using polyN-methylaniline Zr(IV)molybdophosphate cation exchanger columns at room temperature.

S. No	Amount of Cd^{2+} loaded (mg)	Amount of Cd^{2+} found (mg)	% recovery	Eluent Used	Eluent Volume (mL)
1	1.68	1.61	96	0.1M HNO₃	80
2	4.25	4.16	97	0.1M HNO₃	95
3	6.69	6.53	97	0.1M HNO₃	105

Table 7: Selective separations of Cd^{2+} from a synthetic mixture of, Sr^{2+}, Cu^{2+}, Fe^{3+}, Ca^{2+}, Al^{3+}, Zn^{2+} and Cd^{2+} on poly N-methylaniline Zr(IV) molybdophosphate cation exchanger columns.

S. No	Amount of Hg^{2+} loaded (mg)	Amount of Hg^{2+} found (mg)	% recovery	Eluent Used	Eluent Volume (mL)
1	2.43	2.25	92	0.1 M HCl	80
2	5.71	5.53	97	0.1 M HCl	95
3	7.12	7.03	98	0.1 M HCl	105

Table 8: Selective separations of Hg^{2+} from a synthetic mixture of, Sr^{2+}, Cu^{2+}, Fe^{3+}, Ca^{2+}, Cd^{2+} and Hg^{2+} on poly N-methylaniline Zr(IV) molybdophosphate cation exchanger columns.

Conclusions

Novel synthesized semi-crystalline nano-composite cation exchanger shows selective behavior toward heavy metal ions and can withstand fairly high temperature. Thermally stable, it retains significant ion exchange capacity up to 300°C. It can be used for the quantitative separation of metal ions from binary mixture of analytical importance. The material can be explored further for the removal and recovery of important metal ions from industrial effluents. PolyN-methylaniline Zr(IV) molybdate composite cation exchanger thus, exhibits the characteristics of a promising ion-exchanger as well as separating material.

Acknowledgements

Authors are thankful to the Head of Department of Chemistry, Government Degree College Kulgam, for useful discussions. The useful discussions from time to time with the senior faculty members of Department of Chemistry are also gratefully acknowledged.

References

1. Rengaraj S, Joo CK, Kim Y, Yi J (2003) Kinetics of removal of chromium from water and electronic process wastewater by ion exchange resins: 1200H, 1500H and IRN9H, J Hazard Mater B 102: 257-275.

2. Nabi SA, Shahadat M, Bushra R, Shalla AH, Azam A (2011) Synthesis and characterization of nano-composite ion-exchanger; its adsorption behaviour. Colloids Surf B 87: 122-128.

3. Hui KS, Chao CYH, Kot SC (2005) Removal of mixed metal ions in wastewater by zeolite 4A and residual products from recycled coal fly ash. J Hazard Mater B 127: 89-101.

4. Helfferich FG (1995) Ion Exchange. Dover Publications. Inc, New York.

5. Harland CE (1994) Ion Exchange Theory and Practice. 2nd ed. Royal Society of Chemistry, Cambridge.

6. Nabi SA, Shahadat M, Bushra R, Shalla AH, Ahmed F (2010) Development of composite ion-exchange adsorbents for pollutant removal from environmental wastes. Chem Eng J 165: 405-412.

7. Arrad O, Sasson Y (1989) Commercial ion exchange resins as catalysts in solid–solid–liquid reactions. J Org Chem 54: 4993-4998.

8. Niwas R, Khan AA, Varshney KG (1999) Synthesis and ion exchange behavior of polyaniline Sn(IV) arsenophosphate: a polymeric inorganic ion exchanger. Colloids Surf A 150: 7-14.

9. Nabi SA, Shalla AH (2009) Synthesis, characterization and analytical application of hybrid; acrylamide zirconium (IV) arsenate a cation exchanger, effect of dielectric constant on distribution coefficient of metal ions. J Hazard Mater 163: 657-664.

10. Vatutsina OM, Soldatov VS, Sokolova VI, Johann J, Bissen M, Weissenbacher A (2007) A new hybrid (polymer/inorganic) fibrous sorbent for arsenic removal from drinking water. Funct Polym 67: 184-201.

11. Nabi SA, Naushad M, Bushra R (2009) Synthesis and Characterization of a new organic- inorganic Pb^{2+} selective composite cation exchanger, Acrylonitrile stannic(IV) tungstate and its analytical applications. Chem Eng J 152: 80-87.

12. Hafez MA, Kenway MM, Akl MA, Lshein RR (2001) Preconcentration and separation of total mercury in environmental samples using chemically modified chloromethylated polystyrene-PAN (ion-exchanger) and its determination by cold vapour atomic absorptionspectrometry. Talanta 53: 749-760.

13. Ishrat U, Dar AM, Rafiuddin (2014) Synthesis, characterization and electrochemical properties of Cation selective ion exchange composite memberanes. Arabian J Chem.

14. Xin-Guili L, Hung MR, Liu R (2005) Facile synthesis of semi-conducting particles of oxidative melamine/toluidine copolymers with solvatochromism. React Funct Polym 2: 285-294.

15. Topp NE, Pepper KW (1949) Studies on new composite material polyaniline zirconium(IV) tungstophosphate; Th (IV) selective cation exchanger. J Chem Soc 3299-3303.

16. Nabi SA, Shahadat M, Bushra R, Shalla AH (2011) Heavy-metals separation from industrial effluent, natural water as well as from synthetic mixture using synthesized novel composite adsorbent. Chem Eng J 175: 8-16.

17. Dong X, Wang C, Li H, Wu M, Liao S, et al. (2014) The sorption of heavy metals on thermally treated sediments with high organic matter content. Bioresour Technol 160: 123-128

18. Duval C (1963) Inorganic thermogravimetric analysis. Elsevier, Amsterdam p: 315.

19. Rang A, Vangani V, Rakshit K (1997) Synthesis and characterization of some water soluble polymers. J Appl Polym Sci 66: 45-56.

20. Abthagir PS, Saraswathi R, Sivakolunthu S (2004) Aging and thermal degradation of poly (N-methylaniline). J Therm Chim Acta 411: 109-123.

21. Cullity BD (1978) Elements of X-ray Diffraction. 2nd ed, Addison-582 Wesley p: 102.

25

Experimental Study of the Separation of Oil in Water Emulsions by Tangential Flow Microfiltration Process: Analysis of Oil Rejection Efficiency and Flux Decline

Wai Lam Loh[1,2], Thiam Teik Wan[1,2]*, Vivek Kolladikkal Premanadhan[1], Ko Ko Naing[1,2], Nguyen Dinh Tam[1,2], Valente Hernandez Perez[1,2] and Yu Qiao Zhao[1]

[1]*Department of Mechanical Engineering, Faculty of Engineering, National University of Singapore, Singapore*
[2]*Centre for Offshore Research and Engineering (CORE), National University of Singapore, Singapore*

Abstract

Management of produced water is a major issue offshore. Microfiltration has emerged as a useful alternative for treating the oil-water emulsions to meet the regulatory limit for disposal. In this work, tangential flow (cross flow) microfiltration of oil-water mixture was studied. The tangential flow microfiltration process was investigated using a ceramic membrane of 0.5 μm pore size. For this phase of work, medium viscosity paraffin oil was used as substitute to crude oil. Using oily water feed of 500-1000 ppm oil concentration, a microfiltration ceramic membrane of 0.5 μm pore size was proven capable of producing a high purity filtrate lower than the threshold required for offshore produced water effluent, typically 29 mg/l residual oil, in the Gulf of Mexico. However, membrane has a major drawback in the form of fouling. Decline in permeation flux should be expected over time of operation. This limitation has certainly impeded the large scale applications of microfiltration process in the field of oil and gas processing industry. An optimized cleaning process is needed to restore membrane performance.

Keywords: Microfiltration; Oil in water emulsions; Residual oil; Permeability; Oil rejection efficiency

Introduction

Oilfield produced water treating equipment in use today is designed to remove discrete droplets of oil from the water phase. Conventional equipment such as gravity separator and hydro-cyclone separates the dispersed phases from continuous phases according to their density difference under the action of gravity force or centrifugal force induced by swirling flow. As flows move through chokes, valves, pumps, or other constrictions during crude oil production, the droplets can be torn into smaller droplets by the pressure differential across the devices. These small droplets can be further stabilized in the water by surfactants. The addition of excess production chemicals (such as surfactants) forms an encapsulation of hydrophilic and/or hydrophobic particles on the oil droplets, as shown in Figure 1.

It reduces the interfacial tension so that the coalescence and separation of small droplets become extremely difficult. The inherent difficulty of coalescing emulsified oils in the conventional separating equipment, and ineffective separation of finer emulsion droplets slightly lighter than the continuous phase, makes the conventional separation techniques difficult to treat oily effluent which contains fine emulsions. The use of microfiltration technology is becoming an increasingly popular alternation for the separation of emulsified oil in lieu of the conventional separation approaches.

Membrane separation such as microfiltration (MF), ultra-filtration (UF), nano-filtration (NF), and reverse osmosis (RO) can be used to separate different sized materials [1]. A considerable amount of experimental works and theoretical modelling studies [1-7] in the past two decades that have made possible for the use of low pressure driven membranes for MF of membrane pore size between 0.1 to 5 μm. UF with membrane pore size less than 0.1 μm or a combination of MF/UF polymeric or ceramic membranes are suitable for removing oil content of oilfield produced water. This method involves using low pressure to force the continuous phase to permeate through a membrane into the discharge.

The rejection efficiencies are controlled primarily by the choice of the membrane pore size and not by the difference of density in between the dispersed phase and continuous phase. Because of the many unique properties of the membrane technologies such as no phase change, no chemical addition and simple operation, membrane processes usually provide a better option over traditional separation method in oil and gas processing industries.

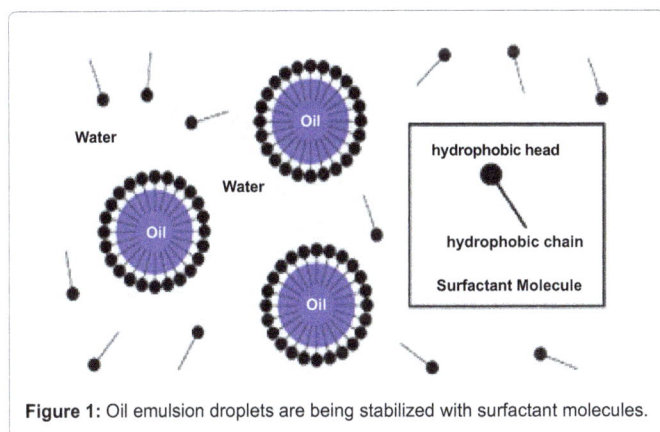

Figure 1: Oil emulsion droplets are being stabilized with surfactant molecules.

*Corresponding author: Wan Thiam Teik, Research Associate, Faculty of Engineering, National University of Singapore, Singapore
E-mail: mpewtt@nus.edu.sg

Koltuniewicz et al. recommended separating oil emulsion from water using ultra/microfiltration performed with ceramic or certain hydrophilic polymeric membranes [5]. Ceramic membranes are often preferred over delicate polymeric membrane because the former have better tolerance to high temperature, high oil contents, foulants, and strong cleaning agents [1]. Mueller et al. proved that microfiltration ceramic membranes at 0.2 and 0.8 µm pore sizes are capable of producing a very high quality permeate from a feed concentrations around 250-1000 ppm of crude oil with droplets size range of 1-10 µm. It was reported to produce a very high quality permeate, containing lower than 6 ppm of total hydrocarbons in the permeate sample [2]. Chen et al. also tested the performance of ceramic tangential flow (cross-flow) microfiltration to separate oil, grease, and suspended solids from produced water. Permeate quality of dispersed oil and grease was 5 mg/L and of suspended solids was less than 1 mg/L [6]. Despite these advances, very few applications of this technology in the separation of oil water mixture have been implemented especially for offshore processing. Although many pilot tests have been conducted using membranes for filtering produced water but with limited success due to their propensity to foul irreversibly with oil and dirt [8].

This research sought to conduct a series of experiments for investigating the characteristics of tangential flow microfiltration of oil-in-water emulsions, and factors determining the membrane performance, as well as the study of fouling control measures. The primary objective of this work is to answer a research question whether tangential flow microfiltration is able to reduce the residual content of the produced water effluent to a level lower than maximum threshold as required for a discharge in sea as claimed in several published works [1-7].

Theory

The permeation flux across the membrane is calculated from $J = V/(A \cdot t)$ where J is the liquid flux across the membrane, A the membrane surface area in contact with the liquid, t the run time of the experiment, and V the volume of permeate collected during time t. The rate of permeate flux through membrane filter is dictated by the resistance of the filter to the flow of fluid and any resistance associated with the trapped particulate. The basic Darcy's Law is defined as:

$$\frac{dp}{dx} = \frac{\mu u}{K} \qquad (2\text{-}1)$$

where dp = ΔP as the pressure drop across the filter, µ as the viscosity of the permeate fluid, dx is the thickness of the filter, K as the permeability, and u=J as permeation flux through the filter. Equation (2-1) is rearranged to give a general equation as:

$$J = \frac{K\Delta P}{\mu \Delta x} \qquad (2\text{-}2)$$

Permeability K refers to the ease with which water can flow through a filter. The resistance to permeation of a membrane is a function of the membrane pore size, feed stream components, and the degree to which gel layer formation and fouling layer formation occur. Increasing the feed stream circulation rate will, as a general rule, reduce gel layer thickness and increase flux.

Darcy's Law states that the flux is directly proportional to the potential pressure drop and inversely proportional to the resistance, R:

$$R = \frac{\Delta P}{J} = \mu R_t \qquad (2\text{-}3)$$

At a constant applied pressure for a given feed with constant composition and flow viscosity, the filtration flux is inversely proportional to the total resistance Rt. The total resistance can be expressed as follows [9]:

$$R_t = R_m + R_c + R_{cp} + R_g \qquad (2\text{-}4)$$

The total resistance consists of resistance by the filter media (R_m) which can include the pore blocking resistance (R_p) and resistance by adsorption (R_a); resistance due to internal colloidal fouling (R_c); resistance due to formation of a highly concentrated layer adjacent to the membrane, concentration polarization (R_{cp}); and resistance caused by the formation of the gel layer (R_g), due to the increasing concentration of particles near the surface of the membrane.

The oil rejection efficiency (R_O) of the filter is defined as the ability to retain dispersed oil phase from flowing across the membrane:

$$R_o = 1 - \frac{C_P}{C_F} \qquad (2\text{-}5)$$

where C_P and C_F are the measured oil concentration of permeate and feed, respectively. For instances, oil rejection efficiency of at least 97% is anticipated for any filter in order to meet a permeation quality of not exceeding 30 ppm of residual oil, assuming a feed stream of 1000 ppm oil concentration.

Methodology

A two phase test rig was built to conduct experiments on the separation of oil-in-water emulsions using tangential flow microfiltration. The experimental set up is shown schematically in Figure 2.

The experimental setup consists of an emulsion feed tank equipped with mixer, clean water tank, centrifugal feed pump, tangential flow microfiltration module, vortex flow meters, pressure gauges, thermo sensors, valves and piping. A ceramic membrane of 0.5 µm pore size was initially back-flushed with clean water. The contaminated oily water was flushed to the emulsion tank to avoid contamination of the clean water tank. After the cleaning, the oil in water feed was pumped into the microfiltration module for the experiments.

The separation of the oil from water is achieved by applying atmospheric pressure to the permeate side of the membrane whilst the other side is exposed to the pressurized liquid to be separated. When the feed stream flows through the membrane bounded channel, it splits into two streams, namely, retentate and permeate (or filtrate). The retentate was returned to the feed tank whilst the filtrate was collected for measurement of residual oil content. The feed flow rate to membrane module was regulated by varying the rate of flows returning to the feed tank, meanwhile, the desirable trans-membrane pressure was obtained by varying both the feed and choke valve on the retentate side.

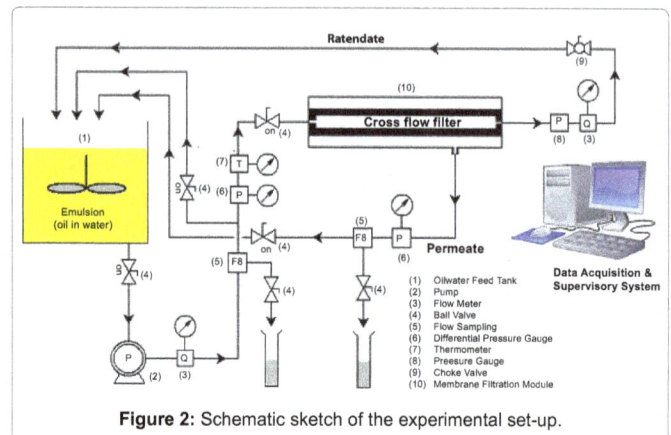

Figure 2: Schematic sketch of the experimental set-up.

Liquid	Density (kg/m³) at 20°C	Viscosity (cP) at 20°C	API
Clean water	998	1.02	10
Paraffin oil	905	87.12	25.54

Table 1: Physical properties of liquids used in the experiments.

When the feed flows through the filter tube, oil droplets tend to deposits on the membrane, but they were continuously removed by the feed flowing tangentially across the surface of the membrane. Constant concentration of the feed is virtually maintained by returning permeate and retentate to the emulsion feed tank. Only a small amount of permeate samples were withdrawn for the measurement of residual content. As the permeate volume compared with the feed volume are small, the feed concentration was considered to be unchanged.

In the experiments, paraffin oil which represents moderate viscosity oil was used as substitute to crude oil. Colored liquid paraffin was used to enable visual inspection during the preparation of emulsion. The dynamic viscosity of paraffin oil used in the experiments was determined using the HAAKE MARS rotational rheometer and the physical properties of the liquids used in the experiments are shown in Table 1.

The oily water mixture was prepared by emulsifying the desirable percentage of paraffin oil in clean water (500 and 1000 ppm). The oil-water mixture was agitated using a mixer for at least five minutes, further re-circulated to the emulsion feed tank by a centrifugal feed pump until a milky white mixture was formed. The result was stable oil in water emulsions which doesn't separate by gravity for days or weeks. However, emulsions is shear sensitive, therefore the retentate was re-circulated to the feed tank, where they were continuously stirred with an emulsifier mixer to prevent oil droplets from coalescing to ensure homogeneity throughout the experiments.

All the microfiltration experiments were conducted in a series of batch runs with trans-membrane pressure of 1.0, 1.5, 2.0, 2.5, 3.0 and 3.5 bar(g), respectively. The starting trans-membrane pressure was initiated from 1 bar(g), and gradually increased by a step of 0.5 bar(g) at each run, until a final pressure of 3 bar(g) was reached during the fifth runs. To this end, the entire cycle was repeated thereafter. Each batch (run) of experiments took a ten minutes interval, sum up to an approximate total time of hundred fifty minutes per fifteen runs of experiments, afterward, then clean water feed was replaced by oil-water feed of 1000 ppm concentration for the study on oil rejection performance. Next, the experiments were repeated with oil-water feed of 500 ppm concentration.

The permeate quality was monitored by collecting samples at different time intervals for the measurement of oil concentration in unit of ppm (part per million) by volume. The sampling approach here involved the collection of representative samples of the emulsion from the pipelines and subsequent analysis of the samples using solvent extraction method. Solvent extraction method is often used to determine the oil content of dilute oil in water emulsion with concentrations as much as 1000 ppm or so [10]. In this work, a TD-500D fluoro-meter was used to measure oil content in water samples by the detection of the fluorescent light emitted by oil that has been extracted into n-hexane. The UV fluorescence value in TD-500D meter was then calibrated to known concentration of oil in water sample to give results equivalent to EPA method 1664 [11]. The fluorescence technique requires calibration by measuring the intensity of fluorescent light that is generated by a known concentration of hydrocarbon. Hence, the fluorescence values in the fluoro-meter were calibrated with respect to a known concentration of oil in water sample, as well as the distilled water sample (two points

calibration, plus the mid-point calibration for a linearity check). The detection range is in between 5-1000 ppm of residual oil content.

During the experiments, all permeate samples collected were immediately photographed for visual comparison. All six samples showed no visible free phase of oil in permeate water. The crystal clear permeate quality gave good indication of effective rejection of oil emulsion after the filtration. However, it was difficult to justify the difference in oil concentration until they were later analyzed by fluoro-meter.

Results and Discussions

The aim of the experiment is to study the oil rejection efficiency under influence of pressure for initial evaluation of the potential application of ceramic microfiltration membrane in filtering oil phase from the oily water. The understanding of oil rejection efficiency under optimum operating conditions was important as a first step in the development of future processing equipment.

Two sets of experiment with a feed solution of different oil concentrations of 500 and 1000 ppm were conducted. Both the experiments were run at approximate similar range of permeability values for fair comparison of results. For evaluation of the process and economic viability of membrane-based filtration applications, flux stability is a significant component which must be taken into consideration.

Fouling, not surprisingly, is frequently cited as the most important factor limiting the utilization of membranes in waste water treatment. It would affect the observation of filtration performance if was not addressed carefully. In the experiments, the first 15 runs uses only clean water as feed for monitoring the effects of trans-membrane pressure on permeability (K=Jμ∆x/∆P), while observing the fouling characteristics of membrane over a history of experiments order. In Figure 4, the ceramic membrane experienced a rapid decline in permeability in the beginning.

The rapid decline in permeability may be connected with the build-up of the concentration polarization layer and pore blocking mechanism. This phenomenon is inherent of all membrane filtration processes. A higher pressure causes the suspended impurities to rapidly deposited on the membrane surface, and the formation of cake increases the resistance to permeate flow and thus reduces the permeate flux and membrane performance, as illustrated in Figure 3.

The irreversible decline of membrane permeability is often due to strong adsorption of particles of contaminants onto the membrane surface and in its pores. In general, flux decline is caused by a decreasing driving force and/or an increased resistance.

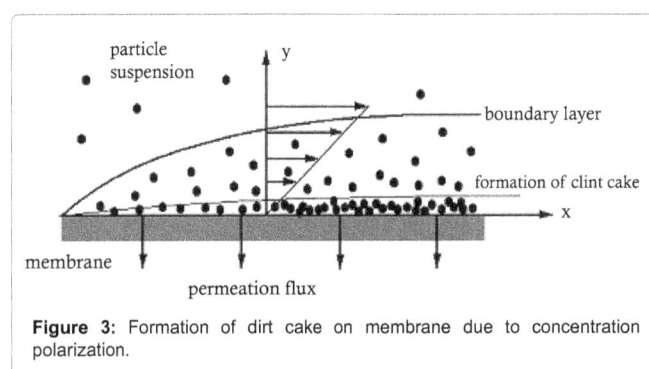

Figure 3: Formation of dirt cake on membrane due to concentration polarization.

Permeability (cm²) versus Experiments Order

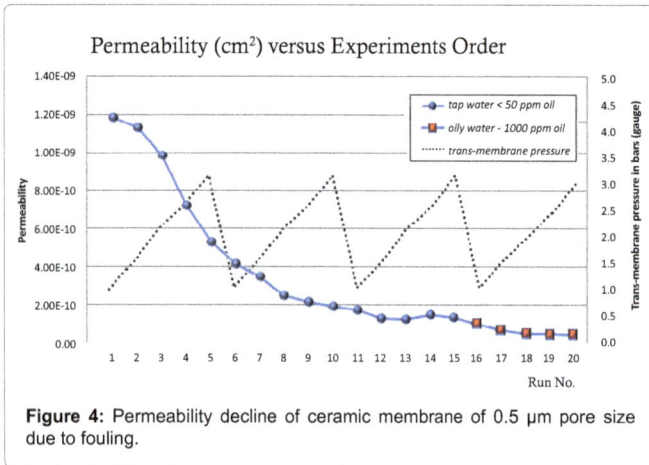

Figure 4: Permeability decline of ceramic membrane of 0.5 μm pore size due to fouling.

Flow Resistance (bars.s/m) Versus Experiments Order

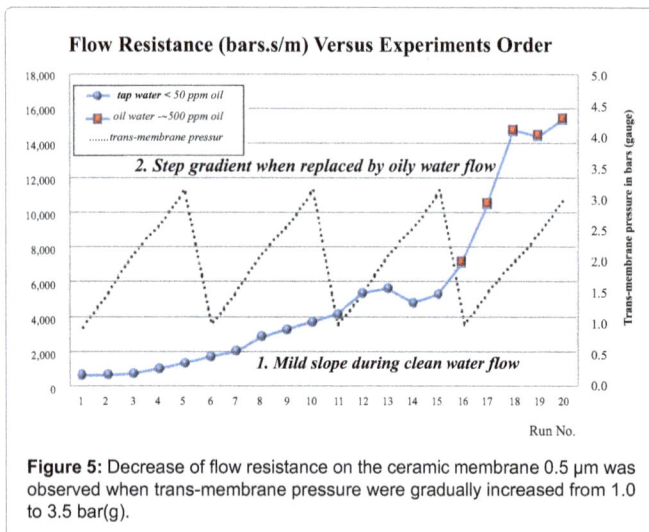

Figure 5: Decrease of flow resistance on the ceramic membrane 0.5 μm was observed when trans-membrane pressure were gradually increased from 1.0 to 3.5 bar(g).

As a series run mode was implemented, trans-membrane pressure was gradually increased from 1 to 3.5 bar(g). The rapid decline in permeability indicates the higher the pressure, the greater decline in permeability were observed. After a nearly 80% decline in flux during first 10 runs of experiments, it finally reach a steady balance between the fouling and hydrodynamic force which constantly sweep away fouling materials from the membrane surface. Hence, a nearly steady permeability or constant permeability stage of approximately $1\sim2\times10$-10 cm² was achieved in the first experiment during the clean water feed phase (run no. 11 to 15), as shown in Figure 4.

With the gradual reduction in fouling, the flow permeability and filtration flux decreased slowly. When a steady permeability pattern is attained, the apparent effect of fouling is negligible. Thus, the flux is linearly dependent on the trans-membrane pressure alone. That's the reason oil-water filtration experiments were conducted from run no. 16 and onward, only then the effects of pressure on filtration quality and flux performance can be reasonably assessed, and a linear relation of pressure on the filtration rate can be established.

The flow permeability and resistance at run no. 15 were 1.33×10-10 cm² and 5292 bar.s/m, respectively, as illustrated in Figures 4 and 5. Once the experiments has transited from the clean water feed phase into the 1000 ppm oily water feed phase, the flow permeability immediately deteriorate thereafter. The resistance curve during the oil-water feed was showing a steeper climb (run no. 16 to 18) in Figure 5, that could

mean an increased rate of fouling due to the existence of a secondary cake layer which may be caused by the organic (oil particles) fouling. The deposition of oil on the surface of membrane has actually trapped more suspended particles into the oil layer, therefore a "mixed" gel layer is formed, increases flow resistance and causes further decline in permeability. The next stable resistance at range around 15 000 bar.s/m was attained afterward, which was almost three times of the previous steady range (Figure 5).

Meanwhile, in another experiment using 500 ppm oily water feed, the flow resistance continues to soar until it reached a peak resistance of around 12000 bar.s/m, as depicted in Figure 6.

This result is 21% lower than the peak resistance value as attained in the first experiment with 1000 ppm of oil in water feed. Afterward, the resistance began to decline when the trans-membrane pressure reached beyond 1.5 bar(g) range. The reason is to be explained later.

The measurements on permeate samples have been repeated five times for each sample and the analysis of statistical mean and standard deviation were carried out. The standard deviation σ shows how much variation or "dispersion" exists from the average (mean, or expected value). In Table 2, a standard deviation of 1.78 means that most measurements (about 68%, assuming a normal distribution) of permeate water samples have a reading within 1.78 ppm of the mean. A relatively low value in the standard deviation indicated that high reproducibility was achieved (Table 2).

Table 3 shows a comparison of the filtrate quality in term of ppm of residual oil content in water after the microfiltration experiment with the clean water feed and the oil in water feed (Table 3).

According to equation (2-5), the results were later converted into oil rejection efficiency as depicted in Table 4.

Flow Resistance (bars.s/m) Versus Experiments Order

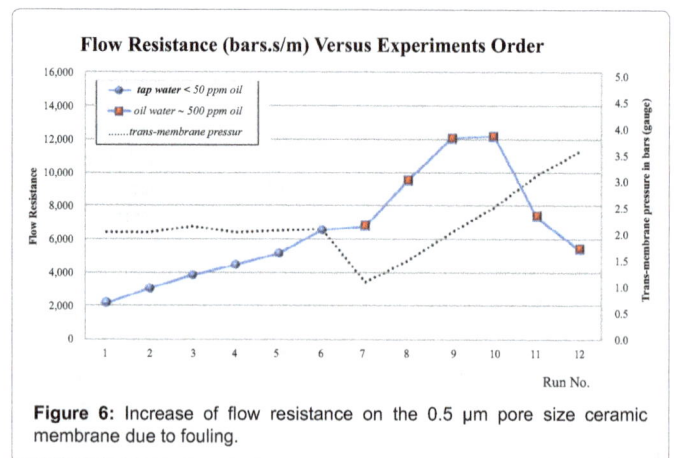

Figure 6: Increase of flow resistance on the 0.5 μm pore size ceramic membrane due to fouling.

Sample Description	Mean	Standard Deviation
S09/P1.0/O1000 Permeate sample @ 1.0 bar(g) TMP	35.9	1.00
S11/P1.5/O1000 Permeate sample @ 1.5 bar(g) TMP	32.8	1.93
S13/P2.0/O1000 Permeate sample @ 2.0 bar(g) TMP	35.7	0.82
S15/P2.5/O1000 Permeate sample @ 2.5 bar(g) TMP	27.8	1.36
S17/P3.0/O1000 Permeate sample @ 3.0 bar(g) TMP	95.3	3.38
S19/P3.5/O1000 Permeate sample @ 3.5 bar(g) TMP	146.1	2.18
Average	62.2	1.78

Table 2: Mean and standard deviation for the measurement of oil concentration (in ppm) for permeate water samples collected from the microfiltration experiment using a ceramic membrane of 0.5 μm pore size and oil-water feed of 1000 ppm.

Feed concentration/TMP	1 bar(g)	1.5 bar(g)	2.0 bar(g)	2.5 bar(g)	3.0 bar(g)	3.5 bar(g)
1000 ppm oily water feed	35.9	32.8	35.7	27.8	95.3	146.1
500 ppm oily water feed	< 10	< 10	< 10	< 10	35.1	72.0

Table 3: Permeate quality (in ppm) from the microfiltration experiment using a ceramic membrane of 0.5 μm pore size and oil-water feed of various concentration at different test pressure.

Oil Rejection Efficiency (%) /TMP in bar(g)	1.0	1.5	2.0	2.5	3.0	3.5
During 500 ppm oily water feed	98.78	>99	>99	>99	92.98	85.6
During 1000 ppm oily water feed	96.17	96.5	96.19	97.03	89.83	84.41

Table 4: Oil rejection efficiency (%) for microfiltration experiment using a ceramic membrane of 0.5 μm pore size and oil-water feed of various concentration at different test pressure.

The permeate quality from the first filtration experiment with 1000 ppm feed concentration has the best results of 27.8 ppm residual oil. The permeate quality were consistent until 2.5 bar(g) pressure. However, the residual content has spiked up to 95.3 ppm oil @ 3 bar(g) pressure, and 146.1 ppm oil @ 3.5 bar(g) pressure. The oil rejection efficiency were found to be maintained around 96~97% before the pressure reached 2.5 bar(g). At higher pressure range, oil phase began to break across the membrane. Therefore, a sudden decline in permeate quality has been observed. A similar finding has been observed when the feed was reduced to 500 ppm oily water mixture in the next experiment. A lower feed concentration is definitely advantageous for attaining higher oil rejection efficiency. For the experiment with 500 ppm oily water feed, the permeate quality reduced to lower than 10 ppm residual oil. Again, oil phase starts to break through the ceramic membrane of 0.5 μm pore size when trans-membrane pressure exceeded 2.5 bar(g).

The results indicated that trans-membrane pressure have negligible effects on the oil rejection efficiency before a critical pressure of 2.5 bar(g). As the suspended oil content of the feed increases, blockage of the pores increases and at a faster rate, which results in the growth of a thicker deposited layer. Therefore lower oil concentrations in the permeate samples should be expected when the feed contained a low concentration of suspended oils.

The previous comparison of peak resistance in both experiments indicates a significant improvement in the flow resistance ($R=\Delta P/J$) at lower feed concentration. Higher feed concentration led to greater problem in fouling, therefore, higher flow resistance. As the trans-membrane pressure increases, the tangential flow velocity has increased as well. Though with increased pressure and flow velocity it may have improved to some extent the degree of fouling, however, it came with a price of having more oil breaking through the membrane when the pressure exceeds 2.5 bar(g). One likely explanation is the theory of deformable drops in capillaries due to pressure.

Nazzal et al. has suggested that the rejection of an oil drop was determined by its ability to deform and flow through a pore as permeate. He applied the Young-Laplace equation to propose a model for calculating the critical pressure required to force the entry of an oil drop into a membrane as follows [12]:

$$P_c = 2\gamma \frac{\cos\theta}{r_p} \left[1 - \left(\frac{2 + 3\cos\theta - \cos^3\theta}{4(r_d/r_p)^3 \cos^3\theta - (2 - 3\sin\theta + \sin^3\theta)} \right)^{1/3} \right] \qquad (4\text{-}1)$$

where γ is the interfacial tension, r_d is the radius of the original drop, r_p is the radius of the pore and θ is the contact angle. In his opinion, this critical pressure can be used for predicting the maximum operating pressure at which oil drops are not longer rejected by the membrane and their penetration leads to membrane fouling. This model is useful to interpret experimental effect of trans-membrane pressure on the microfiltration of oil in water emulsions. Initially, oil drop that does not wet the membrane cannot pass through a pore of smaller diameter. If the applied pressure is great enough beyond the threshold as predicted in equation (4-1), it will deform the drop and overcome surface tension effects within the membrane surface. The higher the applied pressure, it will tend to increase permeate flux and the flow of oil drops to another side of the membrane. The oil drops may either pass through the membrane, or possibly shear deeper into the interconnected pores and stuck inside the dead of the channel. As a results, blockage occur due to passage of oil drops through the membrane, where a constant decline in oil rejection efficiency is clearly revealed as pressure exceeds 2.5 bars(g) and beyond. The threshold could be perceived as the critical pressure where oils will break through the pore in the ceramic membrane.

This observation was further supported by Davis in a simulation study of the micro scale flow of emulsions containing deformable drops through pores neck. He confirmed that whether a drop is able to deform enough the squeeze through the pore neck or not was depending on the capillary number. When the pressure across the membrane was sufficient, it exceeds the critical capillary number (a ratio of viscous deforming force and interfacial restoring forces) and would enable the drop to exceed the average velocity of the continuous phase and squeeze through the pore, otherwise, it would be trapped in a small pore neck [13].

Another explanation to the deterioration in the filtrate quality is due to the corresponding effects of turbulence as tangential flow velocity increases. During the experiment, the tangential flow velocity has increased with the trans-membrane pressure. He has reported that turbulence in flow have remarkably reduced the particle size in the filtration cake [14]. The reduction in particles size is definitely going to affect the quality of filtrate in a negative way. Finer particles and oil drops would find its way to enter and squeeze through the pore and pollute the water in the permeate side. Consequently, more oil would break through the membrane with a constant decline in oil rejection efficiency is clearly revealed.

Conclusions

The performance of microfiltration for the separation of oil-in-water emulsions has been investigated. At 1000 ppm (0.1%) feed concentration, a microfiltration ceramic membrane of 0.5 μm pore size was proven possible of producing a high purity filtrate lower than the threshold required for offshore produced water effluent, typically 29 mg/l (i.e. 33 ppm) residual oil in the Gulf of Mexico [15]. At 500 ppm (0.05%) oil concentration, higher purity filtrate containing lower than 10 ppm residual oil in permeate was proven possible at a trans-membrane pressure not exceeding 2.5 bar(g). The results attained were useful for evaluating the potential of tangential flow microfiltration process in the produced water treatment, with respect to the suitability to fulfill the regulatory requirement for disposal.

The study of the pressure effects on filtration capacity and permeability analysis is a very useful tool for analysis of changes in flux due to complicated influence of membrane pore size, pressure and fouling. The fouling study gives a practical estimation of the filtration capacity especially during the selection appropriate membrane for the need of various industrial treatment processes. The method is also used to optimize the separation efficiency while having the best cleaning ability by hydrodynamic tangential flow.

However, the major challenge of operation is the control of fouling. This research does indeed look at the separation and fouling control as a system instead of an individual subsystem, with the results contributing primarily to the knowledge base of separation with a secondary contribution to the fouling control problem. For the objective of controlling the fouling, a series of studies have been conducted on in-situ cleaning to allow remediation of polluted surface during filtration. The result of the study will be presented in the next paper.

References

1. Fakhru'l-Razi A, Pendashteha A, Abdullaha LC, Awang Biaka DR, Madaenic SS, et al. (2009) Review of technologies for oil and gas produced water treatment. Journal of Hazardous Materials 170: 530 -551.

2. Mueller J, Cen Y, Davis RH (1997) Crossflow microfiltration of oily water. Journal of Membrane Science 129: 221-235.

3. Holdich RG, Cumming IW, Smith ID (1998) Crossflow microfiltration of oil in water dispersions using surface filtration with imposed fluid rotation. Journal of Membrane Science 143: 263-274.

4. Cheryana M, Rajagopalan N (1998) Membrane processing of oily streams. Wastewater treatment and waste reduction, Journal of Membrane Science 151: 13-28.

5. Koltuniewicz, AB, Field RW (1996) Process factors during removal of oil-in-water emulsions with cross-flow microfiltration. Desalination 105: 79-89.

6. Chen ASC, Flynn JT, Cook RG, Casaday AL (1991) Removal of oil, grease, and suspended solids from produced water with ceramic cross flow microfiltration. SPE Prod Eng 6: 131-136.

7. Arnot, TC, Field, RW, Koltuniewicz AB (2000) Cross-flow and dead-end micro filtration of oily water emulsions Part II. Mechanisms and modeling of flux decline. Journal of Membrane Science 169: 1-15.

8. Dejak M (2013) Keeping water soft. Oilfield Technology 6: 35-44.

9. Al-Malack MH, Anderson GK (1997) Use of cross flow microfiltration in wastewater treatment. War Res 31: 3064-3072.

10. Rajinder Pal (1994) Techniques for measuring the composition (oil and water content) of emulsions – a state of the art review. Colloid and Surface A. Physicochemical and Engineering Aspects 84: 141-193.

11. Environment Protection Agency (1999) Method 1664, Rev. A: N-hexane extractable material (HEM; oil and grease) and silica gel treated n-hexane extractable material (SGTHEM; non-polar material) by extraction and gravimetry. EPA-821-R-98-002.

12. Nazzal FF, Wiesner MR (1996) Microfiltration of water-in-oil emulsions. Water Environment Research 68: 1187-1191.

13. Davis RH (2009) Motion of deformable drops through granular media and other confined geometries. Journal of Colloid and Interface Science 334: 113-123.

14. Gaohong He (2012) A comparison of cake properties in traditional and turbulence promoter assisted microfiltration of particulate suspensions. Water Research 46: 2535-2544.

15. Stewart M, Arnold K (2009) Emulsions and Oil Treating Equipment : Selection, Sizing and Troubleshooting. Gulf Professional Publishing, USA, UK. 107-211.

26

Studies on the Radiation Stability of Several Polymeric Flat Sheets used for Actinide Ion Separation from Radioactive Feeds

Mohapatra PK[1]*, Raut DR[1], Shah JG[2] and Bhardwaj YK[3]

[1]*Radiochemistry Division, Bhabha Atomic Research Centre, Trombay, Mumbai-400085, India*
[2]*Process Development Division, Bhabha Atomic Research Centre, Trombay, Mumbai-400085, India*
[3]*Radiation Technology Development Division, Bhabha Atomic Research Centre, Trombay, Mumbai-400085, India*

Abstract

Radiolytic stability of polymeric flat sheet membranes was evaluated from surface morphology (SEM), contact angle and porosity data. The flat sheets made from PES (polyether sulfone), PP (polypropylene), PC (polycarbonate) and PVDF (polyvinylidenefluoride) were irradiated to varying extent using a ^{60}Co gamma ray source and the physical characterization of the membranes was carried out using various techniques mentioned above. Subsequently, the transport efficiency of the irradiated flat sheets was evaluated by studying the mass transfer of Am^{3+} from a feed containing 3 M HNO_3 in to a receiver phase containing 0.01 M HNO_3 as the strippant while 0.1 M TODGA (*N,N,N',N'*-tetraoctyldiglycolamide) +0.5 M DHOA (di-*n*-hexyloctanamide) in *n*-dodecane was used as the carrier extractant. Out of the flat sheet membranes, PC membranes were found to be more promising and transport studies were carried out using PC membranes up to an absorbed dose of 50 MRad which suggested very good transport of Am^{3+} without any deterioration of the membrane.

Keywords: Polymeric membranes; Radiolytic degradation; Transport; Liquid membrane

Introduction

Membrane based separation methods have been used for the remediation of industrial waste water emanating from a large number of industries [1-6]. The major challenge in such studies has been the membrane fouling which arises due to a large number of suspended particles. In view of the high efficiency of the membrane based separation systems, attempts are made, of late, to employ them in the nuclear industry for radioactive waste streams as well. The radioactive wastes are categorized as low, medium and high level wastes depending on their radioactivity content. Though low level wastes have been treated by membrane based separation methods at large scales [7], it is really a challenge to employ membrane separations to high level wastes. This is due to the large amount of radiation dose prevailing in the high level waste mainly due to the high beta gamma containing isotopes such as ^{137}Cs and ^{90}Sr [8]. Secondly, most of the membrane based separation methods employ polymeric membranes which are highly susceptible to radiolytic degradation [9]. Radioactive waste management is one of the major challenges of the nuclear power program [10] and there are reports on the application of membrane based separation methods for the separation of actinide and fission product nuclides from radioactive wastes [11]. Though low level wastes have been treated using polymeric membranes, the latter can be reused and recycled for a significantly long period of time due to the low dose rates (<microcuries per litre of the waste solution) of these wastes. On the other hand, high level waste processing can be done using radiation resistant polymers.

There are limited numbers of reports available in the literature on the radiolytic stability of polymeric membranes. Development of radiation resistant membranes has become important due to the growing interest in membrane based separation methods for radioactive waste water treatment. Also, though commercial membranes are available for gas separation, applications in food processing industries, etc. till recently, there was no need for applying membrane based separation methods for radionuclide separations. Therefore, there is a need to investigate the radiation stability of polymeric membranes. The radiation stability of polysulphonic acid cation exchange membranes was reported by

Ramachandhran and Mishra [12] up to 18.7 MRad absorbed dose. Recently, the authors have reported the transport efficiencies of several polymeric membranes after exposing those to an absorbed dose of 20 M Rad [13]. Though a vague idea was obtained, it was required to carry out a systematic study involving physical characterization of the irradiated membranes as well as metal ion transport. This would enable one to get a direct correlation of the properties such as surface morphology (the surface area has a very relevant role in metal ion transport), hydrophobicity (hydrophobic membranes can retain the organic carrier molecule and solvent for a long time, thereby enhancing the membrane stability), pore size (the pore size may get damaged by exposure to radiation) and tensile strength (overall reusability may be affected if the membranes are damaged) with the metal ion transport over a wide range of absorbed dose.

The authors have used common polymeric membranes like PS (polysulphone), PP (polypropylene), PC (polycarbonate) and PVDF (polyvinylidenefluoride) in the present study. The mass transfer studies were carried out by obtaining the transport behaviour of Am^{3+} using TODGA (*N,N,N'*-tetraoctyl diglycolamide) as the carrier and DHOA (di-*n*-hexyloctanamide) as the modifier. Distilled water was used as the receiver phase in the transport studies. The radiation stability of the polymeric membranes was also evaluated after irradiating them to 50 MRad in a ^{60}Co gamma ray chamber.

---

*Corresponding author: Mohapatra PK, Radiochemistry Division, Bhabha Atomic Research Centre, Trombay, Mumbai-400085, India
E-mail: mpatra@barc.gov.in

Experimental

Materials

PES, PC and PP membranes (with close to 100% polymer content) were prepared at CSMCRI, Bhavnagar while the PVDF membranes were prepared at Desalination Division, BARC. The flat sheet polymeric membranes were irradiated using a ^{60}Co irradiator at a dose rate of 3.54 KGy/h. N,N,N',N'-tetraoctyl diglycolamide (TODGA) and DHOA (di-n-hexyloctanamide) were synthesized by reported methods and were used after ascertaining their purities by NMR and HPLC [14,15]. n-Dodecane (purity: 99%), procured from Lancaster, UK, was used as obtained. All other reagents were of AR grade and were used without further purification. ^{241}Am tracer was taken from a laboratory stock after checking its radiochemical purity [16]. ^{241}Am was assayed radiometrically by gamma ray counting using NaI(Tl) scintillation counter (Para Electronics) interphased with a multi-channel analyzer (ECIL, India). Counting statistics errors were minimized (<1%) by obtaining at least 10,000 counts.

Membrane characterization

SEM images were obtained for surface morphology analysis using a Stereoscan 100 Cambridge model operating at 15/25 kV with a magnification of 50-70X at a working distance of 15 mm at a tilt angle of 45°. The images were obtained after giving a 15 nm coating of gold using a Balzer's coating unit (model: CEA 30). The flat sheet membrane sample was analyzed for contact angle by mounting them on glass plates. The contact angles of the samples were measured by the sessile drop technique using image analysis software. Usually, a drop of MilliQ water (conductivity 0.05 μS.cm^{-1}), was allowed to fall on the flat sheet membrane samples using a software-controlled syringe. An image sequence was obtained through a CCD camera of a goniometer (GBX instruments, France) interfaced with an image capture software (Windrop^{++}, GBX instruments).

The porometer (Porolux 1000 model, Benelux Scientific, Belgium) operated in standard pressure range of 0-35 bar with flow rates up to 200 liters per minute. The sample holder diameter was 10 mm, and data acquisition and analysis was carried out using LabView software supplied by the manufacturer. The "Porefil" solution (γ=16 dyne/cm, vapour pressure=3 mm Hg at 298 K) supplied by Benelux Scientific, Belgium was used as a wetting liquid for capillary flow porometry as it is non-toxic inert fluorocarbon wetting fluid having zero contact angle in membrane or Cos θ=1. Mean flow pore size was calculated from the pressure at which the wet flow was half of the dry flow. The portions of plot of wet run from the point where the first flow of gas was observed up to the point where it meets the dry run plot were used for determining pore size distributions.

Transport studies

Transport of Am^{3+} was monitored using standard procedure reported earlier [17], in which the flat sheet membranes (irradiated as well as pristine membranes) were mounted in between the flanges of a two-compartment Pyrex glass cell for subsequent transport studies. Each compartment of the transport cell has 20 mL capacity and was filled with the feed (3 M HNO$_3$) and the receiver (0.01 M HNO$_3$) solutions. The flat sheet membranes were soaked overnight in the carrier solution (0.1 M TODGA + 0.5 DHOA in n-dodecane) and appeared completely wet before the transport studies. The cumulative % transport as seen from the feed side was determined by the following equation,

$$\% \, T_f = 100 \, \frac{\left(C_{f,o} - C_{f,t}\right)}{C_{f,o}} \tag{1}$$

where, $C_{f,o}$, $C_{f,t}$ and $C_{r,t}$ are the concentration of Am in the feed at 'zero' time, after time 't' and in the receiver compartment after time 't'. Am concentrations in the transport studies were about 10^{-7} M. Permeability coefficient values (P) were calculated from the effective membrane area (Q) and the feed volume (20 mL) using the following expression [18]:

$$\ln\left(C_{f,t} / C_{f,0}\right) = -\left(Q / V_f\right)Pt \tag{2}$$

The transport studies were carried out at ambient temperature (24 ± 1°C) and repeated twice. The results presented are the average of the two measurements.

Results and Discussion

Membrane characterization

Surface morphology analysis: The surface morphology analysis of the flat sheet polymeric membranes was done by taking SEM images of the membranes after exposing to gamma radiation. The flat sheets were exposed to 10, 20, 30 and 50 MRad dose and some membranes were found to disintegrate beyond a particular dose (Table 1). The SEM pictures are shown in Figure 1 for PES and PC flat sheets which were exposed to 30 MRad and 50 MRad doses, respectively. On the other hand, PP membranes were not stable even at 10 MRad absorbed dose and hence, their SEM pictures are not included in Figure 1. PVDF membranes were also irradiated and found to be stable up to an absorbed dose of 30 MRad. However, as good quality SEM pictures of irradiated PVDF membranes were not obtained, the same are not included here. As shown in the figure, the membranes appeared to be

Polymeric membrane	Irradiated up to	Observation
PP	20 MRad	Found to be brittle even after 10 MRad
PC	100 MRad	Integrity of membranes was good up to 50 Mrad. Beyond that membranes appeared brittle
PES	50 MRad	Integrity of membranes was good up to 30 Mrad. Beyond that membranes appeared brittle
PVDF	50 MRad	Integrity of membranes was good up to 30 Mrad. Beyond that membranes appeared brittle

Table 1: Summary of radiation stability data of the polymeric membranes used in this study

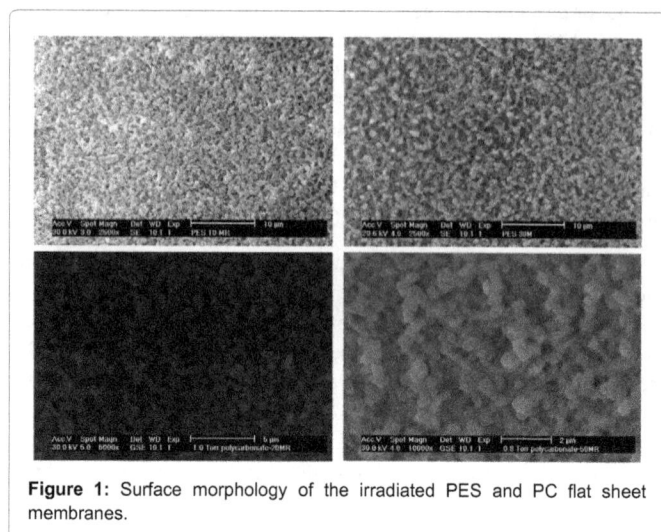

Figure 1: Surface morphology of the irradiated PES and PC flat sheet membranes.

uniform without any visible degradation marks. This is clear from the PES membranes irradiated at 10 MRad and 30 MRad doses and PC membranes irradiated at 20 MRad and 50 MRad absorbed doses. For better clarity, PC membrane irradiated at 50 MRad dose is magnified by a factor of two. The results suggest that PP membranes have much inferior radiation stability and hence, do not hold much promise for application.

Contact angle measurements: The membranes were usually exposed to an estimated absorbed dose up to 50 MRad. However, as some of the membranes were found to be brittle at this high absorbed dose, the data obtained at lower exposure are presented. Actual photographs of the water droplet are included in Table 2 for different membranes. The contact angle data are also presented in Figure 2 as a function of time which can give an indication about the membrane hydrophobicity. The contact angle data indicated that the hydrophobicity of the membranes decreased in the order: PVDF>PP>PC>>PES which was found to get affected with absorbed dose. PES membranes were found to be the least hydrophobic as the water drop was found to get quickly soaked in to the membrane in a few seconds and hence, the photographs are not included in Table 1. On the other hand, PP membranes were not very stable against radiation and became brittle even after exposure to 10 MRad dose (vide supra) suggesting that contact angle measurements was not possible. Out of all the membranes, PC showed better radiation stability while PVDF showed better hydrophobicity. Both the membranes also showed an enhancement in hydrophobicity with increasing absorbed dose. Furthermore, while only 4.4% change in

the contact angle was seen after about 2.5 minutes for the irradiated PVDF membranes, 12.9% change in the contact angle was seen in the same time interval for the irradiated PC membranes. Correlation of the metal ion permeability coefficients with the membrane hydrophobicity are presented below.

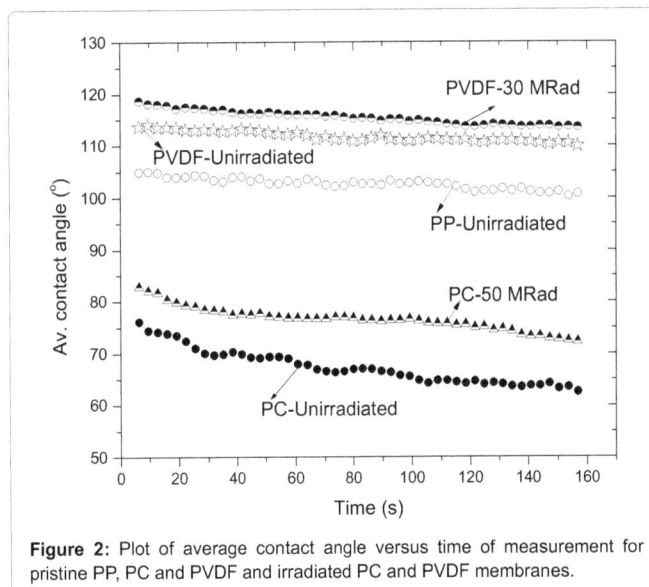

Figure 2: Plot of average contact angle versus time of measurement for pristine PP, PC and PVDF and irradiated PC and PVDF membranes.

Polymeric membrane	Photograph of the water droplet on the membrane at different time	
	Initial	After 157 sec
PC (pristine)		
PP (pristine)		
PVDF (pristine)		
PC (50 MRad)		
PVDF (30 MRad)		

Table 2: Contact angle measurement of the polymeric flat sheet membranes as photographed before and after irradiation.

Porosity measurements: The porosity of the membranes may get changed after exposure to radiation which may affect the transport properties. In view of this, the porosity of the polymeric membranes was measured both with the pristine membranes as well as the irradiated membranes and the data are presented in Table 3. As shown in the table, the errors in the pore size data were much less in the pristine membranes which increased after irradiation suggesting the increase in the average pore size after irradiation. This was particularly seen in case of PC and PVDF membranes while for PS membranes, the pore size did not change significantly after irradiation.

Transport studies

Figure 3 shows the % T data for Am^{3+} transport by the unirradiated PC membrane containing carrier solvent (TODGA + DHOA in n-dodecane) from a feed solution of 3 M HNO_3 into a receiver compartment containing 0.01 M HNO_3. The %T data are for the pristine membranes of all the four polymeric materials are given in Table 4. The %T data presented here are found to be significantly higher than those reported by us earlier with analogous membranes which were attributed to the fabric support used in the previous study which also led to large fraction of the Am^{3+} ion getting trapped in the membrane. In sharp contrast, the present study showed highly efficient Am^{3+} transport with all the four pristine membranes.

Results of the mass transfer studies using the irradiated membranes are also included in Table 4. PP membrane data are not included due

Membrane	Average pore size (μm)
PC (pristine)	0.33 ± 0.01
PC (50 MRad)	0.35 ± 0.03
PES (pristine)	0.36 ± 0.01
PES (30 MRad)	0.36 ± 0.02
PVDF (pristine)	0.25 ± 0.01
PVDF (30 MRad)	0.27 ± 0.03

Table 3: Measurement of pore size of polymeric membranes before and after irradiation. Data for PP membranes are not included as the membranes were found to be brittle even after exposure to even low doses of radiation

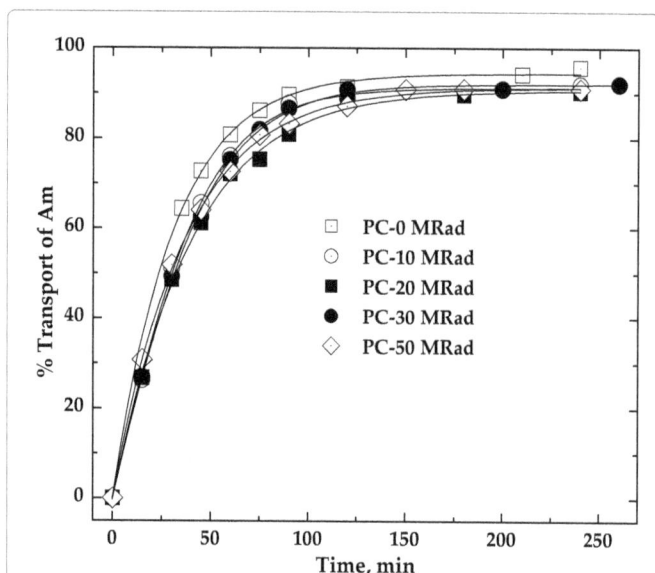

Figure 3: Am^{3+} transport data using polyethersulfone (PES) membrane filter. Feed: 3 M HNO_3; Receiver: 0.01 M HNO_3; Carrier: 0.1 M TODGA+0.5 M DHOA in n-dodecane.

to their poor radiation stability. The mass transfer data (Table 4) for the pristine membranes followed the trend: PP~PC>PES>PVDF which changed marginally to PES>PC>PVDF with the irradiated membranes. With the exception of PC, where significant difference in the Am^{3+} transport data was seen between the pristine and the irradiated membranes, the transport data obtained with the irradiated membranes of the other two membranes were very close to those obtained with the pristine membranes. As mentioned above, the mass transfer rates were found to be much higher as compared to those reported by the authors in a previous report which was attributed to the fabric support used in the membranes studied earlier [13]. Also, the data reported in the present paper are more meaningful as the hold up reported in the previous work is no longer valid [13]. In view of higher radiation stability of PC membranes, subsequent studies were carried out with PC flat sheet membranes exposed to varying amounts of absorbed gamma ray dose. As shown in Figure 3, the transport profiles were found not to deviate significantly from those obtained with the pristine PC membranes suggesting the mass transfer is not significantly affected with the absorbed dose (Table 5). Also, mass transfer studies were carried out using PP membranes which were used in most of the previously reported studies involving hollow fiber contactors [19,20] and the comparable mass transfer data (Table 5) validate PC as a suitable polymeric material for subsequent studies using hollow fiber contactors. All the above studies were carried out in duplicate and the results presented are average of the data. The errors in the mass transfer data were found to be less than ±2% and may be mainly due to non-uniformity of the membrane thickness (±2%) and counting statistics error (<1%). However, the membrane thickness variation averaged out and the counting statistics errors were minimized by long time counting.

In case of other membranes such as PES and PVDF, slight increase in the Am^{3+} transport rate was seen after 2 h with membranes exposed to 30 MRad absorbed dose. Though the membrane hydrophobicity was found to increase after exposing to radiation (Figure 2), this was not reflected clearly in their mass transport results (Table 4).

Polymeric support	% Am transport (2 h)	
	Unirradiated[a]	Irradiated[a]
PP	92.0 ± 0.3	-- [b]
PC	91.4 ± 0.2	87.1 ± 0.7[c]
PES	87.9 ± 0.3	88.2 ± 0.8[d]
PVDF	85.1 ± 0.5	86.3 ± 0.6[d]

Note: [a]Data as seen from the feed side mass balance

[b]Polymer was completely degraded to carry out any experiment

[c] Irradiated to 50 MRad

[d] Irradiated to 30 MRad

Table 4: Transport data after 2 hours with various polymeric solid supports, before and after irradiation. Feed: 3 M HNO_3; Receiver: 0.01 M HNO_3; Carrier: 0.1 M TODGA + 0.5 M DHOA in n-dodecane

Radiation dose (MRad)	Permeability×10³, cm/sec	% Transport of Am (120 min)
Nil	3.548 ± 0.05	91.4 ± 0.2
10	3.645 ± 0.10	88.9 ± 0.4
20	3.004 ± 0.04	88.1 ± 0.4
30	3.200 ± 0.05	90.8 ± 0.7
50	2.992 ± 0.03	87.1 ± 0.7
PP	3.061 ± 0.07*	92.0 ± 0.2*

*Experiments were carried out with PTFE membrane (unirradiated) as support

Table 5: Comparison of Am^{3+} transport performance of PC membranes irradiated at variable doses using ⁶⁰Co irradiator. Feed: 3 M HNO_3; Receiver: 0.01 M HNO_3; Carrier: 0.1 M TODGA + 0.5 M DHOA in n-dodecane

Conclusions

The present study was taken up to evaluate the radiation stability of several polymeric membranes using their physical characterization such as surface morphology, hydrophobicity, pore size etc. and also to study mass transfer for actual application to radioactive waste processing. PP based membranes were found to be highly unstable to radiation while PES membranes were the least hydrophobic suggesting only PVDF and PC based membranes have some amount of resistance against radiation. However, in view of higher radiation resistance of PC membranes, these were further subjected to mass transfer studies which appeared to be quite satisfactory. It is proposed to prepare PC based hollow fibers for large scale mass transfer studies using radioactive solutions.

Acknowledgements

The authors (PKM and DRR) thank Board of Researches in Nuclear Sciences for the DAE-SRC research grant. They also thank Dr. P.K. Pujari, Head, Radiochemistry Division, Bhabbha Atomic Research Centre (BARC) for his keen interest. The authors gratefully acknowledge Dr. A.V.R. Reddy, CSMCRI, Bhavnagar, India for providing samples of PC, PP and PES and Dr. A.K. Ghosh, Desalination Division, BARC for a sample of PVDF membrane.

References

1. Qdaisa HA, Moussa H (2004) Removal of heavy metals from wastewater by membrane processes: a comparative study. Desalination 164: 105-110.

2. Fu F, Wang Q (2011) Removal of heavy metal ions from wastewaters: A review. J Environ Manag 92: 407-e418.

3. Barakat MA (2011) New trends in removing heavy metals from industrial wastewater. Arab J Chem 4: 361-377.

4. Bloecher C, Dorda J, Mavrov V, Chmiel H, Lazaridis NK, et al. (2003) Hybrid flotation—membrane filtration process for the removalof heavy metalions from wastewater. Water Research 37: 4018-4026.

5. Pancharoen U, Somboonpanya S, Chaturabul S, Lothongkum AW (2010) Selective removal of mercury as $HgCl_4^{2-}$ from natural gas well produced waterby TOA via HFSLM. J Alloys Comp 489: 72-79.

6. Saffaj N, Younssi SA, Albizane A, Messouadi A, Bouhria M, et al. (2004) Preparation and characterization of ultrafiltration membranes for toxic removal from wastewater. Desalination 168: 259-263.

7. (2004) Application of membrane technologies for liquid radioactive waste procesing. IAEA Technical Reports Series No. 431. International Atomic Energy Agency, Vienna.

8. Schulz WW, Bray LA (1987) Solvent extraction recovery of by product ^{137}Cs and ^{90}Sr from HNO_3 solutions-a technology review and assessment. Sep Sci Technol 22: 191-214.

9. Clough R, Shalaby SW (1991) Radiation Effects on Polymers, ACS Symposium Series 475, ACS, Washington D.C.

10. Choppin GR, Morgenstern A (2000) Radionuclide separation in radioactive waste disposal. J Radioanal Nucl Chem 243: 45-51.

11. Mohapatra PK, Manchanda VK (2003) Liquid membrane based separations of actinides and fission products. Indian J Chem 42A: 2925-2939.

12. Ramachandhran V, Misra B M (1986) Studies on the radiation stability of ion exchange membranes. J Appl Polym Sci 32: 5743-47.

13. Mohapatra PK, Raut DR, Ghosh AK, Reddy AVR, Manchanda VK (2013) Radiation stability of several polymeric supports used for radionuclide transport from nuclear wastes using liquid membranes. J Radioanal Nucl Chem 298: 807-811.

14. Sasaki Y, Sugo Y, Suzuki S, Tachimori S (2001) The novel extractants, diglycolamides, for the extraction of lanthanides and actinides in HNO_3-n-dodecane system. Solv Extr Ion Exch 19: 91-103.

15. Thiollet G, Musikas C (1989) Synthesis and use of the amide extractants. Solv Extr Ion Exch 7: 813-827.

16. Mohapatra PK (1993) PhD Thesis, University of Bombay.

17. Sriram S, Mohapatra PK, Pandey AK, Manchanda VK, Badheka LP (2000) Facilitated transport of americium(III) from nitric acid media using dimethyldibutyltetradecyl-1,3-malonamide. J Membr Sci 177: 163-171.

18. Danesi PR (1984-85) Separation of metal species by supported liquid membranes. Sep Sci Technnol 19: 857-894.

19. Kandwal P, Ansari SA, Mohapatra PK (2012) A highly efficient supported liquid membrane system for near quantitative recovery of radio-strontium from acidic feeds. Part II: Scale up and mass transfer modeling in hollow fiber configuration. J Membr Sci 406: 85-91.

20. Ansari SA, Mohapatra PK, Raut DR, Adya VC, Thulasidas SK, et al. (2008) Separation of Am(III) and trivalent lanthanides from simulated high-level waste using a hollow fiber-supported liquid membrane. Sep Purif Technol 63: 239-242.

27

Nanofibrous Composite Materials Integrating Nano/Micro Particles between the Fibres

Petr Mikes[1*], **Jiri Chvojka**[1], **Jiri Slabotinsky**[2], **Jiri Pavlovsky**[3], **Eva Kostakova**[1], **Filip Sanetrnik**[1], **Pavel Pokorny**[1] and **David Lukas**[1]

[1]Technical University of Liberec, Czech Republic
[2]National Institute for NBC Protection, Kamenna, Czech Republic
[3]VŠB-Technical University of Ostrava, Czech Republic

Abstract

This article deals with the continual incorporation of particles by the ultrasonic dispersion *in situ* into a nanofibrous matrix produced by the electrospinning process. The new technology is based on the use of the needleless electrospinning method in combination with the ultrasound-enhanced dispersion of sub-micro or micro particles, which are deposited between nanofibres onto the support material. The main advantage of this technology is the independence of particle-incorporation of the electrospinning process. The particles are trapped between the fibres and they remain uncovered by polymer, thus maintaining all their active properties. Such materials can be cut with scissors without the particles being released. In this paper the authors present figures from scans of the electron microscopy of the newly-designed nanocomposite material and its morphological analysis, such as the particle distribution. The material was used as a sorbent of bis(2-chlorethyl) sulfide (mustard gas) with a sorption time greater than 240 minutes. Such material has been developed to be used for protection against chemical warfare agents; yet, it can be employed for several other applications depending on the powder material dispersed onto the nanofibrous layer.

Keywords: Needleless electrospinning; Particles; Ultrasound; Nanocomposites; Mustard gas

Introduction

Nanofibrous materials can be produced by the well-known needle [1] or needleless [2,3] electrospinning processes. Composite materials based on electrospun nanofibres are at the forefront of the modern material science. There are several methods for incorporating particles [4-10] into the nanofibrous materials, e.g. electrospinning of dispersed particles, such as fullerenes [6], nanotubes [8,9] or hydroxyapatite [7] from polymeric solutions, core-shell electrospinning [4,5] or a combination of the electrospinning and electrospraying processes [10]. But all of these methods are somehow limited, mainly by the amount of the powder material that can be incorporated into the nanofibres. The viscosity of the polymeric solution is increased by addition of any powder material, which affects the electrospinning process [6-9].

The task for the presented investigation was to develop a new type of material for sorption of mustard gas and for other applications, depending on the powder used for the process. The achieved outcome was the development of a new method for integration of particles in-between the nanofibres instead of into the fibres themselves.

In contrast to the other methods mentioned above, the particles in this experiment were dispersed by ultrasonic vibrations and then covered with the deposited nanofibres directly after they fell on the collector. This means that the particles are trapped in-between the fibres, which enables a unique incorporation of a higher density of particles. The size of the particles plays an important role in the selection of the process of their incorporation. Furthermore, the applications of particles depend on the material from which they are made. Taking all of that into consideration, the authors prepared membranes containing active carbon particles trapped by nanofibres and tested them for the sorption of the mustard gas.

In the second chapter the methods and materials used in this experiment are described together with the newly-produced materials, their analysis and their usage potential, as revealed by their tests for mustard gas penetration.

Materials and Methods

For the electrospinning process, a 10% aqueous polyvinyl alcohol (PVA) solution was prepared by dissolving the polymer in distilled water. The company Novácké chemické závody, Nováky in Slovakia provided the PVA with a predominant molecular weight of 60.000 amu. Crosslinking of the PVA nanofibres was achieved by adding 3 wt.% of 99% phosphoric acid and 4wt.% of Glyoxal, both of which are produced by Sigma Aldrich. The nanofibrous layer was then heated to 135°C for 5 minutes to activate the crosslinking process. Next, a polymeric solution for the electrospinning process was prepared from two components. Kuraray America, Inc. produced polyvinyl butyral (PVB), Mowital® B 60 H, with an average molecular weight of 60.000 amu; it was dissolved in a 10 wt.% solution in ethanol–water (9:1 v/v).

The nanofibres were collected on the spunbond polypropylene produced by Pegas Nonwovens, Czech Republic. The distribution of the fibre diameter lengths was in the range from 120 to 380 nm with an average of 200 nm. A histogram can be seen in Figure 1A. The NIS Elements software produced by Nikon was used for calculations of the average inter-fibrous distance, which was identified as being (1.67 ± 0.69) μm.

The particles of active carbon Norit GL Extra were milled by RETCH CryoMill with an average diameter at 2.8 μm and deviation of ± 1.5 μm. The particle size distribution depended on the duration of

*Corresponding author: Petr Mikes, Technical University of Liberec, Czech Republic, E-mail: petr.mikes@tul.cz

milling. The average diameter distribution was measured by Zetasizer Nano ZS Malvern UK - model ZEN3601 and can be seen in Figure 1B.

A schematic diagram and photography of the setup used are shown in Figures 2A and 2B, respectively. There are several technological parameters that always have to be controlled for different materials, such as air humidity, concentration of the polymer in the solution for the electrospinning process and the amplitude and frequency for the particle dispersion. The details of this setup are described below.

For the needleless electrospinning process the investigators used positive and negative high voltage sources, a 300 W, high voltage DC power supply with regulators, model number SL 150, manufactured by Spellman High Voltage, Inc. with output parameters of 0-50 kV, 6 mA. The positive voltage of 35 kV was applied to the roller (spinning electrode) and the negative voltage of 5 kV to the belt collector. The positive voltage was set to reach the critical voltage [2] and collector was negatively charged to increase the potential difference and to support the collection of fibers preferentially to the belt collector. The distance between the spinner and the collector was kept at 15 cm which is the optimal distance for fibrous elongation and solvent evaporation.

Nanofibres were collected on the belt collector, which consists of a motor powering two cylinders positioned 40 cm apart and holding the supported non-woven textile. The rotational speed of the gearbox powering the belt collector was 10 rpm and the belt was 80 cm long and 20 cm wide. First of all simple nanofibrous layer without any particles was collected and then the belt collector was twisted three times while both electrospinning and ultrasonic dispersion processes were in progress. Finally the composite was covered by another additional nanofibrous layer. Both inner and outer nanofibrous layer was prepared for better manipulation with the material.

This technology is relatively robust because the processes are independent of each other. The electrospinning parameters, such as voltage, humidity, the distance between the collector and the electrode, the concentration, the molecular weight of the polymer used, etc.,

have already been well described [1,2]. The ultrasonic dispersion is completely independent of the electrospinning process.

The most difficult problem to solve was the formation of a homogeneous distribution of the dispersed particles. Van der Waals forces between particles increase, but the particle diameter decreases and the particles tend to agglomerate [11]. The use of ultrasonic vibrations is necessary for the dispersion of these clusters [12] and for the creation of a more homogeneous type of material. Particles were de-agglomerated by the 18-cm-wide and 1-cm-thick dural sonotrode produced by Ultratech Company. A Sonic Digital ULC generator produced by Weber Ultrasonics with the power of 400 W and frequency of 20 kHz was used to power the sonotrode.

The particles were stored in a reservoir and continually conveyed by a series of vibrations to the edge of the sonotrode. The conveying device was specially designed for this technology. The device is powered by the Afag HLF 12 feeder controlled by the PSG1 unit with the vibrations amplitude with the constant frequency of 220 Hz. The relationship between the amplitude and the amount of the conveyed particles can be seen in Figure 3. The particles were subsequently covered by nanofibres. The temperature during the experiments was $21 \pm 5°C$ and the relative humidity $65 \pm 5\%$. A sieve shaker (Retsch AS 200) was used for testing the particles bonding to the fibres (Figure 3).

The thickness of the produced layer was controlled by the running time of the process and by the size of the particle dosage into the nanofibrous layer.

Adsorption ability of the nanocomposite membrane

A nanofibrous material with incorporated particles of active carbon was tested for penetration of CWA. Ninety-six-per-cent-pure bis(2-chlorethyl) sulfide, or mustard gas was applied in the liquid phase on a membrane made of High Density Polyethylene (HDPE) with the thickness of 30 μm [13]. The mustard gas vapour penetrating through the HDPE was adsorbed by the particles of active carbon. The sorption efficiency on the reverse side of the sample was evaluated by chemical indication factors caused by the degree of penetration and by gas chromatography. The chemical indication was based on the reaction of the penetrating mustard gas with a paper indicator saturated with chloramine (N-chlorobenzoic-o-toluidine) (Figure 4). The change of the pH value was indicated by Congo red [13]. The gas chromatography used the flame-ionizing indication on a Carousel 2000 device to measure the mustard gas permeation [14]. The chemical sensitivity

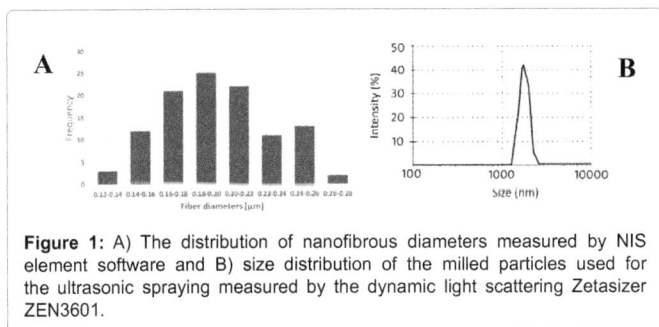

Figure 1: A) The distribution of nanofibrous diameters measured by NIS element software and B) size distribution of the milled particles used for the ultrasonic spraying measured by the dynamic light scattering Zetasizer ZEN3601.

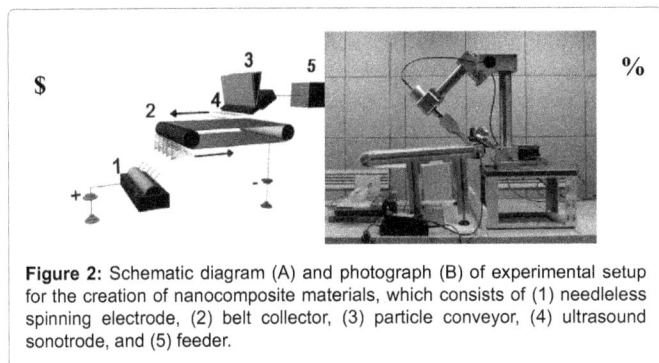

Figure 2: Schematic diagram (A) and photograph (B) of experimental setup for the creation of nanocomposite materials, which consists of (1) needleless spinning electrode, (2) belt collector, (3) particle conveyor, (4) ultrasound sonotrode, and (5) feeder.

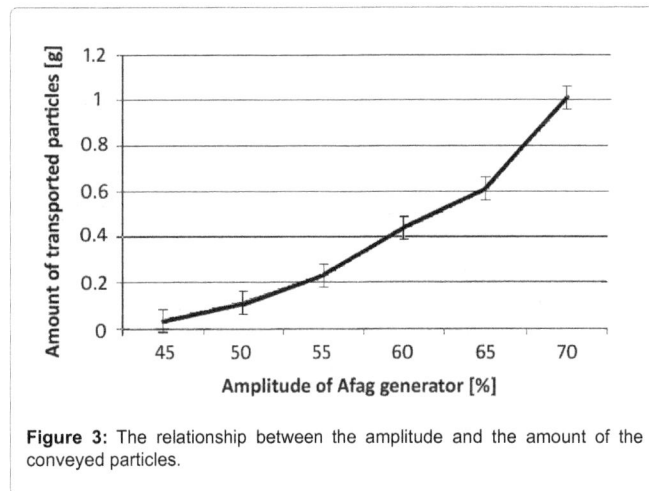

Figure 3: The relationship between the amplitude and the amount of the conveyed particles.

indication was 0.1 µg of mustard gas. The results were measured in terms of the time taken for the vapours to penetrate to the reverse side, which is the breakthrough time (BT) (Table 1). The progress of permeation was characterized by the speed of penetration all the way through to the back side, F, and the total penetrated amount of mustard gas, labelled as Q.

Results and Discussion

The produced nanofibrous composite material has unique properties due to its original and innovative structure. The particles are trapped in-between fibres and do not get covered by any polymer. The material can therefore keep its characteristics, such as, for example, high porosity in the case of active carbon particles [15-18]. The inter-fibrous distances within the nanofibrous matrix enable even sub-micron particles to be incorporated [19].

The average areal density of the nanofibres was 5 g.m^{-2} and the areal density of particles reached up to 20 g.m^{-2} within 10 minutes. These values were measured independently for both processes by weighing the electrospun polymer and the dispersed particles. The amount of particles was controlled by the amplitude of the generator (Figure 3). The scanning electron microscope (SEM) images of the active carbon incorporated in-between the electrospun nanofibres are shown in Figures 5A-5D.

The natural phenomenon of activated carbon particles is their random agglomeration [18-26]. The use of ultrasound proved to be a good method for the dispersion of the dry particles into the nanofibrous matrix [12]; however, some particles remained agglomerated. It was mainly due to the imperfect positioning of the sonotrode and

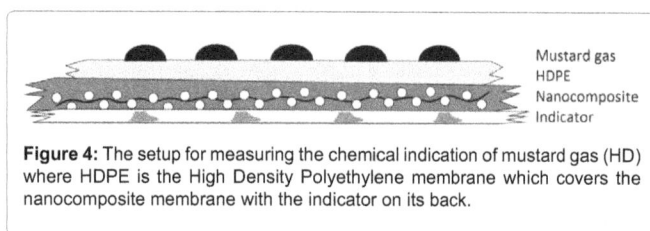

Figure 4: The setup for measuring the chemical indication of mustard gas (HD) where HDPE is the High Density Polyethylene membrane which covers the nanocomposite membrane with the indicator on its back.

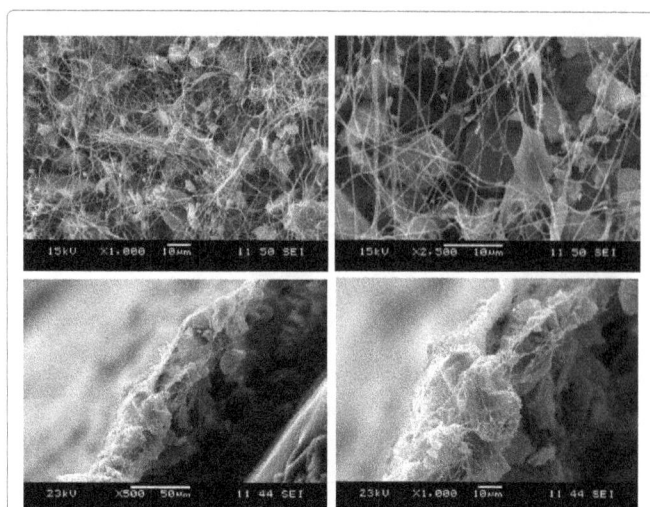

Figure 5: A-D) Scanning electron micrographs of electrospun nanofibrous material with particles of active carbon dispersed into the PVA nanofibrous layer. Figures 5A and 5B are standard SEM pictures of the face of the layer; Figures 5C and 5D are the edges of the layer.

conveyor, which is why some particles fell onto the web undispersed. The homogeneity of particles distribution is important for sorption of mustard gas. The agglomeration creates places with lower amount of active carbon particles where gas can penetrate through the membrane easily.

This technology can be used for many different combinations of polymeric electrospun nanofibres and submicron or micro particles.

Using a bi-component spunbond material made from polypropylene and polyethylene fibres could thermally laminate the produced layer. Polyethylene has a lower melting temperature than polypropylene.

The produced materials were tested for particles released from the nanofibrous matrix by using a Retsch AS 200 sieve shaker. The materials were placed on the sieve with a mesh size of 1 mm. The sample was weighed both before and after shaking with the amplitude of 80% (2.4 mm) for ten minutes. The average amount of loosened particles was 3% from the total weight of all particles.

From the SEM pictures in Figures 5A-5D it is clearly seen that particles of active carbon have been trapped by the fibres. The nanofibrous web prevents their releasing even after it is cut, thus particles stay bonded to the layer. Some particles remained agglomerated due to the above mentioned reason. Samples used for SEM didn't have any additional covering nanofibrous layer for better observation of bonded particles.

Measurement of sorption ability of the resultant material

Table 1 shows the positive effect of active carbon particles compared to the pure HDPE and also to a composite consisting of HDPE, nanofibres and the spunbond material. Not only does the amount of the activated carbon, which is responsible for the efficiency of the protective system, play a role, but also its areal distribution and homogeneity have their significant influence. It can be seen that the maximum breakthrough time (BT) was identified for the HDPE+PVB+AC 16 material. The prolonged time of mustard gas penetration in the fabric will increase its safety.

Even more conclusive results can be seen in Figure 6, where there are almost no differences in the speed of mustard gas penetration. There is a significant decrease in penetration for all composites in comparison to the HDPE membrane. However, the penetration continues even after the gas flow does not influence it any further directly. This is probably caused by the uneven distribution of the sorbent particles in the composite material. Determination of specific surface area at the prepared materials will influence the penetration of mustard gas, as can be seen in Table 2.

It is evident that there is a continuous increase of adsorption of the mustard gas due to the use of the given composite, as shown by the values of Q for the HDPE and the composites in different time intervals. However there is almost no difference between all materials in terms of total penetration Q and penetration rate F. This is probably due to the sorption ability of activated carbon particles trapped by nanofibers. It

Composition	Active carbon content [wt. %]	BT [min]
HDPE	-	4
HDPE+PVA	-	4
HDPE+PVA+AC 8	8	285 ± 45
HDPE+PVA+AC 16	16	313 ± 60
HDPE+PVB+AC 16	16	360 ± 60

Table 1: Breakthrough time (BT) of the composite membrane against mustard gas penetration at 25°C.

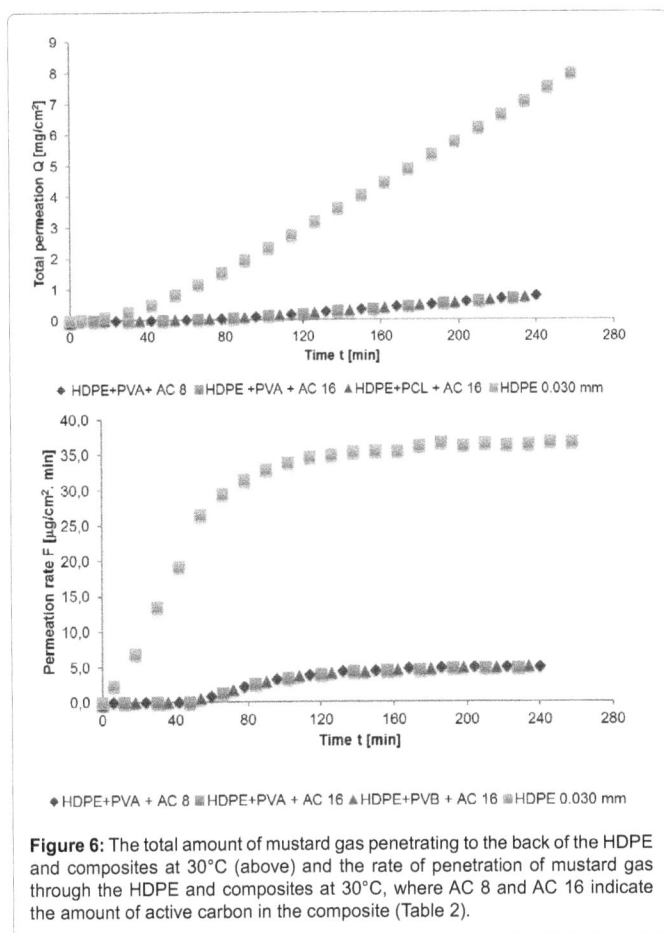

Figure 6: The total amount of mustard gas penetrating to the back of the HDPE and composites at 30°C (above) and the rate of penetration of mustard gas through the HDPE and composites at 30°C, where AC 8 and AC 16 indicate the amount of active carbon in the composite (Table 2).

Specimen	specific surface area [m². g⁻¹]
Active carbon (AC)	1274.0+
HDPE+PVA+AC 8	164.2
HDPE+PVA+AC 16	148.7
HDPE+PVB+AC 16	291.3

Table 2: Specific surface area of active carbon and composites. The specific surface area of microporous material (active carbon) evaluated by using the Langmuir method [16] is labelled by +.

to increase of the BT value [18].

Electrospinning technique has many remarkable characteristics, such as fine diameters ranging from submicron to several nanometers [19], large specific surface area, high porosity and high permeability [20] etc. Simultaneously, recent studies have shown that different properties such as surface area and porosity become more significant when the fibres diameter falls below 20 nm. An ideal fiber diameter is below 20 nm for optimal performances, however electrospun fibers from the traditional process typically have a diameter in the range 100-500 nm [21]. Electrospun nanofibers are featured with very large surface area per unit mass, small pore size a more contacts between individual fibers. Electrospun nanofibers with active carbon have large surface area (Table 2) which is in agreement with Saeed [22].

The objective is a robust method for manufacturing extremely small nanofibers in large quantities and with a uniform size. Recently, the novel two-dimensional (2D) nanofibers were generated in 3D fibrous mats by optimizing processing parameters in electrospinning [23].

In Table 2 the specific surface area studied by the BET [15] theory and for active carbon (AC) by the Langmuir theory [16] is presented.

The final step of the investigation was the performance of the SEM observation of the nanofibrous composite layer after the sorption of mustard gas. Samples were decontaminated by washing in flowing air with the temperature 25-30°C. As can be seen in Figures 7A and 7B, nanofibres slightly swell their volume by sorption of mustard gas but without any visible damage to their structure. The material, which was used for sorption of mustard gas, was finally covered by a thin layer of nanofibres to prevent particles release while being manipulated. Therefore not many particles are visible in Figures 7A and 7B.

As the final result of this study, the use of PVB is a better choice compared to PVA. The amount of activated carbon particles is sufficient with 8 wt.% for time of sorption lower than 240 min. Therefore, the optimal material should combine PVB nanofibers with 8 wt.% of activated carbon.

Conclusions

The materials presented here can be used for a number of technical applications. In the investigation they were successfully tested for the sorption of chemical substances, such as mustard gas. The direct action of the liquid mustard gas through the High Density Polyethylene gave high values of breakthrough time for the composites, but not for the membrane without sorbents. It is already possible to achieve such results with a relatively small amount of active carbon.

The technology developed in the investigation proved to be applicable for different powder materials. Its main advantage is the fact that the particles stay uncovered by the polymeric matrix and they keep their unique properties. They are simply trapped by the adhesion forces of the nanofibres, which enables the integration of even submicron particles depending on the inter-fibrous porosity. It is as well possible to combine particle materials and electrospun polymers and to create

is evident mainly for HDPE+PVA+AC8 and HDPE+PVA+AC16. Both materials showed the same time of sorption as can be seen on Figure 6. The question still remains how it will be for longer time of permeation. But the time of operation of apparatus was limited. Anyway, for its use as a protection membrane against chemical warfare, the time of sorption up to 240 min is sufficient. Therefore the lower amount of activated carbon particles with 8 wt.% is ample.

Prepared materials were also analysed using Sorptomatic series 1990 apparatus, Thermo Quest (CE Instruments), Rodano (Milano) with the method of low-temperature adsorption of nitrogen molecules (boiling point 77.7 K), from a vacuum into atmospheric pressure for determination of specific surface area. It was revealed that the specific surface area was lower for the composite materials in comparison with the activated carbon particles (AC). Also no significant specific surface area differences were measured by the BET analysis for the material HDPE+PVA+AC irrespective to the amount of activated carbon (Table 2). The situation is different for the material HDPE+PVB+AC 16 (16 wt.% of AC), where the specific surface area is approximately two times higher than for HDPE+PVA+AC 16 (16 wt.% of AC). The reason for the difference in higher specific surface of nanofibers with PVB against nanofibers with PVA should be its roughness. Higher content of activated carbon (AC) and its incorporation in-between the electrospun nanofibers thus increases the specific surface area and also more contact points between individual fibers, as shown in Figure 5. Larger specific surface areas can provide sufficient contact between HDPE+PVB+AC 16 and mustard gas (Table 1) and this leads

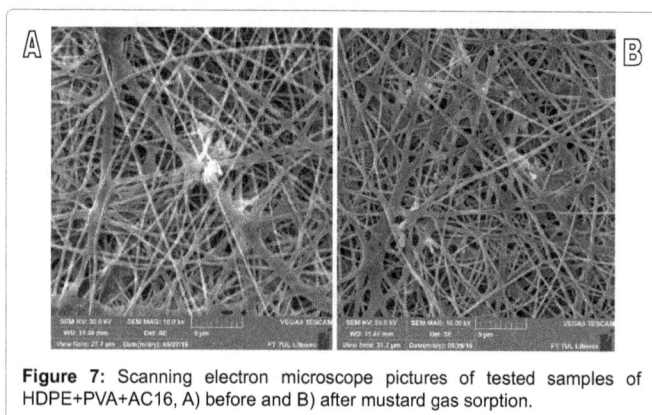

Figure 7: Scanning electron microscope pictures of tested samples of HDPE+PVA+AC16, A) before and B) after mustard gas sorption.

a tailor-made product. Another advantage of these materials is that they can be cut by standard methods (scissors, scalpel, razor blade, etc.), which makes them widely applicable depending on the powder material integrated into the composite.

The composite material with particles was successfully tested for sorption of the mustard gas with time of sorption more than 300 minutes. However, some aspects of the material have to be improved, mainly the homogeneity of the dispersed material. This was evident for the active carbon particles. Active carbon is a very hygroscopic material and therefore its agglomeration increases during the process. The homogeneity can be improved, e.g. by use of some more closed system with air conditioning, which prevents the agglomeration of particles. Another possibility to improve the homogeneity of the particles distribution is the presence of some suction system under the nanofibrous layer.

Acknowledgment

Support for this research was provided by the Ministry of the Interior of the Czech Republic, BV II/2-VS, Nanomaterials for personal protection against CBRN substances". This paper was supported by the project of. Regional Material Technological Research Centre - Sustainability programme", registration number LO1203.

References

1. Reneker DH, Yarin AL (2008) Electrospinning jets and polymer nanofibers. Polymer 49: 2387-2425.

2. Lukas D, Sarkar A, Pokorny P (2008) Self-organization of jets in electrospinning from free liquid surface - a generalized approach. J Appl Phys 103: 1-7.

3. Jirsak O, Sanetrnik F, Lukas D, Kotek V, Martinova L, et al. (2005) A method of nanofibres production from a polymer solution using electrostatic spinning and a device for carrying out the method, US Patent WO2005024101.

4. Moghe AK, Gupta BS (2008) Co-axial electrospinning for nanofiber structures: Preparation and applications. Polym Rev 48: 353-377.

5. Mickova A, Buzgo M, Benada O, Rampichova M, Fisar Z, et al. (2012) Core/Shell Nanofibers with Embedded Liposomes as a Drug Delivery System. Biomacromolecules 4: 952-962.

6. Yang P, Zhan S (2011) The fabrication of PPV/C60 composite nanofibers with highly optoelectric response by optimization solvents and electrospinning technology. Materials Letters 3: 537-539.

7. Kim GM, Asran ASh, Michler GH, Simon P, Kim JS (2005) Electrospun PVA/HAp nanocomposite nanofibers: biomimetics of mineralized hard tissues at a lower level of complexity. Bioinspir Biomim 3: 046003.

8. Kannan P, Eichhorn SJ, Young RJ (2007) Deformation of isolated single-wall carbon nanotubes in electrospun polymer nanofibres. Nanotechnology 18: 235707.

9. Kostakova E, Meszaros L, Gregr J (2009) Composite nanofibers produced by modified needleless electrospinning. Materials Letters 28: 2419-2422.

10. Jaworek A, Krupa A, Lackowski M, Sobczyk AT, et al. (2008) Nanocomposite fabric formation by electrospinning and electrospraying technologies. J Electrostat 67: 435-438.

11. Hartley PA, Parfitt GD, Pollack LB (1985) The role of the van der Waals force in the agglomeration of powders containing submicron particles. Powder Technology 1: 35-46.

12. Sauter C, Emin MA, Schuchmann HP, Tavman S (2008) Influence of hydrostatic pressure and sound amplitude on the ultrasound induced dispersion and de-agglomeration of nanoparticles. Ultrasonics Sonochemistry 15: 517-523.

13. MAZL3/95. Estimation of breakthrough time of mustard drops through protective materials at static conditions. Accredited Method National Institute for NBC Protection of the Czech Republic.

14. MAZL 39-10/Permeatest 4. Determination of resistance membrane material to penetration of mustard by gas chromatography. Accredited Method National Institute for NBC Protection of the Czech Republic.

15. Brunauer S, Emmet PH, Teller EJ (1938) Adsorption of gases in multimolecular layers. J Am Chem Soc 60: 309.

16. Langmuir I (1918) The adsorption of gases on plane surfaces of glass, mica and platinum. J Am Chem Soc 40: 1361 .

17. Dror Y (2003) Carbon nanotubes embedded in oriented polymer nanofibers by electrospinning. Langmuir 19: 7012.

18. March H, Rodriguez-Reinoso F (2006) Actived Carbon. Elsevier, Oxford, UK.

19. Eichhorn SJ, Sampson WW (2005) Statistical geometry of pores and statistics of porous nanofibrous assemblies. J R Soc Interface 2: 309-318.

20. Tingting L, Zhiming Z, Wei L, Ce L, Jianfei W, et al. (2016) H4SiW12O40/polymethylmethacrylate/polyvinyl alcohol sandwichnanofibrous membrane with enhanced photocatalytic activity. Colloids Surf A Physicochem Eng Asp 489: 289-296.

21. Buchko CJ, Chen LC, Shen Y, Martin DC (1999) Processing and microstructural characterization of porous biocompatible protein polymer thin films. Polymer 40: 7397-7407.

22. Saeed K, Haider S, Oh TJ, Park SY (2008) Preparation of amidoxime-modified polyacrylonitrile (PAN-oxime) nanofibers and their applications to metal ions adsorption. J Memb Sci 322: 400-405.

23. Huang CB, Chen SL, Lai CL, Reneker DH, Qiu H, et al. (2006) Electrospun polymer nanofibres with small diameters. Nanotechnology 17: 1558-1563.

24. Li D, McCann JT, Xia Y (2005) Use of electrospinning to directly fabricate hollow nanofibers with functionalized inner and outer surfaces. Small 1: 83-86.

25. Ding B, Li C, Miyauchi Y, Kuwaki O, Shiratori S (2006) Formation of novel 2D polymer nanowebs via electrospinning. Nanotechnology 17: 3685-3691.

26. Anson M, Marchese J, Garis E, Ochoa N, Pagliero C (2004) ABS copolymer-activated carbon mixed matrix membranes for CO2/CH4 separation. J Memb Sci 243: 19-28.

Performance Evaluation of Polyamide Reverse Osmosis Membrane for Removal of Contaminants in Ground Water Collected from Chandrapur District

Vidyadhar V. Gedam[1]*, Jitendra L. Patil[2], Srimanth Kagne[1], Rajkumar S. Sirsam[2] and Pawankumar Labhasetwar[1]

[1]National Environmental Engineering Research Institute, Nagpur- 440020, Maharashtra, India
[2]Department of Chemical Technology, North Maharashtra University, Jalgaon- 425001, Maharashtra, India

Abstract

This paper examines the influence of different operating parameters such as pressure, temperature, pH on the performance of polyamide reverse osmosis membrane. Varying these parameters, intensive trials were undertaken to study the performance of polyamide Reverse Osmosis (RO) membrane. Water samples for experiment were collected from Moradgaon village of Chandrapur district having high concentrations of fluoride, Total dissolved solids (TDS), sulphate and iron. Results indicate that polyamide reverse osmosis membrane can successfully remove 95 to 98% of fluoride, TDS, sulphate, iron and other ground water contaminants under optimized conditions. Different parameters such as pH, pressure and temperature affects RO membrane efficiency. Thus, proper control of these factors is essential for successful operation and maintenance. RO Membrane generates huge quantity of reject water (i.e.65% -75%), which was further passed through RO membrane to study its reuse potential. The results showed that water received from RO membrane after recycling of membrane reject is within the permissible limits of drinking water as prescribed by Bureau of Indian Standards (BIS).

Keywords: Drinking water; Ground water; Polyamide membrane; Operating parameters; Reverse osmosis (RO)

Introduction

India has been well endowed with large freshwater reserves, but increasing population and over-exploitation of surface and groundwater over the past few decades has resulted in water scarcity in most regions. Existing freshwater reserves are being polluted due to inadequate control and unsafe system in view of urbanization, over-exploitation and natural activity. It was estimated that around 37.7 million Indians are affected by waterborne diseases annually, 1.5 million children are estimated to die of diarrhea alone and 73 million working days are lost due to waterborne disease each year [1]. The resulting economic burden is estimated at $600 million a year [2]. The problems of chemical contamination are also prevalent in India with 1,95,813 habitations in the country are affected by poor water quality [3]. The poor quality of raw water sources warrants the application of stringent treatment technologies and proper monitoring to ensure supply of safe drinking water. In India ground water is considered as a safe source of drinking water which is being utilized intensively for drinking, irrigation and industrial purposes. However, due to rapid growth of population, urbanization, industrialization and agricultural activities, Indian ground water resources are under constant stress. There is growing concern on the deterioration of ground water quality due to geogenic and anthropogenic activities. The main ground water quality problems in India are inland salinity, coastal salinity, fluoride, arsenic, iron and nitrate [4].

Many technologies are developed for removal of these contaminants which includes, filtration, chemical treatment, advanced oxidation and membrane separation process. This paper investigates the influence of different operating parameters such as pressure, temperature and pH on performance of polyamide reverse osmosis membrane for removal of fluoride, TDS, sulphate, iron and other ground water contaminants. In addition, its recycling potential for reject water which was generated during above experiments is also studied in detail.

Materials and Methods

The experiment was performed using a thin film composite polyamide spiral wound RO membrane. The module consisted of a Filmtec Spiral wound with composite polyamide membrane module (model no. TW30-1812-75) with effective area of 0.1054 m^2, module length 300 mm and diameter of 40 mm.

A detailed RO membrane module experimental setup is shown in Figure 1. Ground water sample was collected from a hand pump at Moradgaon village, Chandrapur - Maharashtra. The ground water sample had high concentration of TDS, fluoride, chloride, hardness, alkalinity, iron, sulphate and turbidity than the permissible limits as per BIS: 10500. The physico-chemical analysis of water sample collected from Moradgaon village is shown in Table 1.

RO membrane module was operated under various operating parameters such as feed water temperature, pressure and pH. The effect of different operating parameters on performance RO membrane was studied by varying one parameter at a time and keeping others constant. Table 2 shows the operating variables during reverse osmosis.

Recycling potential of membrane reject water

During reverse osmosis process 50 to 65% of membrane reject

*Corresponding author: Vidyadhar V. Gedam, National Environmental Engineering Research Institute, Nagpur- 440020, Maharashtra, India, E-mail: gedmvidyadhar@gmail.com

water is generated [5]. To reutilize such an enormous amount of reject water and to test membrane recycling performance, it was recycled back through RO membrane. During experiment, reject of first run was used as feed for second run and reject of second run was used as feed for third run. This process was continued till RO permeate showed concentration of fluoride, TDS and other contaminants more than the permissible limits as prescribed by [6].

Results and Discussion

Effect of feed water temperature

Temperature is one of the important parameter which affects the performance of RO membrane [7]. The effect of varying temperature keeping other parameters constant on performance of RO membrane is shown in graphs.

It can be observed from Figure 2-6 that as feed water temperature increases from 18 to 40°C, the permeate salinity (TDS) increases from 148 to 288 ppm, permeate flux increased from 1.4 to 2.3 (l/ m².min), fluoride concentration increased from 0.02 to 0.2 ppm and % recovery increased from 21.83 to 35.93 %. But the % salt rejection decreases from 92.43 to 85.23. Increase in TDS, permeate flow, flux, fluoride concentration and % recovery with decrease in % salt rejection is observed because, as temperature increases viscosity decreases and water permeation rate through membrane increases. As temperature increased solubility of solute increased and higher diffusion rate of solute through the membrane is possible [8].

Effect of feed water pressure

Pressure is one of the most important operational parameter

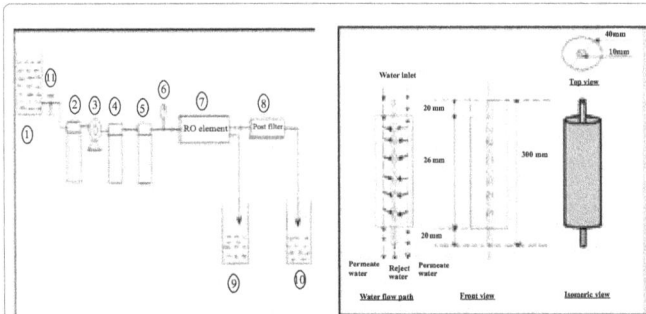

Figure 1: Experimental set up, different views and water flow path for RO membrane module.

Parameters	Water Sample
Temperature°C	22.9
pH	9.2
EC (μs/cm)	3250
TDS (ppm)	1950
Chloride (ppm)	815.35
Fluoride (ppm)	2.13
Hardness (ppm)	140
Calcium (ppm)	120
Magnesium (ppm)	20
Alkalinity (ppm)	28
Iron (ppm)	0.88
Sulphate (ppm)	128.75
Turbidity (NTU)	3.26
Nitrate (ppm)	ND

Table1: Physico-chemical analysis of water sample collected from Moradgaon village.

Experimental Parameters	Operating parameters		
	T (°C)	P (psi)	pH
Feed temp. (T)	Varied	80	8.35
Pressure (P)	30	Varied	8.35
Conc.(TDS)	23	80	8.35
pH	25	80	Varied
Flow rate (F)	25	80	8.00

Table 2: Operating variables of RO Membrane system.

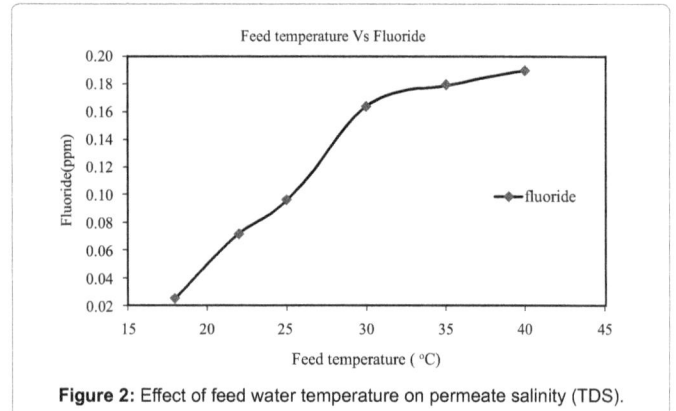

Figure 2: Effect of feed water temperature on permeate salinity (TDS).

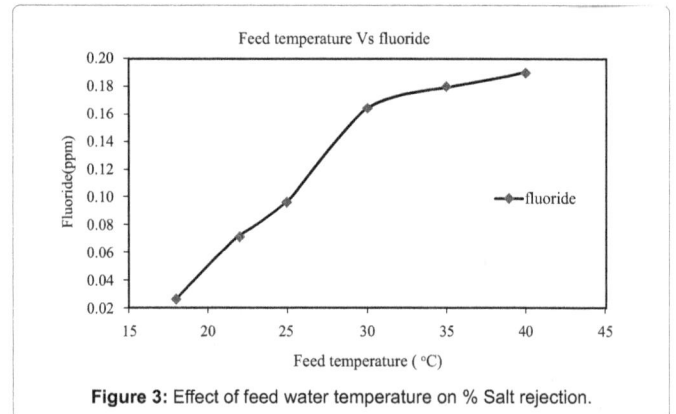

Figure 3: Effect of feed water temperature on % Salt rejection.

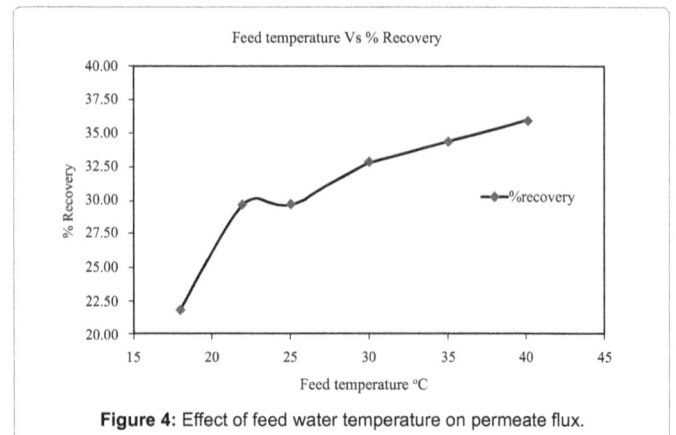

Figure 4: Effect of feed water temperature on permeate flux.

which significantly affects the performance of RO membrane. The effect of pressure on the performance of RO membrane was studied by keeping all other parameter constant and the results are shown below graphically.

Based on experimental data, graphs were plotted for pressure verses % salt rejection, % recovery, permeate concentration (TDS), and fluoride concentration in permeate. Figure 6-9 shows that, as pressure increases from 30 - 80 psi, % recovery increases from 13.12 to 48.43 %, % salt rejection increase from 82.5 to 96.5 but fluoride concentration and permeate TDS decreases from 0.778 to 0.0680 ppm and 195 to 31 ppm respectively. Pressure increases the driving force for the solvent and decrease osmotic pressure hence more amount of water can be passed through the membrane with a high rate of salt rejection [8]. From the graph it was observed that, the optimum value of feed pressure for RO membrane ranges from 70 to 80 psi. At this operating pressure, maximum flux and salt rejection was noticed.

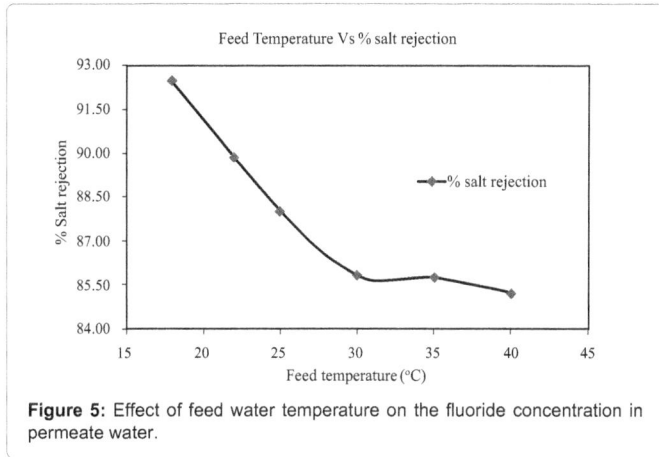

Figure 5: Effect of feed water temperature on the fluoride concentration in permeate water.

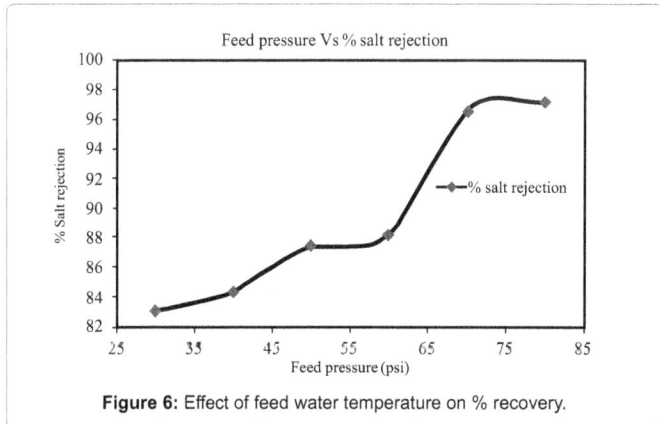

Figure 6: Effect of feed water temperature on % recovery.

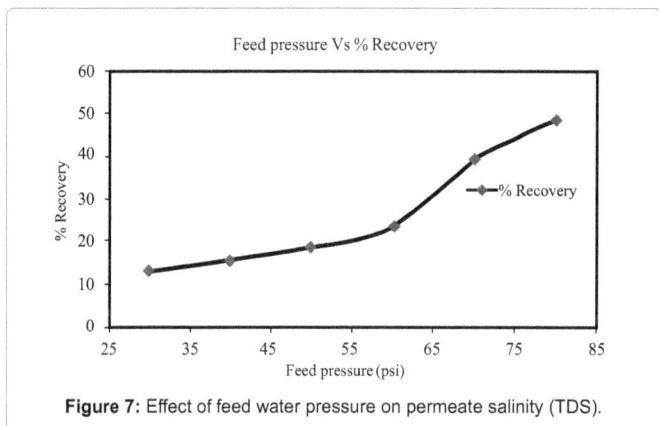

Figure 7: Effect of feed water pressure on permeate salinity (TDS).

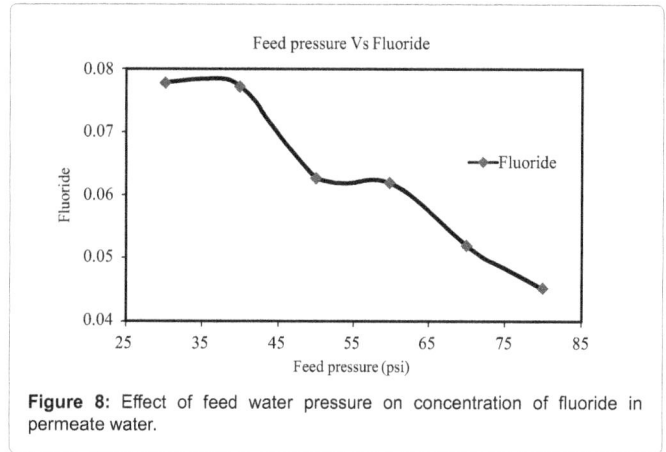

Figure 8: Effect of feed water pressure on concentration of fluoride in permeate water.

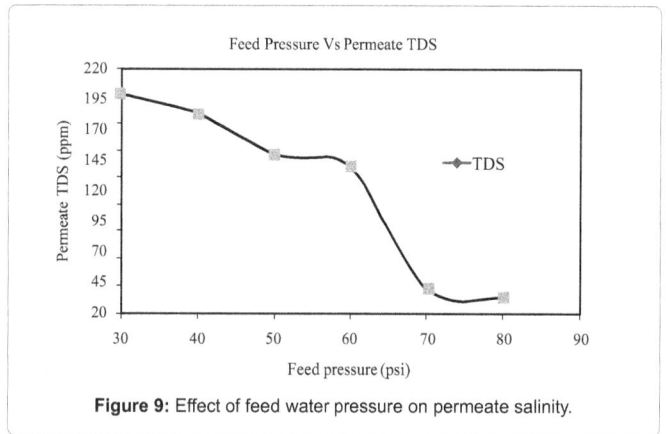

Figure 9: Effect of feed water pressure on permeate salinity.

Parameter	Membrane permeate
Feed pH	8.06
Feed EC (µS/cm)	3380
Feed Conc. -TDS (ppm)	2010
Temperature (°C)	22
Vol. of feed water (L)	5
Chloride (ppm)	921.7
Fluoride (ppm)	2.18
Hardness (ppm)	192
Calcium (ppm)	152
Magnesium (ppm)	40
Alkalinity (ppm)	168
Iron (ppm)	0.90
Sulphate (ppm)	145.6
Turbidity NTU	4.0

Table 3: Physico-chemical Analysis Reject waste water used as feed for RO system.

Effect of feed water pH

Variation in pH affects the performance of RO membrane. Graphs was plotted for feed water pH verses % salt rejection, permeate concentration (TDS), % recovery and fluoride concentration in permeate. Figure 10-13. From Figures 10-13, it was observed that as pH of the feed water increases % salt rejection and % recovery decreases from 91.43 to 89.23% and 39.06 to 20.31% respectively while permeate concentration increases from 167 to 210 ppm.

pH affects the separation performance by affecting the hydration and absorption capacity of solution on membrane [9]. It can be observed from Figure 12 that, as the as pH increases from 3 to 7 fluoride concentration increases from 0.126 to 0.196 ppm but when pH further increases from 7 to 9.5 fluoride concentration decreases from 0.195 to 0.123 ppm. At acidic pH, the fluoride concentration in permeate decreases because of strong hydrogen bonding of fluoride in acidic solution [7,10].

Recycling potential of reject water

Household reverse osmosis units use a lot of water because they have low back pressure. As a result, they recover only 5 to 15 percent of the water entering the system. The remainder is discharged as waste water or rejects water which has no further use; this is one of the disadvantages of RO. To study the reuse potential of such enormous amount of reject water; it was recycled through RO module. The Table 3 and Table 4 show the physico-chemical analysis of reject water used as feed for RO system and recycle reject water after passing from RO membrane.

It can be observed from Table 3 that, during the fifth run of

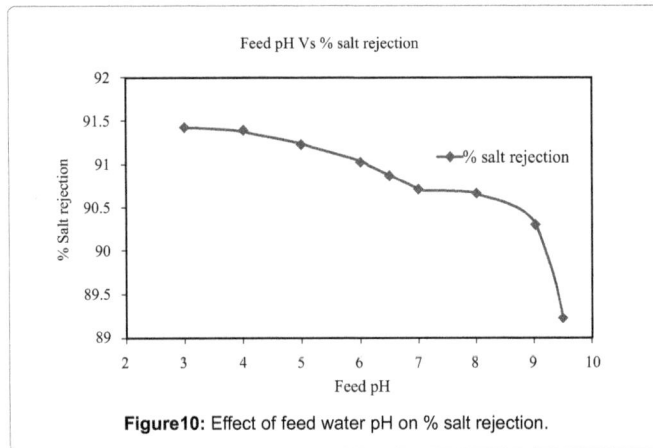

Figure10: Effect of feed water pH on % salt rejection.

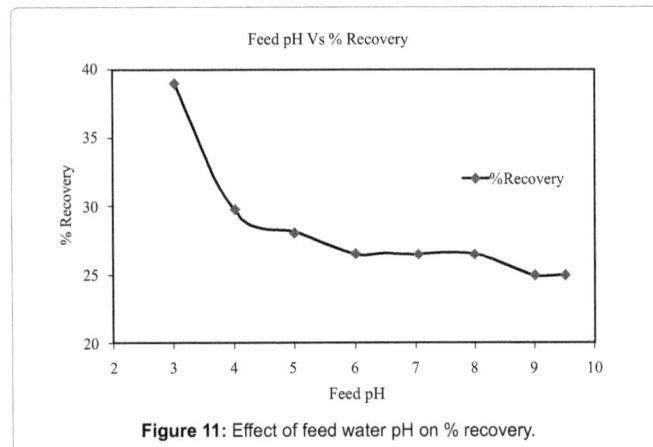

Figure 11: Effect of feed water pH on % recovery.

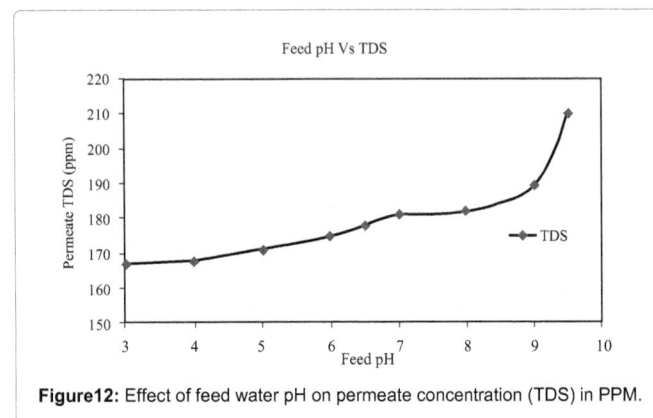

Figure12: Effect of feed water pH on permeate concentration (TDS) in PPM.

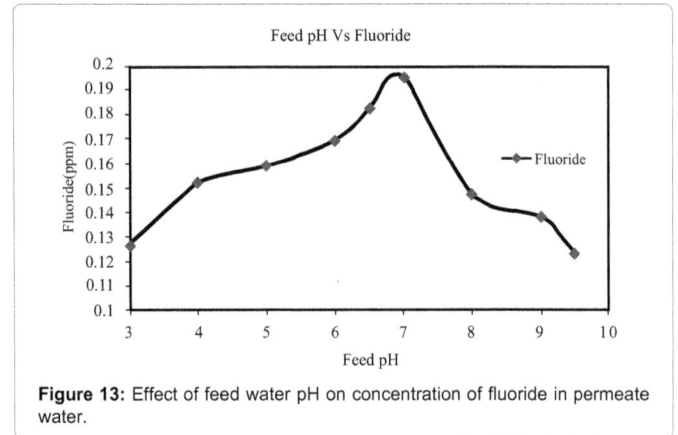

Figure 13: Effect of feed water pH on concentration of fluoride in permeate water.

Parameter/Concentration	Run-1	Run-2	Run-3	Run-4	Run-5
Feed water:					
pH	8.06	8.19	8.5	8.92	7.7
EC (µS/cm)	3380	3390	3720	3820	3950
Concentration -TDS (ppm)	2028	2034	2232	2292	2370
Volume of water (L)	5	3.9	3.1	2.54	2.35
Temperature (°C)	22	23	23	23	23
Permeate water:					
Flow rate(L/min)	0.21	0.18	0.15	0.13	0.13
Temperature (°C)	23.7	24.1	23.3	19.1	21.7
pH	8.5	7.29	8	6.18	7.85
EC (µS/cm)	150.4	311	513	613	819
TDS (ppm)	90.24	186.6	307.8	367.8	491.4
Chloride (ppm)	36.86	65.23	106.35	141.8	170.16
Fluoride (ppm)	0.044	0.0928	0.129	0.158	0.331
Hardness (ppm)	28	24	27	40	48
Calcium (ppm)	2	4	8	8	20
Magnesium (ppm)	2	4	4	8	8
Alkalinity (ppm)	20	16	20	32	40
Iron (ppm)	0.31	0.32	0.28	0.3	0.35
Sulphate (ppm)	2.38	3.21	5.11	5.35	2.94
Turbidity NTU	0.56	0.41	0.39	0.4	0.4
Water collected (lit.)	1.1	0.8	0.56	0.35	0.35
% salt rejection	95.51	90	86.2	83.95	79.26
%Recovery	32.82	28.12	23.44	20.31	20.31
Flux (L/m²min)	2.1	1.8	1.5	1.3	1.3
Membrane permeate:					
Flow rate(l/min)	0.57	0.56	0.56	0.56	0.57
EC (µS/cm)	3390	3720	3820	3950	4030
TDS (ppm)	2034	2232	2292	2370	2418
pH	8.19	8.5	8.92	7.7	7.75
Temperature(°C)	23.1	23	23.5	20.5	23
Water collected (lit.)	3.9	3.1	2.54	2.35	2

Table 4: Physico-chemical analysis of recycle reject water sample for various physico-chemical parameters.

experiment, concentration of permeate water (TDS) increases from 90.24 to 491.4 ppm, feed concentration increases from 2010 ppm to 2370 ppm, % salt rejection decreased from 95.51% to 79.26%, % recovery decreased from 32.82 to 20.31% and fluoride concentration increased from 0.044 to 0.331 ppm. During the experimental run, reject water through RO membrane was within BIS limit of drinking water (i.e. BIS-10500).

Conclusion

The RO membrane was very sensitive to various operating parameters such as feed water temperature, pressure and pH. Increase in temperature increases % recovery, fluoride concentration, permeate concentration (TDS) but decreases the % salt rejection. Increase in pressure, increases % recovery, % salt rejection, but decreases the permeate concentration (TDS) and fluoride concentration. RO is a very efficient process for defluoridation of water as it works at very low pressure and besides fluoride, other inorganic pollutant are also effectively removed. pH has significant effect on the rejection ratio of fluoride and the observed optimum pH was 7. Membrane reject water was recycled through RO membrane for various runs and experimental data shows that recycling membrane reject water through RO is within BIS (10500) limit of drinking water. From these studies, it may be concluded that polyamide reverse osmosis membrane has potential for

membrane reject water recycling but at the same time, feasibility and practicability of reject recycling need to be researched intensively.

References

1. Gupta S (2012) Drinking Water Quality: A Major Concern in Rural India, Journal of Barnolipi 1: 2249-2666.

2. Grail Research (2009) Water–The India Story.

3. Khurana I, Sen R (2010) Drinking water quality in rural India: issues and approaches, Water Aid India.

4. Central Ground Water Board Ministry Of Water Resources Government Of India (2010) Ground Water Quality in Shallow Aquifers of India.

5. Zhao, Shigang, Shi, Weiping (2005) Discussion on Reclaiming and Utilization of Reverse Osmosis Rejected Water. Industrial Water & Wastewater 36: 58-59.

6. Bureau of Indian standard specification for drinking water BIS: 10500 (1991).

7. Mattheus FA, Sablani SS, Al-Maskari SS, Al-Belushi RH, Wilf M (2002) Effect of feed temperature on permeate flux and mass transfer coefficient in spiral-wound reverse osmosis systems. Desalination 144: 367-372.

8. Arora M, Maheshwari RC, Jain SK, Gupta A (2004) Use of membrane technology for potable water production. Desalination 170: 105-112.

9. Berge D, Gad H, Khaled I, Rayan MA (2009) An experimental and analytical study of RO desalination plant. Mansoura Engineering Journal 34: 71-92.

10. Mohammadi T, Moghadam KM, Madaeni SS (2002) Hydrodynamic factors affecting flux and fouling during reverse osmosis of seawater. Desalination 151: 239-245.

Permissions

List of Contributors

El-Toony MM
National center for radiation research and technology, Atomic energy authority, Egypt, 3 Ahmad El Zomr street, Nasr City, Cairo, Egypt. 11370

Hussam Mansour and Wojciech Kowalczyk
Chair of Mechanics and Robotics, University of Duisburg-Essen, Lotharstr, Duisburg, Germany

AT Kassem, N El Said, and HF Aly
Hot Labs. Center, Atomic Energy Authority, P.C. 13759, Cairo, Egypt

Jamel Bouaziz
Laboratory of Industrial Chemistry, National School of Engineering, University of Sfax, BP 1173, 3038 Sfax, Tunisia

Andre Deratani
Institut Européen des Membranes (IEM), Université Montpellier 2, Place E. Bataillon, 34095 Montpellier Cedex 5, France

Semia Baklouti
Laboratory of Materials Engineering and Environment, National School of Engineering, University of Sfax, BP 1173, 3038 Sfax, Tunisia

Sonia Bouzid Rekik
Laboratory of Industrial Chemistry, National School of Engineering, University of Sfax, BP 1173, 3038 Sfax, Tunisia
Laboratory of Materials Engineering and Environment, National School of Engineering, University of Sfax, BP 1173, 3038 Sfax, Tunisia

Maria Visa
Department of Renewable Energy Systems and Recycling, Transilvania University of Brasov, Romania

Nicoleta Popa
Department of Forest District Teliu, National Administration of State Forests Romsilva, Romania

Funamizu Naoyuki
Graduate School of Engineering, Hokkaido University, Japan

Guizani Mokhtar
Graduate School of Engineering, Hokkaido University, Japan
Center for Sustainability Science, Hokkaido University, Japan

Sophie Cerneaux, Vincent Germain, Gil Francisco, David Cornu and André Larbot
Institut Européen des Membranes de Montpellier (IEMM) (UMR 5635), Université Montpellier II (UM2) /Centre National de la Recherche Scientifique (CNRS)/ Ecole Nationale Supérieure de Chimie de Montpellier (ENSCM), Site Halle de Technologie, 276 rue de la Galéra, F-34000 Montpellier, France

Cédric Loubat
Specific Polymers, ZAC Via Domitia, 150 Av. Des Cocardières, 34160 Castries, France

Eric Louradour
CTI-Céramiques Techniques Industrielles, 382 Avenue du Moulinas, F-30340 Salindres, France

Eric Prouzet
University of Waterloo, Department of Chemistry, 200 University Avenue West, Waterloo N2L 3G1, Ontario, Canada

Chamekh Mbareck
Université des Sciences, de Technologie et de Médecine; Faculté des Sciences et Techniques, B.P. 5026, Nouakchott, Mauritanie

Quang Trong Nguyen
P.B.S. UMR 6270 CNRS - Université de Rouen, 76821 Mont-Saint-Aignan, France

Xia Yang, Fang Manquan and Li Jiding
State Key Laboratory of Chemical Engineering, Department of Chemical Engineering, Tsinghua University, Beijing 100084, China

Zhen Wu
Ordos Redbud Innovation Institute, Ordos 017000, China

Shang Han, Shasha Na, Weixing Li and Weihong Xing
College of Chemistry and Chemical Engineering, Nanjing Tech University, Nanjing 210009, China

Vandana Upadhyay
Department of Chemistry, Marwar Business School, Gorakhpur, Uttar Pradesh, India

Shuang Liang and Yubo Zhao
School of Environmental Science and Engineering, Shandong University, Jinan 250100, Shandong, PR China

Haifeng Zhang
Department of Chemical Engineering, Northeast Dianli University, Jilin 132012, Jilin, PR China

Lianfa Song
Department of Civil and Environmental Engineering, Texas Tech University, Lubbock TX 79409 41023, USA

Caravella A
Department of Environmental and Chemical Engineering (DIATIC), University of Calabria, Via P. Bucci, Rende (CS), Italy

Sun Y
Department of Materials Engineering, Hanyang Univeristy, Ansan, Gyeonggi-do, South Korea

Abdel-Hady EE, Abdel-Hamed MO and Gomaa MM
Physics Department, Faculty of Science, Minia University, Minia, Egypt

Haixia Li, Jian Song and Xiaoyao Tan
State Key Laboratory of Separation Membranes and Membrane Processes, Department of Chemical Engineering, Tianjin Polytechnic University, Tianjin 300387, China

Said N El, Kassem AT and Aly HF
Hot Labs, AtomicEnergy Authority, Egypt

Birgit Feketeföldi
Institute for Surface Technologies and Photonics, JOANNEUM RESEARCH Forschungsgesellschaft mbH/ Materials, Franz-Pichler-Straße 30, 8160 Weiz, Austria

Bernd Cermenek, Alexander Schenk, Christoph Grimmer, Merit Bodner and Viktor Hacker
Institute of Chemical Engineering and Environmental Technology, Fuel Cell Systems Group, Graz University of Technology, NAWI Graz, Inffeldgasse 25C, 8010 Graz, Austria

Christina Spirk, Martin Koller and Volker Ribitsch
Institute of Chemistry, University of Graz, Heinrichstraße 28, 8010 Graz, Austria

K. Habib and K. Al-Muhanna
Materials Science Laboratory, Department of Advanced Systems, KISR, P.O.Box 24885 Safat, 1319 Kuwait

Rochd S, Zerradi H, Mizani S, Dezairi A and Ouaskit S
Laboratory of Physics of Condensed Matter (URAC10), Ben M'sik Sciences, University Hassan II Casablanca, Morocco

Song Xiaoxiao and Prince JA
Environmental & Water Technology, Centre of Innovation, Ngee Ann Polytechnic, Singapore

Darren Delai Sun
School of Civil and Environmental Engineering, Nanyang Technological University, Singapore

Panchali Bharali, Somiron Borthakur and Swapnali Hazarika
Chemical Engineering Group, Engineering Science & Technology Division, CSIR-North East Institute of Science and Technology, Jorhat, Assam, India

Mabrouk A
Qatar Environment and Energy Research Institute, Hamad Bin Khalifa University, Qatar Foundation, Qatar

Elhenawy Y and Moustafa G
Port Said University, Egypt

Bogdan C Donose, Greg Birkett and Steven Pratt
The University of Queensland, School of Chemical Engineering, St Lucia, 4072, QLD, Australia

Mir Shabeer Ahmad, Ayaz Mahmood Dar and Shafia Mir
Department of Chemistry, Government Degree College, Kulgam, J&K, India

Shahnawaz Ahmad Mir and Noor Mohd Bhat
Department of Chemistry, Bundelkhand University, Jhansi, UP, India

Vivek Kolladikkal Premanadhan and Yu Qiao Zhao
Department of Mechanical Engineering, Faculty of Engineering, National University of Singapore, Singapore

Wai Lam Loh, Thiam Teik Wan, Ko Ko Naing, Nguyen Dinh Tam and Valente Hernandez Perez
Department of Mechanical Engineering, Faculty of Engineering, National University of Singapore, Singapore
Centre for Offshore Research and Engineering (CORE), National University of Singapore, Singapore

Mohapatra PK and Raut DR
Radiochemistry Division, Bhabha Atomic Research Centre, Trombay, Mumbai-400085, India

Shah JG
Process Development Division, Bhabha Atomic Research Centre, Trombay, Mumbai-400085, India

Bhardwaj YK
Radiation Technology Development Division, Bhabha Atomic Research Centre, Trombay, Mumbai-400085, India

Petr Mikes, Jiri Chvojka, Eva Kostakova, Filip Sanetrnik, Pavel Pokorny and David Lukas
Technical University of Liberec, Czech Republic

Jiri Slabotinsky
National Institute for NBC Protection, Kamenna, Czech Republic

Jiri Pavlovsky
VŠB-Technical University of Ostrava, Czech Republic

Vidyadhar V. Gedam, Srimanth Kagne and Pawankumar Labhasetwar
National Environmental Engineering Research Institute, Nagpur- 440020, Maharashtra, India

Jitendra L. Patil and Rajkumar S. Sirsam
Department of Chemical Technology, North Maharashtra University, Jalgaon- 425001, Maharashtra, India

Index